Studies in Systems, Decision and Control

Volume 373

The series "Studies in Systems, Decision and Control" (SSDC) covers both new developments and advances, as well as the state of the art, in the various areas of broadly perceived systems, decision making and control–quickly, up to date and with a high quality. The intent is to cover the theory, applications, and perspectives on the state of the art and future developments relevant to systems, decision making, control, complex processes and related areas, as embedded in the fields of engineering, computer science, physics, economics, social and life sciences, as well as the paradigms and methodologies behind them. The series contains monographs, textbooks, lecture notes and edited volumes in systems, decision making and control spanning the areas of Cyber-Physical Systems, Autonomous Systems, Sensor Networks, Control Systems, Energy Systems, Automotive Systems, Biological Systems, Vehicular Networking and Connected Vehicles, Aerospace Systems, Automation, Manufacturing, Smart Grids, Nonlinear Systems, Power Systems, Robotics, Social Systems, Economic Systems and other. Of particular value to both the contributors and the readership are the short publication timeframe and the world-wide distribution and exposure which enable both a wide and rapid dissemination of research output.

Indexed by SCOPUS, DBLP, WTI Frankfurt eG, zbMATH, SCImago.

All books published in the series are submitted for consideration in Web of Science.

More information about this series at http://www.springer.com/series/13304

Jagdev Singh · Hemen Dutta · Devendra Kumar · Dumitru Baleanu · Jordan Hristov
Editors

Methods of Mathematical Modelling and Computation for Complex Systems

Springer

Editors
Jagdev Singh
Department of Mathematics
JECRC University
Jaipur, Rajasthan, India

Hemen Dutta
Department of Mathematics
Gauhati University
Guwahati, Assam, India

Devendra Kumar
Department of Mathematics
University of Rajasthan
Jaipur, Rajasthan, India

Dumitru Baleanu
Department of Mathematics and Computer Sciences
Çankaya University
Ankara, Turkey

Jordan Hristov
Department of Chemical Engineering
University of Chemical Technology and Metallurgy
Sofia, Bulgaria

ISSN 2198-4182 ISSN 2198-4190 (electronic)
Studies in Systems, Decision and Control
ISBN 978-3-030-77168-3 ISBN 978-3-030-77169-0 (eBook)
https://doi.org/10.1007/978-3-030-77169-0

This Springer imprint is published by the registered company Springer Nature Switzerland AG
The registered company address is: Gewerbestrasse 11, 6330 Cham, Switzerland

Preface

The book is developed to help understanding and studying various aspects associated with complex systems. The readers will find several new mathematical methods, mathematical models and computational techniques having significance and relevance in studying complex systems. The book consists of 16 chapters and they are organized as follows.

Chapter "On the Diffusion with Decaying Time-Dependent Diffusivity: Formulations and Approximate Solutions Pertinent to Diffusion in Concretes" aims to develop constitutive nonsingular functional relationships of time-dependent diffusivities pertinent to chlorides diffusion in concrete, in particular. It obtained approximated integral-balance solutions (Dirichlet problems) with singular and nonsingular diffusivities to allow minimization of the regression parameters recovered from experimental data and physically more adequate concept to diffusion avoiding the missing causality in the Fickian model applied in modelling of chloride ingress in concretes.

Chapter "Laminar Convection of Power-Law Fluids in Differentially Heated Closed Region: CFD Analysis" aims to analyse characteristics of heat transfer of non-Newtonian fluids in a natural convection application. It studied a 2D square domain containing power-law fluid whose horizontal walls follow adiabatic condition through insulation whereas the vertical walls are differentially heated isothermally. Various parameters viz. Nusselt number, dimensionless vertical velocity and dimensionless temperature have been evaluated to examine the effect of power-law index on heat and mass transfer for different values of Rayleigh number. It further analysed the influence of power-law index and Bingham number on the heat transfer characteristics and proposed the best one with high heat transfer capability for natural convection application. The results obtained are also compared in terms of Nusselt number, velocity and temperature with the help of TECPLOT and ANSYS.

Chapter "Mathematical Perspective of Hodgkin-Huxley Model and Bifurcation Analysis" proposes a modified Hodgkin-Huxley model by considering the higher power (5 and 6) of K activation in potassium ionic currents and studied the comparative behaviour of three models. It observed that the modified Hodgkin-Huxley model with a higher power of potassium activation reached resting state sooner and gains the stability (after oscillatory) at high external current, and that the qualitative behaviour of the modified model (with the higher exponential power) is different as there is shifting of Hopf bifurcation points in comparison with the original Hodgkin-Huxley model. Moreover, a larger periodic region is observed in most of the parameter phase spaces (external current I versus parameters) except against the Na conductance and Na potential. The modified Hodgkin-Huxley model which determines that higher power of K activation is found to be more significant for action potential in neurons.

Chapter "Mathematical Analysis of Two Unequal Collinear Cracks in a Piezo-Electro-Magnetic Media" deals with the problem of two unequal collinear semi-permeable cracks in a piezo-electro-magnetic media and addresses it mathematically using a complex variable technique. It modelled the problem mathematically as a non-homogeneous Riemann-Hilbert problem in terms of unknown complex potential functions. The solutions to the fracture parameters have been obtained in explicit forms by solving the Hilbert problem. A numerical case study has also been included for poled $BaTiO_3$ - $CoFe_2O_4$ ceramic cracked plate to show the effect of volume fractions and inter-crack distance on fracture parameters.

Chapter "Advanced Analysis of Local Fractional Calculus Applied to the Rice Theory in Fractal Fracture Mechanics" aims to include recent results for the analysis of local fractional calculus. It introduced and reviewed the local fractional derivative and the local fractional integral in the fractional (real and complex) sets, the series and transforms involving the Mittag-Leffler function defined on Cantor sets. The uniqueness of the solutions of the local fractional differential and integral equations and the local fractional inequalities has been considered in detail. The local fractional vector calculus has been applied further to describe the Rice theory in fractal fracture mechanics.

Chapter "General Fractional Calculus with Nonsingular Kernels: New Prospective on Viscoelasticity" considers the general fractional derivatives in the different kernel functions, such as Mittag-Lefller, Wiman and Prabhakar functions to model the viscoelastic behaviours in the real materials. It investigated the basic formulas of the fractional calculus in the kernels of the power, Mittag-Lefller, Wiman and Prabhakar functions, and discussed the applications for the general fractional calculus in viscoelasticity. As examples, the Maxwell and Voigt models with the general fractional derivatives have been considered to represent the complexity of the real materials.

Chapter "Group Dynamical Systems on C^*-Algebras Generated by Countable-Infinitely Many Semicircular Elements" aims to characterize starting from a C^*-probability space generated by mutually free, countable-infinitely many semicircular elements, the free distributional data induced by the semicircular elements by joint free moments. It then constructed a certain group acting on the C^*-probability space and studied under a corresponding group dynamical system. From the dynamics, the crossed product C^*-algebra is further constructed from the system, and the free probability on it is considered.

Chapter “Lie Group Theory for Nonlinear Fractional K(m, n) Type Equation with Variable Coefficients” aims to investigate the analytical solution of fractional order $K(m, n)$ type equation with variable coefficient which is an extended type of KdV equations into a genuinely nonlinear dispersion regime. It obtained the Lie point symmetries for this type of time fractional partial differential equations by using the Lie symmetry analysis. It further presented the corresponding reduced fractional differential equations corresponding to the time-fractional $K(m, n)$ type equation.

Chapter “Generalized Rayleigh Wave Propagation in Heterogeneous Substratum Over Homogeneous Half-Space Under Gravity” considers the propagation of Rayleigh waves in an incompressible heterogeneous medium with a general variation of rigidity; resting over another incompressible homogeneous half-space under the effect of gravity. Instead of using Whittaker’s function, the expansion formula proposed by Newland’s has been used to solve the equation of motion for better result in the incompressible half space. The velocity equations have been calculated and presented the results in figures. It observed that except for linear and quadratic variation of rigidity, the relation between the phase velocity of Rayleigh wave and the gravity being directly proportional to each other, and the phase velocity of Rayleigh wave in absence of gravity is smaller than the presence of gravity in all cases except linear and quadratic variation in rigidity.

Chapter “On Defining Trigonometric Box Spline-Like Surface on Type-I Triangulation” aims to define a trigonometric box spline surface on type-I triangulation by introducing a new non-stationary subdivision scheme. The limit surface obtained by the repeated application of this new scheme to an initial regular triangular mesh is a trigonometric box spline-like surface. Besides, having a algorithm, the limit surface is compactly supported, satisfies the convex hull property and is uniformly continuous. It also illustrated the performance of the scheme with some examples.

Chapter “Mathematical Modelling for Perishable Product Supply Chain Under Inflation and Variable Lead Time” presents an inventory model for deteriorating items under a real time situation in which the lead time varies with time. It developed a mathematical model for finding the total cost and order quantity in a finite planning horizon containing m number of cycles. The effects of inflation of currency, shortages and lead time along with the effect of information technology on lead time have been considered. It also obtained certain special cases for complete backlogging and instantaneous deterioration. The study has been illustrated for the error function as demand function. Further, it studied the effect of different parameters like deterioration rate and backlogging parameter on the order quantity and the total supply chain cost. Data and sensitivity analysis have been included in the chapter.

Chapter “Mexican Hat Wavelet Transform and Its Applications” discusses a unique method to time-frequency analysis which gives a centralized way to represent discrete and continuous time-frequency. It considered the Mexican hat wavelet to define the Mexican hat wavelet transform. Further, the theory of Mexican hat wavelet transform has been implemented to obtain the Mexican hat wavelet Stieltjes transform of a bounded variation function. It also included some properties of

Mexican hat wavelet Stieltjes transforms. Further, a standard method has been introduced for representing functions of class $B(m, n)$, and an integral transform has been constructed with the help of the Fourier summation kernel. The construction provided a way to present some conditions that are necessary and sufficient for a function of class $B(m, n)$ to be Mexican hat wavelet and Mexican hat wavelet Stieltjes transform.

Chapter "Fractal Fractional Derivative Operator Method on MCF-7 Cell Line Dynamics" considers the dynamics modelling breast cancer known as MCF-7 cell line by means of a system of ordinary differential equations. The dynamics have been extended to a system of fractal fractional partial differential equations. The well-posed, physiological level and stability conditions for the system of fractal fractional partial differential equations dynamics have been established. Since the extended dynamics are not solvable analytically, a fractal fractional numerical method has been derived, implemented and the results have been presented with respect to the derived stability conditions.

Chapter "The Exponentiated Half Logistic-Topp-Leone-G Power Series Class of Distributions: Model, Properties and Applications" aims to develop a new class of distributions, namely, the exponentiated half logistic-Topp-Leone-G power series class of distributions. It presented some special classes in the proposed distribution. Structural properties have also been derived including moments, entropy and maximum likelihood estimates. Further, it conducted a simulation study to evaluate the consistency of the maximum likelihood estimates, and presented two real data examples to illustrate the usefulness of the new class of distributions.

Chapter "Fixed Points of Multivalued $(\alpha_* - \phi)$-Contractions and Metric Transforms" considers the concept of existence of fixed point sets of multivalued mappings of metric spaces in connection with metric transforms. It considered $(\alpha$-$\phi)$ contractions, multivalued $(\alpha_*$-$\phi)$ contractions, $(\varepsilon - \phi)$ uniform local multivalued contraction and generalized multivalued $(\alpha_*$-$\phi)$ contractions. It aims to extend some fixed point results for multivalued contractions to the case multivalued $(\alpha_*$-$\phi)$ contractions. Further, it used the metrics which are sequentially (D_1), strong semi sequentially (D_2) and semi sequentially (D) equivalent to the Hausdorffmetric (H) on closed and bounded subsets (CB(Y)) to obtain more general fixed point results. It also presented some examples to support some findings.

Chapter "Approximation of Signals Via Different Summability Means with Effects of Gibbs Phenomenon" aims to investigate the notions of the deferred Cesàro, deferred Nörlund and their product summability means of the Fourier series. It estimated the degree of approximation of signal functions belonging to a generalized Lipschitz class by using these notions, and also established some new fundamental approximation theorems in classical sense. Further, it studied the statistical versions of these notions, and demonstrated some Korovkin-type approximation results for trigonometric test functions over a Banach space. It presented some examples demonstrating that the statistical versions of approximation results are stronger than the classical versions. Finally, as regard to the convergence of the Fourier series, the effect of the Gibbs phenomenon has been presented via the proposed means.

The editors are thankful to contributors for their cooperation and patience while the chapters were being reviewed and processed. Reviewers deserve gratitude for offering their kind help in finalizing the book. The editors thankfully acknowledge the encouragement received from many colleagues and friends concerning this book project.

Jaipur, India — Jagdev Singh
Guwahati, India — Hemen Dutta
Jaipur, India — Devendra Kumar
Ankara, Turkey — Dumitru Baleanu
Sofia, Bulgaria — Jordan Hristov
January 2021

Contents

About the Editors

Jagdev Singh is Professor of Mathematics at JECRC University, Rajasthan, India. He did his Master of Science (M.Sc.) and Ph.D. both in Mathematics from University of Rajasthan, India. His fields of research include Mathematical Modelling, Special Functions, Fractional Calculus, Fluid Dynamics, Analytical and Numerical Methods. He has published four books, viz., Fractional Calculus in Medical and Health Science (2020), Engineering Mathematics-I (2008), Engineering Mathematics-II (2013) and Advanced Engineering Mathematics-IV (2008). He has to his credit over 201 research papers published in various Journals of repute and his h-index is 44. He has attended a number of National and International Conferences and presented several research papers. He is on the Editorial Board of various Journals, and reviewer of several Journals. He is on the Clarivate Analytics list of high cited researchers in 2020. He won the Riemann Award for Young Scientists in Testimony of the High Regard of Achievements in the area of Fractional Calculus and its Applications in the First International Conference on Modern Fractional Calculus and Its Applications at Biruni University, Instanbul, Turkey, December 4–6, 2020.

Hemen Dutta is a regular faculty member in the Department of Mathematics at Gauhati University, Guwahati, India. He did his M.Sc, M.Phil and Ph.D. in Mathematics, and also completed Post Graduate Diploma in Computer Application. His current research interests include topics in nonlinear analysis and mathematical modelling. He is recipient of research projects including under SERB-MATRICS scheme. He is a regular and guest-editor of several SCI/SCIE indexed journals. He has also published several thematic issues in leading journals. He has to his credit over 200 publications as research articles, proceedings papers and chapters in books. He has authored and edited several books published by leading publishers. He has also organized 6 academic events so far and served as a speaker in various national and international events.

Devendra Kumar is Assistant Professor in the Department of Mathematics, University of Rajasthan, Rajasthan, India. He did his Master of Science (M.Sc.) and Ph.D. both in Mathematics from University of Rajasthan, India. His areas of interest are Mathematical Modelling, Special Functions, Fractional Calculus, Applied

Functional Analysis, Nonlinear Dynamics, Analytical and Numerical Methods. He has published four books, viz., Engineering Mathematics-I (2008), Engineering Mathematics-II (2013), Fractional Calculus in Medical and Health Science (2020) and Methods of Mathematical Modelling: Fractional Differential Equations (2020). He has to his credit over 218 research papers published in various Journals of repute and his h-index is 44. He has attended a number of National and International Conferences and presented several research papers. He is on the Editorial Board of several Journals of Mathematics. He is also a reviewer of various Journals. He is on the Clarivate Analytics list of high cited researchers in 2020. He won the Bertram Ross Award in Testimony of the High Regard of Achievements in the area of Fractional Calculus and its Applications in the First International Conference on Modern Fractional Calculus and Its Applications at Biruni University, Instanbul, Turkey, December 4–6, 2020.

Dumitru Baleanu is Professor at the Institute of Space Sciences, Magurele-Bucharest, Romania and visiting stuff member at Department of Mathematics, Cankaya University, Ankara, Turkey. His research interests include fractional dynamics and its applications, fractional differential equations, dynamic systems on time scales and Lie symmetries. He has published more than 1000 papers indexed in SCI. He is one of the editors of 5 books published by Springer and one published by AIP Conference Proceedings. He is an editorial board member of the journals indexed in SCI, viz., Mathematical Methods in Applied Sciences, Applied Numerical Mathematics, Mathematics, Journal of Computational and Nonlinear Dynamics, Symmetry and Fractional Calculus and Applied Analysis. He has received more than 15000 citations in journals covered by SCI and his Hirsch index is 65. He was on the Thompson Reuter list of high cited researchers in 2015, 2016, 2017, 2018, 2019 and 2020. He won the International Obada Prize in 2019.

Jordan Hristov is Professor of Chemical Engineering with the University of Chemical Technology and Metallurgy, Sofia, Bulgaria, graduated from the Technical University, Sofia, as an electrical engineer and with a Ph.D. degree in chemical engineering. His Doctor of Sciences thesis on nonlinear and anomalous diffusion models was successfully completed in 2018. Prof. Hristov has more than 39 years' experience in the field of chemical engineering with principle research interests mathematical modelling in complex systems, nonlinear diffusion and fractional calculus applications in modelling, in mechanics, heat transfer, etc. with more than 180 articles published in international journals. He is an editorial board member of Thermal Science, Particuology, Progress in Fractional Differentiation and Applications and Fractional and Fractal. The academic records in Publons and Clarivate indicate more than 2100 citations, H-index 22, and about 23 special issues in scientific journals driven by him.

On the Diffusion with Decaying Time-Dependent Diffusivity: Formulations and Approximate Solutions Pertinent to Diffusion in Concretes

Jordan Hristov

Abstract Constitutive non-singular functional relationship of time-dependent diffusivities pertinent to chlorides diffusion in concrete, in particular, have been conceived. Approximated integral-balance solutions (Dirichlet problems) with singular and non-singular diffusivities have been developed. The approximate solutions developed allow minimization of the regression parameters that should be recovered from experimental data and physically more adequate concept to diffusion avoiding the missing causality in the Fickian model widely applied in modelling of chloride ingress in concretes.

Keywords Diffusion · Chlorides · Concrete · Time-dependent diffusivity · Subdiffusion · Approximate solutions

1 Introduction

The problems of transient diffusions with non-constant diffusivities dependent on the time, spatial coordinate or concentration form a broad and challenging area of research. There is no need to refer here to any works but we have to stress the attention that escaping from the Fickian formulation of the diffusion model we immediately enter in a forest of problems with a variety of solutions. The solution approach applied to a certain non-linear diffusion problem depends mainly on two principle issues:

(i) The nature of the non-linearity, and
(ii) The imagination of the modeler applying suitable solution method.

J. Hristov (✉)
Department of Chemical Engineering, University of Chemical Technology and Metallurgy, Sofia 1756, 8 Kliment Ohridsky, blvd., Sofia, Bulgaria
e-mail: jordan.hristov@mail.bg

J. Singh et al. (eds.), *Methods of Mathematical Modelling and Computation for Complex Systems*, Studies in Systems, Decision and Control 373,
https://doi.org/10.1007/978-3-030-77169-0_1

Last but not least, it is important to recall that in mathematical modelling the solution of a given model is not the sole task, since calculations without physical interpretations of the results do not have much contributions to knowledge.

After these introductory words we like to stress the attention on a problem emerging in many situations when water containing some aggressive substances diffuses (penetrates) into building materials. The problem discussed in the sequel can be found in many articles related to life behaviour of concretes and concrete-based materials exposed to a variety of aqueous solutions (waters) with aggressive chloride ions (in marine waters, for example). Thus, let us see what is the diffusion problem solved and analyzed in this study.

1.1 General Assumptions

The penetration of aggressive substances into concretes provokes corrosion of steel rebars and changes in the cement matrices. Commonly fresh concretes are tested about 28 days (maturity days) [1–12], a time period chosen from practical reasons due to numerous observations [13–16]. The simplest suggestion dominating in the literature is that the chloride (the most aggressive penetrant into concretes) diffuses in accordance with the Fick's law [17, 18]

$$j = -D\frac{\partial C}{\partial x} \tag{1}$$

However there are many facts allowing to revise this assumption because:

- Concretes are too heterogeneous as structures consisting aggregates and cement matrices with voids and defects, and therefore the diffusion coefficients should be, at least spatially dependent, or time-dependent due to existing traps (voids in matrices or cracks at micro or macro levels). That is

$$D = D_0 f(x, t) \tag{2}$$

- The capillarity structures of the concretes changes with the maturity and affect the diffusion coefficient: at constant temperature the diffusion coefficients should be time-dependent.
- There are also situations where the diffusivities of aggressive ions (chloride) are concentration dependent [21], but time-dependence is the dominating case.

The too complex nature of the aggressive diffusion of ions into concretes casts doubts [13–16] about application of the simplest relationship between the flux and gradient (1). If the continuity equation

$$\frac{\partial C}{\partial t} = -\frac{\partial j}{\partial x} \tag{3}$$

and the general assumption of spatially independent but time-dependent diffusivity is applied we get

$$\frac{\partial C(x,t)}{\partial t} = D(t)\frac{\partial^2 C(x,t)}{\partial x^2} \tag{4}$$

and with common Dirichlet boundary condition

$$C(0,t) = C_0, \quad , C(x,0) = 0, \quad C(x \to \infty, t) = 0 \tag{5}$$

The conditions (5) take into account that when a concrete sample is under aggressive attack of ions they come from an infinite bath (i.e., for example, a concrete block in marine water) with practically unchangeable concentration. Moreover, due to the extremely slow diffusion process the model (4)–(5) considers a semi-infinite medium.

1.2 *Common Approaches in Modelling and Solutions*

1.2.1 Linearization

The model (4)–(5) is the dominating in the literature [1–15]. The suggestion that the diffusion coefficient is time-dependent makes the Eq. (4) non-Fickian and as we will see further in this work, the behavior of the diffusion process reveals anomalous subdiffusion character.

A simple way to linearize (4) is to apply the Kirchhoff-type transform [17]

$$I(t) = \int_0^t D(z)dz = D_0 \int_0^t f(z)dz \tag{6}$$

which transforms (4) into a classical diffusion equation

$$\frac{\partial C(x,I)}{\partial I} = D_0 \frac{\partial^2 C(x,I)}{\partial x^2} \tag{7}$$

allowing to apply the classical solution [17, 18]

$$C(x,t) = C_{initial} + (C_{sat} - C_{initial})\, erf\left[\frac{x}{2\sqrt{D_0 I(t)}}\right] \tag{8}$$

The transform (6) was slightly modified here, introducing the functional relationship (2) in the form $D = D_0 f(t)$ where D_0 is the diffusion coefficient of Fickian diffusion. This was especially done in order to preserve the correctness of (7) and

(8) since as it is well-known that in such a case the Boltzmann similarity variable $\eta = x/\sqrt{D_0 t}$ defines the length scale $L = \sqrt{D_0 t}$. Thus, in the linearized equation the new length scale is $L_I = \sqrt{D_0 I}$ and the corresponding similarity variable is $\eta_I = \frac{x}{\sqrt{D_0 I(t)}}$. Further, this approach needs the time-dependent function $f(t)$ to be preliminary known (adequately selected), easily integrable, and the most important point is to allow easy backward transform from the solution in terms of $I(t)$ to a solution expressed in terms of the physical time t.

The same approach was used by Yeih et al. [9]. Very instructive text how this method should be applied step-by-step is provided in the article of Weitsman [19]. The method is widely applied (see further in the text) and the book of Poulsen and Mejlbro (Chap. 2 of [20]) presents it as almost standard approach.

To close these comments, an average diffusion coefficient D_{av} defined by [17, 18, 28, 29]

$$D_{av} = \frac{1}{t}\int_0^t D(z)dz \tag{9}$$

was used for simplification of the solution. Since the integral defining $I(t)$ (6) physically means the area between the axis of the time t and the curve describing the time evolution of $D(t)$, this approach requires correct measurements and post-experimental treatments of many samples; but this will lead to different results as the concrete compositions and experimental conditions change. With this approach the solution is straightforward and similar to (8) [17, 18]

$$C(x,t) = C_{initial} + (C_{sat} - C_{initial})\, erf\left[\frac{x}{2\sqrt{D_{av}t}}\right] \tag{10}$$

1.2.2 Fractional Modelling Approach

Despite the dominating modelling of chloride diffusion by the Fick's model with a time-dependent diffusion coefficient an alternative approach, which corresponds, from a physical basis, to the results obtained with power-law diffusion coefficient (see Sect. 3.1 and the corresponding solutions in Sect. 5.1), is the application of time-fractional sub-diffusion model [30, 31], namely

$$\frac{\partial^\alpha C(x,t)}{\partial t^\alpha} = D_\alpha \frac{\partial^2 C(x,t)}{\partial x^2} \tag{11}$$

Here D_α with dimension $[m^2/s^\alpha]$ is the diffusion coefficient, while the time fractional Caputo derivative of order $0 < \alpha < 1$ represents the time-shift between chloride flux j_a penetrating the concrete and its gradient (see Sect. 6).

Following the same idea, Chen et al. [30], applied the model (11) with variable (time-dependent) fractional order $\alpha(t)$ using the Caputo formulation of fractional derivative [32], namely

$$\frac{\partial^{\alpha(t)}}{\partial^{\alpha(t)}} = \frac{1}{\Gamma(n-\alpha(t))}\int_0^t \frac{1}{(t-z)^{\alpha(t)+1-n}}\frac{df(z)}{dz}dz, \quad n-1 < \alpha(t) \leq 1 \tag{12}$$

where the variable-order fractional parameter is defined as a function of time t [33]

$$\alpha(t) = \alpha_0 + q\frac{1}{T_{max}}, \quad \alpha_0 = 0.6 \tag{13}$$

In the context of this chapter, in (13) T_{max} should equal the time of concrete maturation, denoted here is t_0 (see the sequel); q is a proportional parameter with dimensions $\left[time^{-1}\right]$.

Fractional models will be considered further in this chapter but this will be a natural consequence of the outcomes of the integral-balance solutions with time-dependent diffusivity (power-law functional relationship) (Sects. 3.1 and 5.1) and non-singular $D(t)$ (Sects. 5.2 and 5.3). Readers interested in solutions of time-fractional diffusion equations with time-dependent diffusivities may find detailed information elsewhere [34–38].

2 Motivation of This Study

The main idea of this study is to apply the integral-balance method (see Sect. 4.2) to diffusion equation with time-dependent diffusivity (power-law) and alternative functional forms which do not exhibit singularities (unbounded) at $t = 0$. The method suggested is approximate and does not need initial transformation of the diffusion coefficient and linearization (as commented above). Moreover, the solutions are straightforward for the power-law relationship (solved here as a comparative example) and the new suggested time-dependent diffusivities which are bounded at $t = 0$.

Moreover, we will see that the results of these approximate solutions exhibit subdiffusive behaviours so it is challenging to find how the corresponding time-fractional diffusion model could be build (Sect. 6), as well as to discuss the causality of the original model (4) and time-fractional one.

2.1 Aim and Chapter Organization

2.1.1 Aim

Briefly, the particular aims of this study can be outlined as:

- A brief overview of the existing functional relationships about $D(t)$ and their analysis.

- New non-singular definitions (35) and (36) of time-dependent diffusivities.
- Analytical solutions of diffusion equation in three cases of time-dependent (decaying) diffusivities (the singular paper-law (14)) and non-singular formulations, and analyses of their applicabilities thereof.
- Numerical experiments with the approximate solutions developed.
- Analysis of the subdiffusive behaviour of the approximate solutions and related alternative model constructions with time-fractional derivatives.

2.1.2 Text Organization

The sequel of this chapter is organized as follows: A brief overview of the existing function relationships of $D(t)$ is performed in Sect. 3. Section 4 presents two new forms of non-singular power-law relationship about D(t) (Sect. 4.1); Sect. 4.2 presents the principle integration techniques used of the integral-balance method. Section 5 presents approximate solutions in three cases: with a singular power-law relationship known the literature (Sect. 5.1) and two solutions with newly defined non-singular power-law $D(t)$ (Sects. 5.2 and 5.3). Numerical experiments with the obtained approximate solutions are presented in Sect. 5.4. The outcomes of the approach in modelling with the non-singular $D(t)$ are summarized in Sect. 6 relating them to fractional in time diffusion models. The chapter is closed by Sect. 7 where briefly the main results are outlined.

3 Existing Functional Relationships of D(t): A Brief Overview

The continuous change and refinement in the concrete structure during the maturing stage $0 < t < t_0$ involves several mechanisms of mechanical, rheological and chemical natures acting simultaneously, that finally at macroscopic level results in time-decaying of the chloride diffusivity [1–12, 43].

3.1 Singular (Power-Law) Time Dependence of D(t)

3.1.1 Simple Power-Law D(t)

Generally, the power-law relationship of the time-dependent diffusion coefficient is a consequence of application of the Fickian diffusion model to chloride penetration in concretes and the attempts to adjust the data to the simplified analytical solution [39]. This process of chloride perpetration into concrete is divided into two stages [39]

(i) Maturation stage of 28 days, denoted here as *reference time* t_0, and
(ii) Diffusion beyond the maturation stage where the diffusion coefficient should be constant (time-independent).

In the fist stage we have

$$D_1(t) = \left(\frac{t}{t_0}\right)^k, \quad -1 < k < 0 \tag{14}$$

This functional relationship is unbounded at $t = 0$ and causes problems in calculations. For example, making a formalistic solution of (4)–(5) with $D_1(t)$ it was shown in [12] that the solution is

$$C(x,t) = C_0\left[1 - erf\left(\frac{x}{2\sqrt{D_1(t)t}}\right)\right] \tag{15}$$

This is generally incorrect because introducing the Boltzmann similarity variable $\eta = x/\sqrt{D_0 t}$ and $D = D_0 = const$ we can obtain solutions as either (8) or (9) but not (15); this is easy to check by transforming (4) in terms of $\eta = x/\sqrt{D_0 t}$ when $D(t)$ is defined by (14).

As reported by Wu et al. [39–41] the chloride diffusion is modeled by (14) in two forms

In the first stage for $t < t_0$ (Maturation stage)

$$D_{mat}(t) = D_{ref}\left(\frac{t}{t_0}\right)^{-k} \tag{16}$$

In the second stage for $t > t_0$

$$D_{mat}(t) = D_{ref}\left(\frac{t_R}{t_0}\right)^{-k} \tag{17}$$

where D_{ref} is taken at the beginning of the test lasting 200 days. (corresponds here to the symbol D_0). Moreover, the time t_R is taken 30 years, when it is assumed that the diffusion coefficient is constant. Since the ratio $\left(\frac{t_R}{t_0}\right)$ is constant, we may accept that the time dependence of the diffusivity ends at t_0.

As to the exponent in the power-law relationship, in accordance with the American Life 365 [40–42, 42] the value of k is assumed equal to 0.34 which corresponds to the value used in the earlier study of Mangat and Molloy [2] (see also comments in [39–41]). The tests and modelling approach of Stanish and Thomas [5] revealed that the value of k varied from 0.259 up to 0.565 (similar range 0.32–0.60 was reported by Nokken et al. [6]) and from 0.32 to 0.79 for fly ash based concrete). In addition from [7] (based on the results in [8]) the value of k varies in the range 0.71–0.84. The experiments of Audenaert et al. [11] revealed a range of 0.22–0.43. Hence in the dominating cases $0 < k < 1$ even though in some cases values in the range 1.0–1.5

were determined by data fitting [2]. Despite this, hereafter this study will consider cases with power-law time-dependent diffusivity when $0 < k < 1$.

In general, the power-law form of the diffusion coefficient in disordered medium follows from the short time asymptotics [44]

$$\frac{D(t)}{D_0} \to 1 - A_0\,(D_0 t)^{1/2} + B_0(D_0 t) + O(D_0 t)^{3/2} \tag{18}$$

and

$$\frac{D(t)}{D_0} \to \alpha + \frac{\beta}{t} + \frac{\gamma}{t^{3/2}} + \cdots \tag{19}$$

where the coefficients α,β and γ are assumed to depend on the detailed geometry of porous medium. This functional relationship is thought to be universal [44] but unbounded for $t = 0$. Despite this, the attempts to fit experimental data by the Fickian model (4) defined by different physical methods [45–48] (and the references therein) led to common use of the relationship (14) [1–12, 43]. As example , the plots in Fig.1 show how the power-law varies when the exponent k is in the range $0 < k < 1$.

Some modifications of the simple power-law are defined, with constant or time-dependent boundary condition, for example:

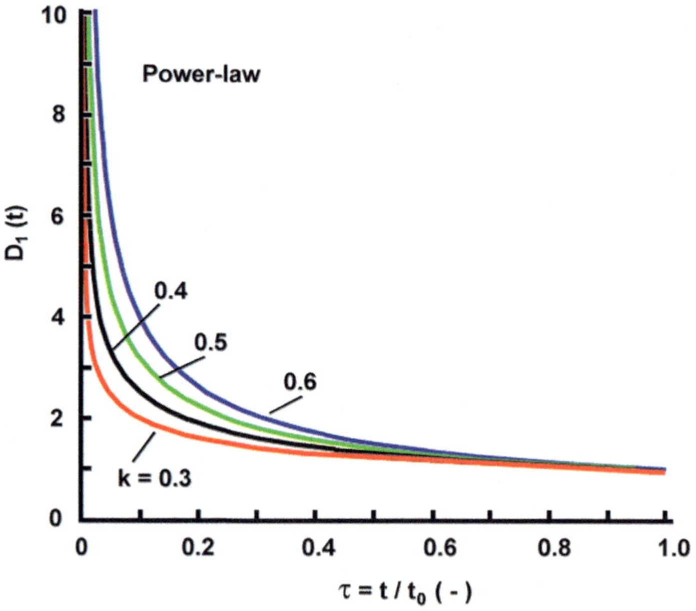

Fig. 1 Power-law $D/D_0 = (t/t_0)^{-k}$ for different values of the exponent k

- Mangat and Molloy [2]

$$D_M(t) = \frac{D_0}{1-m}\left(\frac{t}{t_0}\right)^{-m}, \quad C_0 = C(0,t) = const. \tag{20}$$

- Costa and Appleton [4]

$$D_a = D_R\left(\frac{t_R}{t}\right)^m, \quad D_R = D_\infty, \quad , C(x=0,t) = C_0 t^{n_c} \tag{21}$$

  where D_∞ corresponds to the end of the diffusion test.
- Pack et al. [23]

$$D_a(t) = \frac{D_R}{1-m}\left(\frac{t_R}{t}\right)^m, \quad C(x=0,t) = C_0\left[ln(\beta t+1)\right] + k_p \tag{22}$$

3.1.2 Composite Power-Law D(t)

The research reports encompass wide range of type of concretes (different compositions and technologies of production) and mild or severe conditions in waters containing aggressive ions, mainly chlorides. The data fitting resulted in different relationships, of power-law type but with composite structures. We briefly present them in order to show what is the realistic background in the modelling with time-dependent diffusivities, and mainly to emphasis on the fact that such relationships may fit experimental data in some particular cases, but do not contribute to model build-up.

- Petcherdchoo [7]

$$D_a = \frac{D_R}{1-m}\left[\left(1+\frac{t_I}{t_E}\right)^{1-m} - \left(\frac{t_I}{t_E}\right)^{1-m}\right]\frac{t_R}{t_g}t_E, \quad C(x=0,t) = C_0\sqrt{t} \tag{23}$$

  where t_g is the time to exposure to chlorides; D_R is the diffusion coefficient at the reference time t_R
- Stanish and Thomas [5]

$$D_a = \frac{D_R}{1-m}t_R^m\left(\frac{t^{1-m} - t_I^{1-m}}{t-t_I}\right), \quad C_0 = C(0,t) = const. \tag{24}$$

  where t_I time of first exposure to chlorides.
- Maheswaran and Sanjayan [22]

$$D_a(t) = \frac{D_g}{1-m}\left[\left(\frac{t_R}{t}\right)^m - \frac{t_R}{t}\right], \quad C(c=0,t) = C_0 1 - exp(-n_m t) \tag{25}$$

- **Note**: We can see all these functional relationships contain too many regression parameters such as C_0, m, n_p,n_c, n_m, w, etc, as well no unified symbols allowing direct comparison of the functional relationships of suggested $D(t)$ exist. Moreover, the common idea is to convert the initial equation with suggested functional relationship of $D(t)$ to a diffusion equation with Dirichlet or time-dependent boundary conditions. But, we have to mention, that all theses regression parameters come from fittings of experimental data with solutions of the classical diffusion equation; and then adjusting what functional relationship would fit better the data points (see the book of Poulsen and Mejlbro [20]). Moreover, with too many regression parameters, as mentioned at the beginning of this point, sometimes it appears that they are interrelated; too many regression parameters is not the adequate approach in approximation of experimental data.

3.2 Non-singular Forms of D(t)

3.2.1 Non-singular D(t) of Power-Law Type

Taking into account the singularity of the power-law (14) at $t = 0$, Wang and Fu [12] (see also Yang et al. ([24]) suggested an alternative functional relationship

$$D_{WF}(t) = D_0 \left(\frac{t_0}{1 + t_0} \right)^w \tag{26}$$

where $w = 0.473$ and $D_0 = 4.44 \times 10^{-12}$ m^2/s for chloride ions disillusion in concrete.

The function of $D_{WF}(t)$ can be presented in a more convenient for the present analysis form, namely

$$D_{WF}(t) = D_0 \left(\frac{1}{1 + \frac{t}{t_0}} \right)^w = D_0 \left(1 + \frac{t}{t_0} \right)^{-w} \tag{27}$$

This function is bounded at $t = 0$, since $D_{WF}(t = 0) = 1$ and goes to zero as $t \to \infty$, but at $t = t_0$ we have $D_{WF}(t = t_0) = (1/2)^w = (1/2)^{0.473} \approx 0.720$ (see Fig. 2).

The integration of the time-varying diffusivity with respect to the exposure time (i.e. the linearization approach) defines

$$T = \int_0^t \left(\frac{t_0}{z + t_0} \right)^n dz \Rightarrow D(t) = D_0 \frac{dT}{dt} \tag{28}$$

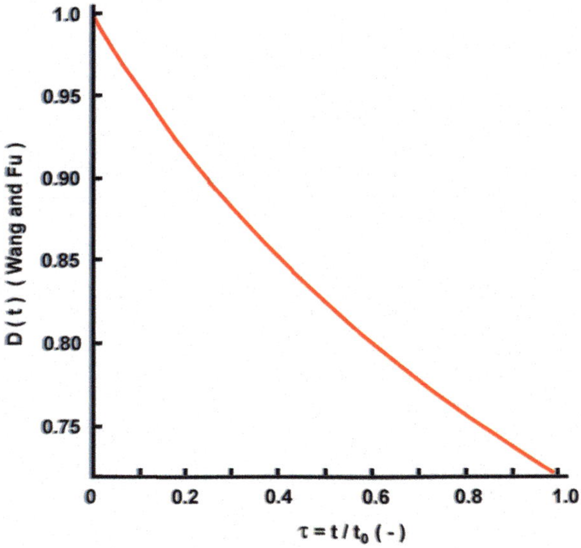

Fig. 2 Non-singular power-law of Wang and Fu

where

$$dz = \frac{t_0^n}{1-n}\left[\left(1+\frac{t_0}{t}\right)^{1-n} - \left(\frac{t_0}{t}\right)^{1-n}\right] t^{1-n} = F(t) \tag{29}$$

This allows the governing equation to be converted to

$$\frac{\partial \widehat{C}(x,T)}{\partial T} = D_0 \frac{\partial \widehat{C}(x,T)}{\partial x^2} \tag{30}$$

with a boundary condition at x_0 as a time-varying function $\widehat{C}_0 = \widehat{C}(x,T)$. This allows to apply the Duhamel theorem and obtain a solution with respect $\widehat{C}(x,T)$ as (we skip here details of the solution available elsewhere [24])

$$C(x,t) = (C_0 - C_\infty) erf \frac{x}{2\sqrt{D_0 F(t)}} + C_0, \quad C_\infty = C(x, t=0) \tag{31}$$

To some extent the transformation (29) coincides with definition of Petcherdchoo [7] of an averaged diffusion coefficient (see Eq. (23) in a slightly modified form

$$D_a = \frac{t_0^n}{1-n}\left[\left(1+\frac{t_0}{t}\right)^{1-n} - \left(\frac{t_0}{t}\right)^{1-n}\right]\left(\frac{t_0}{t}\right)^n \tag{32}$$

and resulting in a solution

$$C(x,t) = C_0\left[erfc\left(\frac{x}{2\sqrt{D_a t}}\right)\right] + k\sqrt{t}\left[exp\left(-\frac{x^2}{4D_0 t}\right) - \left(\frac{x\sqrt{\pi}}{2\sqrt{D_0 t}}\right) erfc\left(\frac{x}{2\sqrt{D_0 t}}\right)\right] \tag{33}$$

Therefore, despite the non-singular formulation (26) the approach is to apply all existing knowledge from the analytical solutions of heat conduction [25] and diffusion [26] as it was already demonstrated in Sect. 1.2.1.

3.2.2 Non-singular D(t) of Exponential Type

Sun et al. [27] considering chloride diffusion in concrete suggested a time-dependent diffusion coefficient

$$D_{sun} = (1 - \phi_0)\left[1 - exp\left(-a_D \frac{t}{t_0}\right)\right] \tag{34}$$

with $t_0 = 730$ days and a_D (depending on chloride concentration) in the range 0.177–1.509, and ϕ_0 is the initial porosity of concrete. This is a function of $D(t)$ bounded at $t = 0$, but the solution is only numerical [27].

4 New Forms of D(t) and Method of Approximate Solutions

Now, after the analysis of the existing situation in modelling of time-dependent diffusivities related to chloride ions ingress in concrete, the present study suggests two simple non-singular power-law functional relationships of $D(t)$ with two principle goals:

- The main reason is the simplicity and reduction of the number of regression parameters that should be recovered from experimental data. Furthermore, a principle issue is to remove the singularity widely spreading in the existing models of $D(t)$ and suggest bounded relationships at $t = 0$ (Sect. 4.1).
- Last but not least, the approximate solutions developed by the integral-balance method (Sect. 4.2) avoids the non-causality of the classical diffusion equation used in all studies commented in the preceding sections; its concept of finite penetration depth is physically based correction of the infinite speed of the parabolic diffusion equation.

4.1 New Non-singular Forms of D(t)

As it follows from the preceding sections the diffusivity changes in time till the point t_0 and for $t > 0$ it is assumed constant (denoted here as D_∞). Two alternative forms of time-dependent diffusion coefficient are suggested in this study, namely

$$D_2(t) = D_0\left[1 - \left(\frac{t}{t_0}\right)^p\right], \quad 0 < p < 1, \qquad 0 \le t \le t_0 \tag{35}$$

$$D_3(t) = D_0\left(\frac{t_0 - t}{t_0}\right)^m = D_0\left(1 - \frac{t}{t_0}\right)^m, \quad 0 < m < 1, \qquad 0 \le t \le t_0 \tag{36}$$

Both functions are bounded at $t = 0$ and go to zero at $t = t_0$. This formulation is not in conflict with aforementioned fact about $D_\infty = const. \neq 0$. A simple vertical shift of the origin of the ordinate in the representation of $D(t)/D_0 = f(t)$ as a function of time moves the zero at D_∞. Hence, in this co-ordinate system at $t = t_0$ the diffusivity becomes zero and the diffusion process ceases (actually ceases the diffusion with the time-dependent diffusivity).

Moreover, at it will be shown in the sequel (the sections devoted to the solution) the depth of the diffusant penetration zone, a measurable value in experiments, allows to establish the values of the exponents p and m. Simulations with (35) and (36) for different values of p and m are shown in Fig. 3. The plots indicate that the function (35) always generates concave profiles, while (36) generates concave profiles for $m > 1$ and convex curves for $0 < m < 1$.

Both functions are distinct (Fig. 4a) or close to each other (Fig. 4b) depending of the selections of the exponents p and m. Actually, the condition both curves to coincide leads to the following equation (with argument $X = t/t_0$)

$$\left(1 - X^p\right) = (1 - X)^m, \quad 0 \le X \le 1, \quad 0 < p < 1, \quad m > 1 \tag{37}$$

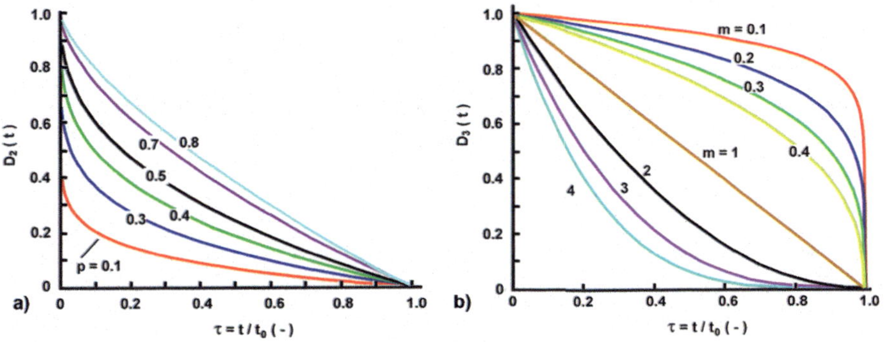

Fig. 3 Non-singular functions of decaying diffusivity for various values of the exponents p and m: **a** $D_2(t)$; **b** $D_3(t)$ used in this study

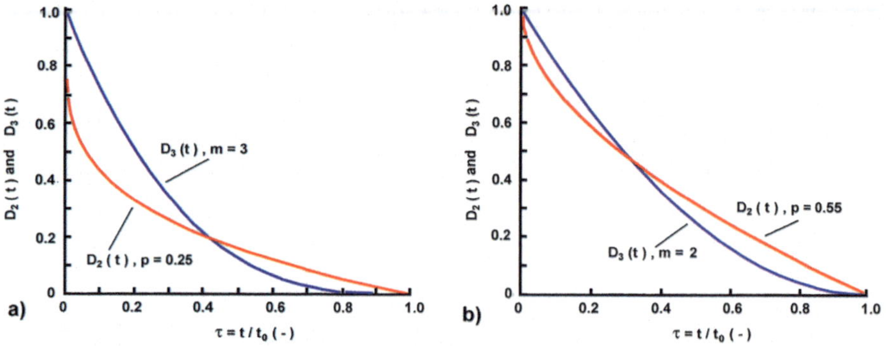

Fig. 4 Non-singular functions $D_2(t)$ and $D_3(t)$ used in this study: **a** Distinct behaviour; **b** Close behaviour

which has two trivial solutions: $X_1 = 0$ and $X_2 = 1$. Therefore, despite the different behaviours the relationships (35) and (36) satisfy the conditions at the boundaries of the interval $0 < X = t/t_0 < 1$.

4.2 Integral-Balance Method

The present chapter addresses approximate solutions of the problem represented by (4)–(5), developed by the integral-balance method in two basic versions: Heat-balance integral method (HBIM) [49, 50] and the Double Integration Method (DIM) [51, 52, 57]. These methods have been commonly applied to diffusion problems (heat and mass) with constant [50], power-law (concentration-dependent) diffusivities (with local [53] and non-local (fractional) derivatives [54]) as well as to time-fractional [55] and space-fractional diffusion problems [56].

The principle assumption of the integral-balance method is the concept of a finite penetration depth of the diffusion substance into the medium. This is actually a physically based correction of the infinite speed of all parabolic models and indirect application of the principle of the causality (see Sect. 6 and the point Sect. 6.3). Mathematically, this concept replaces the last boundary condition in (5) by

$$C(\delta, t) = \frac{\partial}{\partial x} C(\delta, t) = 0 \tag{38}$$

Thus, the concept defines a front $\delta(t)$ separating the medium into two zones: disturbed zone with $C(x, t) > 0$ for $0 \leq x \leq \delta(t)$ and undisturbed zone ($x > \delta(t)$) where the medium is virgin with $C(x, t) = C(x, t = 0)$.

The second condition in (38) means that no diffusion occurs across the front line, while the first one lead to sharp crossing behaviour of the profile (solution) with the coordinate axis. The Integral-balance approach requires the concentration profile to satisfy the boundary conditions imposed at $x = 0$ and $x = \delta$ irrespective of its functional relationships to the spatial coordinate x and the time t.

Details of basic integration techniques of the integral-balance method used in this study are briefly presented next.

4.2.1 HBIM Integration Technique

Integrating (4) over the penetration depth (from 0 to δ) we get

$$\int_0^\delta \frac{\partial C(x,t)}{\partial t} dx = \int_0^\delta \frac{\partial}{\partial x}\left(D(t)\frac{\partial C(x,t)}{\partial x}\right) dx \tag{39}$$

$$\int_0^\delta \frac{\partial C(x,t)}{\partial t} dx = D(t)\frac{\partial C(x,t)}{\partial x}\bigg|_\delta - D(t)\frac{\partial C(x,t)}{\partial x}\bigg|_{x=0} \tag{40}$$

and therefore applying the Leibniz rule we get

$$\frac{d}{dt}\int_0^\delta C(x,t)dx = -D(t)\frac{\partial C(x,t)}{\partial x}\bigg|_{x=0} \tag{41}$$

The relation (41) is the basic equation of the Heat-balance Integral method (HBIM) [49, 50]. Physically it means that the time variation of the accumulated in the medium substance (the integral of $C(t)$ over the diffusion layer $0 \leq< x \leq< \delta$) is controlled by the mass flux at the boundary $x = 0$ (the right-hand side of (41)).

Then, replacing $C(x,t)$ by an assumed profile $C_a(x,t)$ with respect to the spatial co-ordinate (polynomial or exponential [49, 50]) in both sides of (40) the integration yields an ordinary differential equation about δ. The main deficiency of HBIM is that the right-side of (40) contains the gradient $(\partial C(x,t)/\partial x)_{x=0}$ which has to be expressed through the assumed profile.

4.2.2 DIM Integration Technique

The first step of the DIM [53, 55] is the integration from 0 to x (which differs from the classical approach [57, 58]) and the result is

$$\int_0^x \frac{\partial C}{\partial t} dx = D(t)\frac{\partial C}{\partial x}\bigg|_x - D(t)\frac{\partial C}{\partial x}\bigg|_{x=0} \tag{42}$$

Further, representing the integration in left side of (40) as $\int_0^\delta f(\bullet)dx = \int_0^x f(\bullet)dx + \int_x^\delta f(\bullet)dx$ and then subtracting (42) from (40) we get an integral relation in the zone at the vicinity of the front δ, namely

$$\int_x^\delta \frac{\partial C}{\partial t}dx = -D_0(t)\frac{\partial C}{\partial x}\bigg|_x \tag{43}$$

The integration of (43) from 0 to δ yields

$$\int_0^\delta \int_x^\delta \frac{\partial C}{\partial t}dx = D(t)C(0,t) \tag{44}$$

The expression (44) is the principle equation of the Double Integration Method (**DIM**) in case of exponential diffusivity [53, 55].

4.2.3 Assumed Profile

The solutions envisage application of an assumed parabolic profile with unspecified exponent [50, 52, 58], i.e.

$$C_a(x,t) = C_0\left(1 - \frac{x}{\delta}\right)^n \tag{45}$$

With this profile we have $C_a(0,t) = 1$ and $C_a(\delta,t) = 0$, i.e. it satisfies the boundary conditions imposed by the finite penetration depth concept, for any value of the exponent n. The integral-balance method suggests replacement of $C(x,t)$ by $C_a(x,t)$ in the integral relation (40) or (44) thus developing an equation about $\delta(t)$ as it is demonstrated next.

5 Approximate Solutions of (4) with Different D(t)

This section is devoted to approximate (integral-balance) solutions of the diffusion Eq. (4) with three forms of of time-decaying diffusivity: $D_1(t)$ represented by (14), $D_2(t)$ and $D_3(t)$ expressed by (35) and (36), receptively.

5.1 *Solutions with $D_1(t)$*

Denoting $\tau_1 = t/t_0$ we get the following form of Eq. (4), by change of the variables only with respect to the time t, that is

$$\frac{\partial u}{t_0 \partial \tau_1} = D_0 \left(\frac{t}{t_0}\right)^k \frac{\partial^2 u}{\partial x^2} \tag{46}$$

$$\frac{\partial u}{\partial \tau_1} = (D_0 t_0)\, \tau_1^k \frac{\partial^2 u}{\partial x^2} \Rightarrow \frac{\partial u}{\partial \tau_1} = D_e \tau_1^k \frac{\partial^2 u}{\partial x^2} \tag{47}$$

The apparent coefficient D_e in (47) can be easily defined from $D_e = D_0 t_0$ and its dimension should is $\left[m^2\right]$ since the equation is dimensionless with respect the time only.

5.1.1 HBIM Solution

The HBIM technique applied to (47) yields

$$\int_0^\delta \frac{\partial u}{\partial \tau_1} dx = \int_0^\delta D_e \tau_1^k \frac{\partial^2 u}{\partial x^2} dx \tag{48}$$

$$\frac{d}{dt} \int_0^\delta u\,(x,t)\, dx = -D_e \tau_1^k \frac{\partial u\,(0,t)}{\partial x} \tag{49}$$

Now, replacing $C\,(x,t)$ by $C_a = C_0(1 - x/\delta)^n$ in (49) , precisely assuming for simplicity $C_0 = 1$, we get

$$\frac{d}{d\tau_1} \frac{\delta}{n+1} = D_e \tau_1^{-k} \frac{n}{\delta} \tag{50}$$

$$\frac{d}{d\tau_1} \delta^2 = D_e \tau_1^{-k} 2n\,(n+1) \tag{51}$$

For $t = 0$, the physical condition is $\delta\,(t=0) = \delta\,(\tau_1 = 0) = 0$.

$$\delta^2 = D_e \tau_1^{1-k} 2n\,(n+1) \tag{52}$$

Thus, $\delta_{1(HBIM)}$ can be expressed as

$$\delta_{1(HBIM)} = \sqrt{D_e \tau_1^{1-k}} \sqrt{\frac{2n\,(n+1)}{1-k}} \tag{53}$$

or in terms of the physical time t we have

$$\delta_{1(HBIM)}(t_0) = \sqrt{D_0 t_0}\sqrt{\left(\frac{t}{t_0}\right)^{1-k}}\sqrt{\frac{2n(n+1)}{1-k}} \tag{54}$$

The fact that δ is a physical quantity with a dimension of length, the condition $\delta > 0$ requires $1 - k > 0$. This is automatically satisfied because by definition $-1 < k < 0$ and therefore always we have $1 < 1 - k < 2$ for example with $k = -0.5$ we have $1 - k = 1.5$.

As a consequence related to the further calculations in this work we have $0 < (1 - k)/2$. From (53) and more obvious from (54) we can see that the *time evolution of the diffusion front does not follows the Fickian square-root model*, but exhibits *a subdiffusion behaviour* because $\delta_{1(HBIM)} \equiv \tau_1^{\frac{1-k}{2}}$ and $\delta_{1(HBIM)}(t) \equiv t^{\frac{1-k}{2}}$. For instance, with $k = -0.5$, used as example earlier, we have $\delta_{1(HBIM)}(t) \equiv t^{0.75}$. Moreover, since at $t = t_0$ we have $t/t_0 = 1$ we can see directly that the front attains the value

$$\delta_{1(HBIM)}(t_0) = \sqrt{D_0 t_0}\sqrt{\frac{2n(n+1)}{1-k}} \tag{55}$$

Hence, the approximate solution developed by HBIM is

$$C_{1(HBIM)} = \left(1 - \frac{x}{\sqrt{D_0 t_0}\sqrt{\left(\frac{t}{t_0}\right)^{1-k}}\sqrt{\frac{2n(n+1)}{1-k}}}\right)^n \tag{56}$$

5.1.2 DIM Solution with $D_1(t)$

With DIM we have

$$\int_0^{\delta}\left(\int_0^{x} \frac{\partial u}{\partial \tau_1} dx\right) dx = \int_0^{\delta}\left(\int_0^{x} D_e \tau_1^{-k} \frac{\partial^2 u}{\partial x^2} dx\right) dx \tag{57}$$

Replacing $C(x, t)$ by $C_a/C_0 = (1 - x/\delta)^n$ in (57) and after the integration we get the following equation

$$\frac{d}{d\tau_1}\frac{\delta}{(n+1)(n+2)} = D_e \tau_1^{-k} \tag{58}$$

$$\frac{d}{d\tau_2}\delta^2 = D_e \tau_1^{-k}(n+1)(n+2) \tag{59}$$

With the physical condition $\delta(t=0)=\delta(\tau_1=0)=0$ the result is

$$\delta^2 = D_e\tau_1^{1-k}\frac{(n+1)(n+2)}{1-k} \tag{60}$$

$$\delta_{1(DIM)} = \sqrt{D_e\tau_1^{1-k}}\sqrt{\frac{(n+1)(n+2)}{1-k}} \tag{61}$$

In terms of the physical time t the penetration depth $\delta_{1(DIM)}$ can be expressed as

$$\delta_{1(DIM)}(t_0) = \sqrt{D_0 t_0}\sqrt{\left(\frac{t}{t_0}\right)^{1-k}}\sqrt{\frac{(n+1)(n+2)}{1-k}} \tag{62}$$

or as

$$\delta_{1(DIM)}(t_0) = \sqrt{D_0 t}\sqrt{t_0\left(\frac{t}{t_0}\right)^{1-k}}\sqrt{\frac{(n+1)(n+2)}{1-k}} \tag{63}$$

The scaled penetration depth is

$$\frac{\delta_{1(DIM)}(t_0)}{\sqrt{D_0 t_0}} = \sqrt{\left(\frac{t}{t_0}\right)^{1-k}}\sqrt{\frac{(n+1)(n+2)}{1-k}} \tag{64}$$

At $t=t_0$

$$\delta_{1(DIM)}(t_0) = \sqrt{D_0 t_0}\sqrt{\frac{(n+1)(n+2)}{1-k}} \tag{65}$$

$$\frac{\delta_{1(DIM)}(t_0)}{\sqrt{D_0 t_0}} = \sqrt{\frac{(n+1)(n+2)}{1-k}} \tag{66}$$

Since the penetration depth is a physically defined distance we have the condition $\delta \geq 0$, as mentioned in previous point. The result is the same as developed with HBIM, i.e. the front exhibits a subdiffusion behavior, and the only difference appears in the numerical coefficients depending on the exponent n. Moreover, in both solutions (53)–(54) and (61)–(62) we can see that the penetration depth is controlled by the value of the denominator $\sqrt{1-k}$ and bearing on mind that $k<0$ we have $1-k>1$. Therefore, the increase in $|k|$ reduces the length of the penetration zone, and *vice versa*.

To complete this section and comments on the results we have to stress the attention that subdiffusion behavior is direct consequence of the power-law coefficient chosen. In both solutions we may see that the length scale can be defined in two different ways:

- By $\sqrt{D_0 t}$ as in the Fickian diffusion with a dimension $[m]$
- By $\sqrt{D_0 t^{1-k}}$ as in the sub-diffusion processes with a dimension $[m^{\alpha}]$, where $0 < \alpha = (1-k)/2 < 1$.

In the first case,we may use a modified length scale $\sqrt{D_0 t_0}$ (bearing in mind that the dimension of $\sqrt{D_0 t_0}$ is length), and then the scaled penetration depth can be presented as

$$\frac{\delta_{1(DIM)}(t_0)}{\sqrt{D_0 t_0}} = \sqrt{\left(\frac{t}{t_0}\right)^{1-k}} \sqrt{\frac{(n+1)(n+2)}{1-k}} \tag{67}$$

And at $t = t_0$ (see (65)) we have

$$\frac{\delta_{1(DIM)}(t_0)}{\sqrt{D_0 t_0}} = \sqrt{\frac{(n+1)(n+2)}{1-k}} \tag{68}$$

Consequently, the approximate solution is

$$C_{1(DIM)}(t_0) = \left(1 - \frac{x}{\sqrt{D_0 t_0} \sqrt{\left(\frac{t}{t_0}\right)^{1-k}} \sqrt{\frac{(n+1)(n+2)}{1-k}}}\right)^n \tag{69}$$

5.2 Solutions with $D_2(t)$

Denoting $\tau_2 = t/t_0$ we get the following forms of Eq. (4)

$$\frac{\partial C}{t_0 \partial \tau_2} = D(\tau_2) \frac{\partial^2 C}{\partial x^2} \tag{70}$$

$$\frac{\partial C}{\partial \tau_2} = (D_0 t_0)\left(1 - \tau_2^p\right) \frac{\partial^2 C}{\partial x^2} \tag{71}$$

$$\frac{\partial C}{\partial \tau_2} = D_e\left(1 - \tau_2^p\right) \frac{\partial^2 C}{\partial x^2}, \quad D_e = D_0 t_0 \tag{72}$$

5.2.1 HBIM Solution with $D_2(t)$

The HBIM technique applied to (72) yields

$$\int_0^{\delta} \frac{\partial C}{\partial \tau_2} dx = \int_0^{\delta} D_e \tau_2^p \frac{\partial^2 C}{\partial x^2} dx \tag{73}$$

$$\frac{d}{d\tau_2}\int_0^{\delta} u(x,t)\,dx = -D_e\left(1-\tau_2^p\right)\frac{\partial u(0,t)}{\partial x} \tag{74}$$

Now, replacing $C(x,t)$ by $C_a = C_0(1-x/\delta)^n$ in (74) and assuming for simplicity $C_0 = 1$ (Dirichlet problem) we get

$$\frac{d}{d\tau_2}\frac{\delta}{n+1} = D_e\left(1-\tau_2^p\right)\frac{n}{\delta} \tag{75}$$

$$\frac{d}{d\tau_2}\delta^2 = D_e\left(1-\tau_2^p\right)2n(n+1) \tag{76}$$

For $t = 0$, the physical condition is $\delta(t=0) = \delta(\tau_2 = 1) = 0$. Hence, the integration of (76) results in

$$\delta^2 = D_e\left(\tau_2 - \frac{\tau_2^{p+1}}{p+1}\right)2n(n+1) \tag{77}$$

In terms of the physical time t we get

$$\delta_{2(HBIM)} = \sqrt{D_0 t_0}\sqrt{\left(\frac{t}{t_0} - \left(\frac{t}{t_0}\right)^{p+1}\frac{1}{p+1}\right)}\sqrt{2n(n+1)} \tag{78}$$

And, with the length scale $\sqrt{D_0 t_0}$ the dimensionless form is

$$\frac{\delta_{2(HBIM)}}{\sqrt{D_0 t_0}} = \sqrt{\left(\frac{t}{t_0} - \left(\frac{t}{t_0}\right)^{p+1}\frac{1}{p+1}\right)}\sqrt{2n(n+1)} \tag{79}$$

Further, at $t = t_0$ we have

$$\delta_{2(HBIM)}(t_0) = \sqrt{D_0 t_0}\sqrt{\left(1 - \frac{1}{p+1}\right)}\sqrt{2n(n+1)} \tag{80}$$

Hence, the scaled front depth at $t = t_0$ is (see (79))

$$\frac{\delta_{2(HBIM)}(t_0)}{\sqrt{D_0 t_0}} = \sqrt{\left(1 - \frac{1}{p+1}\right)}\sqrt{2n(n+1)} \tag{81}$$

Therefore , the approximate solution is

$$C_{2(HBIM)} = \left(1 - \frac{x}{\sqrt{D_0 t_0}\sqrt{\left(\frac{t}{t_0} - \left(\frac{t}{t_0}\right)^{p+1}\frac{1}{p+1}\right)}\sqrt{2n(n+1)}}\right)^n \tag{82}$$

$$C_{2(HBIM)} = \left(1 - \frac{\eta_0}{\sqrt{F_2(\tau, p) N_{HBIM}}}\right)^n, \quad N_{HBIM} = 2n(n+1) \tag{83}$$

We can see that two dimensionless functions control the solutions : (i) $\eta_0 = x/\sqrt{D_0 t_0}$, that is the classical Boltzmann variable at time $t = t_0$, but actually this a dimensionless distance; (ii) the time-dependent function $F_2 = \left(\frac{t}{t_0} - \left(\frac{t}{t_0}\right)^{p+1}\frac{1}{p+1}\right)$.

5.2.2 DIM Solution with $D_2(t)$

Applying the DIM we get

$$\int_0^\delta \left(\int_0^x \frac{\partial C}{\partial \tau_2} dx\right) dx = \int_0^\delta \left(\int_0^x D_e \tau_2^p \frac{\partial^2 C}{\partial x^2} dx\right) dx \tag{84}$$

Replacing $C(x, t)$ by $C = C_0(1 - x/\delta)^n$ in (84) an after the integration we get the following equation

$$\frac{d}{d\tau_2}\frac{\delta}{(n+1)(n+2)} = D_e\left(1 - \tau_2^p\right) \tag{85}$$

$$\frac{d}{d\tau_2}\delta^2 = D_e\left(1 - \tau_2^p\right)(n+1)(n+2) \tag{86}$$

With the physical condition is $\delta(t = 0) = \delta(\tau_2 = 1) = 0$ the result is

$$\delta^2 = D_e \tau_2 \left(1 - \frac{\tau_2^p}{p+1}\right)(n+1)(n+2) \tag{87}$$

$$\delta_{2(DIM)} = \sqrt{D_e\left(\tau_2 - \frac{\tau_2^{p+1}}{p+1}\right)}\sqrt{(n+1)(n+2)} \tag{88}$$

In terms of the physical time t we can expressed $\delta_{2(DIM)}$ as

$$\delta_{2(DIM)} = \sqrt{D_0 t_0} \sqrt{\left(\frac{t}{t_0} - \left(\frac{t}{t_0} \right)^{p+1} \frac{1}{p+1} \right)} \sqrt{(n+1)(n+2)} \tag{89}$$

At $t = t_0$ we have

$$\delta_{2(DIM)}(t_0) = \sqrt{D_0 t_0} \sqrt{\left(1 - \frac{1}{p+1} \right)} \sqrt{(n+1)(n+2)} \tag{90}$$

Thus, the result is the same as in the solution with HBIM, with only difference emerging in the numerical term dependent on the values of n and p. Thus, the scaled front at $t = t_0$ is

$$\frac{\delta_{2(DIM)}(t_0)}{\sqrt{D_0 t_0}} = \sqrt{\left(1 - \frac{1}{p+1} \right)} \sqrt{(n+1)(n+2)} \tag{91}$$

Hence, the approximate solution is

$$C_{2(DIM)} = \left(1 - \frac{x}{\sqrt{D_0 t_0} \sqrt{\left(\frac{t}{t_0} - \left(\frac{t}{t_0} \right)^{p+1} \frac{1}{p+1} \right)} \sqrt{(n+1)(n+2)}} \right)^n \tag{92}$$

or in dimensionless form as

$$C_{2(DIM)} = \left(1 - \frac{\eta_0}{\sqrt{F_2(\tau, p) N_{DIM}}} \right)^n, \quad N_{DIM} = (n+1)(n+2) \tag{93}$$

Again, two dimensionless functions control the solutions: $\eta_0 = x/\sqrt{D_0 t_0}$ and the time-dependent function $F_2 = \left(\frac{t}{t_0} - \left(\frac{t}{t_0} \right)^{p+1} \frac{1}{p+1} \right)$ which is not affected by the integration technique.

In both solutions the denominators depend on $p + 1$. Taking into account that $-1 < p < 0$ actually we have the same behaviour as in the case of the power-law coefficient where the denominator is controlled by $1 - k$.

5.3 Solutions with $D_3(t)$

Denoting $\tau_3 = t/t_0$ we have $D(\tau_3) = (1 - \tau_3)^m$ and Eq. (4) can be presented as

$$\frac{\partial C}{t_0 \partial \tau_3} = D_0 t_0 (1 - \tau_3)^m \frac{\partial^2 C}{\partial x^2} \tag{94}$$

5.3.1 HBIM Solution with $D_3(t)$

The HBIM technique applied to (94) yields

$$\int_0^\delta \frac{\partial C}{t_0 \partial \tau_3} dx = \int_0^\delta D_0 t_0 (1 - \tau_3)^m \frac{\partial^2 C}{\partial x^2} dx \tag{95}$$

$$\frac{d}{d\tau_3} \int_0^\delta C dx = -D_0 t_0 (1 - \tau_3)^m \frac{\partial C(0, t)}{\partial x} \tag{96}$$

With $C_a = C_0(1 - x/\delta)^n$ and $C_0 = 1$ satisfying the boundary conditions we have

$$\frac{1}{2(n+1)} \frac{d\delta^2}{d\tau_3} = -n D_0 t_0 (1 - \tau_3)^m \tag{97}$$

$$\delta^2 = -D_0 t_0 \frac{2n(n+1)}{m+1} (1 - \tau_3)^{m+1} + P_3 \tag{98}$$

In terms of the physical time t the penetration depth is

$$\delta^2 = -D_0 t_0 \frac{2n(n+1)}{m+1} \left(1 - \frac{t}{t_0}\right)^{m+1} + P_3 \tag{99}$$

For $t = 0$, the physical condition is $\delta(t = 0) = \delta(\tau_3 = 1) = 0$. Hence, the integration constant
P_3 is defined as $P_3 = D_0 t_0 \frac{2n(n+1)}{m+1}$. Further, we get

$$\delta^2 = -D_0 t_0 \left[1 - \left(1 - \frac{t}{t_0}\right)^{m+1}\right] \frac{2n(n+1)}{m+1} \tag{100}$$

$$\delta_{3HBIM}(t) = \sqrt{D_0 t_0} \sqrt{\left[1 - \left(1 - \frac{t}{t_0}\right)^{m+1}\right]} \sqrt{\frac{2n(n+1)}{m+1}} \tag{101}$$

In scaled version we have

$$\frac{\delta_{3HBIM}(t)}{\sqrt{D_0 t_0}} = \sqrt{\left[1 - \left(1 - \frac{t}{t_0}\right)^{m+1}\right]} \sqrt{\frac{2n(n+1)}{m+1}} \tag{102}$$

At $t = t_0$ we have

$$\delta_{3HBIM}(t_0) = \sqrt{D_0 t_0}\sqrt{\frac{2n(n+1)}{m+1}} \tag{103}$$

or in a scaled form as

$$\frac{\delta_{3HBIM}(t_0)}{\sqrt{D_0 t_0}} = \sqrt{\frac{2n(n+1)}{m+1}} \tag{104}$$

Finally, the approximate solution is

$$C_{3HBIM}(t) = \left(1 - \frac{x}{\sqrt{D_0 t_{00}}\sqrt{\left[1-\left(1-\frac{t}{t_0}\right)^{m+1}\right]}\sqrt{\frac{2n(n+1)}{m+1}}}\right)^n \tag{105}$$

or in terms of dimensionless functions as

$$C_{3HBIM}(t) = \left(1 - \frac{\eta_0}{\sqrt{F_3 N_{HBIM}(n,m)}}\right)^n, \quad N_{HBIM}(n,m) = \frac{2n(n+1)}{m+1} \tag{106}$$

where $F_3 = 1 - \left(1 - \frac{t}{t_0}\right)^{m+1}$.

5.3.2 DIM Solution with $D_3(t)$

Applying the DIM we get

$$\int_0^{\delta}\left(\int_0^x \frac{\partial u}{\partial \tau_3}dx\right)dx = -\int_0^{\delta}\left(\int_0^x D_e \tau_3^m \frac{\partial^2 u}{\partial x^2}dx\right)dx \tag{107}$$

$$\int_0^{\delta}\left(\int_0^x \frac{\partial u}{\partial \tau_3}dx\right)dx = -D_e \tau_3^m u(0,t) \tag{108}$$

Replacing $C(x,t)$ in (3–6b) by $C_a = C_0(1 - x/\delta)^n$ and $C_0 = 1$ we get

$$\frac{1}{(n+1)(n+2)}\frac{d\delta^2}{d\tau_3} = -D_e \tau_3^m \tag{109}$$

$$\delta^2 = -D_e \tau_3^{m+1} \frac{(n+1)(n+2)}{m+1} + P_2 \tag{110}$$

The initial condition $\delta(t=0) = \delta(\tau_3 = 1) = 0$ defines $P_2 = [D_e(n+1)(n+2)]/(m+1)$. Hence, in terms of the physical time t the penetration depth can be expressed as

$$\delta_{3DIM}(t) = \sqrt{D_0 t_0} \sqrt{\left[1 - \left(1 - \frac{t}{t_0}\right)^{m+1}\right]} \sqrt{\frac{(n+1)(n+2)}{m+1}} \tag{111}$$

or

$$\frac{\delta_{3DIM}(t)}{\sqrt{D_0 t_0}} = \sqrt{\left[1 - \left(1 - \frac{t}{t_0}\right)^{m+1}\right]} \sqrt{\frac{(n+1)(n+2)}{m+1}} \tag{112}$$

At $t = t_0$ we have

$$\delta_{3DIM}(t) = \sqrt{D_0 t_0} \sqrt{\frac{(n+1)(n+2)}{m+1}} \tag{113}$$

or

$$\frac{\delta_{3DIM}(t)}{\sqrt{D_0 t_0}} = \sqrt{\frac{(n+1)(n+2)}{m+1}} \tag{114}$$

Thus, the approximate solution is

$$C_{3DIM}(t) = \left(1 - \frac{x}{\sqrt{D_0 t_0} \sqrt{\left[1 - \left(1 - \frac{t}{t_0}\right)^{m+1}\right]} \sqrt{\frac{(n+1)(n+2)}{m+1}}}\right)^n \tag{115}$$

or simply in terms of dimensionless functions as

$$C_{3DIM}(t) = \left(1 - \frac{\eta_0}{\sqrt{F_3(\tau, m) N_{DIM}(n, m)}}\right)^n, \quad N_{DIM}(n, m) = \frac{(n+1)(n+2)}{m+1} \tag{116}$$

and $F_3 = 1 - \left(1 - \frac{t}{t_0}\right)^{m+1}$.

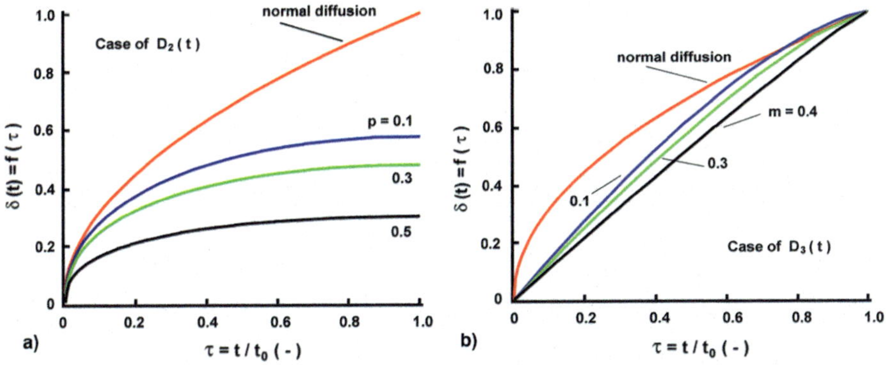

Fig. 5 Subdiffusion behaviour of the fronts (solutions) with non-singular diffusivities $D_2(t)$ **a** and $D_3(t)$ **b** used in this study

5.4 *Numerical Experiments with the Approximate Solutions*

Let now see what is the behaviour of the solutions approximating the diffusion penetration depths and the the concentration profiles thereof.

5.4.1 Penetration Depths

We start with penetration depths since their time-dependencies strongly indicate what type of diffusion process is modelled: *Normal diffusion* as it follows from the acceptance of the second Fick's law or the behaviour is *subdiffusive*?

The solutions developed with the simple form of $D(t)$ as a singular power-law $D_1(t) = (t/t_0)^{-k}$ revealed that $\delta_1 \equiv t^{\frac{1-k}{2}}$, where $1 \leq 1-k \leq 2$, i.e the behaviour is subdiffusive. Similarly the solution with the newly proposed $D_2(t)$ reveals that $\delta_1 \equiv t^{\frac{1+p}{2}}$ (where $0 \leq 1+p \leq 2$), that is the behaviour is subdiffusive. For the case when $D_3(t)$ is used it is hard to see directly from the solutions about $\delta_3(t)$ what is time-dependent behaviour. To check this there is a simple way: to plot $\delta_1(t)$, $\delta_2(t)$ and $\delta_3(t)$ against the time t together with the Gaussian front (normal diffusion propagating with a speed proportional to $\sqrt{t}$. For the sake of simplicity, this test is performed with $n = 2$ as exponent of the assumed profile. The plots in Fig. 5 strongly indicate that the front propagation with $D_2(t)$ and $D_3(t)$ exhibits subdivision behaviour.

In case of $D_2(t)$ the subdivision behaviour is strong. However, in the case of $D_3(t)$ the lines are too close to the Gaussian line $\sqrt{t}$ but with increase in m this difference increases. Close to $t/t_0 = 1$ and small values of m this subdiffusive behaviour is violated. Therefore, based on the numerical experiments performed here with $D_3(t)$, subdiffusive behaviour (the real diffusion mechanism in concretes) can be obtained with $m \geq 0.3$.

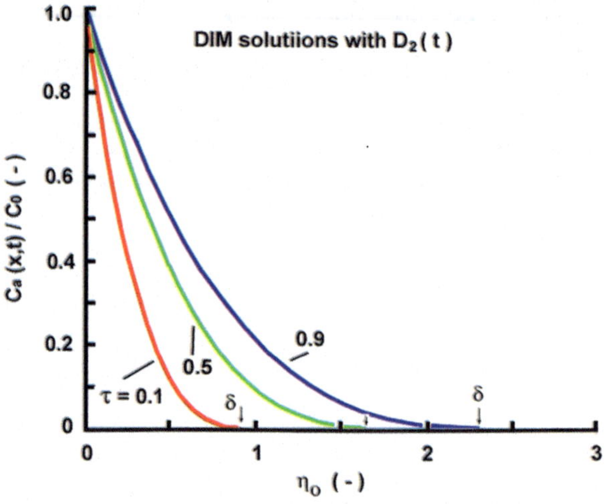

Fig. 6 Approximate profiles (DIM solutions) in depth of the medium at different times (different values of the dimensionless time $\tau = t/t_0$. Case of $D_2(t)$ with a stipulated exponent of the profile $n = 3$ and $m = 0.3$. The space coordinate is dimensionless presented by $\eta_0 = x/\sqrt{D_0 t_0}$. The vertical arrows with labels δ show the points where the profiles end, thus denoting the lengths of the penetration depths

5.4.2 Approximate Profiles with Stipulated Exponents

The plots in Figs. 6 and 7a, at different times (different $\tau = t/t_0$), clearly demonstrate the concept of the penetration depth: with increase in time the front goes deeper and deeper into the medium. Moreover, here η_0 easily shows the dimensionless depths of the diffusion layer.

The three-dimensional profiles in Figs. 7 and 8b actually show the same behaviour; the penetration depths can be easily seen in the plane C, η_0.

5.5 *Comparative Study with Non-singular D(t) From Literature Sources*

Let us now compare the results of the approximate integral-balance method when the two non-singular forms of $D(t)$ commended in Sect. 3.2.1 to the ones developed in this work: (i) The non-singular power-law of Wang and Fu [12] and (ii) The non-singular exponential diffusivity of Sun et al. [27].

- Non-singular power-law of Wang and Fu [12]

With $D_{WF}(t)$ [12] defined by Eq. (26) the application of DIM yields

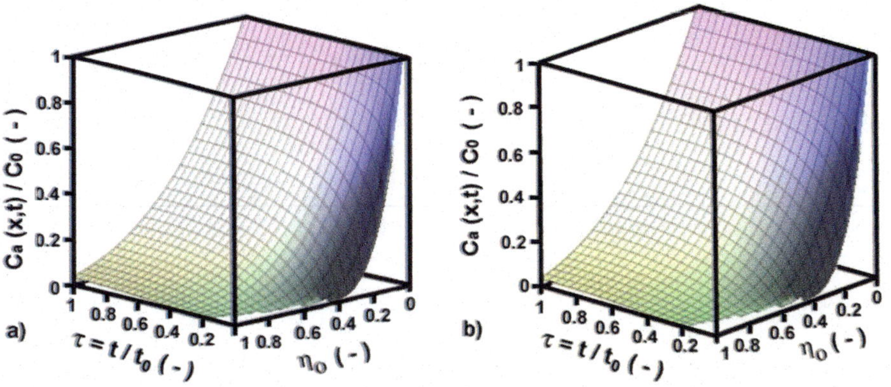

Fig. 7 3D profiles (DIM solutions) in case of $D_2(t)$ with a stipulated exponent of the profile as a function of the similarity variable η_0 and the dimensionless time $\tau = t/t_0$ **a** Case with $p = 0.1$ and $n = 2$ **b** Case with $p = 0.5$ and $n = 3$

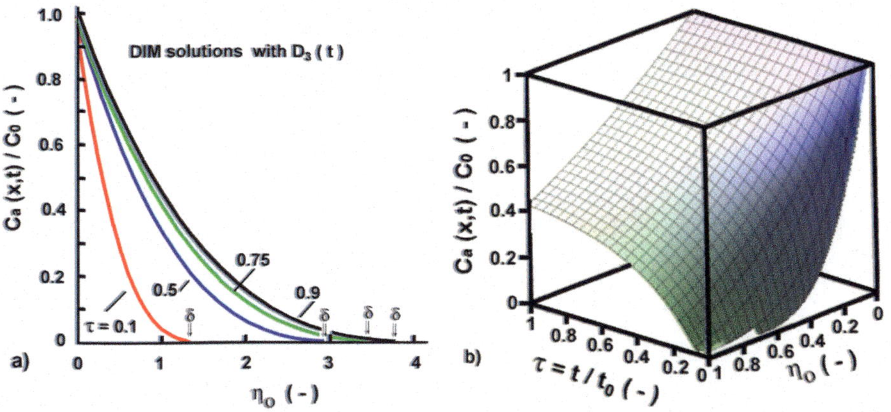

Fig. 8 Approximate profiles (DIM solutions) in case of $D_3(t)$ with a stipulated exponent of the profile $n = 3$ and $m = 0.3$: **a** Two-dimensional profiles in depth of the medium at different times (different values of the dimensionless time $\tau = t/t_0$) **b** $3D$ profile as a function of the similarity variable η_0 and the dimensionless time $\tau = t/t_0$; Note: In Fig. 8a the vertical arrows with labels δ show the points where the profiles end, thus denoting the lengths of the penetration depths

$$\frac{1}{(n+1)(n+2)}\frac{d}{dt}\delta^2 = D_0\left(1 + \frac{t}{t_0}\right)^{-w} C_0(0,t) \tag{117}$$

With $C_0(0,t) = 1$ as in the preceding solutions, and the condition $\delta(t = 0) = 0$, from the integration of (117) we have

$$\delta^2 = D_0\left[\frac{t_0^w (t+t_0)^{1-w}}{1-w}\right]N \Rightarrow \delta(t) = \sqrt{D_0\left[\frac{t_0^w (t+t_0)^{1-w}}{1-w}\right]N}, \quad N = (n+1)(n+2) \tag{118}$$

and in a scaled form as

$$\frac{\delta(t)}{\sqrt{D_0 t}} = \sqrt{\frac{t_0^w}{t^{1-w}}}\sqrt{\left(1+\frac{t_0}{t}\right)^{1-w}}\sqrt{\frac{N}{1-w}} \tag{119}$$

or

$$\frac{\delta(t)}{\sqrt{D_0 t_0}} = \sqrt{\frac{(t+t_0)^{1-w}}{t_0^{2-w}}}\sqrt{\frac{N}{1-w}} \tag{120}$$

Hence, the scaled penetration depth is a growing in time function (more obvious from (118)). Moreover, the important result now is that it scales as $\delta(t) \equiv t^{\frac{1-w}{2}}$, thus exhibiting a subdiffusive behavior of propagation. The plots in Fig. 8a confirm this result since all plots of $\delta_{Sun}(DIM)$ are below the Gaussian line of normal diffusion when $\delta \equiv \sqrt{t}$.

For $t = t_0$ we have

$$\delta_{DIM}(t = t_0) = \sqrt{D_0 t_0}\left(2^{\frac{1-w}{2}}\right)\sqrt{\frac{N}{1-w}} \tag{121}$$

or in a scaled version as

$$\frac{\delta_{DIM}(t = t_0)}{\sqrt{D_0 t_0}} = \left(2^{\frac{1-w}{2}}\right)\sqrt{\frac{N}{1-w}} \tag{122}$$

For $w = 0.5$ as illustrative example and $n = 2$ we get $\frac{\delta_{DIM}(t=t_0)}{\sqrt{D_0 t_0}} = \left(2^{1.25}\right)\sqrt{6} \approx 5.825$

- Non-singular exponential time-dependent $D(t)$ of Sun et al. [27]

  With $D_{Sun}(t)$ (34) and DIM we get

$$\frac{1}{(n+1)(n+2)}\frac{d}{dt}\delta^2 = D_0(1-\phi_0)\left[1 - \exp\left(-a_D\frac{t}{t_0}\right)\right] \tag{123}$$

Since $(1-\phi_0)$ is a dimensionless constant for a given concrete sample, we may denote for the sake of simplicity $D_e = D_0(1-\phi_0)$. Then the integration in (123) yields

$$\frac{\delta^2}{N} = D_e\left[t - \frac{t_0}{a_D}\left(1 - e^{-a_D\frac{t}{t_0}}\right)\right] \Rightarrow \delta = \sqrt{D_e t}\sqrt{1 - \frac{1}{a_D}\frac{t_0}{t}\left(1 - e^{-a_D\frac{t}{t_0}}\right)} \tag{124}$$

In a more convenient form for scaling of the front time propagation we have

$$\delta = \sqrt{D_e t_0}\sqrt{\left[\frac{t}{t_0} - \frac{1}{a_D}\left(1 - exp(-a_D\frac{t}{t_0})\right)\right]'} \quad 0 \leq \frac{t}{t_0} \leq 1 \tag{125}$$

For $t = t_0$

$$\delta\,(t = t_0) = \sqrt{D_e t_0}\sqrt{1 - \frac{1}{a_D}\left(1 - e^{-a_D}\right)} \tag{126}$$

or in scaled forms as

$$\frac{\delta}{\sqrt{D_e t_0}} = \sqrt{1 - \frac{1}{a_D}\frac{t_0}{t}\left(1 - e^{-a_D\frac{t}{t_0}}\right)} \tag{127}$$

Alternatively (from (125))

$$\frac{\delta}{\sqrt{D_0 t_0}} = \sqrt{\left[\frac{t}{t_0} - \frac{1}{a_D}\left(1 - exp(-a_D\frac{t}{t_0})\right)\right]'} \quad 0 \leq \frac{t}{t_0} \leq 1 \tag{128}$$

and

$$\frac{\delta\,(t = t_0)}{\sqrt{D_e t_0}} = \sqrt{1 - \frac{1}{a_D}\left(1 - e^{-a_D}\right)} \tag{129}$$

The results (127) and (128) do not give a direct answer what type of propagation behaviour of the front is: subdiffusive, normal diffusion or superdiffusion ? There is no need of cumbersome equations or function approximating the time dependence of $\delta(t)$ because it is much more easier to see what is its behaviour in comparison to Gaussian time scaling proportional to $\sqrt{t}$. The plots in Fig. 9 show definitively that the behaviour is subdiffusive since all plots with $a_D \in [0.177 - 1.509]$ (the range reported by Sun et al. [27]) are below the Gaussian line of normal diffusion where $\delta \equiv \sqrt{t}$.

5.6 Solution Optimization

The previous examples with approximate solutions were carried out with stipulated exponents of the assumed profile (i.e. $n = 2$ and $n = 3$) only with the idea to demonstrate their behaviors. Now the question is about the accuracy of the approximate solutions. We do not refer to the transformation mentioned at the beginning of this chapter which led to classical solutions in terms of the error function. The main reason for this is that the classical diffusion equation is not causal, precisely the speed of

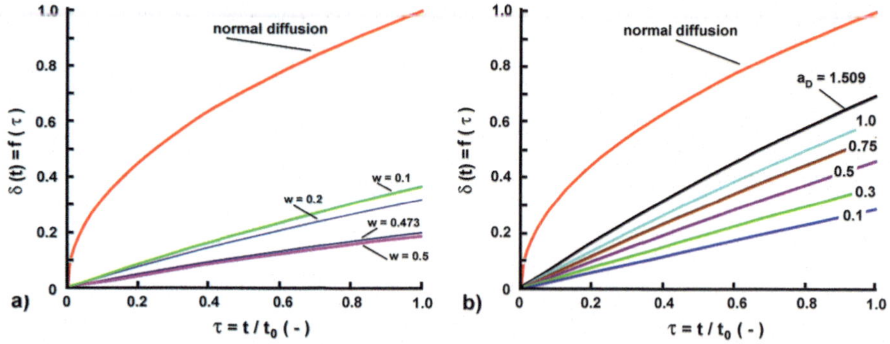

Fig. 9 Time scaling of the front prolongations related to the solutions with non-singular D(t) of Wang and Fu [12] and Sun et al. [27]: **a** The front behaviour (DIM Solution) with non-singular power-law D(t) (26). The curve with $w = 0.473$ corresponds to the study in [12]; **b** The front behaviour (DIM Solution) with non-singular exponential D(t) [27]. The curve with $a_D = 1.509$ corresponds to highest value used in (34)

the solution is infinite, and actually these solutions cannot be accepted as exact thus allowing estimating the accuracy of the approximate solutions from the deviations from them.

Now, recall that the integral-balance method leads to solutions which satisfy its integral-balance relations (of HBIM or of DIM) but not the original model equation (4). Therefore, if we replace the function $C_a(x,t)$ in the original model (4), then the residual function will not be zero, namely

$$R(x,t) = \frac{\partial C_a(x,t)}{\partial t} - D(t)\frac{\partial^2 C_a(x,t)}{\partial x^2} \neq 0 \tag{130}$$

With the assumed profile used in this work, we have

$$\frac{\partial}{\partial t}C_a(x,t) = n\left(1-\frac{x}{\delta}\right)^{n-1}(-x)\frac{1}{\delta^2}\frac{d\delta}{dt} \tag{131}$$

$$\frac{\partial^2}{\partial x^2}C_a(x,t) = \frac{n(n-1)}{\delta^2}\left(1-\frac{x}{\delta}\right)^{n-2} \tag{132}$$

Then, the residual function can be expressed as

$$R(x,t) = n\left(1-\frac{x}{\delta}\right)^{n-1}(-x)\frac{1}{\delta^2}\frac{d\delta}{dt} - D(t)\frac{n(n-1)}{\delta^2}\left(1-\frac{x}{\delta}\right)^{n-2} \tag{133}$$

Simple recasting of (133) yields

$$R(x,t) = \frac{1}{\delta^2}\left[n\left(1-\frac{x}{\delta}\right)^{n-1}\left(-\frac{x}{\delta}\right)\left(\delta\frac{d\delta}{dt}\right) - D(t)\,n(n-1)\left(1-\frac{x}{\delta}\right)^{n-2}\right] \tag{134}$$

Now, we have to see what is the expression of the time-dependent term $\delta\frac{d\delta}{dt}$. In details we have

$$\delta\frac{d\delta}{dt} = \left(\sqrt{D_0 t_0}\sqrt{F}\sqrt{N}\right)\left(\sqrt{D_0 t_0}\sqrt{N}\frac{1}{2}\frac{1}{\sqrt{F}}\frac{dF}{dt}\right) = \frac{1}{2}D_0 t_0 N\frac{dF}{dt} \tag{135}$$

Replacing in (134) and using $D_2(t)$ with corresponding function F_2 we get

$$\delta\frac{d\delta}{dt} = \frac{1}{2}D_0 t_0 N\frac{dF_2}{dt} = \frac{1}{2}D_0 N\left[1-\left(\frac{t}{t_0}\right)^p\right] \tag{136}$$

The time-dependent part of (136) coincides with the definition of $D(t) = D_2(t)$. There is nothing strange in this fact, since the function $F_2(t)$ is a result of integration of $D_2(t)$ with respect to the time t (the same is in the cases of $D_1(t)$ and $D_3(t)$ since this comes from the technology of the integral method- the derivation of the equation about the penetration depth).

Now, the residual function can be presented as a product of two terms: time-dependent function and time independent function depending only on the exponent n and the dimensionless space coordinate $z = x/\delta$, namely

$$R(z,t) = \underbrace{\frac{1}{\delta^2}\left[1-\left(\frac{t}{t_0}\right)^p\right]}_{f_R(t)} D_0 t_0\left[\frac{n(1-z)^{n-1}(-z)\frac{1}{2}N - n(n-1)(1-z)^{n-2}}{N}\right],\ 0 \le z = \frac{x}{\delta} \le 1 \tag{137}$$

With the dimensionless variable $z = x/\delta$ we transform the moving boundary domain of the diffusion layer ($0 \le x \le \delta(t)$) into a fixed boundary domain $0 \le z \le 1$, a step that will facilitate us in the next calculations. The time-dependent term in (137) is

$$f_R(t) = \frac{1}{\delta^2}\left[1-\left(\frac{t}{t_0}\right)^p\right] = \left[1-\left(\frac{t}{t_0}\right)^p\right]\frac{1}{\left(\frac{t}{t_0}\right)\left[1-\left(\frac{t}{t_0}\right)^p\frac{1}{p+1}\right]} \tag{138}$$

With $0 < p < 1$ this function decays faster down to zero at $t/t_0 = 1$ but exhibits a singular behavior at $t/t_0 = 0$ and weak dependence on the parameter p, as it is shown in Fig. 10. Hence, since the residual function is decaying rapidly in time, the main problem in the optimization is to minimize the term dependent only on the exponent n, namely

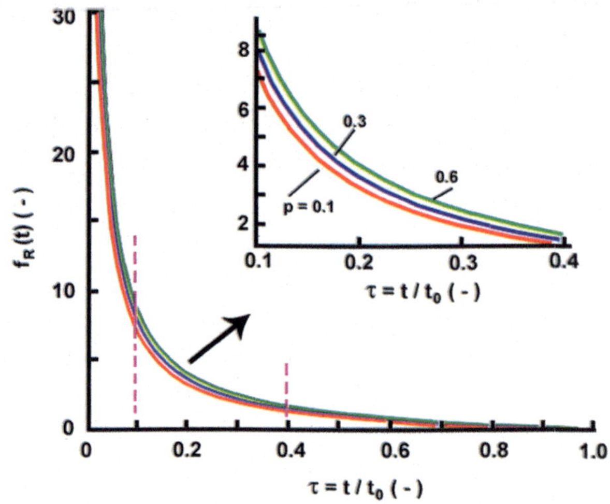

Fig. 10 Behaviour of the time-dependent term of the residual function $f_R(t)$ as a function of $\tau = t/t_0$: Case with $D(t) = D_2(t)$ for three values of the parameter p. Inset: Enlarged sections of the plots demonstrating almost equal decaying behaviours in time

$$R(z, n) \approx \left[\frac{n(1-z)^{n-1}(-z)\frac{1}{2}N - n(n-1)(1-z)^{n-2}}{N} \right] \tag{139}$$

With $N_{HBIM} = 2n(n+1)$, for example, we have

$$R_{HBIM}(z, n) \approx \frac{1}{t} \left[\frac{n(1-z)^{n-1}(-z)\, n(n+1) - n(n-1)(1-z)^{n-2}}{2n(n+1)} \right] \tag{140}$$

The integration over the diffusion layer of the squared residual function yields the mean squared error of approximation, namely

$$E = \int_0^1 [R_{HBIM}(z, t)]^2\, dz \approx \frac{1}{t^2} \int_0^1 [r_{HBIM}(z, n)]^2\, dz \tag{141}$$

Here we have to remember the comments just after Eq. (136), about the relationships between the functions $D(t)$ and $F(t)$. Therefore the function that should be optimized with respect to n in (137), as well as in (140) and in (141), is the same as in the case with when the diffusion equation has a constant (time-independent) diffusivity. This is almost classical case in integral-balance solutions of diffusion equations, studied by Myers [52, 58]; where it was determined that the optimal values for the case of Dirichlet problems are : $n_{opt}(HBIM) \approx 2.233$ and $n_{opt}(DIM) \approx 2.219$.

Here we skip the calculations related to the cases with $D_1(t)$ and $D_3(t)$ but the reader can easy check that the result will be the same. Plots of approximate solutions with the optimal values of n are shown in Figs. 11 and 12.

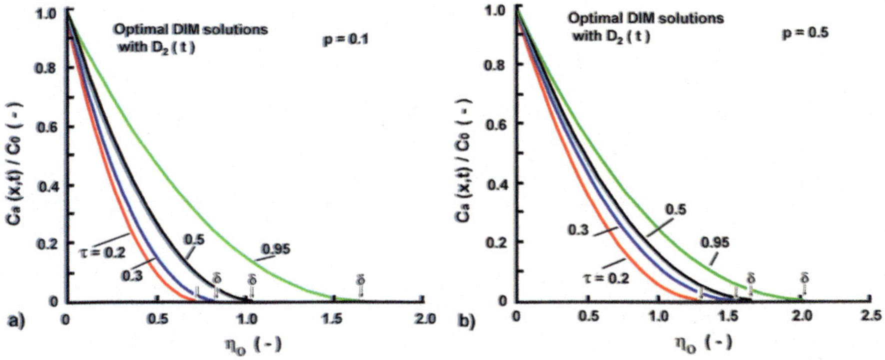

Fig. 11 Two-dimensional profiles with optimal exponents clearly demonstrating the concept of final penetration depth: DIM solutions with $D_2(t)$

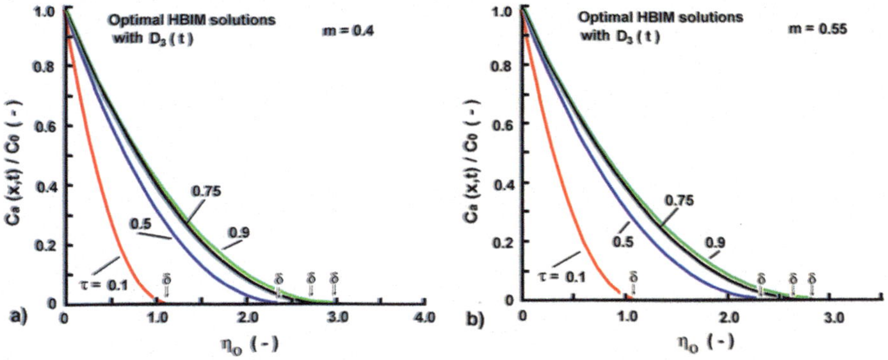

Fig. 12 Two-dimensional profiles with optimal exponents clearly demonstrating the concept of final penetration depth: HBIM solutions with $D_3(t)$

6 Outcomes of the Approximate Solutions

At the end, after the performed approximate solutions of the main models of chlorides diffusion in concretes we stress the attention on two principle problems formulated as

- First of all, we have to sort out the results obtained and see what actually we can say about the modelling approach when the diffusion coefficient is time-dependent (decaying in time). In this context we should detect from where the subdiffusive character of the solutions comes from, and are there alternative approaches leading to similar results?
- Second, we should detect where the causality principle in the models discussed is obeyed and where not.

To find answers to these problems we first have to present briefly the principle of causality. This will facilitate the following analyzes in Sects. 6.2 and 6.3.

6.1 Causality of Dynamic Models: Basic Principles

Here we touch a basic principle in construction of dynamic models, i.e the causality. This term was briefly mentioned several times in the preceding texts, but now, before continuing further in the analysis, we have to define some underlying rules.

Following the analysis of Mittelstaedt and Weingartner [59] in all applied cases *the causal relation satisfies the chronological condition* (i.e. the time-shift between cause and effect), that is no closed time curves exist and always the cause precedes the effect. The main outcome of this condition is that *only a dynamical law describing a time evolution of a certain physical system is a causal relation.* As mentioned by Mittelstaedt and Weingartner [59] (p. 219) under physical realistic situations, the causality of the realistic solutions and the chronological conditions are equivalent.

Actually, the basic problem engaging the attention in this chapter is the causality concept and we will formulate some generalized conditions related to it, among them [60]:

- ***Primitive causality***: *The effect cannot precede the cause.* In such situations the cause and the effects should be correctly defined.
- ***Relativistic causality***: *No signal can propagate with velocity greater than the speed of the light in the vacuum.* It could be considered as a macroscopic causality condition.

However, *the primitive causality condition is more fundamental and general* than the relativistic causality condition.

The causality principle implies that some functions describing transients in dynamical problems should obey some properties: *to vanish over a range of values of its arguments* (see further about the causality of the relaxation functions used in the memory integrals).

If we consider a physical system with a time-dependent input (cause) $x(t)$ and the corresponding output (effect) $y(t)$, and satisfying the following conditions [60].

C1: Linearity. That is, *it obeys the superposition principle* in its simple version implying that the output is a linear functional of the input

$$y(t) = \int_{-\infty}^{\infty} g(t, \tau) x(\tau)\, d\tau \tag{142}$$

where $y(t)$, $g(t)$ and $x(t)$ may represent distributions.

C2: Time-translation invariance. *The system is time-translation invariant* if the input is shifted (forward or backward) by some time interval τ and $x(t+\tau)$ corresponds to $y(t+\tau)$. In this case the function $g(t, \tau)$ should depend only on the difference of the arguments, that is $g(t,\tau) = g(t-\tau)$ and the linear functional can be expressed as

$$y(t) = \int_{-\infty}^{\infty} g(t-\tau)x(\tau)\,d\tau = g(t) * x(t) \qquad (143)$$

Condition **C2** means that the relation (143) is a convolution between the input (cause) $x(t)$ and the output (effect) $y(t)$ and the correlation function (named also memory or kernel) allows to model the time-shift, i.e. that output at time $y(t)$ correspond to an earlier moment of the input $x(t-\tau)$

C3: Primitive causality condition. *The input cannot precede the output*. Therefore, the input $x(t)$ vanishes for $t < T$(T is the moment when the input is applied) that means that the same is valid for $y(t)$, meaning without loss of generality that T can be assumed as zero. As consequence, we obtain that $g(\tau) = 0$ if $\tau < 0$, that is $g(\tau)$ should be a causal function. Moreover, this is equivalent to setting the lower terminal in the (142) and (143) equal to zero.

Now, we can turn on model problems and analyzes of the approximate solutions developed.

6.2 Subdiffusion Front Propagation: Two Formal Explanations

To answer the first question, formulated at the beginning of this section, we present two standpoints about the subdiffusive character of the solutions. Considering the front subdiffusive time evolutions we may interpret formally the results obtained using two existing examples:

(i) Integer-order diffusion model with a power-law coefficient corresponding to Brownian fractional diffusion, and
(ii) Anomalous diffusion, precisely subdiffusion modelled by fractional in time derivatives.

These two explanations allow to see what really can be extracted from the original models of chlorides diffusion in concretes and to see alternatives based on fractional calculus.

6.2.1 Brownian Motion Diffusion Point of View

The results of the approximate solutions strongly indicate that the diffusion has a subdiffusive character. Referring to Henry at al. [61] the easiest way to model anomalous subdiffusion is to construct the diffusion equation with a time-dependent coefficient, that is

$$\frac{\partial C}{\partial t} = D_0 \left(\alpha t^{\alpha-1}\right) \frac{\partial^? C}{\partial x^2} \tag{144}$$

with a Gaussian solution

$$C(x,t) = \frac{1}{2\sqrt{\pi D_0 t^{\alpha}}} exp\left(-\frac{x^2}{4D_0 t^{\alpha}}\right) \tag{145}$$

and the mean square displacement is $\left\langle x^2\right\rangle = 2D_0 t^{\alpha}$.

The solution (145) describes a probability density function of non-Markovian process due to the power-law diffusivity. Now, recall, that such type of result interpretation in the studies on chlorides diffusion in concretes (see all comments in Introduction and commented references) is completely missing. Despite the different functional relationships of $D(t)$ used in the literature and the two new formulations conceived here, we may generalize that they are also of a power-law type. Trying to relate the results to the time-fractional diffusion we may say that the power-law diffusivity in (refeq:Out-Sub-1) can be represented as a time-fractional derivative of a constant [61]

$$D(t) =^{RL} D_t^{1-\alpha} \left[\Gamma(\alpha) D_0\right] \tag{146}$$

where ${}^{RL}D_t^{1-\alpha}$ is a Riemann-Liouville derivative of order $\gamma = 1 - \alpha$ More close to this situation is the definition of $D_1(t) = (t/t_0)^{-k}$, where taking into account (146) it should be $k = 1 - \alpha$ or $\alpha = 1 - k$.

$${}^{RL}D_t^{\gamma} = \frac{1}{\Gamma(1-\gamma)} \frac{d}{dt} \int\limits_0^t D_o \frac{1}{(t-z)^{\gamma-1}} dz \Rightarrow \frac{1}{\Gamma(\alpha)} \frac{d}{dt} \int\limits_0^t D_0 \frac{1}{(t-z)^{\alpha}} dz = \frac{d}{dt} \left[I_t^{\alpha} D_0\right] \tag{147}$$

where $I_t^{\alpha} = \frac{1}{\Gamma(\alpha)} \int\limits_0^t D_0 \frac{1}{(t-z)^{\alpha}} dz$, denoted also as $D_t^{-\alpha}$ is the Riemann-Liouville fractional integral of order $0 < \alpha < 1$.

However, we have to mention here, the constitutive equation of the Fick's first law $j = -D\,(\partial C/\partial x)$ relates the cause (the gradient) and effect (the flux) but *does not satisfy* the *primitive causality* and the *time-translation invariance* conditions. As a consequence, the resulting parabolic diffusion equation with or without time-dependent coefficient is not causal, thus providing solutions with infinite speeds. The integral-valance solutions fix this problem to greater extent (see point Sect. 6.3).

Further, at a glance, despite the previous comments, it is hard to say that the diffusion model (144) as well as the main model studied here (4) are causal since they are parabolic models. Actually the model (144) describes a *fractional Brownian motion diffusion process* which can be also modelled by an evolution Eq. [61]

$$y(t) = y_0 + D_t^{-\alpha} \left[F_G(t)\right] \tag{148}$$

showing the position of a walker at time t starting the motions at y_0. Here $F_G(t)$ is *Gaussian withe noise* with correlation $F(t)F(s) = \delta_D(t-s)$, where δ_D is the Dirac Delta function.

Moreover, in the context of the following analyzes, Eq. (148) may be presented as [61]

$$[y(t) - y_0] = \frac{1}{\Gamma(\alpha)} \int_0^t \frac{F_G(z)}{(t-z)^{\alpha}} dz = D_t^{-\alpha} [F_G(z)] \tag{149}$$

where the right-hand side of (149) is a Riemann-Liouville integral of order $0 < \alpha < 1$.

Here, we can see that causality principles in (149) are satisfied, since *there is a time-shift between the reaction* $[y(t) - y_0]$ *and* $F_G(t)$ *and a cause* (driving force) assured by the memory integral (condition $C2$); and the linearity (condition $C1$), too. Moreover, in all fractional (convolution) operators *the primitive causality condition* $C3$ *is satisfied since the memory kernels are causal functions.*

6.2.2 Power-Law Waiting Time Approach

Concretes are non-nonhomogenous media which so far are not considered as objects of anomalous diffusion from the side of serious physical analysis. The two articles [30, 31] are only attempts applying formally fractional diffusion models, but fortunately fit well experimental data. Therefore, we may suggest that CTRW theory as adequate in such cases and consider the fractional diffusion in terms of the Riemann-Liouville fractional derivative [61, 62] (see the definition in Eq. (147)) as a suitable modelling approach.

With the non-Markovian Pareto waiting time-density [61] $\Psi(t) = \alpha \frac{\tau^{\alpha}}{t^{1+\alpha}}$, $t \in [\tau, \infty]$ and $0 < \alpha < 1$ the cumulative distribution is a power-law $1 - (\tau/t)^{\alpha}$.

In such a case the fractional subdiffusion equation can be presented as [61]

$$\frac{\partial C}{\partial t} = D_{\alpha} \left[{}_0D_t^{1-\alpha} \frac{\partial C^2}{\partial x^2} \right] \tag{150}$$

In the right-hand side of (150) a Riemann-Liouville fractional derivative of order $1 - \alpha$ assures the causality. To be clearer, this is eq. $5b$ developed in [62]. It may be phenomenologically defined from *a flux with a memory* (causal flux-gradient relationship) (which may be considered as a part of the *fading memory approach*-see the last paragraph of this point section) defined as

$$j(x,t) = -D_{\alpha} \left[{}_0D_t^{1-\alpha} \frac{\partial C}{\partial x} \right], \quad 0 < \alpha < 1 \tag{151}$$

where D_α is a fractional diffusivity with a dimension m^2/s^α. This is a causal constitutive equation satisfying both the *linearity* and the *time-translation invariance* conditions.

Then, applying the continuity Eq. (3) we get (150).

$$\frac{\partial C}{\partial t} = D_\alpha \left[{}_0D_t^{1-\alpha} \frac{\partial C}{\partial x} \right] \tag{152}$$

Applying to both sides of (152) the operator $D_t^{\alpha-1}$ and remembering the index law $D^\gamma D^\beta = D^{\gamma+\beta}$ we get the more known form of the time-fractional diffusion equation

$$^{RL}D_t^\alpha C(x,t) = D_\alpha \frac{\partial^2 C(x,t)}{\partial x^2} \tag{153}$$

which a causal equation in contrast to its integer-order counterpart (because the constitutive equation eq:Out-Sub-8 is causal).

The result coincides with the model (11) only formally, because (11) uses the Caputo derivative. However, this is a formal approach and as it was mentioned by Hilfer [62] Eqs. (11) and (152) (and Eq. (153), too) are not equivalent. In the context of the causality of the diffusion models we refer to the integral form of Eq. (152) and Eq. (153), namely

$$\begin{aligned} C(x,t) &= C_{(t=0)}\delta_D(x) + D\alpha \frac{1}{\Gamma(\alpha)} \int_0^t (t-z)^{\alpha-1} \frac{\partial C(x,z)}{\partial z} dz \\ &= C_{(t=0)}\delta_D(x) + D_\alpha I_t^\alpha \left[\frac{\partial C(x,t)}{\partial x} \right] \end{aligned} \tag{154}$$

where the initial condition $C(x, t = 0) = C_{(t=0)}\delta_D(x)$ is incorporated [62]. This is a good example where all conditions of the causality principle are satisfied, plus the superposition principle. Actually, this construction matches the fading memory concept [63–66, 68] for simple materials [67] (simple materials are media where the flux is proportional to the gradient, as in Fick and Fourier laws, and no higher spatial derivatives are needed).

6.3 Finite Penetration Depth and Finite Speed of the Front

Now, let us focus the attention on the basic concept of the integral-balance method. The finite penetration depth actually *corrects the non-causal behaviour of the parabolic diffusion equation* applying the *relativistic causality* principle. That is, the disturbance caused by the boundary conditions should propagate into the medium with a finite speed: that is the finite speed of the solution results in a finite penetration zone of the diffusant into the medium. The finite speed of the front motion for $t > 0$, except the singularity at $t = 0$, obtained by the internal-balance solutions of all examples solved here, satisfies these conditions. With this we end the discussion on the causality of the diffusion models used to model chlorides ingress into concretes, but the problem is open and too much has to be done on modelling approaches in this field.

7 Final Comments

The chapter presented a new approach to the diffusion models with time-dependent diffusivities used in modelling of chlorides ingress into fresh concretes. The solution approach used avoids initial linearization and the straightforward techniques of the integral balance method allow easily to establish the subdiffusion character of the process, a fact not recognized so far in the modelling studies devoted to this important for the building materials problem. The fact that the diffusion process is subdiffusive directly leads to applications of fractional in time models, which comparing to the integer-order versions demonstrated in this chapter, contain lesser parameters that should be established from experimental data. Unfortunately, fractional calculus modelling of aggressive chlorides (or other aggressive substances) in concretes is practically undeveloped, except the two studies mentioned in this text.

Last but not least, the causality principle and the requirement all his conditions to be satisfied are highly demanded in construction of diffusion models in complex media such as concretes. This is the modelling point of view which has direct practical impact when data interpretations should be done, even though for practically oriented engineers this is not so important; we can see the non-causal models with time-dependent diffusivity is widely used, without any question about its correctness, because the Fickian model is the most popular and common accepted as granted for any diffusion problems to be solved. However, the diffusion of chlorides in concretes is not Fickian and all variations in the formulation of the time-dependent diffusivity are actually attempts to fit the non-Fickian behaviour when the ad hoc chosen model is taken from the popular textbooks.

The results reported in this chapter are not only on the application of the integral-balance method to the more widely used diffusion models related to life behaviour of concretes. They emphasis on correctness of the model build-up, the causality principle and what actually is the type of the diffusion process in concrete. The answers are straightforward: (i) The model based on the Fickian diffusion with time-depended diffusivity is not-causal; (ii) Moreover, the diffusion is not Fickian that provides new ideas how the correct modelling should be done. Too much, we believe, has to be done in this direction.

References

1. Berke, N.S., Hicks, M.C.: Predicting chloride profiles in concrete. Corrosion **50**, 234–239 (1994)
2. Mangat, P.S., Molloy, B.T.: Prediction of long term chloride concentration in concrete. Mater. Struct. **27**, 338–346 (1994)
3. Hansen, E.J., Saouma, V.E.: Numeriacl simulation of reinforced concrete deterioration- Part 1: chloride diffusion. ACI Mater. J. **96**, 173–180 (1999)
4. Costa, A.: Appleton J, Chloride penetration into concrete in marine environment-Part 1: main parameters affecting chloride penetration. Mater. Struct. **32**, 252–259 (1999)
5. Stanish, K., Thomas, M.: The use of bulk diffusion tests to establish time-dependent concrete chloride diffusion coefficients. Cem. Concr. Res. **33**, 55–62 (2003)
6. Nokken, M., Boddy, A., Hooton, R.D., Thomas, M.D.A.: Time dependent diffusion in concrete-three laboratory studies. Cem. Concr. Res. **36**, 200–207 (2006)
7. Petcherdchoo, A.: Time dependent models of apparent diffusion coefficients and surface chloride for chloride transport in fly ash concrete. Constr. Build Mater **38**, 487–507 (2013)
8. Chalee, W., Jaturapitakkul, C.: Chindaprasirt P. Predicting the chloride penetration of fly ash concrete in seawater. Marine Struct. **22**, 341–353 (2009)
9. Yeih, W.D., Huang, R., Chang, J.J.: A study of chloride diffusion properties of concrete at early age. J Marine Sci. Technol. **2**, 61–67 (1994)
10. Wang, L., Ueda, T.: Meso-scale modeling of chloride diffusion in concrete with consideration of effects of time and temperature. Water Sci. Eng. **2**, 5870 (2009)
11. Audenaert, K., Yuan, Q., De Schutter, G.: On the time dependency of the chloride migration coefficient in concrete. Constr. Build. Mater. **24**, 396–402 (2010)
12. Wang, Y., Fu, K.: Comparison of instantaneous chloride diffusion coefficients determined by RCM method and chloride natural diffusion test. Constr. Build. Mater. **223**, 595–604 (2019)
13. Boddy, A., Bentz, E., Thomas, M.D.A., Hooton, R.D.: An overview and sensitivity study of a multimechanistic chloride transport model. Cem. Concr. Res. **29**, 827–837 (1999)
14. Patel, R.A., Phung, Q.T., Seetharam, S., Perko, J., Jacques, D., Maes, N., De Schutter, G., Ye, G., Van Breugel, K.: Diffusivity of saturated ordinary Portland cement-based materials: a critical review of experimental and analytical modelling approaches. Cem. Concr. Res. **90**, 52–72 (2016)
15. Sfafikhani, M., Chidiac, S.E.: Quantification of concrete chloride diffusion coefficient-A critical review. Cem. Concr. Compos. **90**, 225–250 (2019)
16. CCAA (Cement Concrete and Aggregates Australia), Chloride resistance of concrete, Report June 2009
17. Poulsen, E.: On a model of Chloride ingress into concrete. In: Proceedings of Nordisk Miniseminarium-Kloridintrangning I Betong. Goteborg (1993)
18. Poulsen, E.: The chloride diffusion characteristics in concrete. Approximate determination by linear regression analysis. Nordic Concrete Research, Publication No 9, Oslo (1990)
19. Weitsman, Y.: Diffusion with time-varying diffusivity, with application to moisture-sorption in composites. J. Comput. Mater. **10**, 193–204 (1976)
20. Poulsen, E., Mejbro, L.: Diffusion of Chloride in Concrete. Taylor and Francis, London and New York (2006)
21. Tang, L.: Concentration dependence of diffusion and migration of chloride ions. Part 1. Theoretical considerations. Cem. Concr. Res. **29**, 1463–1468 (1999)
22. Maheswaran, T., Sanjayan, J.G.: A semi-closed-form solution for chloride diffusion in concrete with time-varying parameters. Mag. Concr. Res. **56**, 359–366 (2004)
23. Pack, S.-W., Jung, M.-S., Song, H.-W., Kim, S.-H., Ann, K.Y.: Prediction of time dependent chloride transport in concrete structures exposed to a marine environment. Cem. Concr. Res. **40**, 302–312 (2010)
24. Yang, L.F., Ma, Q., Yu, B.: Analytical solution and experimental validation for dual time dependent chloride diffusion in concrete. Constr. Build. Mater. **161**, 676–686 (2018)

25. Carslaw, H.S., Jaeger, J.C.: Conduction of Heat in Solids. Oxford University Press, London (1959)
26. Crank, J.: The Mathematics of Diffusion, 2nd edn. UK, Clarendon Press, Oxford (1975)
27. Sun, C., Chen, J., Zhu, J., Zhang, M., Ye, J.: diffusion model of sulfate ions in concrete. Constr. Buid. Mater. **39**, 39–45 (2013)
28. Mertins, H., Pfuff, M.: The dependent diffusion model for fission gas released in fuel rods. J. Nucl. Mater. **92**, 217–220 (1980)
29. Garcia, D.F.: A new proposed moisture diffusion coefficient to transformer paper. Int. J. Heat Mass Transf. **56**, 469–474 (2013)
30. Chen, W., Zhang, J.-J., Zhang, J.-Y.: A varaiable-0rder time-fractional derivative model for chloride ions sub-diffusion in concrete structures. Frac. Calc. Appl. Anal. **16**, 76–92 (2013)
31. Wei, S., Chen, W., Zhang, J.-J.: Time-fractional derivative model for chloride ions sub-diffusion in reinforced concrete. Euro. J. Environ. Civil Eng. (2015). https://doi.org/:10.1080/19648189.2015.111647
32. Coimbra, C.F.M.: Mechanics with variable-order differential operator. Ann. Phys. (Lepzig) **12**, 692–703 (2003)
33. Sun, H., Chen, W., Che, Y.: Variable-order fractional differential operators in anomalous diffusion modeling. Phys. A **388**, 4586–4592
34. Fa, K.S., Lenzi, E.K.: Time-fractional diffusion equation with time dependent diffusion coefficient. Phys. A **72**, article 011107 (2005). https://doi.org/10.1103/PhysRevE.72.011107
35. Fa, K.S., Lenzi, E.K.: Anomalous diffusion, solutions, and the first passage time: influence of diffusion coefficient. Phys. Rev. E **71**, article 012101 (2005). https://doi.org/10.1103/PhysRevE.71.012101
36. Fa, K.S., Lenzi, E.K.: Power law diffusion coefficient and anomalous diffusion: Analysis of solutions and the first passage time. Phys. Rev. E **67**, article 061105 (2003). https://doi.org/10.1103/PhysRevE.67.061105
37. Hristov, J.: Subdiffusion model with time-dependent diffusion coefficient: integral-balance solution and analysis. Thermal Sci. **21**, 69–80 (2017)
38. Le Vot, F., Abad, E., Yuste, S.B.:Continuous time random walk model for anomalous diffusion in expanding media. Phys. Rev. E **96**, article, 96032117 (2017). https://doi.org/10.1103/PhysRevE.96.032117
39. Wu, J., Diao, B., Zhang, W., Ye, Y., Liu, Z., Wang, D.: Chloride diffusivity and service life prediction of fatigue damaged RC beam under sea water wet-dry environment. Constr. Build. Mater. **171**, 942–949 (2018)
40. Moradillo, M.K., Shekarchi, M., Hoseini, M.: Time-dependent performance of concrete surface coatings in tidal zone of marine environment. Constr. Build. Mater. **30**, 198–205 (202)
41. Markeset, G., Kioumarsi, M.: Need for further development in service life modelling of concrete structures in cloride environment. Proc. Eng. **171**, 549–556 (2017)
42. Ehlen, M.A., Thomas, M.D.A., Bentz E.C.: Life-365 service life prediction model [TM version 2.0]. J. Concr. Inst. **31**, 41–46 (2009)
43. Maage, M., Helland, S., Poulsen, E., Vennesland, O., Carl, E.: Service life prediction of existing concrete structures exposed to marine environment. ACI Mater. J. **93**, 602–608 (1996)
44. De Swiet, T.M., Sen, P.N.: Time dependent diffusion coefficient in disordered medium. J. Chem. Phys. **104**, 296–209 (1996)
45. Shi, M., Chen, Z., Sun, J.: Determination of chloride diffusivity in concrete by AC impedance spectroscopy. Cem. Concr. Res. **29**, 1111–1115 (1999)
46. Nugue, F., Yssorche-Cubaynes, M.-P., Olliver, J.P.: Innovative study of non-steady-state tritiated water diffusion test. Cem. Concr. Res. **37**, 1145–1151 (2007)
47. Roa-Rodriguez, G., Aperador, W., Delgado, A.: Calculations of chloride penetration profile in concrete structures. Int. J. Electrochem. Sci. **8**, 5022–5035 (2013)
48. Fleury, M., Berthe, G., Chavlier, Th: Diffusiion of water in industrial cement and concrete. Magn. Reson. Imag. **56**, 32–35 (2019)
49. Goodman, T.R.: Application of integral methods to transient nonlinear heat transfer. In: Irvine, T.F., Hartnett, J.P. (eds.) Advances in Heat Transfer, vol. 1, pp. 51–122. Academic Press, San Diego, CA (1964)

50. Hristov, J.: The heat-balance integral method by a parabolic profile with unspecified exponent: analysis and Benchmark Exercises. Thermal Sci. **13**, 27–48 (2009)
51. Sadoun, N., Si-Ahmed, E.K., Colinet, P.: On the refined integral method for one-phase Stefan problem with time-dependent boundary conditions. Appl. Math. Model. **30**, 531–544 (2006)
52. Mitchell, S.L., Myers, T.G.: Application of standard and refined heat balance integral methods to one-dimensional Stefan problems. SIAM Rev. **52**, 57–86 (2010)
53. Hristov, J.: Integral solutions to transient nonlinear heat (mass) diffusion with a power-law diffusivity: a semi- infinite medium with fixed boundary conditions. Heat Mass Transfer **52**, 635–655 (2016)
54. Hristov, J.: Integral-balance solution to nonlinear subdiffusion equation. In: Bhalekar, S. (ed.) Frontiers in Fractional Calculus, pp.71–106. Bentham Sci. Publ., Sharja (2017)
55. Hristov, J.: Double integral-balance method to the fractional subdiffusion equation: approximate solutions, optimization problems to be resolved and numerical simulations. J. Vib. Contr. **23**, 2795–2818 (2017)
56. Hristov, J.: Space-fractional diffusion with a potential power-law coefficient: transient approximate solution. Progr. Fract. Differ. Appl. **3**, 119–39 (2017)
57. Volkov, V.N., Li-Orlov, V.K.: A Refinement of the integral method in solving the heat conduction equation. Heat Transf. Sov. Res. **2**, 41–47 (1970)
58. Myers, T.G.: Optimizing the exponent in the heat balance and refined integral methods. Int. Commun. Heat Mass Transf. **36**, 143–147 (2009)
59. Mittelstaedt, P., Weingartner, P.A.: Laws of Nature. Springer, Berlin (2005)
60. Nussenzveig, H.: Causality and Dispersion Relations, vol. 95 of Mathematics in Science and Engineering. Academic Press, NY (1972)
61. Henry, B.I., Langlands, T.A.M., Straka, P.: An introduction to fractional diffusion. In: Dewar, R., Detering, F. (eds.) Complex Physical, Biophysical and Econophyscal Systems, vol. 9, pp. 37–89. Canbera, Australian National University (2010)
62. Hilfer, R.: Fractional diffusion based on Riemann-Liouville fractional derivative. J. Phys. Chem. **104**, 3914–2917 (2000)
63. Coleman, B., Noll, W.: Foundations of linear Viscoelasticity. Rev. Modern Phys. **33**, 239–249 (1961)
64. Coleman, B., Gurtin, M.E.: Equipresence and constitutive equations for rigid heat conductors. Z. Angew. Math. Phys. **18**, 188–208 (1967)
65. Gurtin, M.E.: On the thermodynamics of materials with memory. Arch. Rational Mech. Anal. **28**, 40–50 (1968)
66. Gurtin, M.E., Pipkin, A.C.: A general theory of heat conduction with finite wave speeds. Arch. Rational Mech. Anal. **31**, 113–126 (1968)
67. Storm, M.L.: Heat conduction in simple metals. J. Appl. Phys. **22**, 940–951 (1951)
68. Nunciato, J.W.: On heat conduction in materials with memory. Q. Appl. Math. **29**, 187–273 (1971)

Laminar Convection of Power-Law Fluids in Differentially Heated Closed Region: CFD Analysis

Bhuvnesh Sharma, Sunil Kumar, and Carlo Cattani

Abstract In this manuscript, the characteristics of heat transfer of non-Newtonian fluids in a natural convection application is analyzed. A 2D square domain containing power-law fluid is studied whose horizontal walls follow adiabatic condition through insulation whereas the vertical walls are differentially heated isothermally. The provided temperature difference drives the convection current. Various parameters like Nusselt number, dimensionless vertical velocity and dimensionless temperature are evaluated to examine the effect of power-law index on heat and mass transfer for different values of Rayleigh number varying between 10^3 and 10^6. The influence of power-law index and Bingham number on the heat transfer characteristics is analyzed and the best one with high heat transfer capability is proposed for natural convection application.The results thus obtained are compared on the basis of Nusselt number, velocity and temperature with the help of TECPLOT and ANSYS.

Keywords CFD analysis · Heat transfer · Natural convection · Power-law fluids

Nomenculture

k	Flow consistency index (Pa s^n)
n	Flow behavior index (dimensionless)

B. Sharma · S. Kumar (✉)
Department of Mechanical Engineering, National Institute of Technology, Jharkhand 831014, Jamshedpur, India
e-mail: skumar.math@nitjsr.ac.in

B. Sharma
Department of Mechanical Engineering, Vivekananda Institute of Technology Jaipur, 303012 Jaipur, India

C. Cattani
Engineering School (DEIM) University of Tuscia, Largo dell' universetià, 01100 viterbo, Italy

J. Singh et al. (eds.), *Methods of Mathematical Modelling and Computation for Complex Systems*, Studies in Systems, Decision and Control 373,
https://doi.org/10.1007/978-3-030-77169-0_2

c_p	Specific heat at constant pressure (J/kg K)
g	Gravitational acceleration (m/s^2)
h	Heat transfer coefficient (W/m^2 K)
k	Thermal conductivity (W/m K)
$\overline{Nu}$	Mean Nusselt number
Ra	Rayleigh number
Bn	Bingham number
Pr	Prandtl number
L	Length and height of enclosure (m)
T	Temperature
U, V	Dimensionless horizontal and vertical velocity
u, v	Velocity in x and y direction respectively (m/s)
ρ	Density (Kg/m^3)

Greek Symbols

α	Thermal diffusivity (m^2/s)
β	Coefficient of thermal expansion (1/K)
γ	Shear rate (1/s)
μ	Plastic viscosity (Ns/m^2)
μ_{yield}	Yield viscosity (Ns/m^2)
ϕ	General primitive variable
ΔT	Difference between hot and cold wall temperatures i.e. $T_H - T_C$ (K)
τ	Shear stress (N/m^2)
τ_y	Yield stress (N/m^2)

Subscripts

C	Cold wall
H	Hot wall
ref	Refernce value
$wall$	Wall value
eff	Effective value

1 Introduction

Natural convection is common in both nature and technical devices which is due to the variation in density with temperature. Even though this is a simple case of natural convection in the square enclosure it has various engineering applications such as in heating, preserving canned foods and in cooling of electronic components. Most of the cooling processes involve natural convection only. Because of it's simple geometry and its applications this analysis is concentrated on this square enclosure where horizontal walls follow adiabatic condition through insulation while the vertical walls are maintained at different temperatures isothermally as in the case [1]. Previous studies regarding to Newtonian fluids are available in literature [2–4]. While Bejan et al. [5] insulated the vertical walls and maintained the horizontal walls at different temperatures. The present work analyzes the case where the vertical walls are maintained at different temperatures and this difference of temperatures drives the convection current as specified by de Vahl Davies et al. [2]. Even though this described model for Newtonian fluids has been investigated by several researchers but the research regarding to non-Newtonian fluids is limited due to their complex behavior. Lamsaadi has demonstrated the influence of power-law index [6] and shallow enclosures [7] on the temperature distribution where two vertical walls are considered as constant heat flux boundary conditions rather than constant temperature. Barth and Carey [8] used more complex Generalized Newtonian fluid (GNF) models to investigate the 3D model of the same problem and experimental limitations are justified in [9]. Osman et al. [1] analyzed the natural convection in an enclosure filled with Bingham fluid with different Rayleigh numbers and came to know that increasing Bingham number decreases $\overline{Nu}$ values irrespective of the Rayleigh number considered. Also observed that $\overline{Nu}$ rises with increasing Prandtl number (Pr) for Newtonian fluids as well as non-Newtonian fluids at low Bingham number.

In this present research, the outcomes of [1] are carried on to illustrate the effect of power-law fluids on temperature and velocity distribution with Rayleigh number ranging from 10^3 to 10^6. The square cavity of dimension L filled with Bingham plastic, Pseudoplastic and Dilatant fluids is analyzed for heat transfer characteristics and compared with Newtonian fluid results. The viscosity does not remain constant in non-Newtonian fluids. Therefore the relation between tangential stress and velocity gradient or shear rate for non-Newtonian fluids is specified as follows which is presented in Fig. 1.

$$\tau = \tau_y + k\gamma^n, \tag{1}$$

where τ, n, γ, k and τ_y denote the shear stress, power-law index, shear rate, flow consistency index and preliminary or yield stress respectively. For Dilatant and Pseudoplastic fluids the preliminary stress τ_y is zero and for Bingham plastic fluids there exists a preliminary stress. The analysis is being carried out for different Rayleigh number (Ra) values varying from 10^3 to 10^6 at Prandtl number (Pr) = 7. The Rayleigh, Prandtl, Bingham and Nusselt numbers are presented as follows:

$$Ra = \frac{\rho^2 c_p g \beta \Delta T L^3}{\mu k} \tag{2}$$

$$Pr = \frac{\mu c_p}{k}. \tag{3}$$

$$B_n = \frac{\tau_y}{\mu}\sqrt{\frac{L}{g\beta\Delta T}} \tag{4}$$

$$Nu = \frac{hL}{k} \tag{5}$$

where hot and cold wall reference temperatures are considered as T_C and T_H respectively. However, for modeling Bingham Plastic fluid Bi-viscosity model is employed which is described in [10]. The constant plastic viscosity is replaced by effective viscosity expressed as $\mu_{eff} = \frac{\tau_y}{\gamma} + \mu$.

However, the present research assumes that yielded stress and plastic viscosity do not depend on the temperature for the benifit of easiness and dearth of a suitable method for the inclusion of temperature dependent properties. Even though it is not a perfect approach it seems to be a reasonable approximation as the results are in good agreement with preceding investigations on Bingham fluids [11–14]. In addition to this, experimental data [15] for a yield stress system "Carbopo" indicates that the yield stress is constant with respect to temperature and the plastic viscosity is reducing slightly with in the temperature range of 0–90 °C. For modeling Dilatant and Pseudoplastic fluids non-Newtonian power law is considered whose equation is specified below.

$$\eta = k\gamma^{n-1}. \tag{6}$$

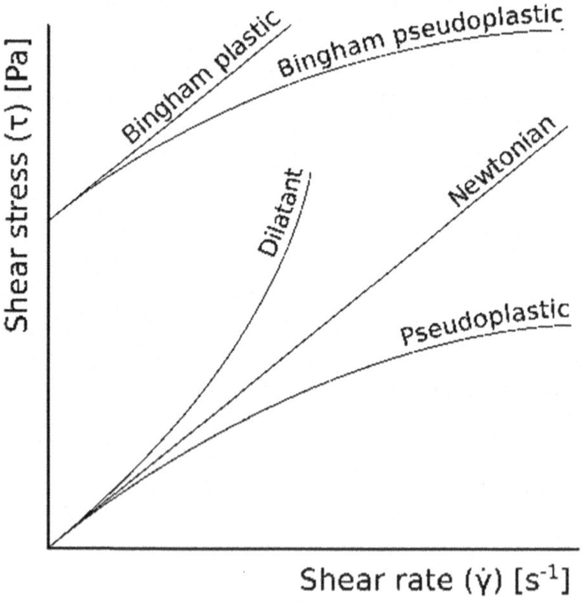

Fig. 1 Shear stress versus Shear rate for different types of Non-Newtonian fluids

This manuscript models the all three non-Newtonian fluids and analyze for heat and mass transfer characteristics in the chosen domain. A comparative study is made between non-Newtonian and Newtonian fluids and analyze the heat dissipation phenomena so obtain highest rate of heat transfer for free convection applications.

2 Numerical Method

The fluid flow postulates the fundamental laws which are established using conservation of mass, energy and momentum. These equations of conservation are evaluated to investigate the fluid flow and heat transfer problems with the help of ANSYS Fluent 18.2. This software utilize central differential scheme and upwind scheme of the second order for diffusive and convective terms respectively. In this analysis, Boussinesq approximation is employed to model the effect of buoyancy forces in the flow field without altering the density. This approximation is valid only for low temperature difference. SIMPLE algorithm is employed to couple velocity and pressure as it accelerates the convergence using fewer resources. The convergence criteria between two consecutive iterations is set to be relative deviation less than $1e-6$ for energy equation and less than $1e-4$ for other variables.

2.1 Governing Equations

The conservation differential equations for chosen fluid are specified below.

$$\text{Continuity Equation}: \quad \frac{\partial u_x}{\partial x_x} = 0, \tag{7}$$

$$\text{Momentum Equation}: \quad \rho u_y \frac{\partial u_x}{\partial x_y} = -\frac{\partial p}{\partial x_x} + \rho g \delta_{x2} \beta (T - T_C) + \frac{\partial \tau_{xy}}{\partial x_y} \tag{8}$$

$$\text{Energy Equation}: \quad \rho u_y c_p \frac{\partial T}{\partial x_y} = \frac{\partial}{\partial x_y}(k \frac{\partial T}{\partial x_y}) \tag{9}$$

Where the right side temperature at wall (T_C) is considered as (T_{ref}) to compute the buoyancy term $\rho g \delta_{x2} \beta (T - T_C)$ in Eq. (8) specified by Boussinesq approximation following several previous studies [2–5, 14].

The Bingham fluid is modeled using bi-viscosity model specified in [10] where properties of fluid are considered independent from temperature. The equations employed for modeling Bingham fluid are specified as

$$\tau = \mu_y \dot{\gamma} \quad \text{for } \dot{\gamma} \leq \frac{\tau_y}{\mu_y} \tag{10}$$

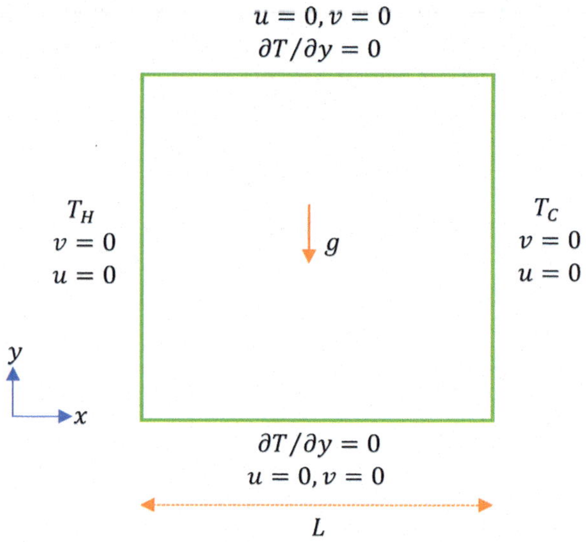

Fig. 2 Appropriate boundary conditions of enclosure

$$\tau = \tau_y + \mu\left[\dot{\gamma} - \frac{\tau_y}{\mu_y}\right] \quad \text{for } \dot{\gamma} \le \frac{\tau_y}{\mu_y} \tag{11}$$

The μ_y is considered 1000 times of μ that estimates the acceptable Bingham model as specified in [10]. In the present work, the ratio of the yielded viscosity (μ_y) and plastic viscosity (μ) is considered as 10000 for better prediction of the effect of Bingham fluid on heat and mass transfer. The default Herschel-Bulkley model present in ANSYS Fluent is used for specifying properties of Bingham fluid. After solving for Bingham fluid the available Non-Newtonian power-law is employed to model both pseudoplastic and dilatant fluids. The relation describing the non-Newtonian power-law is specified in Eq. (6).

2.2 Boundary Conditions

The domain that is to be simulated is shown in Fig. 2 in which the left vertical wall is kept at higher temperature than the right wall temperature and other two horizontal walls are insulated. In addition to this, all the boundaries are assumed to be having non-penetrating and no-slip conditions. The right wall is specified to have significantly lower temperature (i.e. $T(x = 0) = T_C$) and the left wall is specified to have higher temperature (i.e. $T(x = L) = T_H$). Top and bottom walls are specified with insulating boundary condition i.e. $(\partial T/\partial y)_{y=0} = 0$ and $(\partial T/\partial y)_{y=L} = 0$. In order to obtain the Velocity in x and y direction (u, v), density and temperature (ρ, T) at each grid point over the entire domain the governing equations (1 energy + 1 continuity + 2 momentum) are solved numerically by ANSYS Fluent.

Table 1 Grid independence study of the present domain of with Newtonian, Bingham plastic ($Bn = 0.5$), Pseudo plastic ($n = 0.7$) and dilatant fluids ($n = 1.5$) of $Pr = 7$

Fluid type		$\overline{Nu}$				V_{max}			
		M1	M2	M3	M4	M1	M2	M3	M4
Newtonian	ϕ	4.78465	4.7359	4.71826	4.71507	71.5621	72.6246	73.4261	73.4512
	$\Delta\phi$	–	0.04875	0.01764	0.00319	–	1.0625	0.8015	0.0251
	$\frac{\Delta\phi}{\phi_i}\%$	–	1.0189	0.3725	0.0676	–	1.1847	1.1036	0.0342
Bingham plastic	ϕ	6.0189	5.369	5.6627	5.6545	95.843	96.758	97.3168	97.3695
	$\Delta\phi$	–	0.182	0.1742	0.0082	–	0.915	0.5588	0.0527
	$\frac{\Delta\phi}{\phi_i}\%$	–	4.4331	1.8959	0.0721	–	2.7409	0.83	0.2858
Pseudo-plastic	ϕ	6.0189	5.369	5.6627	5.6545	95.843	96.758	97.3168	97.3695
	$\Delta\phi$	–	0.182	0.1742	0.0082	–	0.915	0.5588	0.0527
	$\frac{\Delta\phi}{\phi_i}\%$	–	3.0238	2.9845	0.1448	–	0.9547	0.5775	0.0542
Dilatant	ϕ	4.5139	4.1503	3.9948	3.97505	56.2473	57.2671	57.7153	57.7437
	$\Delta\phi$	–	0.3636	0.1555	0.01975	–	1.0198	0.4482	0.0284
	$\frac{\Delta\phi}{\phi_i}\%$	–	8.0551	3.7467	0.4944	–	1.8131	0.7826	0.0492

2.3 Grid Dependency

To obtain grid independent solution, mesh dependency on the solution is tested by considering four different types of meshes M1 (30 × 30), M2 (60 × 60), M3 (120 × 20), M4 (240 × 240). Two solution variables (i.e. ($\overline{Nu}$) and V_{max}) are considered for testing grid dependency. Numerical simulation is carried out for each type of mesh with same boundary conditions applied. The results obtained are tabulated and shown in Table 1. Note that each mesh has an inflation layer along the four walls for better accuracy.

From above table, we observe that the values of $\overline{Nu}$ obtained from meshes M3 and M4 are very close when compared to other cases and also the percentage change ($\frac{\Delta\phi}{\phi_i}$) in the values obtained is less than 1% for meshes M3 and M4. Also in the case of maximum vertical velocity (V_{max}) the percentage change in the values obtained from meshes M3 and M4 is less than 1%. Even though the number of divisions for mesh M4 is double that of M3 the percentage change in solution variables is less than 1%. Considering the computational efficiency and accuracy the mesh M4 is considered as an optimal mesh to solve the problem. All the results specified in this paper are obtained using mesh M4.

2.4 Benchmark Comparison

The evaluated results of Newtonian fluid with Ra varying from 10^6 to 10^4 and $Pr = 0.71$ are verified with the benchmark results specified in [2]. The comparison between the present simulations and benchmark results are tabulated in Table 2.

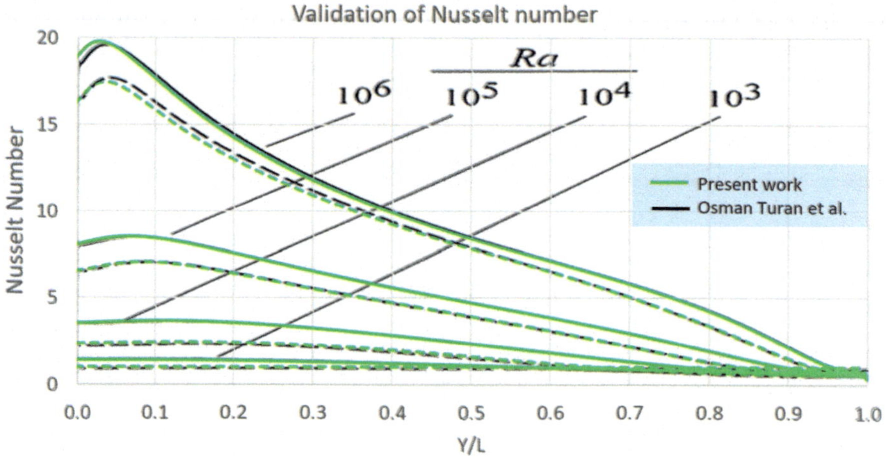

Fig. 3 Nusselt number variation with respect to vertical distance of the hot wall at $Pr = 0.7$

By observing Table 2 we observe that the results are obtained from the present simulations that are very close to the benchmark results [2]. The maximum percentage error between evaluated and benchmark values [2] is less than 2%. In addition to this, the simulation results of Bingham fluid with $Bn = 0.5$ and $Pr = 7$ are compared with the results specified by Osman Turan [1]. The comparison is plotted as a graph and summarized in Fig. 3. No remarkable deviations are found and the results are obtained in satisfactory limits.

The domain is simulated with Bingham fluid of B_n values that varies from 0 to Bn_{max} where Bn_{max} is the Bingham number for $\overline{Nu} = 1$. The solution is not affected by further increase in (B_n) because conduction is merely reason for heat dissipation. Along with Bingham fluid, the domain is simulated with shear-thinning fluid and shear-thickening fluid. The obtained results thus are analyzed and conclusion is drawn for better heat transfer phenomenon.

3 Discussion and Computational Results

3.1 The Effects of Rayleigh Number

The Nusselt number $(\overline{Nu})$ varies with normalized vertical distance (y/L) for all types of fluids (i.e. Newtonian, Bingham plastic, Pseudo plastic and Dilatant) at different Ra values varying from 10^3 to 10^6 with $Pr = 7$ is presented in Fig. 4. It can be observed that out of all fluids Pseudoplastic has the highest $\overline{Nu}$ values and the Bingham plastic has the lowest $\overline{Nu}$ values irrespective of Rayleigh number (Ra) considered. Also,the Dilatant fluid has a lower $\overline{Nu}$ values when compared to the

Table 2 Comparison of present simulation results with the Benchmark results [2] using Newtonian fluid of $Pr = 0.71$

	Particulars	Proposed work	[2]
$Ra = 10^3$	$\overline{Nu}$	1.112	1.118
	Nu_{max}	1.507	1.505
	U_{max}	3.64	3.649
	V_{max}	3.685	3.697
$Ra = 10^4$	$\overline{Nu}$	2.232	2.243
	Nu_{max}	3.543	3.528
	U_{max}	15.91	16.178
	V_{max}	19.517	19.617
$Ra = 10^5$	$\overline{Nu}$	4.518	4.519
	Nu_{max}	7.757	7.717
	U_{max}	34.715	34.73
	V_{max}	68.617	68.59
$Ra = 10^6$	$\overline{Nu}$	8.833	8.8
	Nu_{max}	17.781	17.925
	U_{max}	64.974	64.63
	V_{max}	219.104	219.36

Newtonian fluid at a specific value of Ra. With the increment in the value of Ra, the $\overline{Nu}$ increases for all types of fluids and in all the cases Pseudoplastic has the highest $\overline{Nu}$ values and Bingham plastic has the lowest $\overline{Nu}$ values.

Along with these results, it is always beneficiary to have a look at the variation of θ and V in preceding to define the performance of non-Newtonian fluids in natural convection application. The variation of θ with the horizontal mid-plane for all fluids is depicted in Fig. 5. For $Ra = 10^3$ we can observe that the temperature diffusion is almost linear due to weak buoyancy forces and dominant viscous forces present in the closed region. The heat dissipation is majorly caused by conduction the reason of dominant viscous forces but buoyancy forces convert the heat transfer mode through convection as Ra increases. For a better understanding of this concept, in Fig. 6 the variation of V with the horizontal mid-plane is presented. As the heat transfer through convection is proportional to the vertical velocity we observe that Nu is maximum when V is maximum.

By analyzing the Figs. 4 and 6, we conclude that the V increases as Ra increases. For particular Rayleigh number the vertical velocity V is maximum for Pseudoplastic fluid and minimum for Dilatant fluid. Non–inearity in θ increases as Ra increases because of buoyancy forces. The strengthening of buoyancy-driven flow is shown in Fig. 7a, b in which the temperature and stream function contours are presented for Newtonian fluid at different Rayleigh numbers. By observing them one can clearly confirm that with increasing Ra number the curvature of isotherms is increasing, this proves the concept of increment of buoyancy force with increasing Ra. The

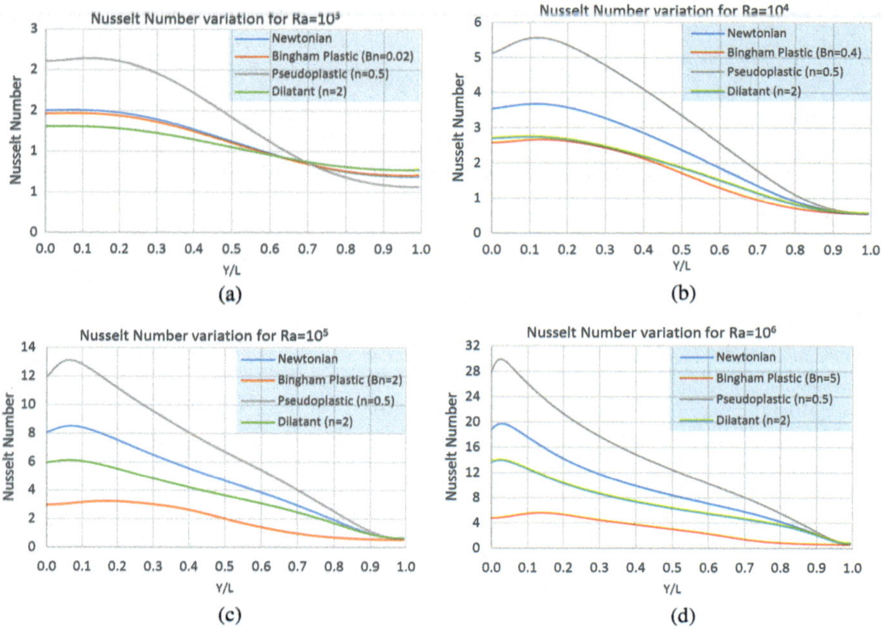

Fig. 4 Nusselt number varies with respect to vertical length of hot wall for $Pr = 7$ with **a** $Ra = 1000$ **b** $Ra = 10000$ **c** $Ra = 100000$ **d** $Ra = 1000000$

isotherms are linear for small Ra due to conduction dominated heat flow taking place in the squared region.

3.2 Bingham Fluids

The $\overline{Nu}$ varies with respect to Bn that is depicted in Fig. 8 with the different values Ra (10^3 to 10^6). Pr is kept constant 7.0, this value seems to be pragmatic for incompressible fluids as described by [1]. we analyze in Fig. 8 that $\overline{Nu}$ decreases as Bn increases then it settles down to unity finally. This type of behavior is valid with earlier studies [14]. It is important to know that $\overline{Nu}$ becomes unity when heat transfers only by conduction because of suppression of buoyancy forces by viscous forces.

This concept can be validated by observing Fig. 9 where the effect of Bn on the variation of θ and V with the horizontal mid-plane are presented for $Ra = 10000$ and 1000000. Figure 9 reveals that V becomes zero and the profile of θ becomes linear for extensive values of Bn. Figure 10 presents the comparison of isotherm contours for different values of Bn. Accordingly the isotherms becomes linear and tends to be a straight line as Bn increases.

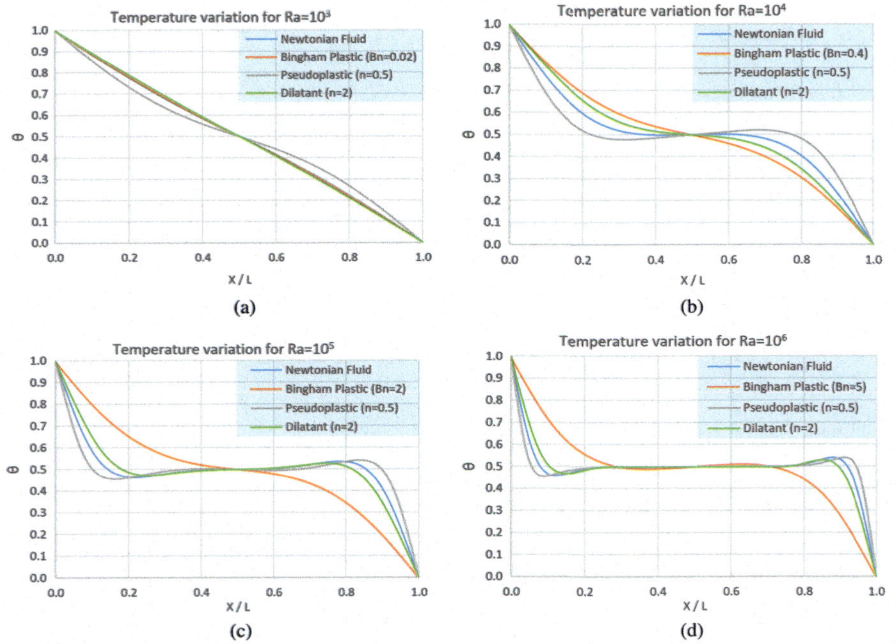

Fig. 5 Temperature (θ) varies along the horizontal mid plane at $Pr = 7$ with **a** $Ra = 1000$ **b** $Ra = 10000$ **c** $Ra = 100000$ **d** $Ra = 1000000$

It is evident from Fig. 9, the temperature profile becomes vertical and linear component of velocity that becomes zero for large values of Bn (i.e. $Bn > Bn_{max}$). This behavior of Bingham fluid can be better explained with the comparison of contours of isotherms by varying Bn for $Ra = 10000$ and 1000000 which are shown in Fig. 10. From this, it is evident with the increment in Bingham number, the variation of isotherms becomes linear and even the higher values of Bn makes the isotherms as straight lines (i.e. Pure conduction case). This clearly shows that, with increasing Bn values the buoyancy effects are dominated by viscous effects resulted that, no powerful flow is induced in the closed area. Both Figs. 9 and 10 concludes that heat transfer due to convection in the domain decreases as Bn increases and moreover, the behavior of fluid tend to be solid as $Bn > Bn_{max}$. The isotherms persist parallel to wall same as the conduction problems in solids in the absence of fluid flow. This is because of the pure conduction taking place and this phenomena is reflected by $\overline{Nu} = 1$ in Fig. 8. As Ra increases the buoyancy force strengthens. The Bingham number Bn at which Nusselt number approaches unity is called as critical Bingham number (Bn_{max}).

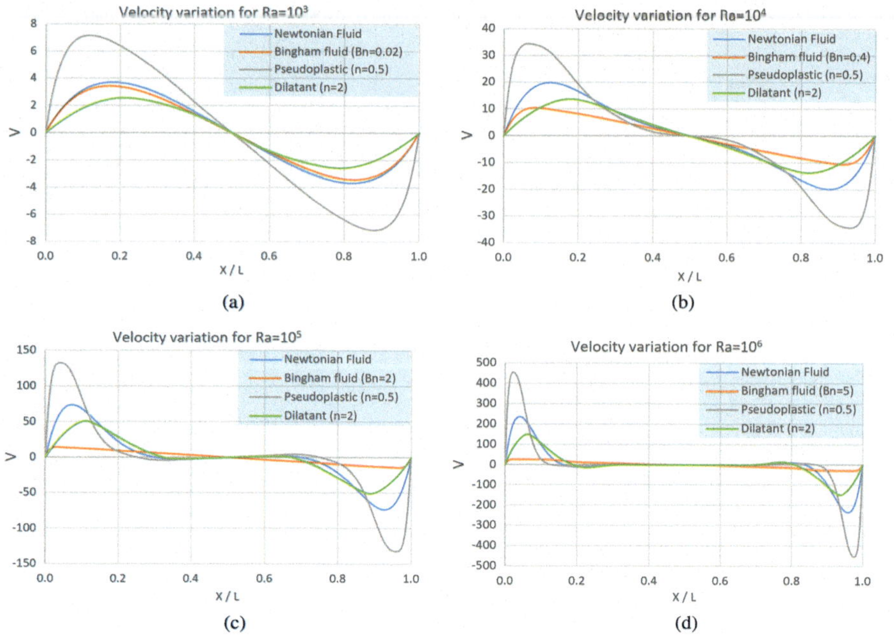

Fig. 6 Vertical velocity (V) variation along horizontal mid plane at $Pr = 7$ for **a** $Ra = 1000$ **b** $Ra = 10000$ **c** $Ra = 100000$ **d** $Ra = 1000000$

3.3 Pseudo Plastic Fluid

The $\overline{Nu}$ varies with respect to n of a Pseudoplastic fluid with having $Ra = 10^6$ and $Pr = 7.0$ is shown in Fig. 11. It is clear that $\overline{Nu}$ increases as n decreases. Decreasing n reduces the viscous forces inside the enclosure. This reduced viscous forces strengthens the buoyancy effect and makes the fluid to flow more rapidly inside the enclosure thereby increasing the convective heat transfer rate. By reason of this increased convective heat transfer the $\overline{Nu}$ is increasing. These results can be validated by observing Fig. 12 in which the effect of n on the variation of V and θ through horizontal mid-planes are specified for $Ra = 10^4$ and 10^6.

Figure 12 revels that the increment in vertical velocity component (V) decreases the power-law index, it proves the concept of strengthening of buoyancy forces. Along with the velocity variation, one can observe the increase in curvature of the dimensionless temperature (θ) because of the increased convective heat transfer rate. This effect is more dominant for high Rayleigh number flows since the buoyancy effect increases with an increase in Ra as shown in Fig. 6. This can be confirmed by observing Fig. 12 where the increase in velocity for $n = 0.5$ is high in the case of $Ra = 10^6$ than that of 10^4 . The Nusselt number depends upon the vertical velocity, the increasment in Nusselt number for $Ra = 10^6$ than that of 10^4. In Fig. 12 the line with $n = 1$ represents the Newtonian fluid results which observe that the Pseudo-

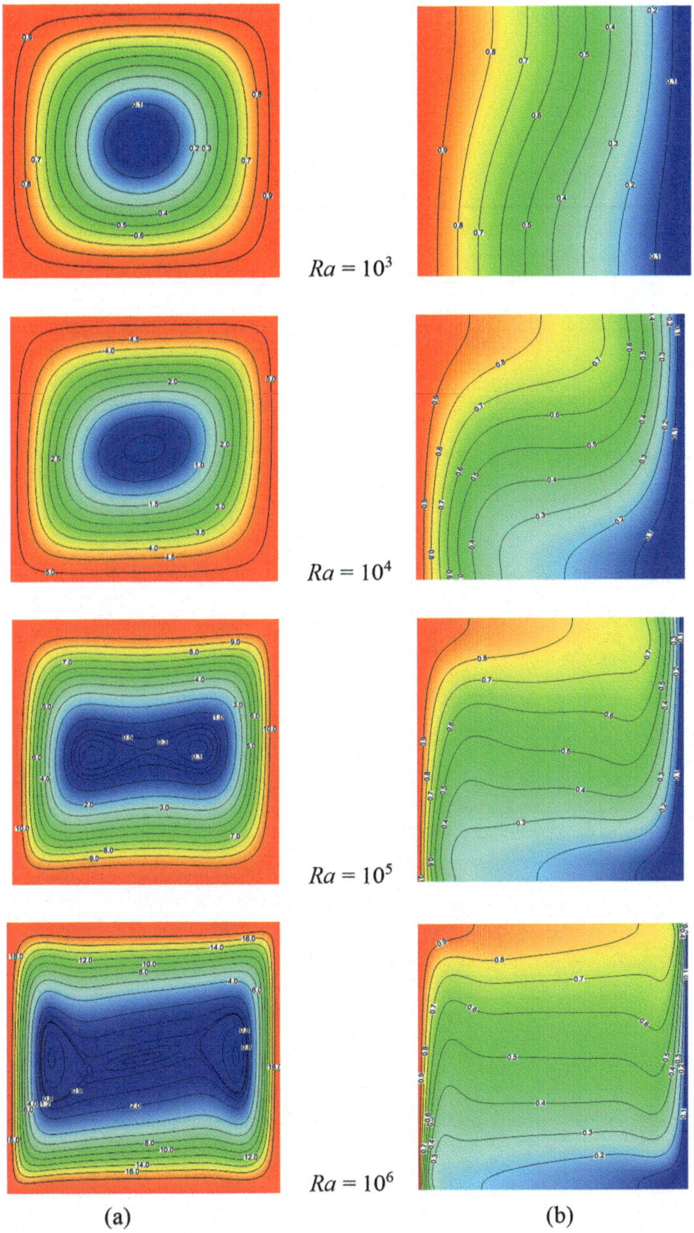

Fig. 7 Contours for Newtonian fluid at $Pr = 7$ for **a** Dimensionless stream function (ψ/α) and **b** Dimensionless temperature (θ)

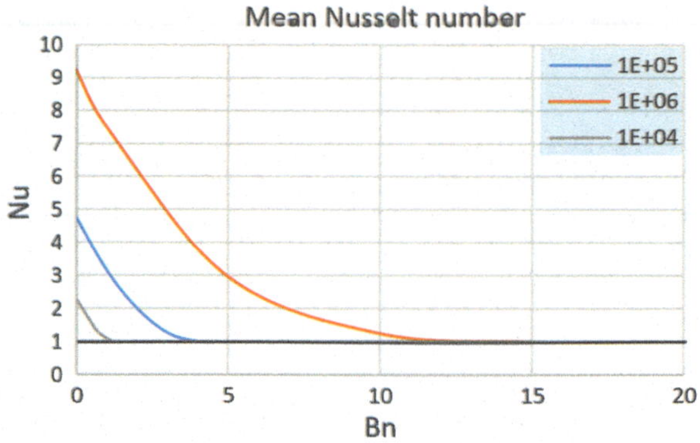

Fig. 8 The interrelation between $\overline{Nu}$ and Bn for different values of Ra numbers at $Pr = 7$

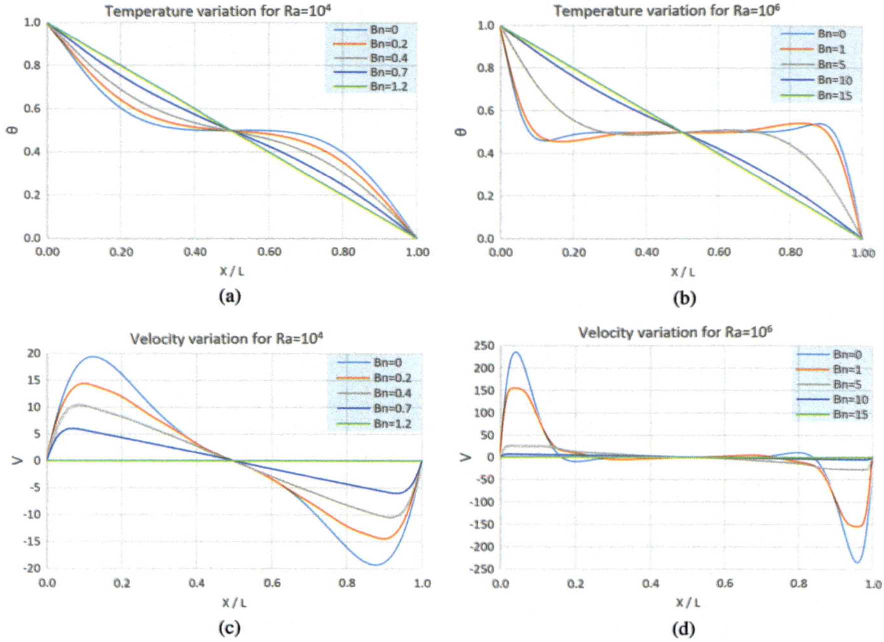

Fig. 9 Temperature and velocity profile with the horizontal mid-plane for various Bn number for $Ra = 10^4$ and 10^6 at $Pr = 7$

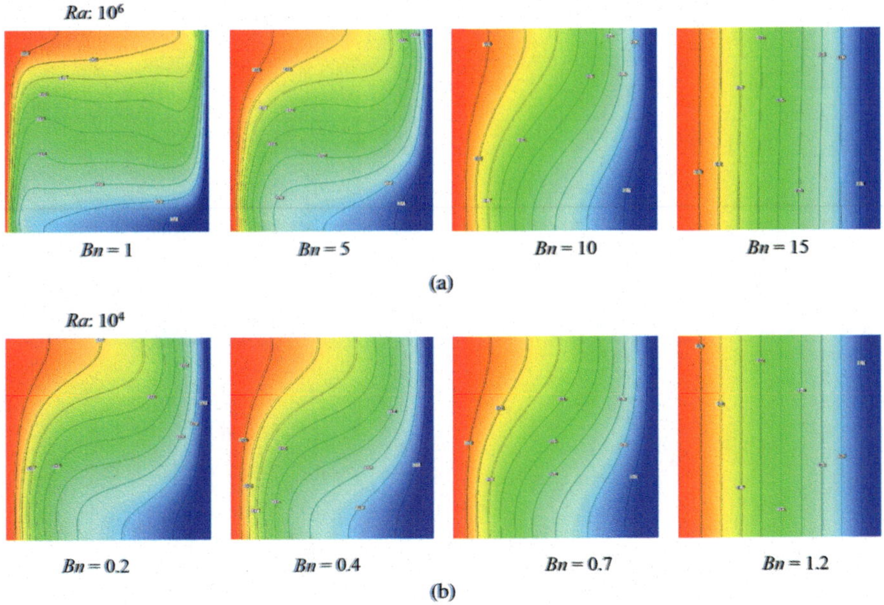

Fig. 10 Contours of dimensionless temperature (θ) with varying Bn at $Pr = 7$ for **a** $Ra = 10^6$ **b** $Ra = 10^4$

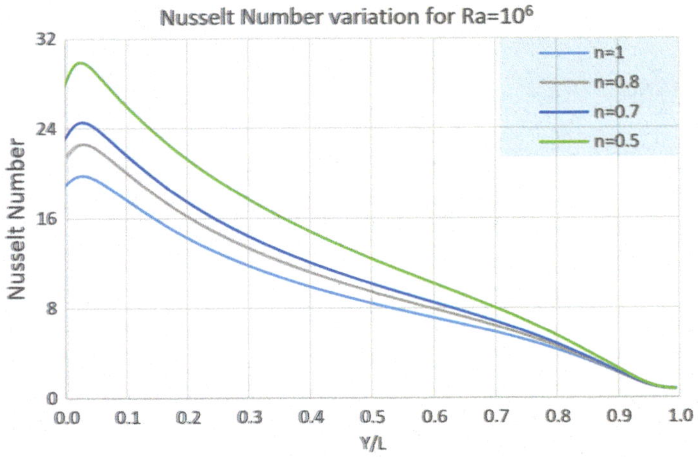

Fig. 11 Nusselt number varies along normalized vertical length of hot wall for distinct values of n with $Ra = 10000$ and 1000000 at $Pr = 7$

plastic fluid has more heat transfer capability than the Newtonian fluid because of the reduced viscous forces.

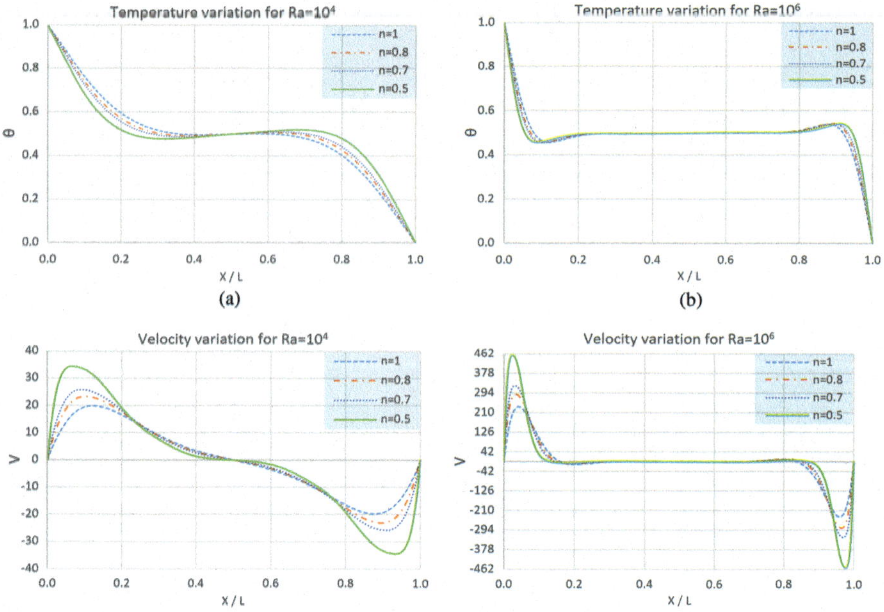

Fig. 12 Vertical velocity profile and temperature through horizontal mid-plane for distinct values of n in case of $Ra = 10^4$ and 10^6 at $Pr = 7$

3.4 *Dilatant Fluids*

Figure 13 presents the $\overline{Nu}$ variation with respect to n of a Dilatant fluid having $(Ra) = 10^6$ and $Pr = 7.0$. It is clear that $\overline{Nu}$ decreases as n increases. This is due to the increased viscous forces with increasing n. The increased viscous forces suppress the buoyancy effect thereby decreasing the convective heat transfer rate. This results can be validated by observing Fig. 14 in which the effect of n on the variation of θ and V through horizontal mid-planes are specified for $Ra = 10^4$ and 10^6.

Figure 14 shows that V decreases as the value of n increases. This proves the concept of reduction in buoyancy effect. Along with the velocity variation, it is clear that the curvature of θ decreases as n increases. The temperature profile becomes linear for increased n due to the reduced convective heat transfer.

From Fig. 14 one can observe that the decrease in V for $n = 2$ is high in the case of $Ra = 10^6$ than that of 10^4. The Nusselt number depends upon the vertical velocity, the decrement in Nusselt number for $Ra = 10^6$ is more than that of 10^4. In Fig. 14, $n = 1$ represents the Newtonian fluid and one can observe that the Dilatant fluid has low heat transfer capability than Newtonian fluid.

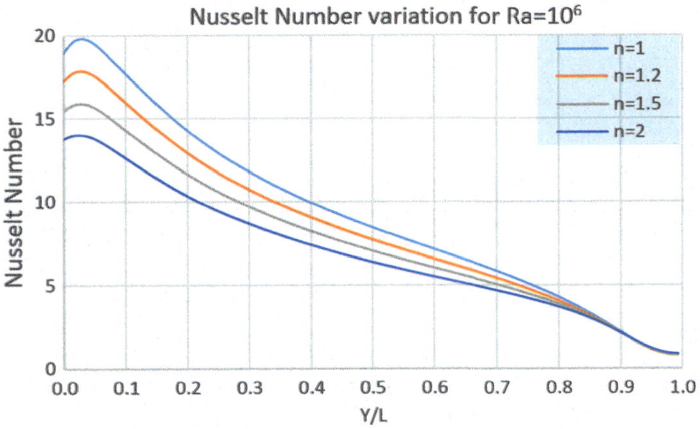

Fig. 13 Nusselt number variation with vertical length of hot wall for distinct values of n with $Ra = 10^4$ and 10^6 at $Pr = 7$

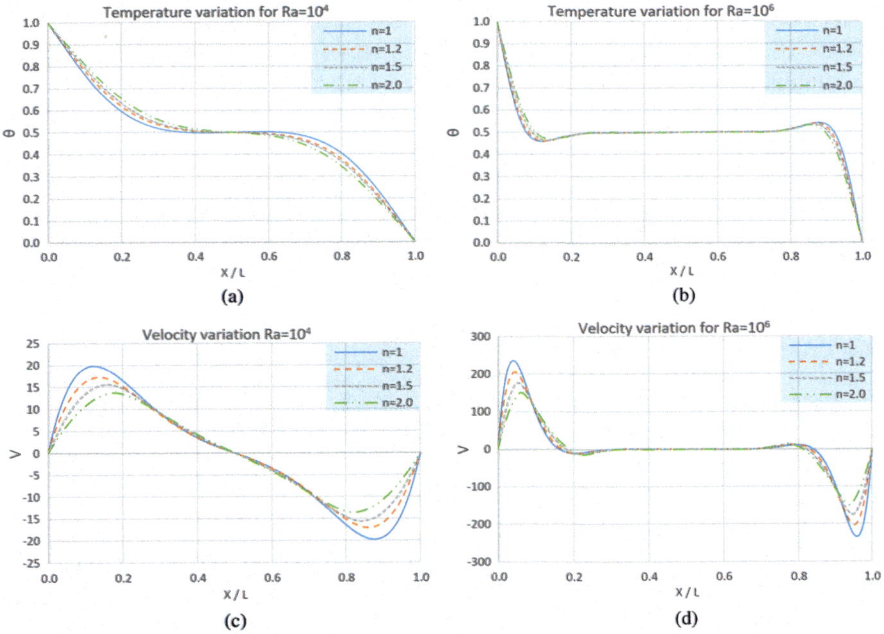

Fig. 14 Temperature and vertical velocity profile through horizontal mid-plane for various values of n with $Ra = 10^4$ and 10^6 at $Pr = 7$

4 Conclusions

The heat transfer characteristics of power-law fluids in a 2D squared domain alongwith the two vertical walls kept at distinct temperature have been investigated. The effect of Ra on momentum and heat transport for different power-law fluids has been evaluated. We noted that $\overline{Nu}$ increases as Ra increases for both power-law and Newtonian fluids. However the $\overline{Nu}$ obtained encase of Pseudo plastic fluids are larger than values obtained for other fluids irrespective of Ra considered. Whereas among all the fluids the values of $\overline{Nu}$ obtained are lower in class of Bingham fluids at considered Ra.

In course of Bingham fluids, $\overline{Nu}$ is found to be decreasing as Bingham number is increasing and further increment in Bingham number ($Bn > Bn_{max}$), $\overline{Nu}$ approaches unity by means of $\overline{Nu} = 1$ as the heat is being transferred only due to conduction because of dominant viscous forces.

For Dilatant fluids, $\overline{Nu}$ is decreasing as the value of n is increasing. This is due to the weakening of buoyancy forces with respect to n. For Pseudo plastic fluids, $\overline{Nu}$ is increasing as the n is decreasing. This is due to the strengthening of buoyancy effects. The relative strengths of viscous, buoyancy forces and the influence of n on momentum and heat transport are investigated.

It is important to notice that in this work yielded stress (τ_y) and plastic viscosity (μ) are considered to be independent of temperature in the ease and better understanding. Even though if the properties with the temperature-dependency are considered for the analysis there will not be any noticeable changes in the present results but it is always beneficiary to include temperature-dependent properties in order to obtain the quantitative results.

Declaration of Conflicting Interests The author(s) declared no potential conflicts of interest with respect to the research, authorship, and/or publication of this article.

Funding This research received no specific grant from any funding agency in the public, commercial, or not-for-profit sectors.

References

1. Turan, O., Chakraborty, N., Poole, R.J.: Laminar natural convection of Bingham fluids in a square enclosure with differentially heated side walls. Journal of Non-Newtonian Fluid Mechanics **165**, 901–913 (2010)
2. de Vahl, D.G.: Natural convection of air in a square cavity: a bench mark numerical solution. Int. J. Numer. Methods Fluids **3**, 249–264 (1983)
3. Emery, A.F., Lee, J.W.: The effects of property variations on natural convection in a square cavity. J. Heat Transfer **121**, 57–62 (1999)
4. Aydin, O., Unal, A., Ayhan, T.: Natural convection in rectangular enclosures heated from one side and cooled from above. Int. J. Heat and Mass Transfer **42**, 2345–2355 (1999)
5. Bejan, A.: Convection Heat Transfer. John Wiley Sons Inc., New York (1984)

6. Lamsaadi, M., Naimi, M., Hasnaoui, M., Mamou, M.: Natural covection in a vertical rectangular cavity filled with a non-Newtonian power law fluid and subjected to a horizontal temperature gradient. Numer. Heat Trans. A **49**, 969–990 (2006)
7. Lamsaadi, M., Naimi, M., Hasnaoui, M.: Natural covection heat transfer in shalow horizontal rectangular enclosures uniformly heated from the side and filled with non-Newtonian power law fluids. Energy Convers. Manage. **47**, 2535–2551 (2006)
8. Barth, W.L., Carey, G.F.: On a natural-convection benchmark problem in a non-Newtonian fluids. Numer. Heat Transfer B **50**, 193–216 (2006)
9. Leung, W.H., Hollands, K.G.T., Brunger, A.P.: On a physically-realizable benchmark problem in internal natural convection. Int. J. Heat Mass Transfer **41**, 3817–3828 (1998)
10. O'Donovan EJ, Tanner RI (1984) Numerical study of the Bingham squeeze film problem. J. Non-Newtonian Fluid Mech. 15;75–83
11. Zhang J, Vola D, Frigaard IA (2006) Yield stress effects on Rayliegh-Benard convection. J. Fluid Mech. 566;389–419
12. Balmforth NJ, Rust AC (2009) Weakly nonlinear viscoplastic convection. J. Non-Newtonian Fluid Mech. 158;36–45
13. Vikhansky, A.: Thermal convection of a viscoplastic liquid with high Rayleigh and Bingham numbers. Phys. Fluids **21**, (2009)
14. Vola D, Boscardin L, Latche JC (2003) Laminar unsteady flows of Bingham fluids: a numerical strategy and some benchmark results. J. Comput. Phys. 187;441–456
15. Peixinho, J., Desaubry, C., Lebouche, M.: Heat transfer of a non-Newtonian fluid (Carbopol aqueous solution) in transitional pipe flow. Int. J. Heat Mass Transfer **51**, 198–209 (2008)

Mathematical Perspective of Hodgkin-Huxley Model and Bifurcation Analysis

Avinita Gautam and Anupam Priyadarshi

Abstract Hodgkin-Huxley model (*HH model*) qualitatively describes the generation of the action potential of squid giant axons. The resting state or oscillatory phase state of the membrane potential (voltage) in the *HH model* depends mainly on applied stimulus (external currents) to neurons. The firing of the action potential in neurons depends upon the depolarization or repolarization of ions. The probability of channel gates (to be open or close) determines the movement of ions across the cell membrane. The term of K^+ ionic currents (related to several activation gates) in external current contains exponential power *4* in *HH model* in which we propose a modified *HH model* by considering the higher power (*5* and *6*) of *K* activation in potassium ionic currents and studied the behavior of all three models comparatively. The modified *HH model* with a *higher power of potassium activation* reached resting-state sooner and gains stability (after oscillatory) at a high external current. The qualitative behavior of the modified model (with the higher exponential power) is different as there is a shifting of Hopf bifurcation points in comparison with the original *HH model*. Moreover, a larger periodic region was observed in most of the parameter phase spaces (external current *I* versus parameters) except against the *Na* conductance and *Na* potential. The modified *HH model* which determines that *higher power of K activation* is more significant for action potential in neurons.

Keywords Potassium activation · Action potential · Hodgkin-Huxley equations · Hopf bifurcation · Resting state

A. Gautam
DST -CIMS, Institute of Science, Banaras Hindu University, Varanasi, India
e-mail: avinita.gautam2@bhu.ac.in

A. Priyadarshi (✉)
Department of Mathematics, Institute of Science, Banaras Hindu University, Varanasi, India

J. Singh et al. (eds.), *Methods of Mathematical Modelling and Computation for Complex Systems*, Studies in Systems, Decision and Control 373,
https://doi.org/10.1007/978-3-030-77169-0_3

1 Neurological Backgrounds

Neuroscience is a branch of biology in which the structure of nervous systems is studied and deals with configuration and chemical processes during the function of the interaction of neurons [1]. The brain is a complicated network of neurons which transmits the information for the functioning of other parts and is remained active continuously (Fig. 1), even in the absence of functions of other body parts. The central nervous system (brain and spinal) and peripheral nervous system (nerves which are made of individual neurons) are two major parts of the nervous systems [1, 2]. Neurons transmit electrochemical signals to communicate within the body [3] and have different structures as compared to other cells such as red cells (i.e. cannot be reproduced) [1, 2].

1.1 How Does Neuron Work?

Nerves are made up of individual neurons, which consist of dendrites, soma, axon, and synapses (Fig. 2) [1]. Dendrites are the small tree-like structure, which accepts information from another cell and passes on to the cell body [3]. The cell body (soma) controls and regulates the functions of the cell and it is a part of the neuron having a roughly rounded shape that contains the nucleus, mitochondria, and other organelles

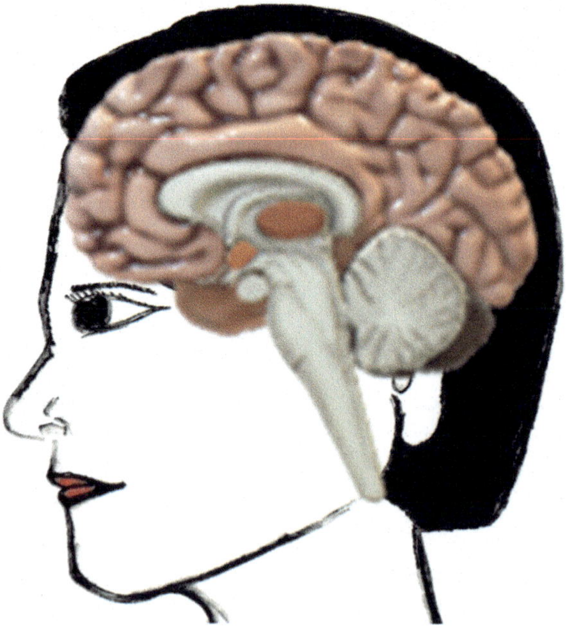

Fig. 1 The most important and complex part of the human body is "Brain"

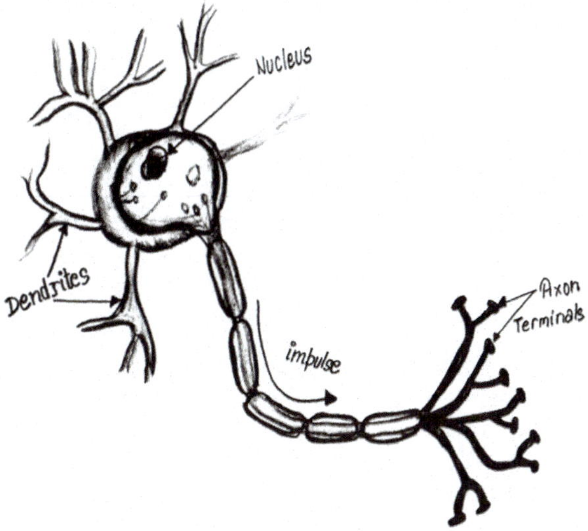

Fig. 2 A single neuron having a complex structure and able to transmit electrochemical signals to communicate within the human body

[1, 2]. An axon is a long slim projection of a nerve cell that transmits information from the cell body to other neurons, muscles, or glands for functioning or movement of other body parts [3]. Neurons have been classified into three parts:

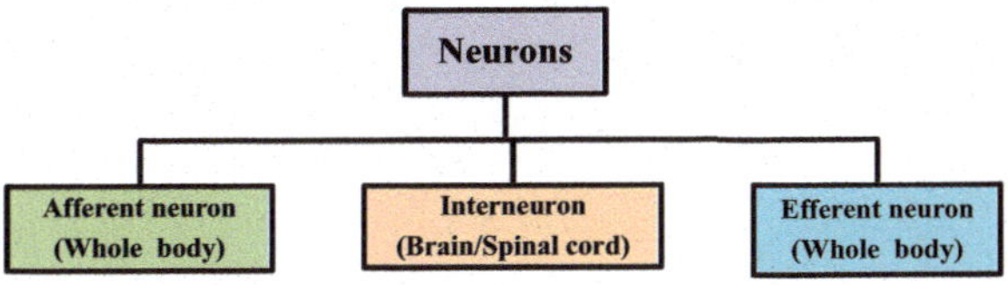

The sensory signal from receptors (vision, touch, hearing, etc.,) to the central nervous system in the body is transmitted by *afferent neurons* [1]. Signals are transmitted with the help of *efferent or motor neurons* from the central nervous system to effectors in the body such as muscles and glands for the functioning of body parts [1, 2]. *Interneuron* is an intermediate between afferent and efferent neurons which helps them to communicate with each other [1]. Afferent and efferent neurons are found throughout the body which is responsible for the functioning of body parts, whereas *interneuron* is found only in the brain and spinal cord [1].

1.2 How Does a Neuron Fire?

There is no movement in the body parts due to the non-transmission of signals from receptors to neurons (at rest). In this resting state, negative charges are more inside the cell relative to the positive charges outside the cell which allows certain ions to

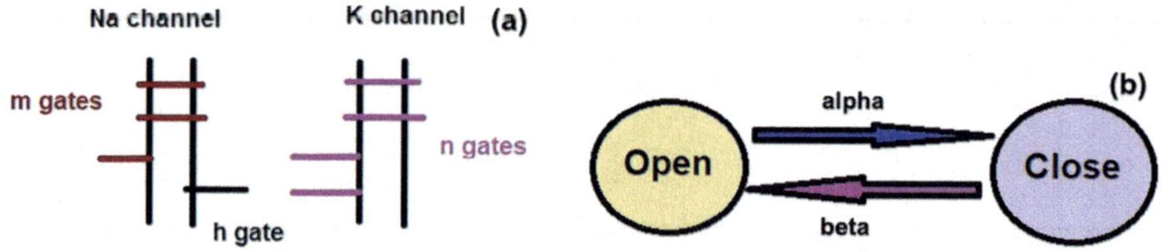

Fig. 3 **a** Structure of channel gates- *3 Na* channel activation gates, *4 K* channel activation gates, and *1 Na* channel inactivation gate, and **b** Opening and closing of the gate in which $\alpha(alpha)$ and β $(beta)$ represents the rate at which gates are close and open

pass across the cell while it prevents other ions from moving into the cell membrane [1]. Sodium ions ($Na+$) and potassium ion ($K+$) cannot easily pass through the membrane; however, potassium ions are free to cross the membrane through ion channels [1]. The negative ions inside the cell are unable to cross the barrier. Therefore, the movement of ions in the cell membrane is dependent on the opening and closing of the gate (Fig. 3).

When an *impulse* is received by dendrites and transmitted to the cell body then neurons become stimulated. Once, this impulse is sent out from the cell body then *Na* channels open and Na^+ ions surge into the cell [4, 5]. Once the membrane potential reaches a certain threshold value, an action potential will fire and an electrical signal is sent down to the axon [6]. The action potential is created by the movement of Na^+ and K^+ ions through the membrane [1]. Action potential follows the *all-or-none law* that is either it does fire or not [7] and hence no partial occurrence of the action potential. After firing an action potential, neurons go to the refractory period, which means no other action potential will occur during the refractory period [1]. Neurons return to its resting potential when the *K* channels reopen and sodium channels closed. After this process, another action potential may occur.

1.3 How Neurons Communicate?

The connections between cells are known as *synapses* in which neurotransmitters are discharged by the neurons to communicate with each other [1, 2]. Neurotransmitters are the chemicals discharged by the neurons into synapses to communicate with each other [2]. It means there is a transmission of an electrical signal within neurons and transmission of chemical signals between neurons (Fig. 4).

Alan Hodgkin [8] started his research on neuron response in 1935 and 1939 (collaboration with Huxley) to work on squid giant fiber, their collaborative research works (*original HH model*) has been published in 1949 [8, 9]. Due to the large diameter of the *Squid giant axon*, Huxley and Hodgkin preferred to experiment on this *Squid giant axon* which resulted in a successful experiment. The study of Hodgkin and Huxley is based on "*How a nerve impulse travels along an axon*?" and the speed of impulse depends on the diameter of the axon. In general, all living cells have a

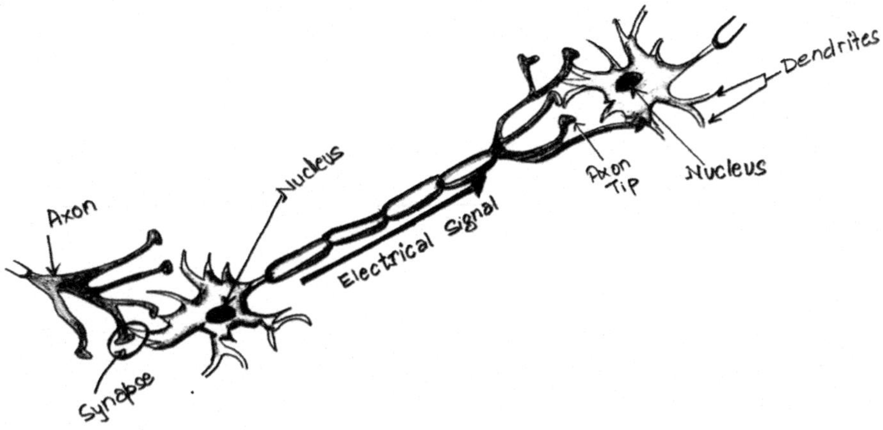

Fig. 4 Neurons communicate by passing a signal between each other and these transmission results as a function of body parts

membrane that separates inside and outside of the cell. The membrane potential [10] is the potential difference created in the cell while resting potential is the potential at rest [11]. Hodgkin and Huxley used the *Voltage Clamp Experiment* for the study of neuron responses in which an electrode is inserted in an axon without disturbing it to measure the potential differences [12–14]. In this experiment, the membrane potential (constant) and the ionic current (time-dependent) are independent of the position on the axon. Many authors like Carl Craver believe that *HH* has knowledge about ions channels only and they have emphasized the whole study around that point and have not introduced the specific reason for the selection of variables in the mathematical model [15].

In this manuscript, we have explored the higher power of the potassium activation gate variable and observed the dynamics which has been shown in the last three sections. We have established a background in neuroscience to connect the mechanism of the Hodgkin-Huxley model and its mathematical description. Time series analysis and bifurcation analysis have been performed to understand the dynamics of the behavior of neurons. These techniques will provide a visual approach for an easy understanding of how neurons become active from resting-state and again how it returns to the resting state after firing an action potential. Also, this study will help to think about the assumptions of the *HH model.*

2 Hodgkin and Huxley Model (*HH Model*)

Hodgkin and Huxley have contributed their major efforts for the transmission of nerve impulse in neural modeling. This model describes the ionic current in form of *Na* and *K* ions. When the neuron is stimulated then the membrane potential changes and causes the opening and closing of channels gates. *Voltage-clamped experiment* was the basis for this model. The Hodgkin-Huxley model described how ionic current initiate and propagate in axon by the following set of equations as follows [16]:

$$\begin{aligned} CdV/dT &= I - [g_K n^4(V - V_K) + g_{Na} m^3 h(V - V_{Na}) + g_L(V - V_L)] \\ dn/dt &= \alpha_n(V)(1-n) - \beta_n(V)n \\ dm/dt &= \alpha_m(V)(1-m) - \beta_m(V)m \\ dh/dt &= \alpha_h(V)(1-h) - \beta_h(V)h \end{aligned} \tag{1}$$

where

$$\begin{aligned} \alpha_m &= 0.1(V+40)/(1-\exp(-(V+40)/10)) \\ \beta_m &= 4\exp(-(V+65)/20) \\ \alpha_h &= 0.07\exp(-(V+65)/20) \\ \beta_h &= 1/(1+\exp(-(V+35)/10)) \\ \alpha_n &= 0.01(V+55)/(1-\exp(-(V+55)/10)) \\ \beta_n &= 0.125\exp(-(V+65)/80) \end{aligned} \tag{2}$$

In the *HH model* (1), *V n, m, and h* are treated as variables for membrane potential, the potential for sodium, potential for potassium, and potential for leakage current respectively, and $c = 1\,\mu\text{F/cm}^2$ is the capacitance of membrane. V_{Na},V_K and V_l are the reversible potentials of sodium ($V_{Na} = 50\,\text{mV}$), potassium ($V_K = -77\,\text{mV}$), and leakage channels ($V_l = -54.4\,\text{mV}$) respectively. g_{Na},g_K and g_L are the maximum conductance of the membrane for the sodium ($g_{Na} = 120\ \text{mS/cm}^2$), potassium ($g_K = 36\ \text{mS/cm}^2$), and leakage currents ($g_L = 0.3\ \text{mS/cm}^2$). $\alpha_n, \alpha_m, \alpha_h, \beta_n, \beta_m$ and β_h are the rate constants (used in the above Eq. (2)) calculated experimentally (Table 1) [17]. *HH* had given the overall

Table 1 Description of Variables and parameters used in the HH Model

Variable and parameters	Representations	Units
V, V_{Na}, V_K and V_L	Membrane potential, the potential for sodium (*Na*), potassium(*K*), and leakage current respectively	mV
C	Membrane capacitance	μF/cm^2
I	External current	μA/cm^2
n, *m*, and *h*	Potassium activation, Sodium activation, and Sodium inactivation	–
g_{Na}, g_K and g_L	Ionic conductance through sodium, potassium, and leakage current	mS/cm^2
α_n, α_m and α_h	Rate constants	–
β_n, β_m and β_h	Rate constants	–
T	Time	ms

structure of the cell membrane and the process of the firing of the action potential [15]. They have used the experimentally fitted values but have not given any descriptive reason for how changes in conductance occur.

HH model is four-dimensional and has complex behavior. Several two-dimensional models depict similar characteristics as given by the *HH model.* There are two-dimensional models such as the *Morris-Lecar model,* FitzHugh-Nagumo *(FHN) model, and* Hindmarsh–*Rose model* which are quite easier to study the behavior of neurons [10]. This study is based on the *HH model* and its mathematical equations.

3 Review of *HH Model*:

The neuron's characteristics are the most influential factor for determining the neuron's response to external electrical stimulations [18]. The stable equilibriums and limit cycles are the most frequent dynamical behavior exhibited by the neuron's response system. At constant (or unvarying) external current, a single spike of action potential has the following two responses of neurons: (i) the resting potential state and (ii) the oscillatory (periodic) state [10]. The display of irregular behavior by electric excitable cell model [19] and the electrical excitation of squid giant axons are extensively studied by Hodgkin-Huxley models or modified Hodgkin-Huxley models which are assumed to be a prototype for excitable cell models [20]. Temperature (control parameter in the *HH model*) may affect the neural activity [21] and variations in it may exhibit chaotic behavior [21, 22]. Global structure of bifurcation (Hopf bifurcation, Saddle-node, Period Doubling, Homoclinic bifurcations) is investigated over the wide range of parameters through non-linear dynamics [23–32].

The dynamics of single-neuron and circuits involved in neural information processing is observed through bifurcations [33, 34]. As the resting potential corresponds to a stable solution in the *HH model* while periodic firing corresponds to the periodic solution which is often characterized by the existence of Hopf bifurcation [23]. The shifting of Hopf bifurcation in such non-linear dynamical models may be possible by changing parameter, this drives us to investigate the shifting of Hopf bifurcation in the *HH model* by changing the exponent in the expression of potassium ionic current. The individual contributions of $K+$, $Na+$, and leakage ionic currents resulted in the total amount of current flow that accounts for the aggregation of other ionic fluxes such as the chloride and bicarbonate ions through the excited membrane in the *HH model* [31, 35]. The membrane currents depend on both the capacitance of the plasma membrane and resistance of the ion channels [36, 37]. As there are *four* activation gates for *potassium* ions in the *HH model*, the potassium activation (n) in the external current is mostly studied for n^4*as* $(g_K n^4(V\text{-}V_K))$ due to the complexity of the *HH model* [30, 38]. There is no valid reason for selecting activation and inactivation variables components. They had chosen the data for mathematical convenience. Even though they had studied for the *4*th exponential power of activation of potassium [15] but higher exponential power (*5* or *6*) of potassium activation may be used to observe the dynamics more clearly [10]. In the present manuscript, for external current, the proportion of the open channels is considered by function n^5 or n^6 instead of n^4 in the *HH model,* and the qualitative behavior of the *modified HH model* is comparatively studied. The results (one-parameter or two parameters bifurcation analysis) are compared and influential parameters are recognized based on the sensitive dependence of the external current and periodic region in the parameter spaces.

4 Modified *HH Models*

In the *HH model,* the conductance change during the depolarizing step of the *voltage clamp experiment* had a sigmoidal shape while it had an exponential shape during the repolarizing step. *HH model* has first-order reactions for the individual channel gates which produce exponential curves, the sigmoidal curves found due to the occurrence of several first-order reactions simultaneously (co-operative process) i.e., the channels contained several gates, *all* of which had to be open simultaneously for the channel itself to be open but for the channel to shut, it is enough to shut *only one* gate. Based on the shape of the experimentally-measured sigmoidal curve Hodgkin-Huxley suggested that *four* would be the best estimate of independent gates within the *K* channel and with a similar analysis of conductance curve shapes for *Na, three* activation gates and *one* inactivation gate (Fig. 3).

The dynamics of neuron's responses (resting-state or oscillatory state) may be explored by assuming higher power (*5* or *6*) of potassium activation (n) in potassium ionic currents [10]. The expression containing potassium activation (n) with exponent power *4* $[g_K n^4(V-V_K)]$, is replaced by exponent powers *5* and *6* which is biologically feasible [7]. Keeping other variables and parameters identical and introducing *K* activations as ($[g_K n^4(V-V_K)]$, $[g_K n^5(V-V_K)]$ *and* $[g_K n^6(V-V_K)]$), the expressions of external current (I) in *HH model* (1) are modified as:

$$I = CdV/dT + g_K n^4(V - V_K) + g_{Na} m^3 h(V - V_{Na}) + g_L(V - V_L) \quad (3)$$

$$I = CdV/dT + g_K n^5(V - V_K) + g_{Na} m^3 h(V - V_{Na}) + g_L(V - V_L) \quad (4)$$

$$I = CdV/dT + g_K n^6(V - V_K) + g_{Na} m^3 h(V - V_{Na}) + g_L(V - V_L) \quad (5)$$

The following parameters are used to observe the global structure of the dynamics of the *HH model* with the above external current (I) for the model cases (3)–(5):

$$\text{para } i = 0, V_{Na} = 50, V_K = -77, V_l = -54.4, g_{Na} = 120, g_K = 36, g_l = 0.3, c = 1$$

We have tried to investigate the patterns and behaviors after modifying the model mathematically. This comparative study will help us to understand the differences with the help of numerical simulations.

5 Comparative Studies of Modified *HH Models*

5.1 *Stability Analysis*

The proposed modified *HH model* Eq. (1) with three cases (3)–(5) have mathematically very complex expressions and they produced an intractable expression of non-trivial equilibrium points. Let $E_1 = (V_1, m_1, h_1, n_1)$, $E_2 = (V_2, m_2, h_2, n_2)$ and $E_3 = (V_3, m_3, h_3, n_3)$ be the existed non-trivial equilibrium points for model cases (3)–(5) respectively which have been obtained by equating the Eq. (1) equals to zero for the corresponding three cases (3)–(3). The stability of equilibrium points E_1, E_2 and E_3 depends upon the eigenvalues of the corresponding Jacobian matrix. The Jacobian matrix can be obtained after linearization of the system (1) around the equilibrium points E_1, E_2 and E_3:

$$\begin{aligned} \frac{dV_i}{dt} &= g^i_{11} V_i + g^i_{12} m_i + g^i_{13} h_i + g^i_{14} n_i \\ \frac{dm_i}{dt} &= g^i_{21} V_i + g^i_{22} m_i \\ \frac{dh_i}{dt} &= g^i_{31} V_i + g^i_{33} h_i \\ \frac{dn_i}{dt} &= g^i_{41} V_i + g^i_{44} n_i \end{aligned} \tag{6}$$

where $i = 1, 2$ and 3 and

$$g^1_{11} = -\frac{g_{Na} m^3 h + g_K n^4 + g_L}{C}, g^1_{12} = -\frac{3 g_{Na} m^2 h (V - V_{Na})}{C},$$

$$g^1_{13} = -\frac{g_{Na} m^3 (V - V_{Na})}{C}, g^1_{14} = -\frac{4 g_K n^3 (V - V_K)}{C}$$

$g^1_{21} = (21m\exp(-v/18 - 65/18))/9 + (m - 1)/(10(\exp(-v - 40)/10 - 1))$
$+ (\exp(-v - 40)(v/10 + 4)(m - 1))/(10(\exp(-v - 40)/10 - 1)^2)$
$g^1_{22} = (v/10 + 4)/(\exp(-v - 40)/10 - 1) - 4\exp(-v/18 - 65/18)$, $g^1_{23} = 0$, $g^1_{24} = 0$
$g^1_{31} = (7\exp(-v/20 - 13/4)(h - 1))/2000 - (h\exp(-v - 35))/(10(\exp(-v - 35)/10 + 1)^2)$
$g^1_{32} = 0$
$g^1_{33} = -(7\exp(-v/20 - 13/4))/100 - 1/(\exp(-v - 35)/10 + 1)$, $g^1_{34} = 0$
$g^1_{41} = (3n\exp(-v/80 - 13/16))/2000 + (n - 1)/(100(\exp(-v - 55)/10 - 1))$
$+ (\exp(-v - 55)(v/100 + 11/20)(n - 1))/(10(\exp(-v - 55)/10 - 1)^2)$
$g^1_{42} = 0$, $g^1_{43} = 0$, $g^1_{44} = (v/100 + 11/20)/(\exp(-v - 55)/10 - 1)$
$-(3\exp(-v/80 - 13/16))/25$

$$g^2_{11} = -\frac{g_{Na}m^3h + g_Kn^5 + g_L}{C}, g^2_{12} = -\frac{3g_{Na}m^2h(V - V_{Na})}{C},$$
$$g^2_{13} = -\frac{g_{Na}m^3(V - V_{Na})}{C}, g^2_{14} = -\frac{5g_Kn^4(V - V_K)}{C}$$

$g^2_{21} = (2m\exp(-v/18 - 65/18))/9 + (m - 1)/(10(\exp(-v - 40)/10 - 1))$
$+ (\exp(-v - 40)(v/10 + 4)(m - 1))/(10(\exp(-v - 40)/10 - 1)^2)$
$g^2_{22} = (v/10 + 4)/(\exp(-v - 40)/10 - 1) - 4\exp(-v/18 - 65/18)$, $g^2_{23} = 0$, $g^2_{24} = 0$
$g^2_{31} = (7\exp(-v/20 - 13/4)(h - 1))/2000 - (h\exp(-v - 35))/(10(\exp(-v - 35)/10 + 1)^2)$,
$g^2_{32} = 0$
$g^2_{33} = -(7\exp(-v/20 - 13/4))/100 - 1/(\exp(-v - 35)/10 + 1)$, $g^2_{34} = 0$
$g^2_{41} = (3n\exp(-v/80 - 13/16))/2000 + (n - 1)/(100(\exp(-v - 55)/10 - 1))$
$+ (\exp(-v - 55)(v/100 + 11/20)(n - 1))/(10(\exp(-v - 55)/10 - 1)^2)$,
$g^2_{42} = 0$, $g^2_{43} = 0$,
$g^2_{44} = (v/100 + 11/20)/(\exp(-v - 55)/10 - 1) - (3\exp(-v/80 - 13/16))/25$

$$g^3_{11} = -\frac{g_{Na}m^4h + g_Kn^6 + g_L}{C}, g^3_{12} = -\frac{3g_{Na}m^2h(V - V_{Na})}{C},$$
$$g^3_{13} = -\frac{g_{Na}m^3(V - V_{Na})}{C}, g^3_{14} = -\frac{6g_Kn^5(V - V_K)}{C}$$

$g^3_{21} = (2m\exp(-v/18 - 65/18))/9 + (m - 1)/(10(\exp(-v - 40)/10 - 1))$
$+ (\exp(-v - 40)(v/10 + 4)(m - 1))/(10(\exp(-v - 40)/10 - 1)^2)$
$g^3_{22} = (v/10 + 4)/(\exp(-v - 40)/10 - 1) - 4\exp(-v/18 - 65/18)$, $g^3_{23} = 0$, $g^3_{24} = 0$
$g^3_{31} = (7\exp(-v/20 - 13/4)(h - 1))/2000 - (h\exp(-v - 35))/(10(\exp(-v - 35)/10 + 1)^2)$
$g^3_{32} = 0 g^3_{33} = -(7\exp(-v/20 - 13/4))/100 - 1/(\exp(-v - 35)/10 + 1)$
$g^3_{34} = 0$
$g^3_{41} = (3n\exp(-v/80 - 13/16))/2000 + (n - 1)/(100(\exp(-v - 55)/10 - 1))$
$+ (\exp(-v - 55)(v/100 + 11/20)(n - 1))/(10(\exp(-v - 55)/10 - 1)^2)$, $g^3_{42} = 0$, $g^3_{43} = 0$
$g^3_{44} = (v/100 + 11/20)/(\exp(-v - 55)/10 - 1) - (3\exp(-v/80 - 13/16))/25$

The *Jacobian matrix* J_{E_i} around the equilibrium point E_i:

$$J_{E_i} = \begin{bmatrix} g^i_{11} & g^i_{12} & g^i_{13} & g^i_{14} \\ g^i_{21} & g^i_{22} & g^i_{23} & g^i_{24} \\ g^i_{31} & g^i_{32} & g^i_{33} & g^i_{34} \\ g^i_{41} & g^i_{42} & g^i_{43} & g^i_{44} \end{bmatrix}$$

from which the *characteristic equation* has been obtained as:

$$\lambda^4 + a_1\lambda^3 + a_2\lambda^2 + a_3\lambda + a_4 = 0 \tag{7}$$

where,

$$a_1 = -(g^i_{11} + g^i_{22} + g^i_{33} + g^i_{44})$$

$$a_2 = g^i_{11}(g^i_{22} + g^i_{33} + g^i_{44}) + g^i_{22}(g^i_{33} + g^i_{44}) + g^i_{33}g^i_{44} - g^i_{12}g^i_{21} - g^i_{13}g^i_{31} - g^i_{14}g^i_{41}$$

$$\begin{aligned} a_3 = {} & g^i_{12}g^i_{21}(g^i_{33} + g^i_{44}) + g^i_{13}g^i_{31}(g^i_{22} + g^i_{44}) + g^i_{14}g^i_{41}(g^i_{22} + g^i_{33}) \\ & - g^i_{11}g^i_{22}(g^i_{33} + g^i_{44}) - (g^i_{11} + g^i_{22})g^i_{33}g^i_{44} \end{aligned}$$

$$a_4 = g^i_{11}g^i_{22}g^i_{33}g^i_{44} - g^i_{12}g^i_{21}g^i_{33}g^i_{44} - g^i_{13}g^i_{22}g^i_{31}g^i_{44} - g^i_{14}g^i_{22}g^i_{33}g^i_{41}$$

Using the Rourth-Hurwitz Criterion equilibrium points are stable under the following conditions:

$$a_1, a_2, a_3 > 0,\ a_1 a_2 > a_3,\ and\ a_1 a_2 a_3 > a_3^2 + a_1^2 a_4^2 \tag{8}$$

Otherwise, the equilibrium points are unstable if the above conditions are not satisfied.

5.2 Comparative Study of Time Series Analysis

Time series analysis is a very helpful technique to understand the dynamics of the model over time. Here in this study different external current values, $I =$ 50,120,154,156 μA/cm^2, with the time evolution of the membrane potential (V) are depicted in Fig. 5. At low external current values periodic behavior (grey shaded region) for all models (3)–(5) is observed but as external current increases, the membrane potential in the model (3) reached a steady-state sooner than the models (4) and (5).

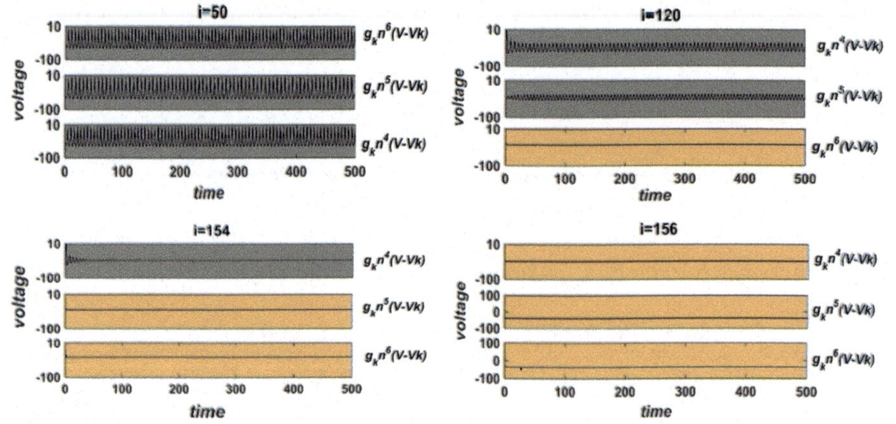

Fig. 5 Time series for model cases (3)-(5) with $[g_K n^4(V-V_K)]$, $[g_K n^5(V-V_K)]$ and $[g_K n^6(V-V_K)]$ at external currents $I = 50$, 120, 154 and 156 μA/cm^2 is drawn (black shaded represents transient oscillatory while orange its resting state)

To understand the impact of external currents (I) on other ion channel gated variables (sodium activation (n) solid lines, potassium activation (m) dotted and sodium inactivation (h) dash-dotted lines) for all three cases (3)–(5) time series is drawn in Fig. 6 and qualitative behavior is described in Table 2.

It has been observed from the Fig. 6a that for the lower value of current, model (3) shows period behavior for all three gated variables (sodium activation (n) solid lines,

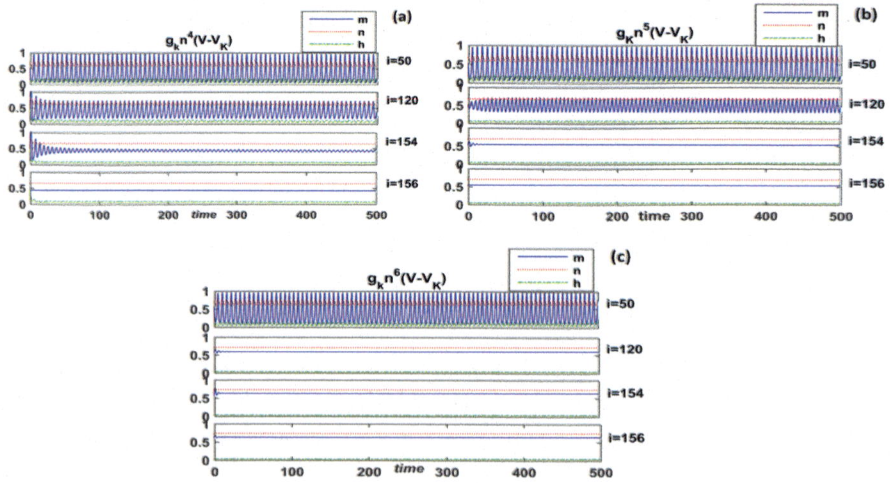

Fig. 6 Comparative study of time series (for m, n and h) for three model cases $[g_K n^4(V-V_K)]$, $[g_K n^5(V-V_K)]$ and $[g_K n^6(V-V_K)]$ at different values of external current at $I = 50$, 120, 154 and 156 μA/cm^2

Table 2 Dynamics (*P/S) of neuron's behavior in three different cases

External current (I)	Membrane potential (V)			Sodium activation (m)			Potassium activation (n)			Sodium inactivation (h)		
	Case (3)	Case (4)	Case (5)	Case (3)	Case (4)	Case (5)	Case (3)	Case (4)	Case (5)	Case (3)	Case (4)	Case (5)
	n^4	n^5	n^6	n^4	n^5	n^6	n^4	n^5	n^6	n^4	n^5	n^6
50 $\mu A/cm^2$	P	P	P	P	P	P	P	P	P	P	P	P
120 $\mu A/cm^2$	P	P	S	P	P	S	P	P	S	P	P	S
154 $\mu A/cm^2$	P	S	S	P	S	S	P	S	S	P	S	S
156 $\mu A/cm^2$	S	S	S	S	S	S	S	S	S	S	S	S

*P ~ Periodic states, S ~ Stable states

potassium activation (m) dotted, and sodium inactivation (h) dash-dotted lines). But as the current values increased to 154 μA/cm^2 its behavior changes from periodic to a stable state. In the case of model (4) and model (5), periodic behavior changes to a stable state for the current values, $I = 154$ and 120 μA/cm^2 respectively as shown in Fig. 6a, b.

5.3 *Comparative Study of One-Parameter Bifurcation*

Based on the time series findings (variations in neuron response i.e. resting or periodic), it is essential to observe the dynamical behavior over the parameter range $I = (0\text{–}200)$ μA/cm^2 (Fig. 7). For the model case (3), the stability curve, Hopf-bifurcations (sub-critical or super-critical), limit cycles (stable or unstable) are explained in the diagram Fig. 7 for a better understanding.

The voltage fluctuations against the external current are drawn in Fig. 7 (stable equilibrium curve (red), unstable equilibrium (black), unstable limit cycles (blue vacant circles), stable limit cycles (green filled circled). The unstable limit cycles (blue circles) emanate from the subcritical-Hopf bifurcation point ($I = 9.779$ μA/cm^2) and undergo *flip bifurcation* to produce stable limit cycles (green circles). At a higher external current value ($I = 154.5$ μA/cm^2) a supercritical-Hopf bifurcation occurs emanating stable limit cycles. Voltage is at resting state for low external currents (0 –9.7) μA/cm2^{nn}while it is periodic for the external current (9.8~154.5) μA/cm^2. The voltage reached its resting state on very high (or very low) external current (I) in the *HH model* while on the low external current it is an oscillatory state.

Similar qualitative behavior over external currents (0–200) μA/cm^2 is obtained for all three cases (3)–(5) in Fig. 8 and Table 3. Similar bifurcations analysis is carried

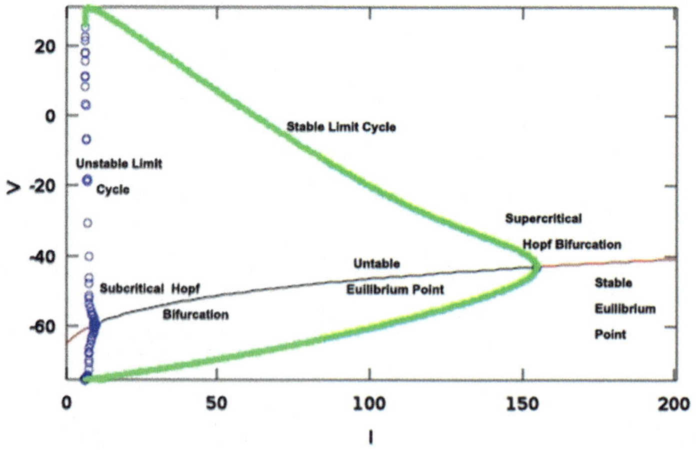

Fig. 7 One parameter bifurcation diagram for membrane potential (V) against external current (I)

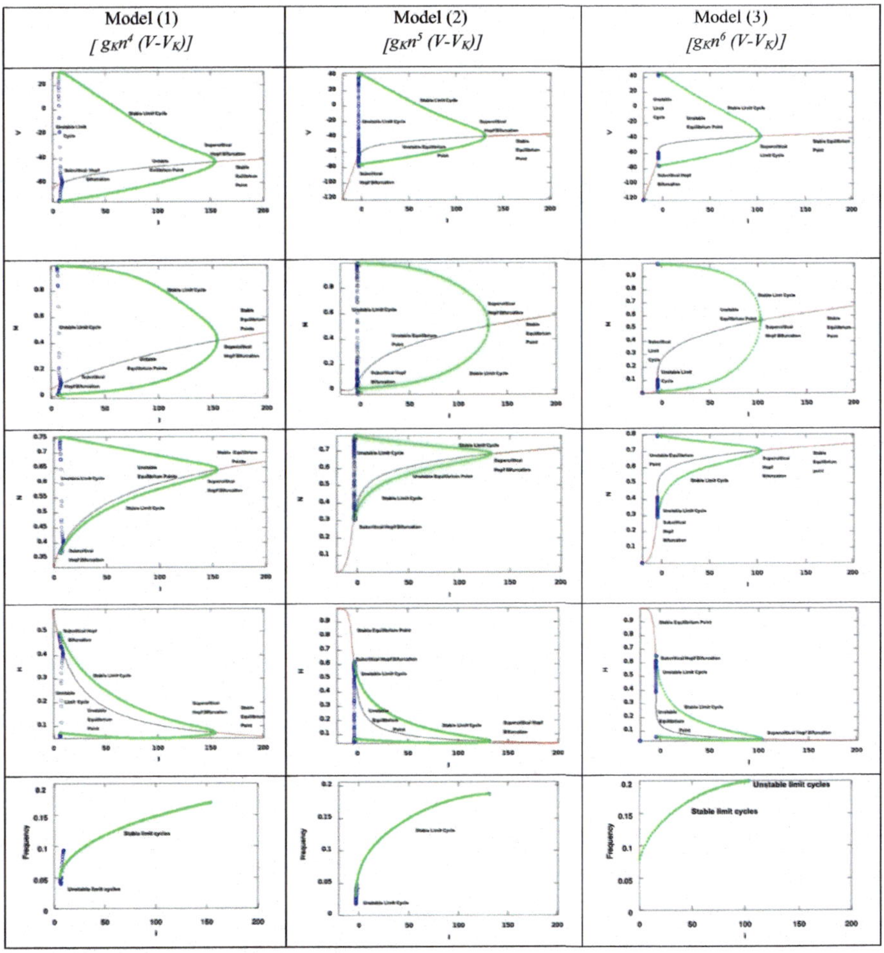

Fig. 8 One-parameter bifurcation diagrams for all variables of the model cases (3) to the case (5) have been drawn. Variation in membrane potential (V), Na activation (m), potassium activation (n) sodium inactivation (h) has been shown in row 1, 2, and 3 respectively

out for all three cases (3)–(5), the one-parameter bifurcation diagrams (Fig. 8), and two-parameter bifurcation diagrams (Fig. 10) are drawn with continuous varying external currents (I = 0–200 μA/cm^2). The qualitative behaviors of all three model cases (3)–(5) are summarized in Table 3 below.

The external current enhances membrane potential (V), *Na* activation *(m)*, and *K* activation *(n)* while it suppresses *K* inactivation *(h)*. The periodic solution for the model case (3) occurred for a bigger range of external currents (9.779~154.5) μA/cm^2 compared to the case (4) (0~131.7) μA/cm^2 and the case (5) (0~103.4) μA/cm^2 (Table 3). This shows that how neuron behaves during communication and action potential produced by the neurons. A neuron can fire spikes periodically as in

Table 3 Dynamic of variables for current values I (in $\mu A/cm^2$)

Variables	Behavior	Case (3) n^4	Case (4) n^5	Case (5) n^6
Membrane potential (V)	Stable	I < 9.779&I > 154.5	I > 131.7	I > 103.4
	Periodic	9.779 < I <154.5	0 < I < 131.7	0 <I < 103.4
Sodium activation (m)	Stable	I < 9.779&I >154.5	I > 131.7	I >103.4
	Periodic	9.779 < I < 54.5	0 < I < 131.7	0 < I <103.4
Potassium activation (m)	Stable	I < 9.779&I > 154.5	I > 131.7	I > 103.4
	Periodic	9.779 <I < 154.5	0 <I < 131.7	0 < I < 103.4
Sodium inactivation (m)	Stable	I < 9.779&I >154.5	I > 131.7	I > 103.4
	Periodic	9.779 < I < 154.5	0 < I < 131.7	0 < I < 103.4

case of high amplitude limit cycles. However, the weak perturbation may reduce the spike of neurons to stable states. With a higher external current, the amplitude of the spike decreases to zero which is said to be a refractory state.

It can be observed that the bifurcation points shift towards left for models (4) and (5) as compared to model (3). Membrane potential increases for models (4) and (5) as the current value increases. If the membrane potential curves have been considered for different current values, then it has been found that the shape of membrane potential curve changes slightly to S-shaped for models (4) and (5). Similar S-shaped patterns have been observed in the case of *Na* activation (m), potassium activation (n) for models (4) and (5). But for (h) sodium inactivation it has been found that the gated variable (h) decreases sharply for models (4) and (5).

5.4 Comparative Study of Two-Parameter Bifurcation

Two-parameter bifurcation analysis helps us to understand the influence of parameters and the relationships among the parameters. Any perturbation against the parameter can be examined and show graphically. Figure 10 shows the comparative graphs to understand the regions of periodic and stable states. The neuron responses as resting (grey shaded region) or periodic state (black shaded region) for all three cases (3)–(5) has been exhibited in two-parameter bifurcation diagrams in which the stable and periodic regions in two-parameter phase spaces (I, g_K), (I, g_L), (I, g_{Na}), (I, V_K),(I, V_L) and (I, V_{Na}) are drawn (Fig. 9). The periodic region in parameter space (I, g_K) is larger for the model case (5) (assuming n^6)) than model case (4) (assuming n^5) and model case (3) (assuming n^4). Similar qualitative behavior (larger periodic region) is found in other three-parameter phase spaces (I, g_L), (I, V_K), and (I, V_L) whereas the smaller periodic region is observed in parameter phase spaces (I, g_{Na}) and (I, V_{Na}).

This understanding will help reader to explore more about the dynamics of the *HH model* mathematically. Since the work of Hodgkin and Huxley were based on

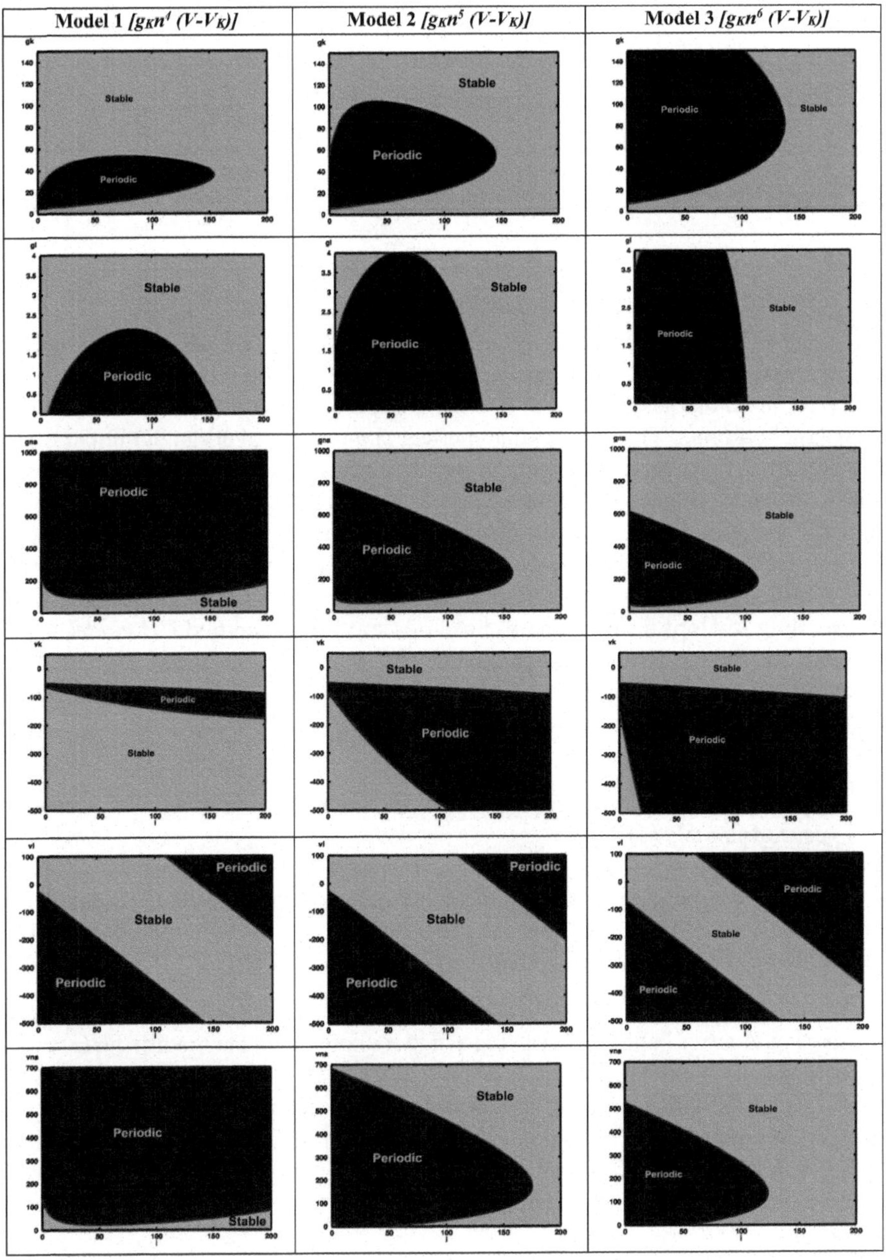

Fig. 9 Two-parameter bifurcation diagrams of ionic conductance for potassium against external current (row 1), ionic conductance for leakage ion against external current (row 2), ionic conductance for sodium against external current (row 3), the potential for potassium against external current (row 4), the potential for leakage ion against external current (row 5) and potential for sodium against external current (row 6) has been drawn

experiments and assumed data for calculations. But mathematically it can be explored more for better understanding and to look into the loopholes of Hodgkin and Huxley studies.

6 Results

In this manuscript, the external current has been applied to study the impacts of depolarizing stimulus to neurons when the exponential power of the potassium activation (n) is assumed to be *5* or *6* (as hinted by [8]). Keeping all other parameters and expression of the variables unchanged, the mathematical models are derived and analyzed based on the exponential power (4, 5 and 6) of the potassium activation (n) in external currents. The extensive numerical simulation suggests that in the modified *HH model*, the threshold voltage which is responsible for the firing of action potential occurred at a low applied external current. The result is quite significant as the low external current may cause the firing of the action potential in neurons. The larger periodic domain in the case of V_K versus I and $V_L\ vs.\ I$ has been obtained in a modified *HH model* which is an advantage in comparison with the *HH model.* The shifting of Hopf bifurcation has been observed in a modified *HH model* which stimulus the neuron firing. One-parameter and two-parameter bifurcation analysis of the present study suggests that in the modified *HH model* with high power of potassium activation that.

i. the resting states reached sooner in the case of models (4) and (5)
ii. shifting of Hopf bifurcation to lower values of externally applied current in case of models (4) and (5)
iii. the larger periodic region is possible in two-parameter phase space with external currents vs. parameters (*K* conductance, leakage conductance, *K* potential, and leakage potential) however smaller periodic region is also observed in external current vs. *Na* conductance and *Na* potential.

The observations are robust as various combinations of parameters also exhibit the same qualitative behavior. By using a higher exponent in the modified *HH model*, the periodic firing can be obtained at a low externally applied current as compared to the original *HH model.* The co-dimension two-parameter bifurcation suggests that a larger domain of periodic region (which refers to periodic firing) in the modified *HH model.* The proposed model gives the flexibility to choose the assumptions of higher power of potassium activation to generate neuron firing at low external applied current.

7 Discussions

In Hodgkin-Huxley's *voltage clamp experiment*, based on the number of independent channel gates (active or inactive) and the experimental data (approximated by sigmoidal function) the exponential power (*4*) for the potassium activation (*n*) in external current has been suggested. Later on, several modified *HH model* has been proposed which produced mathematical results of modified *HH model* based on different terms inclusion or modifications. In the *HH model,* the conductance of leakage ions remains constant which is responsible for resting potential. But the other two channels (*K and Na)* depend on potential and are responsible for action potential in neurons. An action potential occurs when sodium channel gates open during depolarization. Depolarization increases the probability of activation gates and decreases the probability of the inactivation gate. There several types of research that contradict the *HH model.* Since it describes the propagation of an impulse in the form of an electrical signal, even the whole setup was considered an electrical circuit with membrane potential as a capacitor. It suggests that the model is dissipative. But the *HH mod*el fails to brief about the mechanical changes and also about the energy. There is no theory about the experiments and justifications. But the findings of Hodgkin and Huxley was a revolution for neuroscience as well as mathematical modeling in the year 1952. Even in this century, the *HH model* is very important for basic understanding as well as deep study. This manuscript is based on mathematical modeling which includes a set of ordinary differential equations motivated by the work of Hodgkin and Huxley. This is an idea towards the findings of loopholes and limitations of the *HH model* for a better understanding of nerve conduction.

Further, the robustness of the results of mathematically modified *HH models* will certainly attract the researcher for further investigation or more experiments on the neuron's behavior.

Acknowledgements We acknowledge DST-INSPIRE [DST/INSPIRE/03/2016/000597], UGC-BSR, and DST-SERB [SERB/2017/000279] for their financial support.

References

1. Liqun Luo, Principal of Neurobiology, Taylor and Francis Group, ISBN 978–0–8153–4492–6 (1966).
2. Neuroscience: The Science of the Brain, British Neuroscience Association, ISBN: 0–9545204--0–8 (2003).
3. Square, L.R., Bloom, F.E., Spitzer, N.C., Lac, S., Ghosh, A., Berg, D.: Fundamental of Neuroscience, 3rd edn. Academic Press, Elsevier (2008)
4. Rabinovich, M.I., Varona, P., Selverston, A.I., Abarbanel, H.D.I.: Dynamical principles in neuroscience. Rev. Mod. Phys. **78**, 1213–1265 (2006)
5. N. I. Krouchev, F. Rattay, M. Sawan, and A. Vinet, From Squid to Mammals with the HH Model through the Na_v Channels' Half-Activation-Voltage Parameter, PLoS ONE, 2015, 10(12).
6. Bean, B.P.: The action potential in mammalian central neurons. Nat. Rev. Neurosci. **8**, 451–465 (2007)

7. Guckenheimer, J., Oliva, A.: Chaos in the Hodgkin-Huxley model. SIAM J. Appl. Dyn. Syst. **1**, 105–114 (2002)
8. Schwiening, C.J.: A brief historical perspective: Hodgkin and Huxley. J Physiol **590**, 2571–2575 (2012)
9. L. J. Colwell and M. P. Brenner, Action Potential Initiation in the Hodgkin-Huxley Model. PLoS Comput Biol 5(1) (2009).
10. J. Cronin, Mathematical Aspects of Hodgkin–Huxley Neural Theory, Cambridge University Press (1987).
11. Strassberg, A.F., DeFelice, L.J.: Limitation of the Hodgkin-Huxley Formalism: Effects of Single Channel Kinetics on Transmembrane Voltage Dynamics. Neural Comput. **5**, 843–855 (1993)
12. Feudel, U., Neiman, A., Pei, X., Wojtenek, W., Braun, H., Huber, M., Moss, F.: Homoclinic bifurcation in a Hodgkin-Huxley model of thermally sensitive neurons. Chaos **10**, 231 (2000)
13. Traub, R.D., Wong, R.K.S., Miles, R., Michelson, H.: Amodelofa CA3 hippo-campal pyramidal neuron incorporating voltage-clamp data on intrinsic conductances. J. Neurophysiol. **66**, 635–650 (1991)
14. F. Bezanilla and C. M. Armstrong, Inactivation of the sodium channel. I. Sodium current experiments. J. Gen. Physiol. 70(1977), 549–566.
15. Levy, A.: What was Hodgkin and Huxley's Achievement? Br. J. Philos. Sci. **65**, 469–492 (2014)
16. Hodgkin, A.L., Huxley, A.F.: A quantitative description of the membrane and its application to conduction and excitation in nerve. J. Physiol. **117**, 500–544 (1952)
17. Gangal, H., Dar, G.: Mode locking, chaos, and bifurcations in Hodgkin-Huxley neuron forced by sinusoidal current. Chaotic Modeling and Simulation (CMSIM) **3**, 287–294 (2014)
18. Bedrov, Y.A., Akoev, G.N., Dick, O.E.: On the relationship between the number of negative slope regions in the voltage-current curve of the Hodgkin-Huxley model and its parameter values. Biol. Cybern. **73**, 149–154 (1995)
19. Caterall, W.A., Raman, I.M., Robinson, H.P.C., Sejnowski, T.J., Paulsen, O.: The Hodgkin-Huxley Heritage: From channels to circuits. J. Neuroscience **32**(41), 14064–14073 (2012)
20. Fan, Y.S., Chay, T.R.: Generation of periodic and chaotic bursting in an excitable cell model. Biol. Cybern **71**, 417–431 (1994)
21. Shirahata, T.: Numerical Simulation of Bistability between Regular Bursting and Chaotic Spiking in a Mathematical Model of Snail Neurons. International Journal of Theoretical and Mathematical Physics **5**(5), 145–150 (2015)
22. K. Terada, H. Tanaka and S. Yoshizawa, Bifurcations and chaos in the Hodgkin-Huxley equations for muscle. Proc. Nolta '95, Las Vegas, Nevada, U.S.A, pp. (1995) 327–330.
23. H. Fukai, S. Doi, T. Nomura, and S. Sato, Hopf bifurcations in multiple parameter space of the Hodgkin–Huxley equations. I. Global organization of bistable periodic solutions, Biol. Cybern. 82(2000) 215–222.
24. J. Guckenheimer and P. Holmes, Nonlinear Oscillations, Dynamical Systems, and Bifurcation of Vector Fields, Springer (1983).
25. Hassard, B.: Bifurcation of periodic solutions of the Hodgkin-Huxley model for the squid giant axon. J. Theory Biol. **71**, 401–420 (1978)
26. Che, Y., Wang, J., Deng, B., Wei, X., Han, C.: Bifurcation in the Hodgkin- Huxley model exposed to DC electric fields. Neurocomputing **81**, 41–48 (2012)
27. Guckenheimer, J., Labouriau, I.S.: Bifurcation of the Hodgkin and Huxley equations: a new twist. Bull Math Biol **55**, 937–952 (1993)
28. Alexander, J.C., Cai, D.: On the dynamics of bursting systems. J. Math Biol **29**, 405–423 (1991)
29. Fox, R.F.: Stochastic Version of the Hodgkin-Huxley Equations. Biophys. J . **72**, 2068–2074 (1997)
30. P. Pathamanathan and R.A. Gray, Validation and Trustworthiness of Multiscale Models of Cardiac Electrophysiology, Front. Physiol. 2018.
31. F. R. Chavarette, J. M. Balthazar, I. R. Guilherme, and L. Barbanti, Control Design Applied Of The Action Potential Of Membranes Based On Optional Linear Control And Particle Swarm Optimization, Proceedings of the 9th Brazilian Conference On Dynamics Control and their Applications, Serra Negra, 2010, SP-ISSN 2178–3667.

32. Boland, L.M., Jiang, M., Lee, S.Y., Fahrenkrug, S.C., Harnett, M.T., O'Grady, S.M.: Functional properties of a brain-specific NH2-terminally spliced modulator of KV4 channels. Am. J. Physiol. Cell Physiol. **285**, C161–C170 (2003)
33. Lee, S.G., Neiman, A., Kim, S.: Coherence resonance in a Hodgkin-Huxley neuron. Phys. Rev. E **57**, 3 (1998)
34. Kashef, B., Bellman, R.: Solution of the Partial Differential Equation of the Hodgkin-Huxley Model Using Differential Quadrature. Math. Biosci. **19**, 1–8 (1974)
35. Tagluk, M.E., Tekin, R.: The influence of ion concentration on the dynamics behavior of the Hodgkin-Huxley Model-based cortial network. CognNeurodyn **8**, 287–298 (2014)
36. Tasaki, I.: Demonstration of two stable states of the nerve membrane in potassium-rich media. J. Physiol. (London) **148**, 306–331 (1959)
37. Binstock, L., Goldman, L.: Rectification in instantaneous potassium current-voltage relations in Myxicola giant axons. J. Physiol. **217**, 517–531 (1971)
38. Johnston, J., Griffin, S.J., Baker, C., Forsythe, I.D.: Kv4 (A-type) potassium currents in the mouse medial nucleus of the trapezoid body. Eur. J. Neurosci. **27**(6), 1391–1399 (2008)

Mathematical Analysis of Two Unequal Collinear Cracks in a Piezo-Electro-Magnetic Media

Kamlesh Jangid

Abstract In this chapter, we begin our work of studying two unequal collinear semi-permeable cracks in a magneto-electro-elastic media. We employ the Stroh's formalism and complex variable technique to solve the physical problem. We derive the closed form analytic solutions for various fracture parameters, and study the effect of volume fraction and inter-crack distance on these parameters.

Keywords Complex variable · Intensity factor · Piezo-electro-magnetic ceramic · Riemann-Hilbert problem · Semi-permeable cracks

1 Introduction

Piezo-electro-magnetic/Magneto-electro-elastic (MEE) composite materials are widely used in magnetic field probes, acoustic, medical ultrasonic imaging, hydrophones, electronic packaging, electromagnetic sensors, actuators and transducers etc., due to their multi-field-coupled effects. MEE ceramics are brittle in nature and have low fracture toughness. The presence of defects such as cracks, voids leads to the premature failure of these materials under mechanical/electrical/magnetic loadings. Thus fracture study becomes essential for such materials to predict structural integrity and to advance the design of MEE devices.

This chapter reviews extensive work that has been done to better understand the mechanics of MEE materials in the presence of defects such as cracks. As compared to piezoelectric or anisotropic cases, relatively limited work has been done so far in MEE fracture analysis. A large number of publications for a single crack in a MEE materials are available in the literature [1–6]. Further, few work related to multiple cracks in MEE media is available in the literature, also it deserves noting

K. Jangid (✉)
Department of HEAS (Mathematics), Rajasthan Technical University, Kota 324010, India

J. Singh et al. (eds.), *Methods of Mathematical Modelling and Computation for Complex Systems*, Studies in Systems, Decision and Control 373,
https://doi.org/10.1007/978-3-030-77169-0_4

that problems of collinear cracks have been a typical and active topic in fracture mechanics. With the application of MEE ceramics, the collinear-crack problems in them have drawn the attention of many researchers [7–9]. The static and dynamic problems of two collinear interfacial cracks in MEE composites [10–13] have been solved by Zhou and colleagues by using the Schmidt method. Exact solutions for anti-plane collinear cracks in a MEE strip or layer have been derived by Wang et al. [14], Wang and Mai [15], and Singh et al. [16] under different conditions. Most, recently Jangid and Bharagva [17] has derived an analytical solution for two collinear semi-permeable cracks in MEE media using Stroh's formalism and complex variable technique.

The main objective of this chapter is to show the effect of volume fraction, inter-crack distance and prescribed loadings on the collinear semi-permeable cracks. For this, the problem of two unequal collinear semipermeable cracks weakening a MEE media is studied. Only in-plane electro-magnetic and mechanical loading conditions are considered. The problem is formulated employing Stroh's formalism and solved using a complex variable technique (see Sects. 4 and 5). Closed form analytical expressions are derived for various fracture parameters (see Sect. 6).

2 Basic Equations for Piezoelectromagnetic Media

The fundamental equations and the boundary conditions for linear piezo-electro-magnetic media are defined as below:

- *Constitutive Equations*

$$\sigma_{ij} = C_{ijks}\varepsilon_{ks} - e_{sij}E_s - h_{sij}H_s, \tag{1}$$

$$D_i = e_{kis}\varepsilon_{ks} + \kappa_{is}E_s + \beta_{is}H_s, \tag{2}$$

$$B_i = h_{iks}\varepsilon_{ks} + \beta_{is}E_s + \gamma_{is}H_s. \tag{3}$$

- *Kinematic Equations*

$$\varepsilon_{ij} = \frac{1}{2}(u_{i,j} + u_{j,i}), \qquad E_i = \phi_{,i}, \qquad H_i = \varphi_{,i}. \tag{4}$$

- *Equilibrium Equations*

  Equilibrium equations for stresses, electric displacements and magnetic inductions in the absence of body forces, free electric charges and free magnetic currents, may, respectively, be written as

$$\sigma_{ij,j} = 0, \quad D_{i,i} = 0 \text{ and } B_{i,i} = 0, \tag{5}$$

where σ_{ij}, ε_{ij}, D_i, E_i, B_i and H_i denote the components of the stress, strain, electric displacement, electric field, magnetic induction and magnetic field, respectively;

C_{ijks}, e_{iks}, h_{iks} and β_{is} denote the elastic, piezoelectric, piezo-magnetic and electromagnetic constants; κ_{is} and γ_{is} denote the dielectric permittivities and magnetic permeabilities, respectively. Comma denotes partial differentiation with respect to argument following it; ϕ is the electric potential; and φ is the magnetic potential; where i, j, k and $s = 1, 2, 3$.

2.1 *Crack Face Boundary Conditions*

In the literature, mainly three crack face boundary conditions for MEE ceramics are available. These are represented mathematically as:

- *Impermeable boundary conditions* (proposed by Deeg [18])
  The crack faces are assumed to be traction-free, electrically and magnetically impermeable

$$\sigma_{ij}n_j = 0; \quad D_2^+ = D_2^- = 0 \text{ and } B_2^+ = B_2^- = 0; \tag{6}$$

- *Permeable boundary conditions* (proposed by Parton [19])
  In this case, crack is traction-free and does not obstruct any electric field from conduction

$$\sigma_{ij}n_j = 0; \quad \phi^+ = \phi^-; \quad \varphi^+ = \varphi^-; \quad D_2^+ = D_2^- \neq 0 \text{ and } B_2^+ = B_2^- \neq 0; \tag{7}$$

- *Semi-permeable boundary conditions*
  This condition, gives a more realistic boundary condition for a open cracks, its modification are proposed by Hao and Shen [20] for piezoelectric solids. These assumption establishes that medium between the crack surfaces partially conducts the electric and magnetic fields and can be expressed as

$$\sigma_{ij}n_j = 0; \; D_2^+ = D_2^- = D_2^c = -\kappa_c \frac{\Delta\phi(x_1)}{\Delta u(x_1)} \text{ and } B_2^+ = B_2^- = B_2^c = -\gamma_c \frac{\Delta\varphi(x_1)}{\Delta u(x_1)}, \tag{8}$$

where superscripts + and – represent, respective, values on the upper and lower crack surfaces, considering crack along x_1-axis; $\kappa_c = \kappa_r \kappa_o (\kappa_o = 8.85 \times 10^{-12} F/m)$, κ_r is electric permittivity and $\gamma_c = \gamma_r \gamma_o (\gamma_o = 1.26 \times 10^{-6} Ns^2/C^2)$, γ_r is magnetic permeability of the medium between the crack faces, respectively; $\Delta\phi$, $\Delta\varphi$ and Δu are the jumps of electric potential, magnetic potential and crack opening displacement across the crack, respectively.

3 Fundamental Formulation and Solution Methodology

According to Stroh's formulation [21] the general solution satisfying Eqs. (1)–(5) may be written as (solution methodology is recapitulated from Jangid and Bhargava [17])

$$\mathbf{u}_{,1} = \mathbf{AF}(z) + \overline{\mathbf{AF}(z)}, \tag{9}$$

$$\Phi_{,1} = \mathbf{BF}(z) + \overline{\mathbf{BF}(z)}, \tag{10}$$

where, $\mathbf{A} = (\mathbf{a}_1, \mathbf{a}_2, \mathbf{a}_3, \mathbf{a}_4, \mathbf{a}_5)$, $\mathbf{B} = (\mathbf{b}_1, \mathbf{b}_2, \mathbf{b}_3, \mathbf{b}_4, \mathbf{b}_5)$, $\mathbf{u} = [u_1, u_2, u_3, \phi, \varphi]^T$, $\mathbf{F}(z) = \dfrac{d\mathbf{f}(z)}{dz}$, $\mathbf{f}(z_\alpha) = [f_1(z_1), f_2(z_2), f_3(z_3), f_4(z_4), f_5(z_5)]^T$ and $z_\alpha = x_1 + p_\alpha x_2$, where p_α is a non-real root of

$$|\mathbf{W} + p(\mathbf{R} + \mathbf{R}^T) + p^2\mathbf{Q}| = 0. \tag{11}$$

The matrices **W**, **R** and **Q** are given by

$$\mathbf{W} = \begin{bmatrix} C_{1jk1} & e_{1j1} & h_{1j1} \\ e^T_{1k1} & -\kappa_{11} & -\beta_{11} \\ h^T_{1k1} & -\beta_{11} & -\gamma_{11} \end{bmatrix}, \mathbf{R} = \begin{bmatrix} C_{1jk1} & e_{2j1} & h_{2j1} \\ e^T_{1k2} & -\kappa_{12} & -\beta_{12} \\ h^T_{1k2} & -\beta_{12} & -\gamma_{12} \end{bmatrix},$$
$$\mathbf{Q} = \begin{bmatrix} C_{2jk2} & e_{2j2} & h_{2j2} \\ e^T_{2k2} & -\kappa_{22} & -\beta_{22} \\ h^T_{2k2} & -\beta_{22} & -\gamma_{22} \end{bmatrix}, j, k = 1, 2, 3. \tag{12}$$

The column vectors of matrix $\mathbf{B} = (\mathbf{b}_1, \mathbf{b}_2, \mathbf{b}_3, \mathbf{b}_4, \mathbf{b}_5)$ are related to the column vectors of matrix $\mathbf{A} = (\mathbf{a}_1, \mathbf{a}_2, \mathbf{a}_3, \mathbf{a}_4, \mathbf{a}_5)$ in the following form

$$\mathbf{b}_k = (\mathbf{R}^T + p_k\mathbf{Q})\mathbf{a}_k, \quad k = 1, 2, 3, 4, 5$$

and Φ is the generalized stress function such that

$$\text{œ}_2 = [\sigma_{2j}, D_2, B_2]^T = \Phi_{,1}, \quad \text{œ}_1 = [\sigma_{1j}, D_1, B_1]^T = -\Phi_{,2}. \tag{13}$$

4 Statement of the Problem

An infinite transversely isotropic piezo-electro-magnetic 2D domain is considered for the analysis in the ox_1x_2-plane. Two unequal collinear cracks L_1 and L_2 are taken along the x-axis occupying the intervals $[d, c]$ and $[b, a]$, respectively. The traction free crack face and semipermeable boundary condition are taken for the analysis. The remote boundary of the plate is prescribed in-plane mechanical load σ^∞_{22}, electric displacement D^∞_2, and magnetic induction B^∞_2. The entire configuration

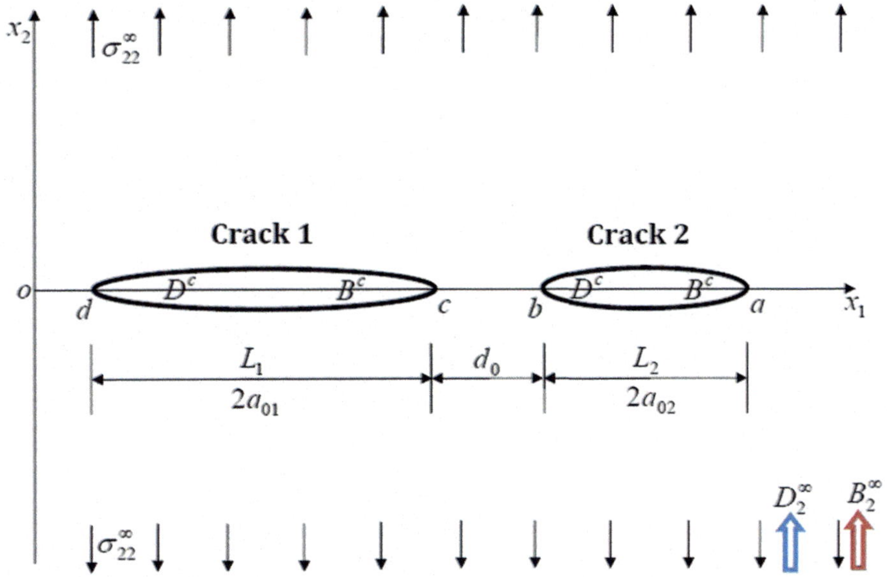

Fig. 1 Schematic representation of the problem

is schematically presented in Fig. 1. The physical boundary conditions stated above may be written as

(i) $\sigma_{2j}^{+} = \sigma_{2j}^{-} = 0,\ \ D_2 = D^c,\ \ B_2 = B^c$ on $L = \bigcup_{1}^{2} L_i$

(ii) $\sigma_{22} = \sigma_{22}^{\infty},\ \ D_2 = D_2^{\infty},\ \ B_2 = B_2^{\infty}$ for $|x_2| \to \infty$

(iii) $u_j^{+} = u_j^{-},\ \sigma_{2j}^{+} = \sigma_{2j}^{-},\ D_2^{+} = D_2^{-},\ B_2^{+} = B_2^{-},\ \phi^{+} = \phi^{-},\ \varphi^{+} = \varphi^{-}$ for $|x_1| < d,\ c < |x_1| < b,\ |x_1| > a$

(iv) $\mathbf{\Phi}_{,1}^{+} = \mathbf{\Phi}_{,1}^{-} = -\mathbf{V},\ \mathbf{V} = [0\ \sigma_{22}^{\infty}\ 0\ D_2^{\infty}\ B_2^{\infty}]^T$ for $d < |x_1| < c, b < |x_1| < a$.

where D^c and B^c are the electric and magnetic fluxes through the crack regions (d, c) and (b, a), which can be determined with the help of the Eq. (8).

5 Solution of the Problem

The continuity of $\mathbf{\Phi}_{,1}(x_1)$ on the whole real axis implies that

$$[\mathbf{BF}(x_1) - \overline{\mathbf{BF}}(x_1)]^{+} - [\mathbf{BF}(x_1) - \overline{\mathbf{BF}}(x_1)]^{-} = \mathbf{0}. \tag{14}$$

According to Muskhelishvil [22] its solution may be written as

$$\mathbf{BF}(z) = \overline{\mathbf{BF}}(z) = h(z)(say) \tag{15}$$

Boundary condition (iv) together with Eqs. (10, 15) yield following vector Hilbert problem

$$\mathbf{h}^{+}(x_1) + \mathbf{h}^{-}(x_1) = \mathbf{V}^0 - \mathbf{V},\ \mathbf{V}^0 = [0, 0, 0, D^c, B^c]^T \quad \text{on } L \tag{16}$$

Introducing a new complex function vector $\mathbf{\Omega}(z) = [\Omega_1(z), \Omega_2(z), \Omega_3(z), \Omega_4(z), \Omega_5(z)]^T$ as

$$\mathbf{\Omega}(z) = \mathbf{H}^R\mathbf{BF}(z).$$

Which using Eq. (15) gives the relation

$$\mathbf{h}(z) = \mathbf{\Lambda\Omega}(z), \tag{17}$$

where $\Lambda = [\mathbf{H}^R]^{-1},\ \mathbf{H}^R = 2Re\mathbf{Y},\ \mathbf{Y} = i\mathbf{AB}^{-1}$.

Consequently Eq. (16) may be written in component form for $\Omega_2(z)$, $\Omega_4(z)$ and $\Omega_5(z)$, yield following scalar Hilbert problems

$$\Lambda_{22}[\Omega_2^+(x_1) + \Omega_2^-(x_1)] + \Lambda_{24}[\Omega_4^+(x_1) + \Omega_4^-(x_1)] + \Lambda_{25}[\Omega_5^+(x_1) + \Omega_5^-(x_1)] = -\sigma_{22}^{\infty}, \tag{18}$$

$$\Lambda_{42}[\Omega_2^+(x_1) + \Omega_2^-(x_1)] + \Lambda_{44}[\Omega_4^+(x_1) + \Omega_4^-(x_1)] + \Lambda_{45}[\Omega_5^+(x_1) + \Omega_5^-(x_1)] = D^c - D_2^{\infty}, \tag{19}$$

$$\Lambda_{52}[\Omega_2^+(x_1) + \Omega_2^-(x_1)] + \Lambda_{54}[\Omega_4^+(x_1) + \Omega_4^-(x_1)] + \Lambda_{55}[\Omega_5^+(x_1) + \Omega_5^-(x_1)] = B^c - B_2^{\infty}. \tag{20}$$

The solution of above Hilbert problems written using According to Muskhelishvil [22] as

$$\Omega_2(z) = \frac{\Delta_1}{2\Delta}\left\{\frac{P_1(z)}{(a_{11}a_{22} - a_{12}a_{21})X_1(z)} - 1\right\}, \tag{21}$$

$$\Omega_4(z) = \frac{\Delta_2}{2\Delta}\left\{1 - \frac{P_1(z)}{(a_{11}a_{22} - a_{12}a_{21})X_1(z)}\right\}, \tag{22}$$

$$\Omega_5(z) = \frac{\Delta_3}{2\Delta}\left\{1 - \frac{P_1(z)}{(a_{11}a_{22} - a_{12}a_{21})X_1(z)}\right\}. \tag{23}$$

where $X_1(z)$, $P_1(z)$ etc. are given in "Appendix A".

6 Applications

In this section, closed form analytical expressions are derived for crack opening displacement (COD), crack opening potential drop (COPD), crack opening induction drop (COID), stress intensity factor (SIF), electric displacement intensity factor (EDIF) and magnetic induction intensity factor (MIIF).

6.1 Crack Opening Displacement (COD)

The jump displacement vector $\Delta\mathbf{u}_{,1}$ may be given as

$$i\Delta\mathbf{u}_{,1} = \Omega^{+}(x_1) - \Omega^{-}(x_1). \tag{24}$$

Taking the second component of the above equation, we get

$$\Delta u_{2,1}(x_1) = -i[\Omega_2^{+}(x_1) - \Omega_2^{-}(x_1)]. \tag{25}$$

Substituting value of $\Omega_2(z)$ from Eq. (21) and integrating one obtains

$$\Delta u_2(x_1) = \frac{\Delta_1}{(a_{11}a_{22} - a_{12}a_{21})\Delta}\{C_0 S_3 + C_1 S_4 + C_2 S_5\}, \text{ on } d < |x_1| < c \tag{26}$$

$$\Delta u_2(x_1) = \frac{-\Delta_1}{(a_{11}a_{22} - a_{12}a_{21})\Delta}\{C_0 S_6 + C_1 S_7 + C_2 S_8\}, \text{ on } b < |x_1| < a \tag{27}$$

where the symbol Δ indicates the difference between the values on the upper and lower crack surfaces and S_3, S_4 etc. are given in "Appendix B".

6.2 Crack Opening Potential Drop (COPD)

Comparing the fourth component from Eq. (24) and using the value of $\Omega_4(x_1)$ from Eq. (22) and integrating one obtains the COP drop, $\Delta\phi(x_1)$, between the two faces of the crack as

$$\Delta u_4(x_1) = \frac{-\Delta_2}{(a_{11}a_{22} - a_{12}a_{21})\Delta}\{C_0 S_3 + C_1 S_4 + C_2 S_5\}, \text{ on } d < |x_1| < c \tag{28}$$

$$\Delta u_4(x_1) = \frac{\Delta_2}{(a_{11}a_{22} - a_{12}a_{21})\Delta}\{C_0 S_6 + C_1 S_7 + C_2 S_8\}, \text{ on } b < |x_1| < a. \tag{29}$$

6.3 Crack Opening Induction Drop (COID)

Comparing the fifth component from Eq. (24) and using the value of $\Omega_5(x_1)$ from Eq. (23) and integrating one obtains the COI drop, $\Delta\varphi(x_1)$, between the two faces of the crack as

$$\Delta u_5(x_1) = \frac{-\Delta_3}{(a_{11}a_{22} - a_{12}a_{21})\Delta}\{C_0 S_3 + C_1 S_4 + C_2 S_5\}, \text{ on } d < |x_1| < c \quad (30)$$

$$\Delta u_5(x_1) = \frac{\Delta_3}{(a_{11}a_{22} - a_{12}a_{21})\Delta}\{C_0 S_6 + C_1 S_7 + C_2 S_8\}, \text{ on } b < |x_1| < a. \quad (31)$$

The values of electric and magnetic fluxes, D^c and B^c, respectively, are obtained by substituting the required values from Eqs. (26), (28), (30) into Eq. (8) simplifying and solving the system of non-linear equations

$$m_1 D^{c2} + D^c(m_4\sigma_{22}^{\infty} - m_1 D_2^{\infty} - m_5 B_2^{\infty} + m_2\kappa_c) + B^c D^c m_5 + B^c m_3\kappa_c = -\kappa_c(m_1\sigma_{22}^{\infty} - m_2 D_2^{\infty} - m_3 B_2^{\infty}), \quad (32)$$

$$m_5 B^{c2} + B^c(m_4\sigma_{22}^{\infty} - m_1 D_2^{\infty} - m_5 B_2^{\infty} + m_6\gamma_c) + B^c D^c m_1 + D^c m_3\gamma_c = -\gamma_c(m_5\sigma_{22}^{\infty} - m_3 D_2^{\infty} - m_6 B_2^{\infty}), \quad (33)$$

where,

$$m_1 = \Lambda_{42}\Lambda_{55} - \Lambda_{45}\Lambda_{52}, \quad m_2 = \Lambda_{22}\Lambda_{55} - \Lambda_{25}\Lambda_{52}, \quad m_3 = \Lambda_{25}\Lambda_{42} - \Lambda_{22}\Lambda_{45},$$
$$m_4 = \Lambda_{44}\Lambda_{55} - \Lambda_{45}\Lambda_{54}, \quad m_5 = \Lambda_{25}\Lambda_{44} - \Lambda_{24}\Lambda_{45}, \quad m_6 = \Lambda_{22}\Lambda_{44} - \Lambda_{24}\Lambda_{42}.$$

6.4 Stress Intensity Factor (SIF)

Open mode stress intensity factor K_I at the crack tips $x_1 = d, c, b$, and a is obtained using following formulae

$$K_I(d) = \lim_{x_1 \to d^-} \sqrt{2\pi(d - x_1)}\sigma_{22}(x_1), \tag{34}$$

$$K_I(c) = \lim_{x_1 \to c^+} \sqrt{2\pi(x_1 - c)}\sigma_{22}(x_1), \tag{35}$$

$$K_I(b) = \lim_{x_1 \to b^-} \sqrt{2\pi(b - x_1)}\sigma_{22}(x_1), \tag{36}$$

$$K_I(a) = \lim_{x_1 \to a^+} \sqrt{2\pi(x_1 - a)}\sigma_{22}(x_1). \tag{37}$$

Substituting $\sigma_{22}(x_1)$ obtained from Eqs. (10), (15), (17) and (20) into above equations and simplifying we obtain

$$K_I(d) = \frac{-\sqrt{2\pi}\,(\Lambda_{25}\Delta_3 + \Lambda_{24}\Delta_2 - \Lambda_{22}\Delta_1)}{\Delta(a_{11}a_{22} - a_{12}a_{21})} \left\{ \frac{C_0 d^2 + C_1 d + C_2}{\sqrt{(a-d)(b-d)(c-d)}} \right\}, \tag{38}$$

$$K_I(c) = \frac{\sqrt{2\pi}\,(\Lambda_{25}\Delta_3 + \Lambda_{24}\Delta_2 - \Lambda_{22}\Delta_1)}{\Delta(a_{11}a_{22} - a_{12}a_{21})} \left\{ \frac{C_0 c^2 + C_1 c + C_2}{\sqrt{(a-c)(b-c)(c-d)}} \right\}, \tag{39}$$

$$K_I(b) = \frac{\sqrt{2\pi}\,(\Lambda_{25}\Delta_3 + \Lambda_{24}\Delta_2 - \Lambda_{22}\Delta_1)}{\Delta(a_{11}a_{22} - a_{12}a_{21})} \left\{ \frac{C_0 b^2 + C_1 b + C_2}{\sqrt{(a-b)(b-c)(b-d)}} \right\}, \tag{40}$$

$$K_I(a) = \frac{-\sqrt{2\pi}\,(\Lambda_{25}\Delta_3 + \Lambda_{24}\Delta_2 - \Lambda_{22}\Delta_1)}{\Delta(a_{11}a_{22} - a_{12}a_{21})} \left\{ \frac{C_0 a^2 + C_1 a + C_2}{\sqrt{(a-b)(a-c)(a-d)}} \right\}. \tag{41}$$

6.5 Electric Displacement Intensity Factor (EDIF)

Similarly, Open mode EDIF, K_{IV}, at the crack tips $x_1 = d, c, b,$ and a may be obtain as

$$K_{IV}(d) = \frac{-\sqrt{2\pi}\,(\Lambda_{45}\Delta_3 + \Lambda_{44}\Delta_2 - \Lambda_{42}\Delta_1)}{\Delta(a_{11}a_{22} - a_{12}a_{21})} \left\{ \frac{C_0 d^2 + C_1 d + C_2}{\sqrt{(a-d)(b-d)(c-d)}} \right\}, \tag{42}$$

$$K_{IV}(c) = \frac{\sqrt{2\pi}\,(\Lambda_{45}\Delta_3 + \Lambda_{44}\Delta_2 - \Lambda_{42}\Delta_1)}{\Delta(a_{11}a_{22} - a_{12}a_{21})} \left\{ \frac{C_0 c^2 + C_1 c + C_2}{\sqrt{(a-c)(b-c)(c-d)}} \right\}, \tag{43}$$

$$K_{IV}(b) = \frac{\sqrt{2\pi}\,(\Lambda_{45}\Delta_3 + \Lambda_{44}\Delta_2 - \Lambda_{42}\Delta_1)}{\Delta(a_{11}a_{22} - a_{12}a_{21})} \left\{ \frac{C_0 b^2 + C_1 b + C_2}{\sqrt{(a-b)(b-c)(b-d)}} \right\}, \tag{44}$$

$$K_{IV}(a) = \frac{-\sqrt{2\pi}\,(\Lambda_{45}\Delta_3 + \Lambda_{44}\Delta_2 - \Lambda_{42}\Delta_1)}{\Delta(a_{11}a_{22} - a_{12}a_{21})} \left\{ \frac{C_0 a^2 + C_1 a + C_2}{\sqrt{(a-b)(a-c)(a-d)}} \right\}. \tag{45}$$

6.6 Magnetic Induction Intensity Factor (MIIF)

Analogously, MIIF, K_V, at the crack tips $x_1 = d,\ c,\ b$, and a may be obtain as

$$K_V(d) = \frac{-\sqrt{2\pi}\,(\Lambda_{55}\Delta_3 + \Lambda_{54}\Delta_2 - \Lambda_{52}\Delta_1)}{\Delta(a_{11}a_{22} - a_{12}a_{21})}\left\{\frac{C_0 d^2 + C_1 d + C_2}{\sqrt{(a-d)(b-d)(c-d)}}\right\}, \tag{46}$$

$$K_V(c) = \frac{\sqrt{2\pi}\,(\Lambda_{55}\Delta_3 + \Lambda_{54}\Delta_2 - \Lambda_{52}\Delta_1)}{\Delta(a_{11}a_{22} - a_{12}a_{21})}\left\{\frac{C_0 c^2 + C_1 c + C_2}{\sqrt{(a-c)(b-c)(c-d)}}\right\}, \tag{47}$$

$$K_V(b) = \frac{\sqrt{2\pi}\,(\Lambda_{55}\Delta_3 + \Lambda_{54}\Delta_2 - \Lambda_{52}\Delta_1)}{\Delta(a_{11}a_{22} - a_{12}a_{21})}\left\{\frac{C_0 b^2 + C_1 b + C_2}{\sqrt{(a-b)(b-c)(b-d)}}\right\}, \tag{48}$$

$$K_V(a) = \frac{-\sqrt{2\pi}\,(\Lambda_{55}\Delta_3 + \Lambda_{54}\Delta_2 - \Lambda_{52}\Delta_1)}{\Delta(a_{11}a_{22} - a_{12}a_{21})}\left\{\frac{C_0 a^2 + C_1 a + C_2}{\sqrt{(a-b)(a-c)(a-d)}}\right\}. \tag{49}$$

7 Case Study

In this section, the effect of inter-crack distance and volume fraction are shown on the intensity factors (discussed in Sect. 5).

Piezo-electro-magnetic composite $BaTiO_3$-$CoFe_2O_4$ is selected for numerical case study considering $BaTiO_3$ as inclusion and $CoFe_2O_4$ as matrix. The volume fraction of the inclusion is denoted by V_f. The proportion of the two phases can be varied by adjusting the volume fraction of inclusion and the matrix. The elastic constants, dielectric permittivities and magnetic permeabilities, as well as piezoelectric and piezo-magnetic constants, are obtained by fraction rule {taken from Wang and Mai [23]}

$$\kappa_{is}^{c} = V_f.\kappa_{is}^{i} + (1 - V_f).\kappa_{is}^{m} \tag{50}$$

where the superscripts c, i and m represent composite, inclusion and matrix, respectively. κ_{is} denotes the dielectric permittivities.

We assume the crack faces as semi-permeable ($\kappa_r = \gamma_r = 1$). And the length of bigger crack, L_1, smaller crack, L_2, prescribed mechanical load, electric displacement and magnetic induction are $2a_{01}(= 5$ mm), $2a_{02}(= 4$ mm), $\sigma_{22}^{\infty} = 5$ MPa, $D_2^{\infty} = 2(e_{33}/c_{33})\sigma_{22}^{\infty}$ and $B_2^{\infty} = 2(h_{33}/c_{33})\sigma_{22}^{\infty}$, respectively, till otherwise specified. Material constants for $BaTiO_3$-$CoFe_2O_4$ for different volume fraction are given in Table 1, taken from Zhong [24].

Table 1 Material constants for $BaTiO_3 - CoFe_2O_4$ for different volume fraction

Material constants	$V_f(0.25)$	$V_f(0.50)$	$V_f(0.75)$
$c_{11}(10^9\ \text{N/m}^2)$	245	215	186
$c_{12}(10^9\ \text{N/m}^2)$	145	125	115
$c_{13}(10^9\ \text{N/m}^2)$	144	112	90
$c_{33}(10^9\ \text{N/m}^2)$	235	210	181
$c_{44}(10^9\ \text{N/m}^2)$	46	50	51
$e_{31}(\text{C/m}^2)$	–1.5	–2.8	–3.8
$e_{33}(\text{C/m}^2)$	4.2	8.7	13.2
$e_{15}(\text{C/m}^2)$	0.0	0.2	0.3
$h_{31}(\text{N/Am})$	380	220	90
$h_{33}(\text{N/Am})$	475	290	135
$h_{15}(\text{N/Am})$	335	180	75
$\kappa_{11}(10^{-9}\ \text{C}^2/\text{Nm}^2)$	0.1	0.25	0.5
$\kappa_{33}(10^{-9}\ \text{C}^2/\text{Nm}^2)$	3.2	6.3	9.4
$\gamma_{11}(10^{-6}\ \text{Ns}^2/\text{C}^2)$	–3.55	–2.00	–0.90
$\gamma_{33}(10^{-6}\ \text{Ns}^2/\text{C}^2)$	1.2	0.8	0.45
$\beta_{11}(10^{-9}\ \text{Ns/VC})$	3.1	5.3	6.8
$\beta_{33}(10^{-9}\ \text{Ns/VC})$	2350	2750	1800

7.1 *Effect of Inter-Crack Distance*

Figure 2a, b show the variation of stress intensity factors (SIFs) versus normalized inter-crack distance for different volume fractions. It may be seen, that due to the mutual interactions of two cracks, the SIFs at the crack tips are increased as the inter-crack distance decreases. Also it may be seen, that SIF at the inner crack tips (at $x_1 = c$ and $x_1 = b$) is higher as compare to that at the outer crack tips (at $x_1 = d$ and $x_1 = a$), which implies that the cracks will open more at the inner tips as compared to that at outer tips. Moreover, K_I stabilizes for $d_0/a_{02} \geq 3$. Also, SIF is decreased as the volume fraction increases. Similarly, Figs. 3 and 4 show the variations of EDIF and MIIF versus inter-crack distance for different volume fractions.

7.2 *Effect of Crack Length*

Effect of crack length a_{02} on stress intensity factor (SIF), K_I, for different volume fractions is shown in Fig. 5. It may be seen from the figure that at the interior and exterior tips of the longer crack, K_I increases at both the tips as the crack length is increased. Increase in K_I at interior tip is more steep vis-a-vis than at exterior tip. The similar variation is observed at the interior and exterior tips of the shorter crack.

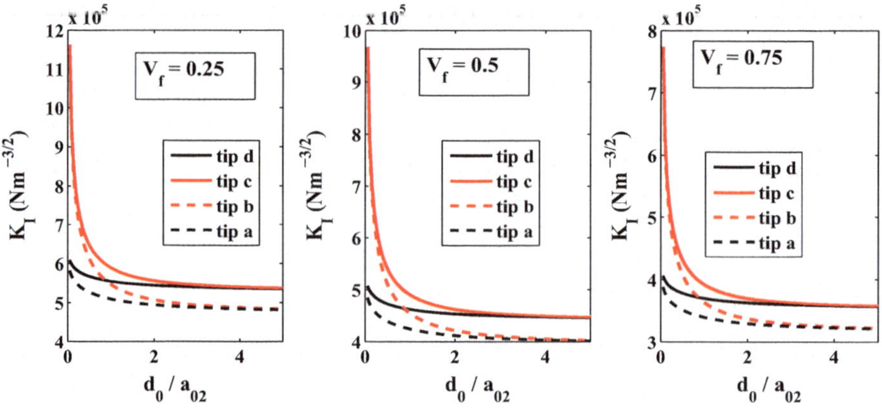

Fig. 2 Effect of normalized inter-crack distance d_0/a_{02} on SIF for different volume fractions

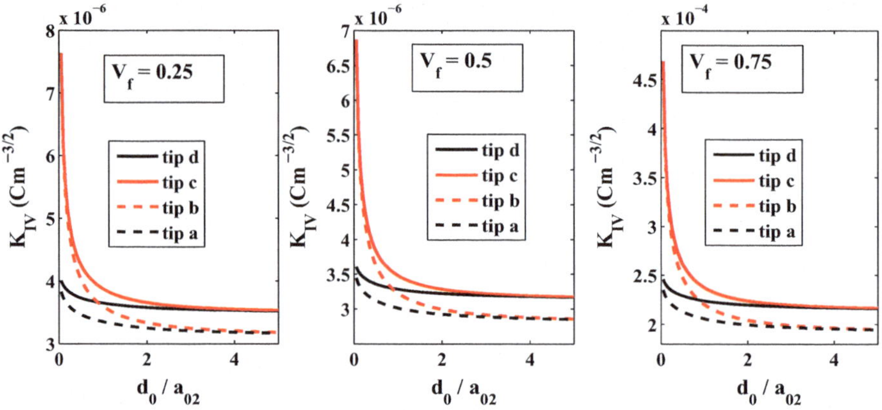

Fig. 3 Effect of normalized inter-crack distance d_0/a_{02} on EDIF for different volume fractions

It is to be noted that for half length of the crack equal to 2.5 mm (i.e., the length of the both cracks is equal), the curves for K_I at the interior tips of both cracks and exterior tips of the cracks become equal. Figures 6 and 7 show the same variations for EDIF and MIIF, respectively.

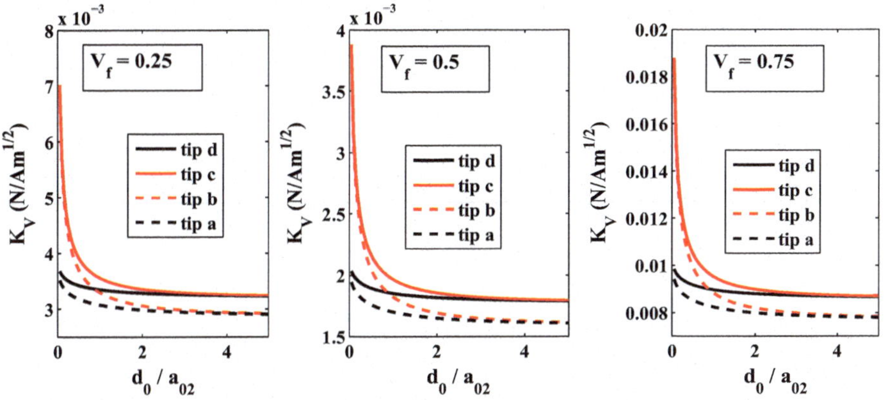

Fig. 4 Effect of normalized inter-crack distance d_0/a_{02} on MIIF for different volume fractions

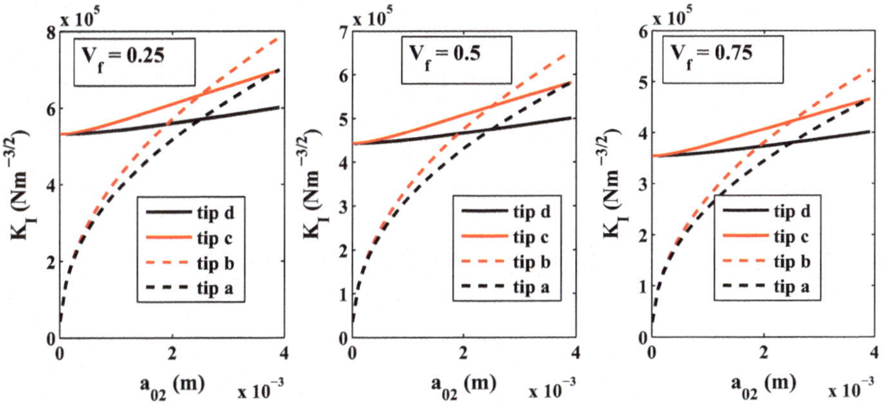

Fig. 5 Effect of crack length a_{02} on SIF for different volume fractions

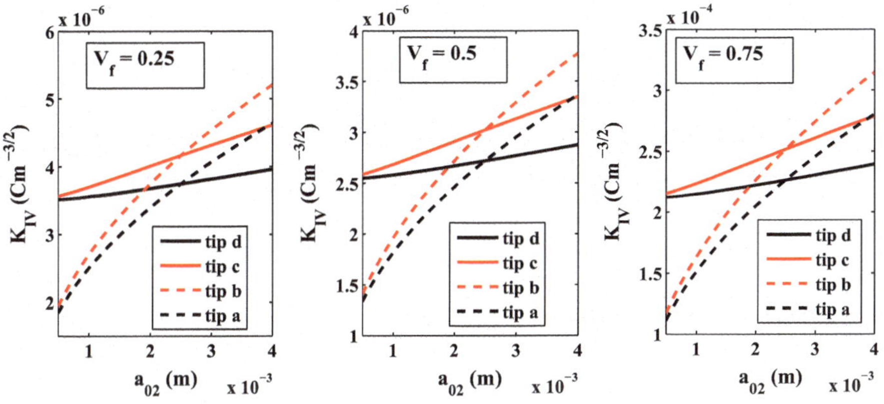

Fig. 6 Effect of crack length a_{02} on EDIF for different volume fractions

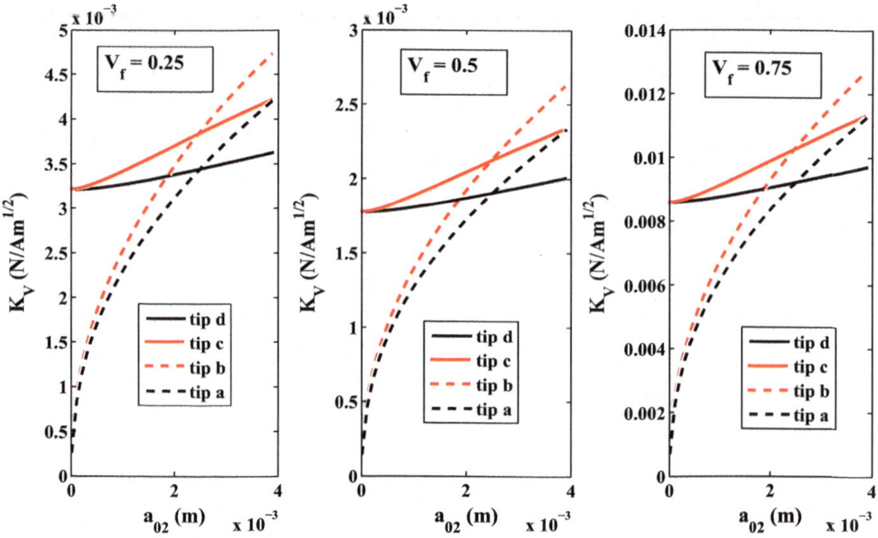

Fig. 7 Effect of crack length a_{02} on MIIF for different volume fractions

8 Conclusions

Considering the aforementioned analytical and numerical studies done on the proposed model, the following points are concluded.

(i) A complex variable and Stroh's formalism technique is successfully applied to study the two unequal collinear semi-permeable cracks in a piezo-electro-magnetic media.
(ii) The closed form analytic expressions are derived for the COD, COPD, COID, SIF, EDIF and the MIIF for the proposed model.
(iii) Two non-linear equations are derived, to obtain the electric displacement and magnetic induction inside the crack gap media.
(iv) The effect of volume fraction is observed on the intensity factors(IFs). All the IFs are decreased with the increase in the volume fraction.
(v) The effect of the inter-crack distance is observed on the IFs. All the IFs are increased with the decrease in the inter-crack distance.
(vi) The effect of crack length is observed on the IFs. All the IFs are increased with the increase in the crack length.

Appendix (A)

$X_1(z) = \sqrt{(z-a)(z-b)(z-c)(z-d)}, \qquad P_1(z) = C_0z^2 + C_1(z) + C_2;$
$\Delta = \Lambda_{22}(\Lambda_{44}\Lambda_{55} - \Lambda_{45}\Lambda_{54}) - \Lambda_{24}(\Lambda_{42}\Lambda_{55} - \Lambda_{45}\Lambda_{52}) + \Lambda_{25}(\Lambda_{42}\Lambda_{54} - \Lambda_{44}\Lambda_{52});$
$\Delta_1 = -\sigma_{22}^{\infty}(\Lambda_{44}\Lambda_{55} - \Lambda_{45}\Lambda_{54}) - (D^c - D_2^{\infty})(\Lambda_{24}\Lambda_{55} - \Lambda_{25}\Lambda_{54}) + (B^c - B_2^{\infty})$
$(\Lambda_{25}\Lambda_{44} - \Lambda_{24}\Lambda_{45});$
$\Delta_2 = \sigma_{22}^{\infty}(\Lambda_{42}\Lambda_{55} - \Lambda_{45}\Lambda_{52}) + (D^c - D_2^{\infty})(\Lambda_{22}\Lambda_{55} - \Lambda_{25}\Lambda_{52}) + (B^c - B_2^{\infty})$
$(\Lambda_{25}\Lambda_{42} - \Lambda_{22}\Lambda_{45});$
$\Delta_3 = \sigma_{22}^{\infty}(\Lambda_{44}\Lambda_{52} - \Lambda_{42}\Lambda_{54}) + (D^c - D_2^{\infty})(\Lambda_{24}\Lambda_{52} - \Lambda_{22}\Lambda_{54}) + (B^c - B_2^{\infty})$
$(\Lambda_{22}\Lambda_{44} - \Lambda_{24}\Lambda_{42});$
$C_0 = a_{11}a_{22} - a_{12}a_{21}, \qquad C_1 = a_{20}a_{12} - a_{10}a_{22}, \qquad C_2 = a_{21}a_{10} - a_{11}a_{20},$
$k^2 = \dfrac{(a-b)(c-d)}{(a-c)(b-d)};$
$g = \dfrac{2}{\sqrt{(a-c)(b-d)}}, \quad \alpha^2 = \dfrac{d-c}{a-c}, \quad \beta^2 = \dfrac{a-b}{a-c}, \quad a_{11} = g[aF(k) + (d-a)\Pi(\alpha^2, k)];$
$a_{12} = gF(k), \qquad a_{21} = g[cF(k) + (b-c)\Pi(\beta^2, k)], \qquad a_{22} = gF(k);$
$a_{10} = g\left[a^2F(k) + 2a(d-a)\Pi(\alpha^2, k) + (d-a)^2V_2\right];$
$a_{20} = g\left[c^2F(k) + 2c(b-c)\Pi(\beta^2, k) + (b-c)^2V_3\right];$
$V_2 = \dfrac{1}{2(\alpha^2-1)(k^2-\alpha^2)}\left\{\alpha^2E(k) + (k^2-\alpha^2)F(k) + (2\alpha^2k^2 + 2\alpha^2 - \alpha^4 - 3k^2)\Pi(\alpha^2, k)\right\};$
$V_3 = \dfrac{1}{2(\beta^2-1)(k^2-\beta^2)}\left\{\beta^2E(k) + (k^2-\beta^2)F(k) + (2\beta^2k^2 + 2\beta^2 - \beta^4 - 3k^2)\Pi(\beta^2, k)\right\};$
where $F(k)$, $E(K)$ and $\Pi(\alpha^2, k)$ are the complete elliptic integrals of the first, second and third kind, respectively.

Appendix (B)

$\alpha_1^2 = \dfrac{a}{d}\alpha^2, \quad \beta_1^2 = \dfrac{c}{b}\beta^2, \quad \nu = sin^{-1}\sqrt{\dfrac{(a-c)(y-d)}{(d-c)(a-y)}}, \quad \psi = sin^{-1}\sqrt{\dfrac{(a-c)(y-b)}{(a-b)(y-c)}};$
$S_1 = \alpha^2E(\nu, k) + (k^2-\alpha^2)F(\nu, k) + (2\alpha^2k^2 + 2\alpha^2 - \alpha^4 - 3k^2)\Pi(\nu, \alpha^2, k) - \dfrac{\alpha^4 \text{sn}u\text{cn}u\text{dn}u}{1-\alpha^2\text{sn}^2u};$
where sn$u$, cn$u$ and dn$u$ are the Jacobian elliptic functions.

$S_2 = \beta^2E(\psi, k) + (k^2-\beta^2)F(\psi, k) + (2\beta^2k^2 + 2\beta^2 - \beta^4 - 3k^2)\Pi(\psi, \beta^2, k) - \dfrac{\beta^4 \text{sn}u\text{cn}u\text{dn}u}{1-\beta^2\text{sn}^2u};$
$S_3 = d^2g\dfrac{\alpha_1^4}{\alpha^4}\left\{F(\nu, k) + \dfrac{2(\alpha^2-\alpha_1^2)}{\alpha_1^2}\Pi(\nu, \alpha^2, k) + \dfrac{(\alpha^2-\alpha_1^2)^2}{2\alpha_1^4(\alpha^2-1)(k^2-\alpha^2)}S_1\right\};$
$S_4 = dg\dfrac{\alpha_1^2}{\alpha^2}\left\{F(\nu, k) + \dfrac{\alpha^2-\alpha_1^2}{\alpha_1^2}\Pi(\nu, \alpha^2, k)\right\}, \qquad S_5 = gF(\nu, k);$

$$S_6 = b^2 g \frac{\beta_1^4}{\beta^4}\left\{F(\psi,k) + \frac{2(\beta^2-\beta_1^2)}{\beta_1^2}\Pi(\psi,\beta^2,k) + \frac{(\beta^2-\beta_1^2)^2}{2\beta_1^4(\beta^2-1)(k^2-\beta^2)}S_2\right\};$$

$$S_7 = bg\frac{\beta_1^2}{\beta^2}\left\{F(\psi,k) + \frac{\beta^2-\beta_1^2}{\beta_1^2}\Pi(\psi,\beta^2,k)\right\}, \qquad S_8 = gF(\psi,k);$$

where $F(,k)$, $E(,k)$ and $\Pi(,\ k)$ are the incomplete elliptic integrals of first, second and third kinds, respectively.

References

1. Wang, B.L., Han, J.C.: Discussion on electromagnetic crack face boundary conditions for the fracture mechanics of magneto-electro-elastic materials. Acta Mechanica Sinica **22**, 233–242 (2006)
2. Zhong, X.C., Li, X.F.: Closed-form solution for an eccentric anti-plane shear crack normal to the edges of a magnetoelectroelastic strip. Acta Mechanica **186**, 1–15 (2006)
3. Wang, B.L., Mai, Y.W.: Exact and fundamental solution for an anti-plane crack vertical to the boundaries of a magnetoelectroelastic strip. Int. J. Damage Mech. **16**, 77–94 (2007)
4. Li, Y.D., Lee, K.Y.: Fracture analysis and improved design for a symmetrically bonded smart structure with linearly non-homogeneous magnetoelectroelastic properties. Eng. Fracture Mech. **75**, 3161–3172 (2008)
5. Zheng, R.F., Wu, T.H., Li, X.Y., Chen, W.Q.: Analytical and numerical analyses for a penny-shaped crack embedded in an infinite transversely isotropic multi-ferroic composite medium: semi-permeable electro-magnetic boundary condition. Smart Mater. Struct. **27**, 065020 (2018)
6. Jangid, K., Bhargava, R.R.: Influence of polarization on two unequal semi-permeable cracks in a piezoelectric media. Strength Fracture Complexity **10**(3–4), 129–144 (2017)
7. Chyanbin, H.: Explicit solutions for collinear interface crack problems. Int. J. Solids Struct. **30**, 301–312 (1993)
8. Li, Y.D., Lee, K.Y., Dai, Y.: Dynamic stress intensity factors of two collinear mode-III cracks perpendicular to and on the two sides of a bi-FGM weak discontinuous interface. Euro. J. Mech. Solid **27**, 808–823 (2008)
9. Zhong, X.C., Liu, F., Li, X.F.: Transient response of a magnetoelectroelastic solid with two collinear dielectric cracks under impacts. Int. J. Solids Struct. **46**, 2950–2958 (2009)
10. Zhou, Z.G., Wang, B., Sun, Y.G.: Two collinear interface cracks in magneto-electro-elastic composites. Int. J. Eng. Sci. **42**, 1155–1167 (2004)
11. Zhou, Z.G., Wu, L.Z., Wang, B.: The dynamic behavior of two collinear interface cracks in magneto-electro-elastic materials. Euro. J. Mech. Solid **24**, 253–262 (2005)
12. Zhou, Z.G., Zhang, P.W., Wu, L.Z.: Solutions to a limited-permeable crack or two limited-permeable collinear cracks in piezoelectric/piezomagnetic materials. Arch. Appl. Mech. **77**, 861–882 (2007)
13. Zhou, Z.G., Tang, Y.L., Wu, L.Z.: Non-local theory solution to two collinear limited-permeable mode-I cracks in a piezoelectric/piezomagnetic material plane. Sci. China Phys. Mech. Astronout. **55**, 1272–1290 (2012)
14. Wang, B.L., Han, J.C., Mai, Y.W.: Exact solution for mode-III cracks in a magnetoelectroelastic layer. Int. J. Appl. Electromag. Mech. **24**, 33–44 (2006)
15. Wang, B.L., Mai, Y.W.: An exact analysis for mode-III cracks between two dissimilar magnetoelectroelastic layers. Mech. Compos. Mater. **44**, 533–548 (2008)
16. Singh, B.M., Rokne, J., Dhaliwal, R.S.: Closed-form solutions for two anti-plane collinear cracks in a magnetoelectroelastic layer. Euro. J. Mech. Solid **28**, 599–609 (2009)
17. Jangid, K., Bhargava, R.R.: Complex variable-based analysis for two semi-permeable collinear cracks in a piezoelectromagnetic media. Mech. Adv. Mater. Struct. **24**, 1007–1016 (2017)

18. Deeg, W.E.F.: Tha analysis of dislocation, Crack and inclusion problems in piezoelectric solids, Ph.D. thesis, Stanford University (1980)
19. Parton, V.Z.: Fracture mechanics of piezoelectric materials. Acta Astronout. **3**, 671–683 (1976)
20. Hao, T.H., Shen, Z.Y.: A new electric boundary condition of electric fracture mechanics and its applications. Eng. Fracture Mech. **47**, 793–802 (1994)
21. Stroh, A.N.: Dislocations and cracks in anisotropic elasticity. Philoso. Mag. **3**, 625–646 (1958)
22. Muskhelishvili, N.I.: Some Basic Problems of Mathematical Theory of Elasticity. Noordhoff Leyden (1975)
23. Wang, B., Mai, Y.: Applicability of the crack-face electromagnetic boundary conditions for fracture of magnetoelectroelastic materials. Int. J. Solids Struct. **44**, 387–398 (2007)
24. Zhong, X.C.: Analysis of a dielectric crack in a magnetoelectroelastic layer. Int. J. Solids Struct. **46**, 4221–4230 (2009)

Advanced Analysis of Local Fractional Calculus Applied to the Rice Theory in Fractal Fracture Mechanics

Xiao-Jun Yang, Dumitru Baleanu, and H. M. Srivastava

Abstract In this chapter, the recent results for the analysis of local fractional calculus are considered for the first time. The local fractional derivative (LFD) and the local fractional integral (LFI) in the fractional (real and complex) sets, the series and transforms involving the Mittag-Leffler function defined on Cantor sets are introduced and reviewed. The uniqueness of the solutions of the local fractional differential and integral equations and the local fractional inequalities are considered in detail. The local fractional vector calculus is applied to describe the Rice theory in fractal fracture mechanics.

Keywords Local fractional calculus · Local fractional derivative · Local fractional integral · Mittag-leffler function · Local fractional vector calculus · Local fractional partial differential equation · Local fractional integral transform · Local fractional integral equation · Local fractional inequality · Rice theory · Fractal fracture mechanics · Fractals.

X.-J. Yang (✉)
State Key Laboratory for Geomechanics and Deep Underground Engineering, China University of Mining and Technology, Xuzhou 221116, Jiangsu, People's Republic of China
e-mail: dyangxiaojun@163.com; xjyang@cumt.edu.cn

D. Baleanu
Department of Mathematics and Computer Sciences, Faculty of Arts and Sciences, Çankaya University, Ankara 06790, Turkey

Institute of Space Sciences, Magurele-Bucharest, Romania

H. M. Srivastava
Department of Mathematics and Statistics, University of Victoria, V8W 3R4, Victoria, BC, Canada

Department of Medical Research, China Medical University Hospital China Medical University, Taichung 40402, Taiwan, People's Republic of China

J. Singh et al. (eds.), *Methods of Mathematical Modelling and Computation for Complex Systems*, Studies in Systems, Decision and Control 373,
https://doi.org/10.1007/978-3-030-77169-0_5

1 Introduction

The theory of fractional calculus (FC) has successfully been utilized to describe the fractal problems in engineering practices [1–4]. The important examples are the fractal Fokker-Planck equations [5] and fractal description of stress and strain in elasticity [6–8]. There are several alternative approaches for handling the complex and fractal behaviors in nature [9–12].

The theory of the local fractional calculus (LFC) is a mathematical tool for handling the non-differentiable problems under the consideration of the complex and fractal behaviors of the real-world problems [13–19]. The local fractional derivative (LFD) and the local fractional integral (LFI) were used to present the approaches for describing the fractal phenomena in mathematical physics (see [20–23]). For the details of the applications of the LFC, we see as follows: the LFC to model the shallow water surfaces [24, 25], LC-electric circuit [26–28], local fractional partial differential equations (PDEs) [29–32], local fractional ordinary differential equations (ODEs) [33, 34], and so on. The special inequalities via LFI, such as the Ostrowski type [35], Steffensen type [36, 37], and Pompeiu type [38] inequalities for the LFIs and other inequalities were considered (see [39–49]).

The local fractional integral transforms via LFC were proposed in [9, 10, 50] and developed in [12, 16]. The local fractional Fourier type integral transform was investigated in [51–54]. The local fractional Laplace type integral transform was investigated in [54–61]. These integral transforms were applied to find the non-differentiable solutions for the local fractional PDEs (see [12, 62]). From the functional analysis point of view, the uniqueness of the solutions of the local fractional ODE and local fractional integral equations were considered in [9, 10] for the first time. The existence and uniqueness of the solutions of some local fractional abstract differential equations were presented in [63]. The existence and uniqueness of solutions for local fractional differential equations and its applications were reported in [9, 10, 62, 64]. The local fractional vector calculus and applications in the fractal heat conduction problems were presented in [2, 11].

The main aim of this chapter is to investigate the properties of the LFC, the series and transforms involving the Mittag-Leffler function defined on Cantor sets, analysis of the local fractional differential and integral equations, local fractional inequalities and local fractional vector calculus, and to present the applications of the extended version of the Rice theory in fractal fracture mechanics.

The structure of the chapter is as follows. In Sect. 2, the theory of the LFD and LFI in the fractional (real and complex) sets is presented. In Sect. 3, the analysis of the local fractional differential and integral equations is derived. In Sect. 4, the local fractional inequalities are discussed in detail. In Sect. 5, the series and transforms involving the Mittag-Leffler function defined on Cantor sets are reported. In Sect. 6, the local fractional vector calculus and its application in fractal fracture mechanics are considered in detail. Finally, the conclusions are given in Sect. 7.

2 The LFD and LFI in the Fractional Real and Complex Sets

In this section, we introduce the LFC of the real and complex variables and consider α as the fractal dimension in the chapter.

Let $\mathbb{N}$, $\mathbb{R}$ and $\mathbb{C}$ be sets of the natural numbers, real numbers and complex numbers.

Let $\mathbb{N}^\alpha$, $\mathbb{R}^\alpha$ and $\mathbb{C}^\alpha$ be the fractional sets of the natural numbers, real numbers and complex numbers (For more details of the notations of the fractional sets, see [9–11, 35, 36, 38, 40–42, 44–49]).

Definition 1 The complex number defined on the fractal set $\mathbb{C}^\alpha$ is given as follows [9–13]:

$$z^\alpha = x^\alpha + i^\alpha y^\alpha \qquad (x, y \in \mathbb{R}) \tag{1}$$

and its conjugate by

$$\overline{z^\alpha} = x^\alpha - i^\alpha y^\alpha \qquad (\overline{z^\alpha} \in \mathbb{C}^\alpha, x, y \in \mathbb{R}), \tag{2}$$

with its fractional modulus defined as follows [9–13]:

$$\left|\overline{z^\alpha}\right| = |z^\alpha| = \sqrt{\overline{z^\alpha} \cdot z^\alpha} = \sqrt{x^{2\alpha} + y^{2\alpha}}. \tag{3}$$

The complex number defined on the fractal set $\mathbb{C}^\alpha$ is represented in the form:

$$z^\alpha = \Re\left(z^\alpha\right) + i^\alpha Im\left(z^\alpha\right) = x^\alpha = x^\alpha + i^\alpha y^\alpha,$$

where $\Re\left(z^\alpha\right) = x^\alpha$ is the purely real part and $Im\left(z^\alpha\right) = y^\alpha$ is the purely imaginary part, which can be expressed as follows [9–13]:

$$z^\alpha = x^\alpha + i^\alpha y^\alpha = \sqrt{x^{2\alpha} + y^{2\alpha}}\left(\cos_\alpha\left(x^\alpha\right) + i^\alpha \sin_\alpha\left(x^\alpha\right)\right),$$

with

$$\cos_\alpha\left(x^\alpha\right) = \frac{x^\alpha}{\sqrt{x^{2\alpha} + y^{2\alpha}}},$$

$$\sin_\alpha\left(x^\alpha\right) = \frac{y^\alpha}{\sqrt{x^{2\alpha} + y^{2\alpha}}},$$

where

$$\cos_\alpha\left(z^\alpha\right) := \sum_{k=0}^{\infty} (-1)^k \frac{z^{2\alpha k}}{\Gamma\left(1 + 2\alpha k\right)}, \tag{4}$$

$$\sin_\alpha (z^\alpha) := \sum_{k=0}^{\infty} (-1)^k \frac{z^{(2k+1)\alpha}}{\Gamma[1+\alpha(2k+1)]}. \tag{5}$$

Definition 2 The complex Mittag-Leffler function on the fractal set $\mathbb{C}^\alpha$ is defined as follows (see [9–13]; see also [65]):

$$E_\alpha (z^\alpha) := \sum_{k=0}^{\infty} \frac{z^{\alpha k}}{\Gamma(1+k\alpha)}, \tag{6}$$

where $z^\alpha \in \mathbb{C}^\alpha$, which leads to the formulation in the form given by [9–13]

$$\begin{aligned} & z^\alpha \\ & = x^\alpha + i^\alpha y^\alpha \\ & = \sqrt{x^{2\alpha} + y^{2\alpha}} \left(\cos_\alpha (x^\alpha) + i^\alpha \sin_\alpha (x^\alpha)\right) \\ & = \sqrt{x^{2\alpha} + y^{2\alpha}} E_\alpha (i^\alpha z^\alpha), \end{aligned}$$

where

$$E_\alpha (i^\alpha z^\alpha) := \cos_\alpha (z^\alpha) + i^\alpha \sin_\alpha (z^\alpha). \tag{7}$$

2.1 *The LFD and LFI in the Fractional Real Set*

Definition 3 A function $f(x)$ is said to be local fractional continuous at $x = x_0$ if for each $\varepsilon > 0$ there exists for $\delta > 0$ such that (see [9–13, 35, 36, 38, 40–42, 44–49])

$$|f(x) - f(x_0)| < \varepsilon^\alpha, \tag{8}$$

whenever $0 < |x - x_0| < \delta$.

It is to say that

$$\lim_{x \to x_0} f(x) = f(x_0). \tag{9}$$

If $f(x)$ is local fractional continuous in the domain $I = (a, b)$, then we write it as [22–30]

$$f(x) \in H_\alpha (a, b). \tag{10}$$

Definition 4 Let $f(x) \in H_\alpha (a, b)$. The LFD of the function $f(x)$ of order αat $x = x_0$, denoted as $f^{(\alpha)}(x_0)$ or $\frac{d^\alpha f(x)}{dx^\alpha}\big|_{x=x_0}$, is defined as follows (see [9–13]):

$$D^{(\alpha)} f(x) = f^{(\alpha)}(x_0) = \frac{d^\alpha f(x)}{dx^\alpha}\Big|_{x=x_0} = \lim_{x \to x_0} \frac{\Delta^\alpha (f(x) - f(x_0))}{(x - x_0)^\alpha}, \tag{11}$$

where

$$\Delta^{\alpha}\left(f\left(x\right)-f\left(x_{0}\right)\right)\cong\Gamma\left(1+\alpha\right)\Delta\left(f\left(x\right)-f\left(x_{0}\right)\right).$$

Let

$$f\left(x\right),g\left(x\right)\in H_{\alpha}\left(a,b\right).$$

The properties of the LFD are presented as follows [9–13]:

(1)

$$\frac{d^{\alpha}}{dx^{\alpha}}\left(f\left(x\right)\pm g\left(x\right)\right)=\frac{d^{\alpha}f\left(x\right)}{dx^{\alpha}}\pm\frac{d^{\alpha}g\left(x\right)}{dx^{\alpha}};$$

(2)

$$\frac{d^{\alpha}\left(f\left(x\right)g\left(x\right)\right)}{dx^{\alpha}}=g\left(x\right)\frac{d^{\alpha}f\left(x\right)}{dx^{\alpha}}+f\left(x\right)\frac{d^{\alpha}g\left(x\right)}{dx^{\alpha}};$$

(3)

$$\frac{d^{\alpha}}{dx^{\alpha}}\left(\frac{f\left(x\right)}{g\left(x\right)}\right)=\frac{1}{g\left(x\right)^{2}}\left(g\left(x\right)\frac{d^{\alpha}f\left(x\right)}{dx^{\alpha}}+f\left(x\right)\frac{d^{\alpha}g\left(x\right)}{dx^{\alpha}}\right),$$

where $g\left(x\right)\neq 0$;

(4)

$$\frac{d^{\alpha}\left(hf\left(x\right)\right)}{dx^{\alpha}}=h\frac{d^{\alpha}f\left(x\right)}{dx^{\alpha}},$$

where h is a constant;

(5) If $y\left(x\right)=\left(f\circ u\right)\left(x\right)$, where $u\left(x\right)=g\left(x\right)$, then we have

$$\frac{d^{\alpha}y\left(x\right)}{dx^{\alpha}}=f^{(\alpha)}\left(g\left(x\right)\right)\left(g^{(1)}\left(x\right)\right)^{\alpha}.$$

The LFDs of the elementary functions defined on fractal sets are given as follows [9–13]:

(1)

$$\frac{d^{\alpha}}{dx^{\alpha}}\frac{x^{k\alpha}}{\Gamma(1+k\alpha)}=\frac{x^{(k-1)\alpha}}{\Gamma(1+(k-1)\,\alpha)};$$

(2)

$$\frac{d^{\alpha}E_{\alpha}\left(x^{\alpha}\right)}{dx^{\alpha}}=E_{\alpha}\left(x^{\alpha}\right);$$

(3)

$$\frac{d^{\alpha}E_{\alpha}\left(kx^{\alpha}\right)}{dx^{\alpha}}=kE_{\alpha}\left(kx^{\alpha}\right),$$

where k is a constant.

(4)

$$\frac{d^{\alpha}\sin_{\alpha}\left(x^{\alpha}\right)}{dx^{\alpha}}=\cos_{\alpha}\left(x^{\alpha}\right);$$

(5)

$$\frac{d^{\alpha}\cos_{\alpha}\left(x^{\alpha}\right)}{dx^{\alpha}}=-\sin_{\alpha}\left(x^{\alpha}\right).$$

Theorem 1 (The mean value theorem for the LFD) *If* $f(x)\in H_{\alpha}[a,b]$, *then there exists a point* $x_0\in(a,b)$ *such that (see [9–13])*

$$f(b)-f(a)=f^{(\alpha)}\left(x_0\right)\frac{(b-a)^{\alpha}}{\Gamma(1+\alpha)}.$$

Definition 5 Let $f(x)\in H_{\alpha}[a,b]$. The LFI of the function $f(x)$ of order α $(0<\alpha\leqq 1)$ is defined as follows (see [9–13]):

$${}_aI_b^{(\alpha)}f(x)=\frac{1}{\Gamma(1+\alpha)}\int_a^b f(x)(dx)^{\alpha}=\frac{1}{\Gamma(1+\alpha)}\lim_{\Delta x_k\to 0}\sum_{k=0}^{N-1}f\left(x_k\right)\left(\Delta x_k\right)^{\alpha},$$

where $\Delta x_k=x_{k+1}-x_k$ with

$$x_0=a<x_1<\cdots<x_{N-1}<x_N=b.$$

Let $f(x), g(x)\in H_{\alpha}(a,b)$. The properties of the LFI are presented as follows [9–13]:

(1)

$${}_aI_b^{(\alpha)}\left(f(x)\pm g(x)\right)={}_aI_b^{(\alpha)}f(x)\pm{}_aI_b^{(\alpha)}g(x);$$

(2)

$${}_aI_b^{(\alpha)}\left(hf(x)\right)=h\,{}_aI_b^{(\alpha)}f(x),$$

where h is a constant.

The LFIs of the elementary functions defined on fractal sets are given as follows [9–13]:

(1)

$$\frac{1}{\Gamma(1+\alpha)}\int_a^b E_{\alpha}\left(x^{\alpha}\right)(dx)^{\alpha}=E_{\alpha}\left(b^{\alpha}\right)-E_{\alpha}\left(a^{\alpha}\right);$$

(2)

$$\frac{1}{\Gamma(1+\alpha)}\int_a^b\frac{x^{k\alpha}}{\Gamma(1+k\alpha)}(dx)^{\alpha}=\frac{a^{(k+1)\alpha}}{\Gamma(1+(k+1)\alpha)}-\frac{b^{(k+1)\alpha}}{\Gamma(1+(k+1)\alpha)};$$

(3)

$$\frac{1}{\Gamma(1+\alpha)}\int_a^b \sin_\alpha(x^\alpha)(dx)^\alpha = \cos_\alpha(a^\alpha) - \cos_\alpha(b^\alpha);$$

(4)

$$\frac{1}{\Gamma(1+\alpha)}\int_a^b \cos_\alpha(x^\alpha)(dx)^\alpha = \sin_\alpha(b^\alpha) - \sin_\alpha(a^\alpha)$$

Theorem 2 (The mean value theorem for the LFI) *If $f(x) \in H_\alpha[a,b]$, then there exists a point $\xi \in (a,b)$ such that [9–13]*

$${}_aI_b^{(\alpha)} f(x) = f(\xi)\frac{(b-a)^\alpha}{\Gamma(\alpha+1)}.$$

Theorem 3 *If $f(x) \in H_\alpha[a,b]$, then there exists a point $\xi \in (a,b)$ such that [9–13]*

$$f(b) - f(a) = \frac{f^{(\alpha)}(\xi)(b-a)^\alpha}{\Gamma(1+\alpha)}.$$

Theorem 4 *Suppose that $f(x) \in H_\alpha[a,b]$, then there is a function [9–13]*

$$\Pi(x) = {}_aI_x^{(\alpha)} f(x),$$

such that it has the LFD,

$$\frac{d^\alpha \Pi(x)}{dx^\alpha} = f(x), a \leqq x \leqq b.$$

Theorem 5 (The LFI is anti-differentiation) *If $f(x) = g^{(\alpha)}(x) \in C_\alpha[a,b]$, then we have [9–13]*

$${}_aI_b^{(\alpha)} f(x) = g(b) - g(a).$$

Theorem 6 (The LFI by parts) *If $f^{(\alpha)}(x), g^{(\alpha)}(x) \in C_\alpha[a,b]$, then [9–13]*

$${}_aI_b^{(\alpha)} f(t)\, g^{(\alpha)}(t) = [f(t)\, g(t)]_a^b - {}_aI_b^{(\alpha)} f^{(\alpha)}(t)\, g(t).$$

Theorem 7 (The local fractional Taylor' theorem) *Suppose that $f^{((k+1)\alpha)}(x) \in C_\alpha(a,b)$ for $k = 0, 1, ..., n$, then [9–13]*

$$f(x) = \sum_{k=0}^{n} \frac{f^{(k\alpha)}(x_0)}{\Gamma(1+k\alpha)}(x-x_0)^{k\alpha} + \frac{f^{((n+1)\alpha)}(\xi)}{\Gamma(1+(n+1)\alpha)}(x-x_0)^{(n+1)\alpha}$$

$$with\, a < x_0 < \xi < x < b\, and\, \forall x \in (a,b)\,,\, where\ f^{((k+1)\alpha)}(x) = \overbrace{D_x^{(\alpha)}...D_x^{(\alpha)}}^{k+1\ times} f(x).$$

2.2 The LFD and LFI in the Fractional Complex Set

Let the complex function $f(z)$ be defined in a neighborhood of a point z_0.

Definition 6 The LFD of $f(z)$ at the point z_0, denoted by ${}_{z_0}D_z^{\alpha} f(z)$, $\frac{d^\alpha}{dz^\alpha} f(z)\big|_{z=z_0}$ or $f^{(\alpha)}(z_0)$, is defined as follows [9–11]:

$$ {}_{z_0}D_z^{\alpha} f(z) =: \lim_{z \to z_0} \frac{\Delta^\alpha f(z)}{(z - z_0)^\alpha},\ 0 < \alpha \leqq 1, \tag{12}$$

where

$$\Delta^\alpha f(z) = \Gamma(1+\alpha)\left[f(z) - f(z_0)\right].$$

If this limit exists, then the function $f(z)$ is said to be local fractional analytic at z_0.

If this limit exists for all $z_0$in a region $\aleph^\alpha \in \mathbb{C}^\alpha$, then the function $f(z)$ is said to be local fractional analytic in a region $\aleph^\alpha \in \mathbb{C}^\alpha$.

Let $f(z)$ and $g(z)$ be local fractional analytic functions. Then there is as follows [9–11]:

(1)

$$\frac{d^\alpha(f(z) \pm g(z))}{dz^\alpha} = \frac{d^\alpha f(z)}{dz^\alpha} \pm \frac{d^\alpha g(z)}{dz^\alpha};$$

(2)

$$\frac{d^\alpha(f(z)g(z))}{dz^\alpha} = g(z)\frac{d^\alpha f(z)}{dz^\alpha} + f(z)\frac{d^\alpha g(z)}{dz^\alpha};$$

(3)

$$\frac{d^\alpha}{dz^\alpha}\left(\frac{f(z)}{g(z)}\right) = \frac{1}{g(z)^2}\left(g(z)\frac{d^\alpha f(z)}{dz^\alpha} + f(z)\frac{d^\alpha g(z)}{dz^\alpha}\right)$$

where $g(z) \neq 0$;

(4)

$$\frac{d^\alpha(hf(z))}{dz^\alpha} = h\frac{d^\alpha f(z)}{dz^\alpha}$$

where h is a constant.

Definition 7 Let $f(z)$ be defined, single-valued and local fractional continuous in a region $\aleph^\alpha \in \mathbb{C}^\alpha$. The LFI of the complex function $f(z)$ along the contour C in $\aleph^\alpha \in \mathbb{C}^\alpha$ from point z_p to point z_q is defined as follows (see [9–11]):

$$I_C^{\alpha} f(z) = \frac{1}{\Gamma(1+\alpha)} \lim_{\Delta z \to 0} \sum_{i=0}^{n-1} f(z_i)(\Delta z)^{\alpha} = \frac{1}{\Gamma(1+\alpha)} \int_C f(z)(dz)^{\alpha}, \quad (13)$$

where $\Delta z_i = z_i - z_{i-1}$, $z_0 = z_p$, $z_n = z_q$ and $i \in \mathbb{N}$.

Our main theorems for the LFC of the complex variables are presented below.

Theorem 8 *If the contour C have the end points z_p and z_q with the orientation z_p to z_q, then we have [9–11]*

$$\frac{1}{\Gamma(1+\alpha)} \int_C f(z)(dz)^{\alpha} = F(z_q) - F(z_p) \quad (14)$$

where the function $f(z)$ has the primitive $F(z)$ on the contour C.

Theorem 9 *Let the function $f(z)$ be a primitive on C, where C is a simple closed contour. Then we have [9–11]*

$$\frac{1}{\Gamma(1+\alpha)} \oint_C f(z)(dz)^{\alpha} = 0 \quad (15)$$

Theorem 10 *If $f(z)$ is local fractional analytic on C_1 , C_2 and between them, and the contours C_1 and C_2 have same end points, then we have [9–11]*

$$\frac{1}{\Gamma(1+\alpha)} \int_{C_1} f(z)(dz)^{\alpha} = \frac{1}{\Gamma(1+\alpha)} \int_{C_2} f(z)(dz)^{\alpha} \quad (16)$$

Theorem 11 *If the closed contours C_1 and C_2 are such that C_2 lies inside C_1, then we have [9–11]*

$$\frac{1}{\Gamma(1+\alpha)} \int_{C_1} f(z)(dz)^{\alpha} = \frac{1}{\Gamma(1+\alpha)} \int_{C_2} f(z)(dz)^{\alpha}, \quad (17)$$

where $f(z)$ is local fractional analytic on C_1 , C_2 and between them.

Theorem 12 *If $f(z)$ is local fractional analytic within and on a simple closed contour C and z_0 is any point interior to the contour C, then we have [9–11]*

$$\frac{1}{(2\pi)^{\alpha} i^{\alpha}} \cdot \frac{1}{\Gamma(1+\alpha)} \oint_C \frac{f(z)}{(z-z_0)^{\alpha}} (dz)^{\alpha} = f(z_0) \quad (18)$$

Theorem 13 *If $f(z)$ is local fractional analytic within and on a simple closed contour C and z_0 is any point interior to the contour C, then we have [9–11]*

$$\frac{1}{(2\pi)^{\alpha} i^{\alpha}} \cdot \frac{1}{\Gamma(1+\alpha)} \oint_C \frac{f(z)}{(z-z_0)^{(n+1)\alpha}} (dz)^{\alpha} = f^{(n\alpha)}(z_0). \tag{19}$$

Theorem 14 *If $f(z)$ is local fractional analytic within and on a simple closed contour C and z_0 is any point interior to the contour C, then we have [9–11]*

$$\frac{1}{(2\pi)^{\alpha}} \cdot \frac{1}{\Gamma(1+\alpha)} \oint_C \frac{(dz)^{\alpha}}{(z-z_0)^{\alpha}} = i^{\alpha} \tag{20}$$

Theorem 15 *If $f(z)$ is local fractional analytic within and on a simple closed contour C and z_0 is any point interior to the contour C, then we have [9–11]*

$$\frac{1}{\Gamma(1+\alpha)} \oint_C \frac{(dz)^{\alpha}}{(z-z_0)^{n\alpha}} = 0, \tag{21}$$

where $n > 1$.

Definition 8 Let $f(z) = \varphi(z) / (z-z_0)^{n\alpha}$ and $\varphi(z) \neq 0$, where $\varphi(z)$ is local fractional analytic everywhere in a region including $z = z_0$. There are given as follows [9–11]:

(1) If n is a positive integer, then $f(z)$ has an isolated singularity at $z = z_0$, the point is called as a pole of order n, where n is a positive integer.
(2) If $n = 1$, the pole is often called a simple pole;
(3) if $n = 2$, it is called as a double pole.

Theorem 16 *If $f(z)$ has a pole of order n at $z = z_0$ but is local fractional analytic at every other point inside and on a contour C with the center at the point z_0, then $(z-z_0)^{n\alpha} f(z)$ is local fractional analytic at all points inside and on the contour C and has a local fractional Laurent type series about $z = z_0$ so that $f(z) = \sum_{i=-\infty}^{\infty} a_k (z-z_0)^{k\alpha}$, $0 < \alpha \leqq 1$ where [9–11]*

$$a_k = \frac{1}{(2\pi)^{\alpha}} \cdot \frac{1}{i^{\alpha}} \cdot \frac{1}{\Gamma(1+\alpha)} \oint_C \frac{f(z)}{(z-z_0)^{(k+1)\alpha}} (dz)^{\alpha} \tag{22}$$

for the contour $C : |z-z_0|^{\alpha} \leqq R^{\alpha}$.

Theorem 17 *If $f(z)$ is local fractional analytic within and on the boundary C of a region $\aleph^{\alpha} \in \mathbb{C}^{\alpha}$ except at a number of poles a within $\mathfrak{R}$, then (see [9–11])*

$$\frac{1}{(2\pi)^{\alpha} i^{\alpha} \Gamma(1+\alpha)} \oint_C f(z) (dz)^{\alpha} = \underset{z=z_0}{Res}\, f(z) = a_{-1}. \tag{23}$$

where $\underset{z=z_0}{Res}\, f(z) = a_{-1}$ is the residue of the function $f(z)$.

3 Analysis of the Local Fractional Differential and Integral Equations

Here, in this section, we introduce the local fractional continuity, convergence, and completeness in a generalized metric space (For the notations of the fractional sets, we used them in [9–12]).

Definition 9 A metric space on a fractal set E is a map $\rho_\alpha : E \times E \to \mathbb{R}^\alpha$ such that, for all $x^\alpha, y^\alpha, z^\alpha \in E$, the following rules hold true (see [9–12]):

(1) $\rho_\alpha (x^\alpha, y^\alpha) \geq 0$ with the equality $\rho_\alpha (x^\alpha, y^\alpha) = 0$ if $x^\alpha = y^\alpha$;
(2) $\rho_\alpha (x^\alpha, y^\alpha) = \rho_\alpha (y^\alpha, x^\alpha)$;
(3) $\rho_\alpha (x^\alpha, z^\alpha) \leqq \rho_\alpha (x^\alpha, y^\alpha) + \rho_\alpha (y^\alpha, z^\alpha)$. The pair (E, ρ_α) is a generalized metric space in the fractal space with the fractal dimension α.

Let E is a generalized metric space and $a^\alpha, b^\alpha, c^\alpha \in E$. Then we have

$$|\rho_\alpha (a^\alpha, b^\alpha) - \rho_\alpha (b^\alpha, c^\alpha)| \leqq \rho_\alpha (a^\alpha, c^\alpha) . \tag{24}$$

Definition 10 Suppose that X and Y are generalized metric spaces and f is a mapping of X into Y. If, for each $\varepsilon > 0$, there exists $\delta > 0$ such that

$$\rho_\alpha (f (a) , f (x)) < \varepsilon^\alpha ,$$

whenever $x^\alpha \in X$ and $\rho (a, x) < \delta$, then f is said to be local fractional continuous at the point $a^\alpha \in X$, which is noted as follows [9–11]:

$$\lim_{x \to a} f (x) = f (a) . \tag{25}$$

Definition 11 Let X be a generalized metric space. A sequence $\left\{x_n^\alpha\right\}_{n=1}^\infty$ in a generalized metric space X is called a Cauchy sequence if, for each $\varepsilon > 0$, there exists a positive integer N such that (see [9–11])

$$\rho_\alpha \left(x_m^\alpha, x_n^\alpha\right) < \varepsilon^\alpha , \tag{26}$$

whenever $m, n \geqq N$. This is equivalent to the requirement that

$$\lim_{m,n \to \infty} \rho_\alpha \left(x_m^\alpha, x_n^\alpha\right) = 0. \tag{27}$$

Definition 12 Let X be a generalized metric space. If each Cauchy sequence in the space X converges in X, the generalized metric space X is complete (see [9–11]).

We notice that $\mathbb{R}_n^\alpha$ and $\mathbb{C}_n^\alpha$ are complete.

Definition 13 (X, ρ_α) is a generalized metric space and $T : X \to X$, if there exists a number $\beta \in (0, 1)$ such that [9–11]

$$\rho_\alpha \left(T\left(x^\alpha\right), T\left(y^\alpha\right)\right) \leqq \beta^\alpha \rho_\alpha \left(x^\alpha, y^\alpha\right) \tag{28}$$

for all $x^\alpha, y^\alpha \in X$.

We say that T is a contraction mapping on the generalized metric space X.

Definition 14 (*see* [9–11]) Let (X, ρ_α) be a generalized metric space. If $x^\alpha \in X$ and $Tx^\alpha = x^\alpha$, then we say that x^α is a fixed point of T.

Theorem 18 (see [9–11]) *Let X be a generalized metric space. A convergent sequence in the fractal space X may have more than one limit in X.*

Theorem 19 (Contraction Mapping Theorem) (see [9–11]) *A contraction mapping T defined on the complete generalized metric space (X, ρ_α) has a unique fixed point.*

Theorem 20 (Generalized Contraction Mapping Theorem) (see [9–11]) *Let $T : X \to X$ be a map on the complete metric space (X, ρ_α). Then, for some $m \geqq 1$, T^m is a contraction and*

$$\rho_\alpha \left(T^m \left(x^\alpha\right), T^m \left(x^\alpha\right)\right) \leqq \beta^\alpha \rho_\alpha \left(x^\alpha, y^\alpha\right) \tag{29}$$

for all $x^\alpha, y^\alpha \in X$.

3.1 The Uniqueness of the Solutions of the Local Fractional Differential Equations

In this subsection, we discuss the uniqueness of the solutions of the local fractional differential equations.

Theorem 21 *Suppose that $x_0 \in [a, b]$ and $y_0 \in \mathbb{R}^\alpha$, $F : [a, b] \times \mathbb{R}_1^\alpha \to \mathbb{R}_1^\alpha$ is local fractional continuous. For all $x \in [a, b]$, there is a continuous condition given as (see [9–11, 64])*

$$|F(x, y_1) - F(x, y_2)| \leqq k^\alpha |y_1 - y_2|^\alpha . \tag{30}$$

where $1 > k > 0$ and $1 \geqq \alpha > 0$.

Then the following local fractional differential equation:

$$\frac{d^\alpha y}{dx^\alpha} = F(x, y) \tag{31}$$

subject to the initial condition $y_0 = y(x_0)$ has a unique solution in the space $C_\alpha [a, b]$.

Proof We consider the map $T : C_\alpha [a, b] \to C_\alpha [a, b]$ defined as

$$Tf(x) = y_0 + \frac{1}{\Gamma(1+\alpha)} \int_{x_0}^{x} F(t, f(t)) (dt)^{\alpha}.$$

We claim that for all n,

$$\left| T^n f_1(x) - T^n f_2(x) \right| \leqq k^{n\alpha} \frac{|x - x_0|^{n\alpha}}{\Gamma(1+n\alpha)} \rho_\alpha(f_1, f_2).$$

The proof is by the induction on n.
The case $n = 0$ is trivial (and $n = 1$ is already done).
The induction step is as follows:

$$\begin{aligned}
&\left| T^{n+1} f_1(x) - T^{n+1} f_2(x) \right| \\
&= \left| \frac{1}{\Gamma(1+\alpha)} \int_{x_0}^{x} F(t, T^n f_1(x)) - F(t, T^n f_2(x)) (dt)^{\alpha} \right| \\
&\leqq \left| \frac{1}{\Gamma(1+\alpha)} \int_{x_0}^{x} k^{\alpha} \left| F(t, T^n f_1(x)) - F(t, T^n f_2(x)) \right| (dt)^{\alpha} \right| \\
&\leqq \left| \frac{1}{\Gamma(1+\alpha)} \int_{x_0}^{x} \frac{k^{(n+1)\alpha} |x - x_0|^{n\alpha}}{\Gamma(1+n\alpha)} \rho_\alpha(f_1, f_2) (dt)^{\alpha} \right| \\
&\leqq \left| \frac{1}{\Gamma(1+\alpha)} \int_{x_0}^{x} k^{(n+1)\alpha} \frac{|x - x_0|^{n\alpha}}{\Gamma(1+n\alpha)} \rho_\alpha(f_1, f_2) (dt)^{\alpha} \right| \\
&\leqq k^{(n+1)\alpha} \frac{|x - x_0|^{(n+1)\alpha}}{\Gamma(1+(n+1)\alpha)} \rho_\alpha(f_1, f_2) \\
&\leqq k^{(n+1)\alpha} \frac{|b - a|^{(n+1)\alpha}}{\Gamma(1+(n+1)\alpha)} \rho_\alpha(f_1, f_2).
\end{aligned}$$

We have

$$k^{(n+1)\alpha} \frac{|b - a|^{(n+1)\alpha}}{\Gamma(1+(n+1)\alpha)} \rho_\alpha(f_1, f_2) \to 0$$

as $n \to 0$.

If n is sufficiently large, we have

$$0 < k^{(n+1)\alpha} \frac{|b - a|^{(n+1)\alpha}}{\Gamma(1+(n+1)\alpha)} < 1$$

such that T^n is a contraction on the space $C_\alpha[a, b]$.

Hence, T has a unique fixed point in the space $C_\alpha[a, b]$, which gives a unique solution to the local fractional differential equation.

Example 1 The local fractional differential equation

$$\frac{d^{\alpha} f(x)}{dx^{\alpha}} + f(x) = 0,$$

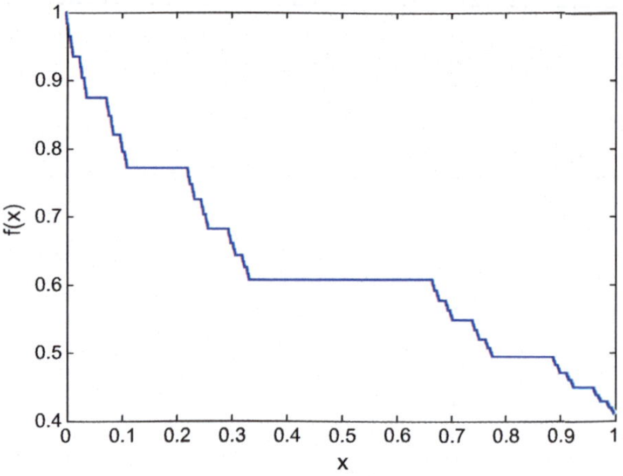

Fig. 1 The plot of the solution of the local fractional differential equation when $\alpha = \ln 2/\ln 3$

has the unique solution given as $f(x) = E_\alpha(-x^\alpha)$, where $f(0) = 1$, and its graph is shown in Fig. 1.

3.2 *The Uniqueness of the Solutions of the Local Fractional Integral Equations*

In this subsection, we discuss the uniqueness of the solutions of the local fractional integral equations.

Theorem 22 *Let $C_\alpha[a, b] = \{x(t) : x(t)$ be local fractional continuous on the interval $[a, b]$. The metric on the space $C_\alpha[a, b]$ is defined as (see [9–11])*

$$\rho_\alpha(x, y) = \{\max|x(t) - y(t)| : t \in [a, b], x, y \in C_\alpha[a, b]\}. \tag{32}$$

Let us consider that the local fractional integral equation

$$f(x) = \frac{\lambda^\alpha}{\Gamma(1+\alpha)} \int_a^x F(x, y) f(y) (dy)^\alpha + \varphi(x), \tag{33}$$

has a unique solution in $C_\alpha[a, b]$, where $\lambda^\alpha \in \mathbb{R}^\alpha$, $\varphi \in C_\alpha[a, b]$ and $F(x, y) \in C_\alpha[a, b] \times C_\alpha[a, b]$.

Proof We define $T : C_\alpha[a, b] \to C_\alpha[a, b]$ by

$$Tf(x) = \frac{\lambda^\alpha}{\Gamma(1+\alpha)} \int_a^x F(x, y) f(y) (dy)^\alpha + \varphi(x).$$

Let $f_1, f_2 \in C_\alpha[a,b]$. Then

$$\begin{aligned}
&\rho_\alpha(Tf_1, Tf_2)\\
&= \max_{x\in[a,b]} |Tf_1 - Tf_2|\\
&= \max_{x\in[a,b]} \left| \frac{\lambda^\alpha}{\Gamma(1+\alpha)} \int_a^x F(x,y)(f_1(y) - f_2(y))(dy)^\alpha \right|\\
&\leqq \frac{|\lambda|^\alpha M}{\Gamma(1+\alpha)} \left[\max_{x\in[a,b]} |f_1(x) - f_2(x)| \right] \left| \int_a^x (dy)^\alpha \right|\\
&\leqq \frac{|\lambda|^\alpha M \rho_\alpha(f_1,f_2)}{\Gamma(1+\alpha)} \left| \int_a^x (dy)^\alpha \right|\\
&= \frac{|\lambda|^\alpha M \rho_\alpha(f_1,f_2)}{\Gamma(1+\alpha)} |x-a|^\alpha\\
&\leqq \frac{|\lambda|^\alpha M |b-a|^\alpha}{\Gamma(1+\alpha)} \rho_\alpha(f_1, f_2)
\end{aligned}$$

where $M = \max \leqq \{|F(x,y)| : x, y \in [a,b]\}$.

We claim that for all n,

$$\rho_\alpha(T^n f_1, T^n f_2) \leqq \frac{|\lambda|^{n\alpha} M^n |x-a|^{n\alpha}}{\Gamma(1+n\alpha)} \rho_\alpha(f_1, f_2) \leqq \frac{|\lambda|^{n\alpha} M^n |b-a|^{n\alpha}}{\Gamma(1+n\alpha)} \rho_\alpha(f_1, f_2).$$

The induction step is as follows:

$$\begin{aligned}
&\rho_\alpha\left(T^{n+1} f_1, T^{n+1} f_2\right) = \max_{x\in[a,b]} \left|T^{n+1} f_1 - T^{n+1} f_2\right|\\
&= \max_{x\in[a,b]} \left| \frac{\lambda^\alpha}{\Gamma(1+\alpha)} \int_a^x F(x,y)(T^n f_1(y) - T^n f_2(y))(dy)^\alpha \right|\\
&\leqq \frac{|\lambda|^{(n+1)\alpha} M^{n+1}}{\Gamma(1+n\alpha)} \left[\max_{x\in[a,b]} |f_1(x) - f_2(x)| \right] \left| \frac{1}{\Gamma(1+\alpha)} \int_a^x (x-a)^{n\alpha} (dy)^\alpha \right|\\
&\leqq \frac{|\lambda|^{(n+1)\alpha} M^{n+1} \rho_\alpha(f_1,f_2)}{\Gamma(1+(n+1)\alpha)} |x-a|^{(n+1)\alpha}\\
&\leqq \frac{|\lambda|^{(n+1)\alpha} M^{n+1} |b-a|^{(n+1)\alpha}}{\Gamma(1+(n+1)\alpha)} \rho_\alpha(f_1, f_2).
\end{aligned}$$

For each $\lambda^\alpha \in \mathbb{R}^\alpha$, there exists $N \in \mathbb{N}$ such that

$$0 < \frac{|\lambda|^{n\alpha} M^n |b-a|^{n\alpha}}{\Gamma(1+n\alpha)} \rho_\alpha(f_1, f_2) < 1,$$

where $n > N$.

It is to say that T^n is a contraction mapping and has a unique fixed point f.

Thus, f provides the unique local fractional continuous solution to the local fractional integral equation.

Example 2 The local fractional integral equation

$$f(x) - \frac{\lambda}{\Gamma(1+\alpha)} \int_0^x f(x)(dx)^\alpha = 1,$$

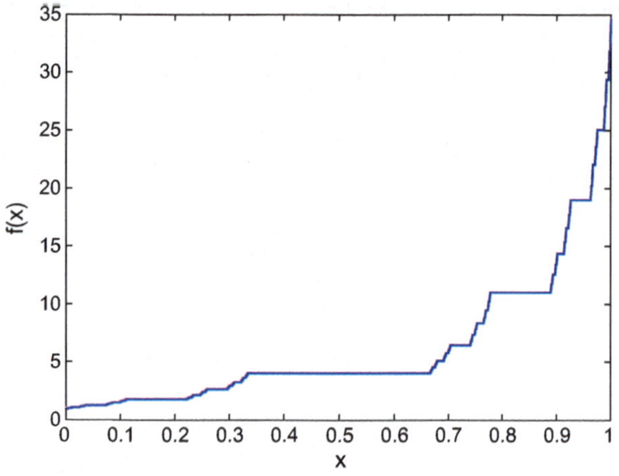

Fig. 2 The plot of the solution of the local fractional integral equation when $\lambda = 2$ and $\alpha = \ln 2/\ln 3$

has the unique solution given as $f(x) = E_\alpha(\lambda x^\alpha)$ and its graph is shown in Fig. 2.

4 Local Fractional Inequalities

In this chapter, we present the inequalities within local fractional integral, such as the Hölder type, Cauchy-Schwarz type and Minkowski type inequalities.

Let E be a fractal set.

The Hölder type, Cauchy-Schwarz type and Minkowski type inequalities in the fractal finite series are presented as follows:

Theorem 23 (Generalized Hölder type inequality) (see [9–11]) *Let* $|x_i^\alpha| > 0$, $|y_i^\alpha| > 0$, $p > 0$, $q > 0$, $i \in \mathbb{N}$ *and* $1/p + 1/q = 1$. *Then we have*

$$\sum_{i=1}^{n} |x_i^\alpha| \, |y_i^\alpha| \leqq \left(\sum_{i=1}^{n} |x_i^\alpha|^p\right)^{\frac{1}{p}} + \left(\sum_{i=1}^{n} |y_i^\alpha|^q\right)^{\frac{1}{q}}, \tag{34}$$

where $p > 1$, $q > 1$ *and* $0 < \alpha \leqq 1$.

Theorem 24 (Generalized Cauchy-Schwarz type inequality) (see [9–11])
Let $|x_i^\alpha| > 0$, $|y_i^\alpha| > 0$ *and* $i \in \mathbb{N}$. *Then we have*

$$\sum_{i=1}^{n} |x_i^\alpha| \, |y_i^\alpha| \leqq \left(\sum_{i=1}^{n} |x_i^\alpha|^2\right)^{\frac{1}{2}} + \left(\sum_{i=1}^{n} |y_i^\alpha|^2\right)^{\frac{1}{2}}. \tag{35}$$

Theorem 25 (Generalized Minkowski type inequality) (see [9–11])

$$\left(\sum_{i=1}^{n}\left|x_i^\alpha - y_i^\alpha\right|^p\right)^{\frac{1}{p}} \leqq \left(\sum_{i=1}^{n}\left|x_i^\alpha\right|^p\right)^{\frac{1}{p}} + \left(\sum_{i=1}^{n}\left|y_i^\alpha\right|^p\right)^{\frac{1}{p}}, \tag{36}$$

where $p > 1$ *and* $0 < \alpha \leqq 1$.

For the linear space of bounded infinite sequences, denoted as $E = l_{p,\alpha}$, the generalized normed linear space on E is defined by (see [9–11]):

$$\|x^\alpha\|_{p,\alpha} =: \left(\sum_{i=1}^{\infty}\left|x_i^\alpha\right|^p\right)^{\frac{1}{p}} < \infty, \tag{37}$$

where $1 \leqq p < \infty$.

Theorem 26 (The infinite version of the generalized Minkowski type inequality) *The infinite version of generalized Minkowski type inequality can be write as [9–11]:*

$$\left(\sum_{i=1}^{\infty}\left|x_i^\alpha - y_i^\alpha\right|^p\right)^{\frac{1}{p}} \leqq \left(\sum_{i=1}^{\infty}\left|x_i^\alpha\right|^p\right)^{\frac{1}{p}} + \left(\sum_{i=1}^{\infty}\left|y_i^\alpha\right|^p\right)^{\frac{1}{p}},$$

where $\infty > p \geqq 1$ *and* $0 < \alpha \leqq 1$.

Let $E = L_{p,\alpha}[a, b]$. Then the normed space with the $p-$norm is given as ([9–11]):

$$\|f\|_{p,\alpha} =: \left(\frac{1}{\Gamma(1+\alpha)}\int_a^b |f(t)|^p (dt)^\alpha\right)^{\frac{1}{p}} < \infty, \tag{38}$$

where $0 < \alpha \leqq 1$ and $\infty > p \geqq 1$.

The following rules hold ([9–11]):

1. If $\|f\|_{1,\alpha} = 0$, then $f(x) = 0$;
2. $\|ag\|_{1,\alpha} = |a|^\alpha \|f\|_{1,\alpha}$;
3. $\|f + g\|_{1,\alpha} \leqq \|f\|_{1,\alpha} + \|g\|_{1,\alpha}$.

Theorem 27 (The integral form of the generalized Hölder type inequality) *Let* $f, g \in L_{p,\alpha}[\mathbb{R}]$, $1 \leqq p < \infty$. *Then (see [9–11])*

$$\|fg\|_{1,\alpha} \leqq \|f\|_{p,\alpha} \|g\|_{q,\alpha}, \tag{39}$$

where $p \geqq 1$, $q \geqq 1$ *and* $1/q + 1/p = 1$.

Theorem 28 (The integral form of the generalized Minkowski type inequality) *Let* $f, g \in L_{p,\alpha}[\mathbb{R}]$, $1 \leqq p < \infty$. *Then (see [9–11])*

$$\|f + g\|_{p,\alpha} \leqq \|f\|_{p,\alpha} + \|g\|_{p,\alpha}. \tag{40}$$

For more details of the Hölder type, Cauchy-Schwarz type and Minkowski type inequalities, which are defined on the fractal domain, see [9–11].

5 The Series and Transforms Involving the Mittag-Leffler Function Defined on Cantor Sets

In this section, we consider the concepts and theorems of the series and transforms involving the Mittag-Leffler function defined on Cantor sets.

5.1 The Fourier Type Series Via the Mittag-Leffler Function Defined On Cantor Sets

In this subsection, we introduce the concepts and theorems of the series involving the Mittag-Leffler function defined on Cantor sets.

Definition 15 Let $f(x)$ be 2π-periodic. For $n \in \mathbb{Z}$, the complex Mittag-Leffler form of the local fractional Fourier type series of $f(x)$ involving the Mittag-Leffler function defined on Cantor sets is defined as follows (see [9–11, 13])

$$f(x) \sim \sum_{k=-\infty}^{\infty} C_n E_\alpha \left(i^\alpha (nx)^\alpha\right), \tag{41}$$

where the Fourier coefficients are represented as follows (see [9–11, 13]):

$$C_n = \frac{1}{(2\pi)^\alpha} \int_{-\pi}^{\pi} f(x) E_\alpha \left(-i^\alpha (nx)^\alpha\right) (dx)^\alpha. \tag{42}$$

Theorem 29 *Suppose that $f(x)$ is 2π-periodic, bounded and local fractional integral on $[-\pi, \pi]$. Then, the local fractional series of the function $f(x)$ involving the Mittag-Leffler function defined on Cantor sets converges to $f(x)$ at $x \in [-\pi, \pi]$, and (see [9–11, 13])*

$$\frac{f(x+0)+f(x-0)}{2} = \sum_{k=-\infty}^{\infty} C_n E_\alpha \left(i^\alpha (nx)^\alpha\right), \tag{43}$$

where the Fourier type coefficients are expressed by

$$C_n = \frac{1}{(2\pi)^\alpha} \int_{-\pi}^{\pi} f(x) E_\alpha (\pi^\alpha i^\alpha (nx)^\alpha) (dx)^\alpha. \tag{44}$$

Definition 16 Let $f(x)$ be $2l$-periodic. For $n \in \mathbb{Z}$, the complex generalized Mittag-Leffler form of the local fractional Fourier type series of the function $f(x)$ involving the Mittag-Leffler function defined on Cantor sets is defined as follows (see [9–11, 13]):

$$f(x) \sim \sum_{k=-\infty}^{\infty} C_n E_\alpha \left(\frac{\pi^\alpha i^\alpha (nx)^\alpha}{l^\alpha} \right), \tag{45}$$

where the Fourier type coefficients are given by

$$C_n = \frac{1}{(2l)^\alpha} \int_{-l}^{l} f(x) E_\alpha \left(\frac{-\pi^\alpha i^\alpha (nx)^\alpha}{l^\alpha} \right) (dx)^\alpha. \tag{46}$$

Theorem 30 *Suppose that $f(x)$ is $2l$-periodic, bounded and local fractional integral on $[-l, l]$. Then, the local fractional series of the function $f(x)$ involving the Mittag-Leffler function defined on Cantor sets converges to $f(x)$ at $x \in [-l, l]$, and (see [9–11, 13])*

$$\frac{f(x+0) + f(x-0)}{2} = \sum_{k=-\infty}^{\infty} C_n E_\alpha \left(\frac{\pi^\alpha i^\alpha (nx)^\alpha}{l^\alpha} \right), \tag{47}$$

where the Fourier type coefficients are represented as

$$C_n = \frac{1}{(2l)^\alpha} \int_{-l}^{l} f(x) E_\alpha \left(\frac{-\pi^\alpha i^\alpha (nx)^\alpha}{l^\alpha} \right) (dx)^\alpha. \tag{48}$$

5.2 The Fourier Type Transform Via Mittag-Leffler Function Defined on Cantor Sets

In this subsection, we introduce the concepts and theorems of the Fourier type transform involving the Mittag-Leffler function defined on Cantor sets.

Definition 17 The local fractional Fourier type transform of the function $f(x)$ involving the Mittag-Leffler function defined on Cantor sets is defined as follows (see [9–11, 13]):

$$F_\alpha\{f(x)\} = f_\omega^{F,\alpha}(\omega) := \frac{1}{\Gamma(1+\alpha)} \int_{-\infty}^{\infty} E_\alpha(-i^\alpha \omega^\alpha x^\alpha) f(x) (dx)^\alpha,$$

whenever the latter integral converges. The sufficient condition for convergence is given by (see [9–11, 13])

$$\left| \frac{1}{\Gamma(1+\alpha)} \int_{-\infty}^{\infty} f(x) E_\alpha\left(-i^\alpha \omega^\alpha x^\alpha\right) (dx)^\alpha \right| \leq \frac{1}{\Gamma(1+\alpha)} \int_{-\infty}^{\infty} |f(x)| (dx)^\alpha = \|f\|_{1,\alpha} < \infty,$$

which can be written as $f \in L_{1,\alpha}[\mathbb{R}]$. If $f \in L_{1,\alpha}[\mathbb{R}]$, then local fractional Fourier type transform of the function $f(x)$ exists. Moreover, the inverse local fractional Fourier type transform involving the Mittag-Leffler function defined on Cantor sets is defined as follows (see [9–11, 13]):

$$f(x) = F_\alpha^{-1}\left(f_\omega^{F,\alpha}(\omega)\right) := \frac{1}{(2\pi)^\alpha} \int_{-\infty}^{\infty} E_\alpha(i^\alpha \omega^\alpha x^\alpha) f_\omega^{F,\alpha}(\omega) (d\omega)^\alpha,$$

whenever the latter integral converges.

Definition 18 The local fractional convolution of the functions $f_1(x)$ and $f_2(x)$ is defined as follows (see [9–11, 13]):

$$f_1(x) * f_2(x) = \frac{1}{\Gamma(1+\alpha)} \int_{-\infty}^{\infty} f_1(t) f_2(x-t) (dt)^\alpha.$$

There are the equalities as follows (see [9–11, 13]):

$$f_1(x) * f_2(x) = f_2(x) * f_1(x),$$

$$f_1(x) * (f_2(x) + f_3(x)) = f_1(x) * f_2(x) + f_1(x) * f_3(x).$$

The theorems for the local fractional Fourier type transform are presented as follows (see [9–11, 13]):

Let $f, f_1, f_2 \in L_{1,\alpha}[\mathbb{R}]$, $F_\alpha\{f(x)\} = f_\omega^{F,\alpha}(\omega)$, $F_\alpha\{f_1(x)\} = f_{\omega,1}^{F,\alpha}(\omega)$ and $F_\alpha\{f_2(x)\} = f_{\omega,2}^{F,\alpha}(\omega)$. Then, we have the following:

(1) $F_\alpha\{f_1(x) + f_2(x)\} = F_\alpha\{f_1(x)\} + F_\alpha\{f_2(x)\}$;
(2) $F_\alpha\{f_1(x) * f_2(x)\} = f_{\omega,1}^{F,\alpha}(\omega) f_{\omega,2}^{F,\alpha}(\omega)$;
(3) $F_\alpha\left\{f^{(\alpha)}(x)\right\} = i^\alpha \omega^\alpha F_\alpha\{f(x)\}$, where $\lim_{|x|\to\infty} f(x) = 0$;
(4) $F_\alpha\left\{{}_{-\infty}I_x^{(\alpha)} f(x)\right\} = F_\alpha\{f(x)\} / (i^\alpha \omega^\alpha)$, where $\lim_{x\to\infty} {}_{-\infty}I_x^{(\alpha)} f(x) \to 0$.

6 Local Fractional Vector Calculus with an Application in Fractal Fracture Mechanics

In this chapter, we introduce the theory of the local fractional vector calculus and present an application to the Rice theory in the fractal fracture mechanics.

6.1 Local Fractional Vector Calculus

In this subsection, we introduce the basic theory and theorems of the local fractional vector calculus.

Definition 19 For $1 > \alpha > 0$, the local fractional line integral of the function $\mathbf{u}(x_P, y_P, z_P)$ along a fractal line l^α is defined as follows (see [2, 11]):

$$\int_{l^{(\alpha)}} \mathbf{u}(x_P, y_P, z_P) \cdot d\mathbf{l}^{(\alpha)} = \lim_{N\to\infty} \sum_{P=1}^{N} \mathbf{u}(x_P, y_P, z_P) \cdot \Delta\mathbf{l}_P^{(\alpha)} \tag{49}$$

where the elements of line $\Delta\mathbf{l}_P^{(\alpha)}$ are so required that all $\left|\Delta l_P^\alpha\right| \to 0$ as $N \to \infty$ and $\beta = 2\alpha$.

Definition 20 For $\gamma = \frac{3}{2}\beta = 3\alpha$, $1 > \alpha > 0$, the local fractional surface integral of $u(r_P)$ is defined as follows (see [2, 11]):

$$\iint u(r_P)\, d\mathbf{S}^{(\beta)} = \lim_{N\to\infty} \sum_{P=1}^{N} u(r_P)\, \mathbf{n}_P \Delta S_P^{(\beta)}, \tag{50}$$

where $d\mathbf{S}^{(\beta)}$ are N elements of area with a unit normal local fractional vector n_P, $\Delta S_P^{(\beta)} \to 0$ as $N \to \infty$.

Definition 21 For $\gamma = \frac{3}{2}\beta = 3\alpha$, $1 > \alpha > 0$, the local fractional volume integral of the function $\mathbf{u}(r_P)$ is defined as follows (see [2, 11]):

$$\iiint \mathbf{u}(r_P)\, dV^{(\gamma)} = \lim_{N\to\infty} \sum_{P=1}^{N} \mathbf{u}(r_P)\, \Delta V_P^{(\gamma)}, \tag{51}$$

where $\Delta V_P^{(\gamma)}$ are the elements of the volume $\Delta V_P^{(\gamma)} \to 0$ as $N \to \infty$.

Basic operators of the local fractional vector integrals are as follows (see [2, 11]):

$$\int_{l^{(\alpha)}} (\mathbf{u}_1 + \mathbf{u}_2) \cdot d\mathbf{l}^{(\alpha)} = \int_{l^{(\alpha)}} \mathbf{u}_1 \cdot d\mathbf{l}^{(\alpha)} + \int_{l^{(\alpha)}} \mathbf{u}_2 \cdot d\mathbf{l}^{(\alpha)},$$

$$\int_{l^{(\alpha)}} \mathbf{u}\cdot d\mathbf{l}^{(\alpha)} = \int_{l_1^{(\alpha)}} \mathbf{u}\cdot d\mathbf{l}^{(\alpha)} + \int_{l_2^{(\alpha)}} \mathbf{u}\cdot d\mathbf{l}^{(\alpha)},$$

$$\iint_{S^{(\beta)}} (\mathbf{u}_1 + \mathbf{u}_2)\cdot d\mathbf{S}^{(\beta)} = \iint_{S^{(\beta)}} \mathbf{u}_1\cdot d\mathbf{S}^{(\beta)} + \iint_{S^{(\beta)}} \mathbf{u}_2\cdot d\mathbf{S}^{(\beta)},$$

$$\iint_{S^{(\beta)}} \mathbf{u}\cdot d\mathbf{S}^{(\beta)} = \iint_{S_1^{(\beta)}} \mathbf{u}\cdot d\mathbf{S}^{(\beta)} + \iint_{S_2^{(\beta)}} \mathbf{u}\cdot d\mathbf{S}^{(\beta)},$$

$$\iiint_{V^{(\gamma)}} (\mathbf{u}_1 + \mathbf{u}_2)\cdot dV^{(\gamma)} = \iiint_{V^{(\gamma)}} \mathbf{u}_1\cdot dV^{(\gamma)} + \iiint_{V^{(\gamma)}} \mathbf{u}_2\cdot dV^{(\gamma)},$$

$$\iiint_{V^{(\gamma)}} (\mathbf{u}_1 + \mathbf{u}_2)\cdot dV^{(\gamma)} = \iiint_{V_1^{(\gamma)}} \mathbf{u}\cdot dV^{(\gamma)} + \iiint_{V_2^{(\gamma)}} \mathbf{u}\cdot dV^{(\gamma)},$$

where $\mathbf{l}^{(\alpha)} = \mathbf{l}_1^{(\alpha)} + \mathbf{l}_2^{(\alpha)}$, $\mathbf{S}^{(\beta)} = \mathbf{S}_1^{(\beta)} + \mathbf{S}_2^{(\beta)}$ and $V^{(\gamma)} = V_1^{(\gamma)} + V_2^{(\gamma)}$.

Definition 22 For

$$\gamma = \frac{3}{2}\beta = 3\alpha,\ 1 > \alpha > 0,$$

the local fractional gradient of the scale function φ is defined as follows (see [2, 11]):

$$\nabla^\alpha \varphi = \lim_{dV^{(\gamma)}\to 0} \left(\frac{1}{dV^{(\gamma)}} \oiint_{S^{(\beta)}} \varphi d\mathbf{S}^{(\beta)} \right) = \frac{\partial^\alpha \varphi}{\partial x_1^\alpha} e_1^\alpha + \frac{\partial^\alpha \varphi}{\partial x_2^\alpha} e_2^\alpha + \frac{\partial^\alpha \varphi}{\partial x_3^\alpha} e_3^\alpha, \tag{52}$$

where $V^{(\gamma)}$ is a small fractal volume enclosing P, $S^{(\beta)}$ is its bounding fractal surface, and ∇^α is a local fractional Hamilton operator.

Definition 23 For $\gamma = \frac{3}{2}\beta = 3\alpha$, $1 > \alpha > 0$, the local fractional divergence of the vector function $\mathbf{u}$ is defined by (see [2, 11])

$$\nabla^\alpha \bullet \mathbf{u} = \lim_{dV^{(\gamma)}\to 0} \left(\frac{1}{dV^{(\gamma)}} \oiint_{S^{(2\alpha)}} \mathbf{u} \bullet d\mathbf{S}^{(\beta)} \right) = \frac{\partial^\alpha u_1}{\partial x_1^\alpha} + \frac{\partial^\alpha u_2}{\partial x_2^\alpha} + \frac{\partial^\alpha u_3}{\partial x_3^\alpha}, \tag{53}$$

where

$$\mathbf{u} = u_1 e_1^\alpha + u_2 e_2^\alpha + u_3 e_3^\alpha.$$

Definition 24 For $\gamma = \frac{3}{2}\beta = 3\alpha$, $1 > \alpha > 0$, the local fractional curl of the vector function $\mathbf{u}$ is defined by (see [2, 11]):

$$\nabla^{\alpha} \times \mathbf{u} = \lim_{dS^{(\beta)} \to 0} \left(\frac{1}{dS^{(\beta)}} \oint_{l^{(\alpha)}} \mathbf{u} \cdot d\mathbf{l}^{(\alpha)} \right) \mathbf{n}_P$$
$$= \left(\frac{\partial^{\alpha} u_3}{\partial x_2^{\alpha}} - \frac{\partial^{\alpha} u_2}{\partial x_3^{\alpha}} \right) e_1^{\alpha} + \left(\frac{\partial^{\alpha} u_1}{\partial x_3^{\alpha}} - \frac{\partial^{\alpha} u_3}{\partial x_1^{\alpha}} \right) e_2^{\alpha} + \left(\frac{\partial^{\alpha} u_2}{\partial x_1^{\alpha}} - \frac{\partial^{\alpha} u_1}{\partial x_2^{\alpha}} \right) e_3^{\alpha}, \tag{54}$$

where

$$\mathbf{u} = u_1 e_1^{\alpha} + u_2 e_2^{\alpha} + u_3 e_3^{\alpha}.$$

Theorem 31 (Local fractional Gauss theorem) *For* $\gamma = \frac{3}{2}\beta = 3\alpha$, $1 > \alpha > 0$, *the local fractional Gauss theorem of the fractal vector field states that (see [2, 11])*

$$\iiint_{V^{(\gamma)}} \nabla^{\alpha} \cdot \mathbf{u} dV^{(\gamma)} = \oiint_{S^{(\beta)}} \mathbf{u} \cdot d\mathbf{S}^{(\beta)}. \tag{55}$$

Theorem 32 (Local fractional Stokes' theorem) *For* $\beta = 2\alpha$, $1 > \alpha > 0$, *the local fractional Stokes' theorem of the fractal field states that (see [2, 11])*

$$\oint_{l^{(\alpha)}} \mathbf{u} \cdot d\mathbf{l}^{\alpha} = \iint_{S^{(\beta)}} (\nabla^{\alpha} \times \mathbf{u}) \cdot d\mathbf{S}^{(\beta)}.$$

For more details of the local fractional vector calculus, see [2, 11].

6.2 An Application to Rice Theory in Fractal Mechanics

Let us consider the work of the traction in fractal boundary, the elastic energy in fractal medium and the fractal losing energy be

$$W_1 = \iint_{S^{(\beta)}} \mathbf{p} \cdot \mathbf{u} d\mathbf{S}^{(\beta)}, \ W_2 = -\iiint_{V^{(\gamma)}} w dV^{(\gamma)}$$

and

$$W_3 = \int_{l^{(\alpha)}} D \cdot d\mathbf{l}^{(\alpha)},$$

respectively, where $\mathbf{p}$ is the traction in the fractal boundary, $\mathbf{u}$ is the fractal displacement, w is the fractal elastic energy density, and D is the fractal losing energy in unit fractal line.

The energy in fractal medium can be written as

$$W = \int_{l^{(\alpha)}} p_i u_i d\mathbf{l}^{(\alpha)} - \iint_{S^{(\beta)}} w d\mathbf{S}^{(\beta)}, \tag{56}$$

where p_i and u_i are components of both traction in the fractal boundary and the fractal displacement.

Consider the fractal losing energy and finding the LFD, we give

$$\frac{\partial^\alpha W}{\partial t^\alpha} = \frac{\partial^\alpha}{\partial t^\alpha}\int\limits_{l^{(\alpha)}} p_i u_i d\mathbf{l}^{(\alpha)} - \frac{\partial^\alpha}{\partial t^\alpha}\iint\limits_{S^{(\beta)}} w d\mathbf{S}^{(\beta)} - \frac{\partial^\alpha D}{\partial t^\alpha}. \tag{57}$$

With the use of

$$\frac{\partial^\alpha}{\partial t^\alpha}\int\limits_{l^{(\alpha)}} p_i u_i d\mathbf{l}^{(\alpha)} = \int\limits_{l^{(\alpha)}} p_i \frac{\partial^\alpha u_i}{\partial t^\alpha} d\mathbf{l}^{(\alpha)} = \int\limits_{l^{(\alpha)}} p_i \frac{\partial^\alpha u_i}{\partial a^\alpha}\left(\frac{\partial a}{\partial t}\right)^\alpha d\mathbf{l}^{(\alpha)},$$

$$\frac{\partial^\alpha}{\partial t^\alpha}\iint\limits_{S^{(\beta)}} w d\mathbf{S}^{(\beta)} = \iint\limits_{S^{(\beta)}} \frac{\partial^\alpha w}{\partial t^\alpha} d\mathbf{S}^{(\beta)} = \iint\limits_{S^{(\beta)}} \frac{\partial^\alpha w}{\partial a^\alpha}\left(\frac{\partial a}{\partial t}\right)^\alpha d\mathbf{S}^{(\beta)},$$

$$\frac{\partial^\alpha D}{\partial t^\alpha} = \frac{\partial^\alpha D}{\partial a^\alpha}\left(\frac{\partial a}{\partial t}\right)^\alpha,$$

where a is the length of crack, we obtain from Eq. (25) that

$$\begin{aligned} &\frac{\partial^\alpha W}{\partial t^\alpha} \\ &= \int\limits_{l^{(\alpha)}} p_i \frac{\partial^\alpha u_i}{\partial a^\alpha}\left(\frac{\partial a}{\partial t}\right)^\alpha d\mathbf{l}^{(\alpha)} - \iint\limits_{S^{(\beta)}} \frac{\partial^\alpha w}{\partial a^\alpha}\left(\frac{\partial a}{\partial t}\right)^\alpha d\mathbf{S}^{(\beta)} - \frac{\partial^\alpha D}{\partial a^\alpha}\left(\frac{\partial a}{\partial t}\right)^\alpha \\ &= \left(\frac{\partial a}{\partial t}\right)^\alpha\left(\int\limits_{l^{(\alpha)}} p_i \frac{\partial^\alpha u_i}{\partial a^\alpha} d\mathbf{l}^{(\alpha)} - \iint\limits_{S^{(\beta)}} \frac{\partial^\alpha w}{\partial a^\alpha} d\mathbf{S}^{(\beta)} - \frac{\partial^\alpha D}{\partial a^\alpha}\right). \end{aligned} \tag{58}$$

When $\partial^\alpha W/\partial t^\alpha = 0$, we have from Eq. (26) that

$$\int\limits_{l^{(\alpha)}} p_i \frac{\partial^\alpha u_i}{\partial a^\alpha} d\mathbf{l}^{(\alpha)} - \iint\limits_{S^{(\beta)}} \frac{\partial^\alpha w}{\partial a^\alpha} d\mathbf{S}^{(\beta)} - \frac{\partial^\alpha D}{\partial a^\alpha} = 0. \tag{59}$$

The J-integral in fractal medium is defined as

$$J_\alpha = \frac{\partial^\alpha D}{\partial a^\alpha}.$$

From Eq. (27), we obtain that

$$J_\alpha = \int\limits_{l^{(\alpha)}} p_i \frac{\partial^\alpha u_i}{\partial a^\alpha} d\mathbf{l}^{(\alpha)} - \iint\limits_{S^{(\beta)}} \frac{\partial^\alpha w}{\partial a^\alpha} d\mathbf{S}^{(\beta)}. \tag{60}$$

As an extended version of the Rice's theory, we give that

$$\frac{\partial^{\alpha} W}{\partial t^{\alpha}} \geq 0. \tag{61}$$

From Eq. (29), there are two cases:

Case 1. When the crack tip is super-static, there is $\partial^{\alpha} W/\partial t^{\alpha} > 0$;

Case 2. When the crack tip is sub-static, there is $\partial^{\alpha} W/\partial t^{\alpha} = 0$.

When the crack length has is greater and the horizontal coordinate value is smaller, there is relationship of both increment of crack length and increment of horizontal coordinate value given as

$$(dx)^{\alpha} = -(da)^{\alpha} \tag{62}$$

which leads to

$$J_{\alpha} = \int\limits_{l^{(\alpha)}} p_i \frac{\partial^{\alpha} u_i}{\partial a^{\alpha}} d\mathbf{l}^{(\alpha)} - \iint\limits_{S^{(\beta)}} \frac{\partial^{\alpha} w}{\partial a^{\alpha}} d\mathbf{S}^{(\beta)} = \int\limits_{l^{(\alpha)}} w\,(dy)^{\alpha}\, d\mathbf{l}^{(\alpha)} - \int\limits_{l^{(\alpha)}} p_i \frac{\partial^{\alpha} u_i}{\partial a^{\alpha}} d\mathbf{l}^{(\alpha)}. \tag{63}$$

By using the traction on the fractal boundary given as

$$\mathbf{P} = \mathbf{N} \cdot \mathbf{œ}, \tag{64}$$

we have

$$(dx)^{\alpha} = (N_1) \cdot d\mathbf{l}^{(\alpha)},\ (dy)^{\alpha} = N_2 d\mathbf{l}^{(\alpha)}, \tag{65}$$

where

$$N_1 = \frac{(dx)^{\alpha}}{\sqrt{(dx)^{\alpha} + (dy)^{\alpha}}},\ N_2 = -\frac{(dy)^{\alpha}}{\sqrt{(dx)^{\alpha} + (dy)^{\alpha}}}. \tag{66}$$

Suppose that $w = \int_0^{\varepsilon_{ij}} \sigma_{ij} d\left(\varepsilon_{ij}\right)^{\alpha}$, where $\sigma_{ij} = \partial^{\alpha} w/\partial\left(\varepsilon_{ij}\right)^{\alpha}$ and $\varepsilon_{ij} = \partial^{\alpha} u_i/\partial x_j^{\alpha}$, we have

$$\int\limits_{l^{(\alpha)}} w\,(dy)^{\alpha}\, d\mathbf{l}^{(\alpha)} = \iint\limits_{S^{(\beta)}} \frac{\partial^{\alpha} w}{\partial x^{\alpha}} d\mathbf{S}^{(\beta)} = \iint\limits_{S^{(\beta)}} \sigma_{ij} \frac{\partial^{\alpha} \varepsilon_{ij}}{\partial x^{\alpha}} d\mathbf{S}^{(\beta)} = \oint\limits_{l^{(\alpha)}} \sigma_{ij} N_j \frac{\partial^{\alpha} u_i}{\partial x^{\alpha}} d\mathbf{l}^{(\alpha)}, \tag{67}$$

which leads to

$$J_{\alpha} = \oint\limits_{l^{(\alpha)}} \left(\sigma_{ij} N_j - p_i\right) \frac{\partial^{\alpha} u_i}{\partial x^{\alpha}} d\mathbf{l}^{(\alpha)} = 0, \tag{68}$$

where $l^{(\alpha)}$ is the closed circle.

The result states the crack tip is always super-static or sub-static in the real materials and the two cases always take place in the real crack progression in the differential fractal dimension of the material surface (see [11, 12] and the cited references).

7 Conclusion

In the present work, we introduce the analysis of the LFC for the first time. The concepts and properties of the LFD and LFI in the fractional (real and complex) sets and of the series and transforms involving the Mittag-Leffler function defined on Cantor sets were investigated in detail. The uniqueness of the solutions of the local fractional differential and integral equations and local fractional inequalities were also discussed. The local fractional vector calculus were used to describe the extended version of the Rice theory in fractal fracture mechanics with aid of the LFC operator. The results are accurate and efficient for handling a family of the fractal problems by using the local fractional differential and integral equations from the functional analysis point of view.

References

1. Tarasov, V.E.: Fractional Dynamics: Applications of Fractional Calculus to Dynamics of Particles, Fields and Media. Springer Science & Business Media (2011)
2. Cattani, C., Srivastava, H.M., Yang, X.J.: Fractional Dynamics. De Gruyter Open, Berlin (2015)
3. West, B., Bologna, M., Grigolini, P.: Physics of Fractal Operators. Springer Science & Business Media (2012)
4. Baskin, E., Iomin, A.: Electrostatics in fractal geometry: fractional calculus approach. Chaos, Solitons Fractals **44**(4–5), 335–341 (2011)
5. Tarasov, V.E.: Fractional Fokker-Planck equation for fractal media. Chaos Interdiscip. J. Nonlinear Sci. **15**(2), 023102 (2005)
6. Carpinteri, A., Cornetti, P., Kolwankar, K.M.: Calculation of the tensile and flexural strength of disordered materials using fractional calculus. Chaos, Solitons Fractals **21**(3), 623–632 (2004)
7. Carpinteri, A., Cornetti, P.: A fractional calculus approach to the description of stress and strain localization in fractal media. Chaos, Solitons Fractals **13**(1), 85–94 (2002)
8. Carpinteri, A., Chiaia, B., Cornetti, P.: Static-kinematic duality and the principle of virtual work in the mechanics of fractal media. Comput. Methods Appl. Mech. Eng. **191**(1–2), 3–19 (2001)
9. Yang, X.J.: Local fractional integral transforms. Progress Nonlinear Sci. **4**(1), 1–225 (2011)
10. Yang, X.J.: Local Fractional Functional Analysis and Its Applications. Asian Academic Publisher Limited, Hong Kong (2011)
11. Yang, X.J.: Advanced Local Fractional Calculus and Its Applications. World Science Publisher, New York (2012)
12. Yang, X.J., Baleanu, D., Srivastava, H.M.: Local Fractional Integral Transforms and Their Applications. Academic, New York (2015)
13. Yang, X.J., Baleanu, D., Srivastava, H.M.: Local fractional similarity solution for the diffusion equation defined on Cantor sets. Appl. Math. Lett. **47**, 54–60 (2015)
14. Liu, H.Y., He, J.H., Li, Z.B.: Fractional calculus for nanoscale flow and heat transfer. Int. J. Numer. Methods Heat Fluid Flow **24**(6), 1227–1250 (2014)
15. Jafari, H., Jassim, H.K., Tchier, F., Baleanu, D.: On the approximate solutions of local fractional differential equations with local fractional operators. Entropy **18**(4), 150 (2016)
16. Yang, X.J., Baleanu, D., Machado, J.A.T.: Mathematical aspects of the Heisenberg uncertainty principle within local fractional Fourier analysis. Boundary Value Problems **2013**(1), 131 (2013)

17. Debbouche, A., Antonov, V.: Finite-dimensional diffusion models of heat transfer in fractal mediums involving local fractional derivatives. Nonlinear Stud. **24**(3), 527–535 (2017)
18. Yang, X.J., Machado, J.A.T., Baleanu, D.: Exact traveling-wave solution for local fractional Boussinesq equation in fractal domain. Fractals **25**(04), 1740006 (2017)
19. Hemeda, A.A., Eladdad, E.E., Lairje, I.A.: Local fractional analytical methods for solving wave equations with local fractional derivative. Math. Methods Appl. Sci. **41**(6), 2515–2529 (2018)
20. Yang, X.J., Gao, F., Srivastava, H.M.: A new computational approach for solving nonlinear local fractional PDEs. J. Comput. Appl. Math. **339**, 285–296 (2018)
21. Yang, X.J., Baleanu, D.: Fractal heat conduction problem solved by local fractional variation iteration method. Thermal Sci. **17**(2), 625–628 (2013)
22. Kumar, D., Singh, J., Baleanu, D.: A hybrid computational approach for Klein-Gordon equations on Cantor sets. Nonlinear Dyn. **87**(1), 511–517 (2017)
23. Yang, X.J., Srivastava, H.M., He, J.H., Baleanu, D.: Cantor-type cylindrical-coordinate method for differential equations with local fractional derivatives. Phys. Lett. A **377**(28–30), 1696–1700 (2013)
24. Yang, X.J., Machado, J.A.T., Baleanu, D., Cattani, C.: On exact traveling-wave solutions for local fractional Korteweg-de Vries equation. Chaos Interdiscip. J. Nonlinear Sci. **26**(8), 084312 (2016)
25. Ye, S.S., Mohyud-Din, S.T., Belgacem, F.B.M.: The Laplace series solution for local fractional Korteweg-de Vries equation. Thermal Sci. **20**(3), S867–S870 (2016)
26. Yang, X.J., Machado, J.A.T., Cattani, C., Gao, F.: On a fractal LC-electric circuit modeled by local fractional calculus. Commun. Nonlinear Sci. Numer. Simul. **47**, 200–206 (2017)
27. Zhao, X.H., Zhang, Y., Zhao, D., Yang, X.J.: The RC circuit described by local fractional differential equations. Fundamenta Informaticae **151**(1–4), 419–429 (2017)
28. Yang, X.J., Machado, J.A.T., Gao, F., Carlo, C.: On linear and nonlinear electric circuits: A local fractional calculus approach, Chapter 11. In: Azar, A.T., Radwan, A., Vaidyanathan, S. (eds.) Fractional Order Systems: Optimization, Control, Circuit Realizations and Applications. Academic, NY, USA (2018)
29. Yang, X.J., Machado, J.A.T., Hristov, J.: Nonlinear dynamics for local fractional Burgers' equation arising in fractal flow. Nonlinear Dyn. **84**(1), 3–7 (2016)
30. Singh, J., Kumar, D., Nieto, J.J.: A reliable algorithm for a local fractional tricomi equation arising in fractal transonic flow. Entropy **18**(6), 206 (2016)
31. Zhang, Y., Srivastava, H.M., Baleanu, M.C.: Local fractional variational iteration algorithm II for non-homogeneous model associated with the non-differentiable heat flow. Adv. Mech. Eng. **7**(10), 1–7 (2015)
32. Jafari, H., Tajadodi, H., Johnston, J.S.: A decomposition method for solving diffusion equations via local fractional time derivative. Thermal Sci. **19**(suppl.1), 123–129 (2015)
33. Yang, X.J., Machado, J.A.T.: A new insight into complexity from the local fractional calculus view point: modelling growths of populations. Math. Methods Appl. Sci. **40**(17), 6070–6075 (2017)
34. Yang, X.J., Gao, F., Srivastava, H.M.: Non-differentiable exact solutions for the nonlinear ODEs defined on fractal sets. Fractals **25**(04), 1740002 (2017)
35. Sarikaya, M.Z., Budak, H.: Generalized Ostrowski type inequalities for local fractional integrals. Proc. Am. Math. Soc. **145**(4), 1527–1538 (2017)
36. Tunç, T., Sarikaya, M.Z., Srivastava, H.M.: Some generalized Steffensen's inequalities via a new identity for local fractional integrals. Int. J. Anal. Appl. **13**(1), 98–107 (2017)
37. Wu, S.H., Srivastava, H.M.: Some improvements and generalizations of Steffensen's integral inequality. Appl. Math. Comput. **192**, 422–428 (2007)
38. Erden, S., Sarikaya, M.Z.: Generalized Pompeiu type inequalities for local fractional integrals and its applications. Appl. Math. Comput. **274**, 282–291 (2016)
39. Chen, G.S., Srivastava, H.M., Wang, P., Wei, W.: Some further generalizations of Hölder's inequality and related results on fractal space. Abstract Appl. Anal. Article ID **832802**, 1–7 (2014)

40. Sarikaya, M.Z., Tunc, T., Budak, H.: On generalized some integral inequalities for local fractional integrals. Appl. Math. Comput. **276**, 316–323 (2016)
41. Budak, H., Sarikaya, M.Z., Yildirim, H.: New inequalities for local fractional integrals. Iranian J. Sci. Technol. Trans. Sci. **41**(4), 1039–1046 (2017)
42. Mo, H., Sui, X.: Hermite-Hadamard-type inequalities for generalized s-convex functions on real linear fractal set $\mathbb{R}^{\alpha}$ $(0 < \alpha < 1)$. Math. Sci. **11**(3), 241–246 (2017)
43. Srivastava, H.M., Zhang, Z.H., Wu, Y.D.: Some further refinements and extensions of the Hermite-Hadamard and Jensen inequalities in several variables. Math. Comput. Model. **54**, 2709–2717 (2011)
44. Kılıçman, A., Saleh, W.: Notions of generalized s-convex functions on fractal sets. J. Inequalities Appl. **2015**(1), 312 (2015)
45. Choi, J., Set, E., Tomar, M.: Certain generalized Ostrowski type inequalities for local fractional integrals. Commun. Korean Math. Soc. **32**(3), 601–617 (2017)
46. Liu, Q., Sun, W.: A Hilbert-type fractal integral inequality and its applications. J. Inequalities Appl. **2017**(1), 83 (2017)
47. Liu, Q., Chen, D.: A Hilbert-type integral inequality on the fractal space. Integral Transf. Special Funct. **28**(10), 772–780 (2017)
48. Kiliçman, A., Saleh, W.: Generalized Convex Functions and their Applications. Selected Topics, Mathematical Analysis and Applications, pp. 77–99 (2018)
49. Lara, T., Meretes, N., Rosales, E., Sanchez, R.: Convexity on fractal sets. UPI J. Math. Biostat. **1**(1), 22–31 (2018)
50. Srivastava, H.M., Golmankhaneh, A.K., Baleanu, D., Yang, X.J.: Local fractional Sumudu transform with application to IVPs on Cantor sets. Abstract Appl. Anal. **2014**, 1–7 (2014)
51. He, J.H.: Asymptotic methods for solitary solutions and compactons. Abstract Appl. Anal. **2012**, 1–130 (2012)
52. He, J.H.: A tutorial review on fractal spacetime and fractional calculus. Int. J. Theor. Phys. **53**(11), 3698–3718 (2014)
53. Yang, X.J., Liao, M.K., Chen, J.W.: A novel approach to processing fractal signals using the Yang-Fourier transforms. Procedia Eng. **29**, 2950–2954 (2012)
54. Yang, A.M., Zhang, Y.Z., Long, Y.: The Yang-Fourier transforms to heat-conduction in a semi-infinite fractal bar. Thermal Sci. **17**(3), 707–713 (2013)
55. Zhong, W.P., Gao, F.: Application of the yang laplace transforms to solution to nonlinear fractional wave equation with local fractional derivative. In: International Conference on Computer Technology and Development, 3rd (ICCTD), ASME Press (2011)
56. Yan, S.P.: Local fractional Laplace series expansion method for diffusion equation arising in fractal heat transfer. Thermal Sci. **19**(suppl.1), 131–135 (2015)
57. Liu, C.F., Kong, S.S., Yuan, S.J.: Reconstructive schemes for variational iteration method within Yang-Laplace transform with application to fractal heat conduction problem. Thermal Sci. **17**(3), 715–721 (2013)
58. Jassim, H.K.: The analytical solutions for volterra integro-differential equations within local fractional operators by yang-laplace transform. Sahand Commun. Math. Anal. **6**(1), 69–76 (2017)
59. Zhang, Y.Z., Yang, A.M., Long, Y.: Initial boundary value problem for fractal heat equation in the semi-infinite region by Yang-Laplace transform. Thermal Sci. **18**(2), 677–681 (2014)
60. Zhao, C.G., Yang, A.M., Jafari, H., Haghbin, A.: The Yang-Laplace transform for solving the IVPs with local fractional derivative. Abstract Appl. Anal. **2014**, 1–5 (2014)
61. Jassim, H.K., Ünlü, C., Moshokoa, S.P., Khalique, C.M.: Local fractional Laplace variational iteration method for solving diffusion and wave equations on Cantor sets within local fractional operators. Math. Problems Eng. **2015**, 1–9 (2015)
62. Hassan, K.J.: Analytical Solutions of Partial Differential Equations on Cantor Sets Within Local Fractional Derivative Operators. University of Mazandaran, Babolsar, Ph.D. Thesis (2016)
63. Zhong, W.P., Yang, X.J., Gao, F.: A Cauchy problem for some local fractional abstract differential equation with fractal conditions. J. Appl. Funct. Anal. **8**(1), 92–99 (2013)

64. Jafari, H., Jassim, H.K., Qurashi, M.A., Baleanu, D.: On the existence and uniqueness of solutions for local fractional differential equations. Entropy **18**(11), 420 (2016)
65. Kilbas, A.A., Srivastava, H.M., Trujillo, J.J.: Theory and Applications of Fractional Differential Equations, vol. 204. North-Holland Mathematical Studies. Elsevier (North-Holland), Amsterdam (2006)

General Fractional Calculus with Nonsingular Kernels: New Prospective on Viscoelasticity

Xiao-Jun Yang, Feng Gao, and Yang Ju

Abstract In the chapter, the general fractional derivatives in the different kernel functions, such as Mittag-Lefller, Wiman and Prabhakar functions are considered to model the viscoelastic behaviors in the real materials. We investigate the basic formulas of the fractional calculus (FC) in the kernels of the power, Mittag-Lefller, Wiman and Prabhakar functions. We discuss the applications for the general fractional calculus (GFC) in viscoelasticity. As the examples, the Maxwell and Voigt models with the general fractional derivatives (GFD) are considered to represent the complexity of the real materials.

Keywords Mittag-Lefller function · Wiman function · Prabhakar function · General fractional derivative · General fractional integral · General fractional calculus · Viscoelasticity

1 Introduction

Fractional calculus (FC) within the singular power-law kernel in the Riemann–Liouville and Liouville-Caputo types (see [1–9]) has been the increasing interests for scientists and engineers to represent the mathematical models in areas of a great many of the applications in engineering practices, such as the electric circuit [10], control theory [11], physics [12], mechanics [13], heat transfer [14], mathematical economy

X.-J. Yang (✉) · F. Gao · Y. Ju
State Key Laboratory for Geomechanics and Deep Underground Engineering, China University of Mining and Technology, XuzhouJiangsu 221116, China
e-mail: dyangxiaojun@163.com; xjyang@cumt.edu.cn

X.-J. Yang · F. Gao
School of Mechanics and Civil Engineering, China University of Mining and Technology, Xuzhou 221116, China

Y. Ju
State Key Laboratory of Coal Resources and Safe Mining, China University of Mining and Technology At Beijing, Beijing 100083, China

J. Singh et al. (eds.), *Methods of Mathematical Modelling and Computation for Complex Systems*, Studies in Systems, Decision and Control 373,
https://doi.org/10.1007/978-3-030-77169-0_6

and finance (see [15, 16]), complex population dynamics [17], mathematical biology [18] and many others (see [20] and the cited references therein).

From mathematical and physical point of view, there may exist some of the new prospective of the applications of the operators involving the special functions and power-law functions to linear viscoelasticity (see [21–33]). With the use of the Nutting's observation [28], the laws of deformation with the operators involving the Riemann–Liouville [23, 24], Liouville-Caputo [25] and Caputo-Fabrizio [28] types, local FD [29], general FDs [30], and others [31–33] were reported in detail. The hereditary elastic rheological models, represented as the Volterra integral equation, were reported in [20, 21, 33]. The Maxwell and Voigt models involving the different fractional and fractal operators were proposed in [20, 21, 28–36].

Nowadays, there may exist the new unsolved problems including the Nutting equation [37] and anomalous Nutting equation in the real materials, such as rock and mining rock. Motivated by the above ideas, the brief targets of the chapter are to investigate the general fractional derivatives (GFDs) and the general fractional integrals (GFIs) with the nonsingular power-law kernel to describe the real material with the power-law phenomena by using the general fractional-order Maxwell and Voigt models.

The structure of the present chapter is suggested as follows. In Sect. 2, we introduce the FC and GFC operators with the power-law kernel. In Sect. 3, we investigate the recent applications of the GFDs to the general fractional-order viscoelasticity in the real materials. Finally, the conclusion is given in Sect. 4.

2 Mathematical Tools

In order to discuss the GFC, we introduce the special functions and the FC operator of the Riemann–Liouville and Liouville-Caputo types in this section. Meanwhile, we present the recent results on the GFC operators in the kernels of the special functions. Finally, the Laplace transforms of the FC and GFC operators are considered in detail (see [38–51]).

2.1 The Special Functions with Power Law

Let $\mathbb{C}$, $\mathbb{R}$, $\mathbb{R}_0^+$, $\mathbb{N}$ and $\mathbb{N}_0$ be the sets of complex numbers, real numbers, non-negative real numbers, positive integers and $\mathbb{N}_0 = \{0\} \cup \mathbb{N}$, respectively.

The Mittag–Leffler function, introduced by Swedish mathematician Gosta Mittag–Leffler in 1903, is defined as [47]:

$$E_\nu(\eta) = \sum_{\kappa=0}^{\infty} \frac{\eta^\kappa}{\Gamma(\kappa\nu + 1)}, \tag{1}$$

where $\eta, \nu \in \mathbb{C}, \Re(\nu) \in \mathbb{R}_0^+, \kappa \in \mathbb{N}$, and $\Gamma(\cdot)$ is the familiar Gamma function [3].

As first extension of the Mittag–Leffler function, the extended Mittag–Leffler function, structured by Wiman in 1905, is defined as [48]:

$$E_{\nu,\upsilon}(\eta) = \sum_{\kappa=0}^{\infty} \frac{\eta^{\kappa}}{\Gamma(\kappa\nu + \upsilon)}, \tag{2}$$

where $\eta, \nu, \upsilon \in \mathbb{C}, \Re(\nu), \Re(\upsilon) \in \mathbb{R}_0^+$, and $\kappa \in \mathbb{N}$.

As further extension of the Mittag–Leffler function, the extended Mittag–Leffler function, introduced by Prabhakar in 1971, is given as [49]:

$$E_{\nu,\upsilon}^{\phi}(\eta) = \sum_{\kappa=0}^{\infty} \frac{(\phi)_{\kappa}}{\Gamma(\kappa\nu + \upsilon)} \frac{\eta^{\kappa}}{\Gamma(\kappa + 1)}, \tag{3}$$

where $\eta, \nu, \upsilon, \phi \in \mathbb{C}, \Re(\nu), \Re(\upsilon), \Re(\phi) \in \mathbb{R}_0^+, \kappa \in \mathbb{N}$, and the familiar Pochhammer symbol is expressed as [50]:

$$(\phi)_{\kappa} = \begin{cases} 1, \ \kappa = 0, \\ \dfrac{\Gamma(\phi + \kappa)}{\Gamma(\phi)}, \kappa \in \mathbb{N}. \end{cases} \tag{4}$$

For $\lambda \in \mathbb{C}$, the Laplace transforms of the functions with power law are given as [38, 40, 49]:

$$\mathrm{L}\left[\frac{t^{-\nu}}{\Gamma(1-\nu)}\right] = s^{\nu}, \tag{5}$$

$$\mathrm{L}\left[\frac{t^{\nu}}{\Gamma(1+\nu)}\right] = s^{-\nu}, \tag{6}$$

$$\mathrm{L}\left[t^{\upsilon-1} E_{\nu,\upsilon}^{\phi}(\lambda t^{\nu})\right] = \frac{1}{s^{\upsilon}(1 - \lambda s^{-\nu})^{\varphi}}, \tag{7}$$

$$\mathrm{L}\left[t^{\upsilon-1} E_{\nu,\upsilon}(\lambda t^{\nu})\right] = \frac{1}{s^{\upsilon}(1 - \lambda s^{-\nu})}, \tag{8}$$

$$\mathrm{L}\left[E_{\nu}(\lambda t^{\nu})\right] = \frac{1}{1 - \lambda s^{-\nu}}, \tag{9a}$$

$$\mathrm{L}[\delta(t)] = 1, \tag{9b}$$

where $\delta(t)$ is the Dirac delta (see [38]) and the Laplace transform is defined as [38]:

$$L[\Phi(t)] = \Phi(s) := \int_0^\infty e^{-st}\Phi(t)dt. \tag{9c}$$

2.2 GFC in a Kernel Function

The GFD of the Riemann–Liouville type is defined as [38, 40, 41, 45, 46]:

$$\left(D^{RL}_{(\Xi)}\Theta\right)(\tau) = \frac{d}{d\theta}\int_a^\tau \Xi(\tau - t)\Theta(t)dt \left(\tau \in \mathbb{R}_0^+\right), \tag{10}$$

where $\Theta \in AC\left(\mathbb{R}_0^+\right)$, and $\Xi(\tau)$ is the kernel function.

The GFD of the Liouville-Caputo type is defined as [38, 40, 41, 45, 46]:

$$\left(D^{LC}_{(\Xi)}\Theta\right)(\theta) = \int_0^\tau \Xi(\tau - t)\Theta^{(1)}(t)dt \left(\tau \in \mathbb{R}_0^+\right), \tag{11}$$

where $\Theta^{(1)}(\tau) = d\Theta(\tau)/d\tau$, $\Theta^{(1)} \in L_1^{loc}\left(\mathbb{R}_0^+\right)$, and $\Xi(\tau)$ is the kernel function.

The relationship between Eqs. (11) and (10) is given as [40, 41]:

$$\left(D^{C}_{(\Xi)}\Theta\right)(\tau) = \left(D^{RL}_{(\Xi)}\Theta\right)(\tau) - \Xi(\tau)\Theta(0). \tag{12}$$

2.3 FC Within the Singular Power-Law Kernel

With the use of the kernel $\Xi(\tau) = \tau^{-\nu}/\Gamma(1-\nu)$, the Riemann–Liouville FD of the function $\Theta(\tau)$ of order $(0 < \nu < 1)$ is given by [1, 2, 4, 5]:

$$\left({}_0^{RL}D_\tau^{(\nu)}\Theta\right)(\tau) = \frac{1}{\Gamma(1-\nu)}\frac{d}{d\tau}\int_0^\tau \frac{\Theta(t)}{(\tau - t)^\nu}dt \ (\tau > 0). \tag{13}$$

where $\Theta \in AC\left(\mathbb{R}_0^+\right)$, and the Liouville-Caputo FD of the function $\Theta(\tau)$ by [1, 2, 4–9]

$$\left({}_0^C\mathrm{D}_\tau^{(\nu)}\Theta\right)(\tau) = \frac{1}{\Gamma(1-\nu)}\int\limits_0^t \frac{\Theta^{(1)}(t)}{(\tau-t)^\nu}dt \ (\tau > 0), \tag{14}$$

where $\Theta^{(1)}(\tau) = d\Theta(\tau)/d\tau$ and $\Theta^{(1)} \in L_1^{loc}(\mathbb{R}_0^+)$.

The relationship between Eqs. (13) and (14) is given as [2]:

$$\left({}_0^{RL}\mathrm{D}_\tau^{(\nu)}\Theta\right)(\tau) = \left({}_0^C\mathrm{D}_\tau^{(\nu)}\Theta\right)(\tau) + \frac{\tau^{-\nu}}{\Gamma(1-\nu)}\Theta(0). \tag{15}$$

Suppose that $\mathbb{N}$ is the set of positive integers, $m \in \mathbb{N}$ and $m-1 < \nu < m$. Equations (13) and (14) yield [2, 4, 7, 8]:

$$\left({}_0^{RL}\mathrm{D}_\tau^{(\nu)}\Theta\right)(\tau) = \frac{1}{\Gamma(m-\nu)}\frac{d^m}{d\tau^m}\int\limits_0^\tau \frac{\Theta(t)}{(\tau-t)^{m-\nu-1}}dt \ (\tau > 0), \tag{16}$$

$$\left({}_0^{RL}\mathrm{D}_\tau^{(\nu)}\Theta\right)(\tau) = \frac{1}{\Gamma(m-\nu)}\int\limits_0^\tau \frac{\Theta^{(m)}(t)}{(\tau-t)^{m-\nu-1}}dt \ (\tau > 0), \tag{17}$$

respectively.

The Laplace transforms of the FC operators in the nonsingular power-law kernel are given as [2]:

$$\mathrm{L}\left[\left({}_0^{RLT}\mathrm{D}_\tau^{(\nu)}\Theta\right)(\tau)\right] = s^\nu\Theta(s), \tag{18a}$$

$$\left({}_0\mathrm{I}_t^{(\nu)}\Omega\right)(\tau) = \frac{1}{\Gamma(\nu)}\int\limits_0^\tau \frac{\Omega(t)}{(\tau-t)^{1-\nu}}dt. \tag{18b}$$

The inverse operator (the Riemann–Liouville fractional integral) is given as [2]:

$$\left({}_0\mathrm{I}_t^{(\nu)}\Omega\right)(\tau) = \frac{1}{\Gamma(\nu)}\int\limits_0^\tau \frac{\Omega(t)}{(\tau-t)^{1-\nu}}dt. \tag{19}$$

The Laplace transforms of the GFC operators in the nonsingular power-law kernel are given as [2]:

$$\left({}_0^{RL}\mathrm{D}_\tau^{(\nu)}\Theta\right)(t) = s^\nu\Theta(s),$$

$$\left({}_0^{LC}\mathrm{D}_\tau^{(\nu)}\Omega\right)(t) = s^{\nu-1}(s\Omega(s) - \Omega(0)),$$

where $\Theta(s)$ is the Laplace transform of the function $\Theta(s)$.

The properties of the GFD in the nonsingular power-law kernel are given as [2]:

$$\left({}_{a}^{RL}\mathrm{D}_{\tau}^{(\nu)}(\Theta_1+\Theta_2)\right)(\tau)=\left({}_{a}^{RL}\mathrm{D}_{\tau}^{(\nu)}\Theta_1\right)(\tau)+\left({}_{a}^{RL}\mathrm{D}_{\tau}^{(\nu)}\Theta_2\right)(\tau),$$

$${}_{a}^{RL}\mathrm{D}_{\tau}^{(\nu)}1 = \frac{(\tau-a)^{-\nu}}{\Gamma(1-\nu)},$$

$${}_{a}^{LC}\mathrm{D}_{\tau}^{(\nu)}1=0,$$

$$\left({}_{a}^{LC}\mathrm{D}_{\tau}^{(\nu)}(\Theta_1+\Theta_2)\right)(\tau)=\left({}_{a}^{LC}\mathrm{D}_{\tau}^{(\nu)}\Theta_1\right)(\tau)+\left({}_{a}^{LC}\mathrm{D}_{\tau}^{(\nu)}\Theta_2\right)(\tau).$$

Remark 1 Liouville derived the fractional derivative formula (see [4]).

$${}_{0}^{C}\mathrm{D}_{\infty}^{(\nu)}\Theta(\tau)=\frac{1}{(-1)^{\nu}\Gamma(\nu)}\int_{0}^{\infty}\Theta^{(m)}(\tau+t)t^{\nu-1}dt,$$

and the formula (see [4])

$$h\int_{0}^{\tau}(\tau-t)^{-\frac{1}{2}}\Theta^{(1)}(t)dt=m(\tau),$$

where $h=1/\sqrt{2\mathrm{g}}$ is the constant, though not quite rigorously from the modern point of view.

So nine introduced the following fractional derivative given as (see [6])

$${}_{a}^{C}\mathrm{D}_{\tau}^{(\nu)}\Theta(\tau)=\frac{1}{\Gamma(p-\nu+1)}\int_{a}^{\tau}(\tau-t)^{p-\nu}\Theta^{(1)}(t)dt,\ \mathrm{Re}(n)<\nu<\mathrm{Re}(n+1).$$

Caputo and Smit and De Vries introduced the fractional derivative in the form (see [7, 8])

$${}_{a}^{C}\mathrm{D}_{\tau}^{(\nu)}\Theta(\theta)=\frac{1}{\Gamma(n-\nu)}\int_{a}^{\tau}\frac{1}{(\tau-t)^{\nu}}\Theta^{(n)}(t)dt.$$

In 1968, Dzhrbashyan and Nersesyan introduced the fractional derivative (see[9])

$${}_{0}^{C}\mathrm{D}_{\infty}^{(\nu)}\Theta(\theta)=\frac{1}{\Gamma(n-\nu)}\int_{0}^{\infty}\frac{1}{(\tau-t)^{n-\nu}}\Theta^{(n)}(t)dt.$$

Theorem 1 (see [49]).

Let $\tau \in \mathbb{R}_0^+$, $\nu \in (0,1)$, $\Xi \in L(\mathbb{R}_0^+)$ and $\Omega^{(1)} \in L_1^{loc}(\mathbb{R}_0^+)$. Then, there is an Abel integral of the form

$$\frac{1}{\Gamma(\nu)} \int_0^\tau \frac{\Xi(t)}{(\tau - t)^{1-\nu}} dt = \Omega(\tau), \tag{20a}$$

with the solution given as

$$\Xi(t) = \frac{1}{\Gamma(1+\nu)} \int_0^\tau (\tau - t)^\nu \Omega^{(1)}(t) dt + \frac{\tau^\nu}{\Gamma(1+\nu)} \Omega(0), \tag{20b}$$

where $\Omega(\tau = 0) = \Omega(0)$.

2.4 GFC with the Nonsingular Power-Law Kernel

When the kernel in Eq. (1) is given as $\Xi(\tau) = \tau^\nu / \Gamma(1+\nu)$, the Riemann–Liouville-type GFD of the function $\Theta(\tau)$ of order $(0 < \nu < 1)$ in the nonsingular power-law kernel is defined as [21, 38]

$$\left({}_0^{RLT}\mathrm{D}_\tau^{(\nu)}\Theta\right)(\tau) = \frac{1}{\Gamma(1+\nu)} \frac{d}{d\theta} \int_0^\tau (\tau - t)^\nu \Theta(t) dt \ (\tau > 0), \tag{21}$$

where $\Theta \in AC(\mathbb{R}_0^+)$, and the Liouville-Caputo-type GFD of the function $\Theta(\tau)$ of order $(0 < \nu < 1)$ in the nonsingular power-law kernel as [21, 38]

$$\left({}_0^{CT}\mathrm{D}_\tau^{(\nu)}\Theta\right)(\tau) = \frac{1}{\Gamma(1+\nu)} \int_0^\tau (\tau - t)^\nu \Theta^{(1)}(t) dt \ (\tau > 0), \tag{22}$$

where $\Theta^{(1)}(\tau) = d\Theta(\tau)/d\tau$ and $\Theta(1) \in L_1^{loc}(\mathbb{R}_0^+)$.

The relationship between Eq. (21) and Eq. (22) is presented as [21, 38]:

$$\left({}_0^{RLT}\mathrm{D}_\tau^{(\nu)}\Theta\right)(\tau) = \left({}_0^{CT}\mathrm{D}_\tau^{(\nu)}\Theta\right)(\tau) + \frac{\theta^\nu \Theta(0)}{\Gamma(1+\nu)}. \tag{23}$$

Similarly, for $m - 1 < \nu < m$, Eqs. (23) and (24) yield:

$$\left({}_{0}^{RLT}\mathrm{D}_{\tau}^{(\nu)}\Theta\right)(\tau)=\frac{1}{\Gamma(m+\nu)}\frac{d^m}{d\tau^m}\int\limits_0^{\tau}(\tau-t)^{m-\nu-1}\Theta(t)dt\ (\tau>0), \tag{24}$$

$$\left({}_{0}^{CT}\mathrm{D}_{\tau}^{(\nu)}\Theta\right)(\tau)=\frac{1}{\Gamma(m+\nu)}\int\limits_0^{\tau}(\tau-t)^{m-\nu-1}\Theta^{(m)}(t)dt\ (\tau>0). \tag{25}$$

The Laplace transforms of Eqs. (21) and (22) are presented as follows [21, 38]:

$$\mathrm{L}\left[\left({}_{0}^{RLT}\mathrm{D}_{\tau}^{(\nu)}\Theta\right)(\tau)\right]=\frac{1}{s^{\nu}}\Theta(s), \tag{26}$$

$$\mathrm{L}\left[\left({}_{0}^{CT}\mathrm{D}_{\tau}^{(\nu)}\Theta\right)(\tau)\right]=\frac{1}{s^{1+\nu}}(s\Theta(s)-\Theta(0)). \tag{27}$$

Its inverse operator (the general fractional integral) is defined as [21, 38]:

$$\left({}_{a}^{L}\mathrm{I}_{\tau}^{(\nu)}\Omega\right)(\tau)=\frac{1}{\Gamma(-\nu)}\int\limits_a^{\tau}\frac{1}{(\tau-t)^{1+\nu}}\Omega(t)dt. \tag{28}$$

The Laplace transforms of the GFC operators in the nonsingular power-law kernel are given as [21, 38]:

$$\left({}_{0}^{RL}\mathrm{D}_{\tau}^{(\nu)}\Theta\right)(t)=s^{\nu}\Theta(s),$$

$$\left({}_{0}^{LC}\mathrm{D}_{\tau}^{(\nu)}\Omega\right)(t)=s^{\nu-1}(s\Omega(s)-\Omega(0)).$$

The properties of the GFD in the nonsingular power-law kernel are given as [21, 38]:

$$\left({}_{a}^{RL}\mathrm{D}_{\tau}^{(\nu)}(\Theta_1+\Theta_2)\right)(\tau)=\left({}_{a}^{RL}\mathrm{D}_{\tau}^{(\nu)}\Theta_1\right)(\tau)+\left({}_{a}^{RL}\mathrm{D}_{\tau}^{(\nu)}\Theta_2\right)(\tau),$$

$${}_{a}^{RL}\mathrm{D}_{\tau}^{(\nu)}1=\frac{(\tau-a)^{-\nu}}{\Gamma(1-\nu)},$$

$${}_{a}^{LC}\mathrm{D}_{\tau}^{(\nu)}1=0,$$

$$\left({}_{a}^{LC}\mathrm{D}_{\tau}^{(\nu)}(\Theta_1+\Theta_2)\right)(\tau)=\left({}_{a}^{LC}\mathrm{D}_{\tau}^{(\nu)}\Theta_1\right)(\tau)+\left({}_{a}^{LC}\mathrm{D}_{\tau}^{(\nu)}\Theta_2\right)(\tau).$$

Theorem 2 (see [21, 38]).

Let $\tau\in\mathbb{R}_0^+$, $\nu\in(0,1)$, $\Xi\in L\left(\mathbb{R}_0^+\right)$ and $\Omega(1)\in L_1^{loc}\left(\mathbb{R}_0^+\right)$. Then, there is an Abel type integral

$$\frac{1}{\Gamma(-\nu)}\int_{0}^{\tau}\frac{\Xi(t)}{(\tau-t)^{1+\nu}}dt=\Omega(\tau), \tag{29a}$$

with the solution given as

$$\Xi(t)=\frac{1}{\Gamma(1-\nu)}\int_{0}^{\tau}(\tau-t)^{-\nu}\Omega^{(1)}(t)dt+\frac{\tau^{-\nu}}{\Gamma(1-\nu)}\Omega(0), \tag{29b}$$

where $\Omega(\tau=0)=\Omega(0)$.

2.5 GFC with the Nonsingular Mittag–Leffler Function Kernel

When the kernel in Eq. (1) is given as: $\Xi(\tau)=E_{\nu}(-\tau^{\nu})$, the GFD of Riemann–Liouville type in the kernel of the Mittag–Leffler function is defined by [21, 38, 41]:

$$\left({}_{a}^{RLT}\mathrm{D}_{\tau}^{(\nu)}\Theta\right)(\tau)=\frac{d}{d\tau}\int_{a}^{\tau}E_{\nu}(-(\tau-t)^{\nu})\Theta(t)dt\ (\tau>a), \tag{30}$$

where $\Theta\in AC\left(\mathbb{R}_{0}^{+}\right)$, and the GFD of the Liouville-Caputo type in the kernel of the Mittag–Leffler function by [21, 38, 41]:

$$\left({}_{a}^{CT}\mathrm{D}_{\tau}^{(\nu)}\Theta\right)(\tau)=\int_{a}^{\tau}E_{\nu}(-(\tau-t)^{\nu})\Theta^{(1)}(t)dt\ (\tau>a), \tag{31}$$

where $\Theta^{(1)}(\tau)=d\Theta(\tau)/d\tau$ and $\Theta^{(1)}\in L_{1}^{loc}\left(\mathbb{R}_{0}^{+}\right)$.

The relationship between Eqs. (30) and (31) becomes [21, 38, 41]:

$$\left({}_{0}^{CT}\mathrm{D}_{\tau}^{(\nu)}\Theta\right)(\tau)=\left({}_{0}^{RLT}\mathrm{D}_{\tau}^{(\nu)}\Theta\right)(\tau)-E_{\nu}(\tau^{\nu})\Theta(0). \tag{32}$$

Similarly, for $m-1<\nu<m$, Eqs. (13) and (14) yield:

$$\left({}_{a}^{RLT}\mathrm{D}_{\tau}^{(\nu)}\Theta\right)(\tau)=\frac{d^{m}}{d\tau^{m}}\int_{a}^{\tau}E_{\nu}(-(\tau-t)^{\nu})\Theta(t)dt\ (\tau>0), \tag{33}$$

$$\left({}_{a}^{CT}\mathrm{D}_{\tau}^{(\nu)}\Theta\right)(\tau) = \int_{a}^{\tau} E_{\nu}(-(\tau-t)^{\nu})\Theta^{(m)}(t)dt \ (\tau > 0). \tag{34}$$

Its inverse operator (the general fractional integral) is defined as

$$\left({}_{0}\mathrm{I}_{t}^{(\nu)}\Omega\right)(\tau) = \Omega(\tau) - \frac{1}{\Gamma(\nu)}\int_{0}^{\tau}\frac{\Omega(t)}{(\tau-t)^{1-\nu}}dt$$

Remark 2 Hille and Tamarlcin proposed the Abel type integral equation of the second kind (see [50]).

$$\Omega(\tau) - \frac{\tau}{\Gamma(\nu)}\int_{0}^{x}\frac{\Omega(t)}{(\tau-t)^{1-\nu}}dt = \Theta(\tau), 0 < \alpha < 1,$$

with the solution given as

$$\Omega(\tau) = \frac{d}{d\tau}\int_{0}^{\tau} E_{\nu}\left[\lambda(\tau-t)^{\nu}\right]\Theta(t)dt.$$

Hille introduced the following fractional differential operator (see [51])

$$\left({}_{a}^{RLT}\mathrm{D}_{\tau}^{(\nu)}\Theta\right)(\tau) = \lambda\frac{d}{d\tau}\int_{0}^{\tau} E_{\nu}\left[\lambda(t-x)^{\nu}\right]f(t)dt.$$

Atangana and Baleanu introduced the general fractional derivative with the Mittag–Leffler function involving the normalization parameter (see [43])

$$\left({}_{a}^{CT}\mathrm{D}_{\tau}^{(\nu)}\Theta\right)(\tau) = \frac{\Im(\nu)}{1-\nu}\int_{a}^{\tau}\mathrm{E}_{\nu}\left(-\frac{\nu}{1-\nu}(\tau-t)^{\nu}\right)\Theta^{(1)}(t)dt,$$

where $\Im(\nu)$ is the normalization parameter.

The Laplace transforms of the GFC operators in the nonsingular Mittag–Leffler kernel are given as [21, 38, 41]:

$$\left({}_{0}^{RL}\mathrm{D}_{\tau}^{(\nu)}\Theta\right)(t) = \left(1-s^{-\nu}\right)^{-1}\Theta(s),$$

$$\left({}_{0}^{LC}\mathrm{D}_{\tau}^{(\nu)}\Omega\right)(t) = s^{-1}\left(1-s^{-\nu}\right)^{-1}(s\Omega(s)-\Omega(0)).$$

The properties of the GFD in the nonsingular Mittag–Leffler kernel are given as [21, 38, 41]:,

$$\left({}_{\alpha}^{RL}\mathrm{D}_{\tau}^{(v)}(\Theta_1+\Theta_2)\right)(\tau)=\left({}_{\alpha}^{RL}\mathrm{D}_{\tau}^{(v)}\Theta_1\right)(\tau)+\left({}_{\alpha}^{RL}\mathrm{D}_{\tau}^{(v)}\Theta_2\right)(\tau),$$

$${}_{a}^{RL}\mathrm{D}_{\tau}^{(v)}1=E_{\nu}(-\tau^{\nu}),$$

$${}_{\alpha}^{LC}D_{\tau}^{(v)}1=0,$$

$$\left({}_{\alpha}^{LC}\mathrm{D}_{\tau}^{(v)}(\Theta_1+\Theta_2)\right)(\tau)=\left({}_{\alpha}^{LC}\mathrm{D}_{\tau}^{(v)}\Theta_1\right)(\tau)+\left({}_{\alpha}^{LC}\mathrm{D}_{\tau}^{(v)}\Theta_2\right)(\tau).$$

2.6 GFC with the Nonsingular Wiman Kernel

When the kernel in Eq. (1) is given as: $\Xi(\tau)=\tau^{\upsilon-1}E_{\nu,\upsilon}(-\tau^{\nu})$, the GFD of Riemann–Liouville type in the kernel of the Wiman function is defined by [21, 38, 41]:

$$\left({}_{a}^{RLT}\mathrm{D}_{\tau}^{(v)}\Theta\right)(\tau)=\frac{d}{d\tau}\int_{a}^{\tau}(\tau-t)^{\upsilon-1}E_{\nu,\upsilon}(-(\tau-t)^{\nu})\Omega(t)dt\ (\tau>a), \tag{35}$$

where $\Theta\in AC\left(\mathbb{R}_0^{+}\right)$, and the GFD of the Liouville-Caputo type in the kernel of the Wiman function by:

$$\left({}_{E_{\nu,\upsilon}(-)}^{C}\mathrm{D}_{a}^{(v)}\Omega\right)(\tau)=\int_{a}^{\tau}(\tau-t)^{\upsilon-1}E_{\nu,\upsilon}(-(\tau-t)^{\nu})\Omega^{(1)}(t)dt\ (\tau>a), \tag{36}$$

where $\Theta^{(1)}(\tau)=d\Theta(\tau)/d\tau$ and $\Theta^{(1)}\in L_1^{loc}\left(\mathbb{R}_0^{+}\right)$.

The relationship between Eqs. (35) and (36) is [21, 38, 41]:

$$\left({}_{a}^{CT}\mathrm{D}_{\tau}^{(v)}\Theta\right)(\tau)=\left({}_{a}^{RLT}\mathrm{D}_{\tau}^{(v)}\Theta\right)(\tau)-\tau^{\upsilon-1}E_{\nu,\upsilon}(-\tau^{\nu})\Omega(0). \tag{37}$$

Similarly, for $m-1<\nu<m$, Eqs. (13) and (14) yield:

$$\left({}_{a}^{RLT}\mathrm{D}_{\tau}^{(v)}\Theta\right)(\tau)=\frac{d^m}{d\tau^m}\int_{a}^{\tau}(\tau-t)^{\upsilon-1}E_{\nu,\upsilon}(-(\tau-t)^{\nu})\Omega(t)dt\ (\tau>0), \tag{38}$$

$$\left({}_{a}^{CT}\mathrm{D}_{\tau}^{(\nu)}\Theta\right)(\tau) = \int_{a}^{\tau} (\tau - t)^{\upsilon-1} E_{\nu,\upsilon}(-(\tau - t)^{\nu})\Theta^{(m)}(t)dt \ (\tau > 0). \quad (39)$$

Its inverse operator (the general fractional integral) is defined as [21, 38, 41]

$$\left({}_{0}\mathrm{I}_{t}^{(\nu)}\Omega\right)(\tau) = \int_{0}^{\tau} (\tau - t)^{-\upsilon} E_{\nu,1-\upsilon}^{-1}(-(\tau - t)^{\nu})\Omega(t)dt$$

The Laplace transforms of the GFC operators in the nonsingular Wiman kernel are given as [21, 38, 41]:

$$\left({}_{0}^{RL}\mathrm{D}_{\tau}^{(\upsilon)}\Theta\right)(t) = s^{1-\upsilon}\left(1 + s^{-\nu}\right)^{-1}\Theta(s),$$

$$\left({}_{0}^{LC}\mathrm{D}_{\tau}^{(\upsilon)}\Omega\right)(t) = s^{-\upsilon}\left(1 + s^{-\nu}\right)^{-1}(s\Omega(s) - \Omega(0)).$$

The properties of the GFD in the nonsingular Wiman kernel are given as [21, 38, 41]:

$$\left({}_{\alpha}^{LC}D_{\tau}^{(\nu)}(\Theta_1 + \Theta_2)\right)(\tau) = \left({}_{\alpha}^{LC}D_{\tau}^{(\nu)}\Theta_1\right)(\tau) + \left({}_{\alpha}^{LC}D_{\tau}^{(\nu)}\Theta_2\right)(\tau).$$

$${}_{a}^{RL}\mathrm{D}_{\tau}^{(\upsilon)}1 = \tau^{\upsilon-1}E_{\nu,\upsilon}(-\tau^{\nu}),$$

$${}_{\alpha}^{LC}D_{\tau}^{(\nu)}1 = 0,$$

$$\left({}_{\alpha}^{LC}D_{\tau}^{(\nu)}(\Theta_1 + \Theta_2)\right)(\tau) = \left({}_{\alpha}^{LC}D_{\tau}^{(\nu)}\Theta_1\right)(\tau) + \left({}_{\alpha}^{LC}D_{\tau}^{(\nu)}\Theta_2\right)(\tau).$$

2.7 GFC with the Nonsingular Prabhakar Kernel

When the kernel in Eq. (1) is given as: $\Xi(\tau) = \tau^{\upsilon-1}E_{\nu,\upsilon}^{\phi}(-\tau^{\nu})$, the GFD of Riemann–Liouville type in the kernel of the Prabhakar function is defined as [21, 38, 41]:

$$\left({}_{a}^{RLT}\mathrm{D}_{\tau}^{(\upsilon)}\Theta\right)(\tau) = \frac{d}{d\tau}\int_{a}^{\tau} (\tau - t)^{\upsilon-1} E_{\nu,\upsilon}^{\phi}(-(\tau - t)^{\nu})\Omega(t)dt \ (\tau > a), \quad (40)$$

where $\Theta \in AC\left(\mathbb{R}_0^+\right)$, and the GFD of the Liouville-Caputo type in the kernel of the Prabhakar function as [21, 38, 41]:

$$\left({}^{C}_{E_{\nu,\upsilon}(-)}\mathrm{D}^{(\nu)}_{a}\Omega\right)(\tau) = \int\limits_{a}^{\tau} (\tau - t)^{\upsilon-1} E^{\phi}_{\nu,\upsilon}(-(\tau - t)^{\nu})\Omega^{(1)}(t)dt \ (\tau > a), \tag{41}$$

where $\Theta^{(1)}(\tau) = d\Theta(\tau)/d\tau$ and $\Theta^{(1)} \in L_1^{loc}(\mathbb{R}_0^+)$.

The relationship between Eqs. (40) and (41) is [21, 38, 41]:

$$\left({}^{CT}_{a}\mathrm{D}^{(\nu)}_{\tau}\Theta\right)(\tau) = \left({}^{RLT}_{a}\mathrm{D}^{(\nu)}_{\tau}\Theta\right)(\tau) - \tau^{\upsilon-1}E^{\phi}_{\nu,\upsilon}(-\tau^{\nu})\Omega(0). \tag{42}$$

Similarly, for $m - 1 < \nu < m$, Eqs. (13) and (14) yield:

$$\left({}^{RLT}_{a}\mathrm{D}^{(\nu)}_{\tau}\Theta\right)(\tau) = \frac{d^m}{d\tau^m}\int\limits_{a}^{\tau} (\tau - t)^{\upsilon-1} E^{\phi}_{\nu,\upsilon}(-(\tau - t)^{\nu})\Omega(t)dt \ (\tau > 0), \tag{43}$$

$$\left({}^{CT}_{a}\mathrm{D}^{(\nu)}_{\tau}\Theta\right)(\tau) = \int\limits_{a}^{\tau} (\tau - t)^{\upsilon-1} E^{\phi}_{\nu,\upsilon}(-(\tau - t)^{\nu})\Theta^{(m)}(t)dt \ (\tau > 0). \tag{44}$$

Its inverse operator (the general fractional integral) is defined as [[21, 38, 41]]

$$\left({}_{0}\mathrm{I}^{(\nu)}_{t}\Omega\right)(\tau) = \int\limits_{0}^{\tau} (\tau - t)^{-\upsilon} E^{-\phi}_{\nu,1-\upsilon}(-(\tau - t)^{\nu})\Omega(t)dt$$

Remark 3 Kilbas, Saigo and Saxena introduced the following general fractional derivative (see [42]).

$$\left({}^{RLT}_{a}\mathrm{D}^{(\nu)}_{\tau}\Theta\right)(\tau) = \frac{d^m}{d\tau^m}\int\limits_{a}^{\tau} (\tau - t)^{\mu+m-\beta-1} E^{\phi}_{\nu,\mu+m-\beta}(\lambda(\tau - t)^{\nu})\Theta(t)dt,$$

which is called the Kilbas-Saigo-Saxena GFD.

The Laplace transforms of the GFC operators in the nonsingular Prabhakar kernel are given as [21, 38, 41]:

$$\left({}^{RL}_{0}\mathrm{D}^{(\nu)}_{\tau}\Theta\right)(t) = s^{1-\upsilon}\left(1 + s^{-\nu}\right)^{-\phi}\Theta(s),$$

$$\left({}^{LC}_{0}\mathrm{D}^{(\nu)}_{\tau}\Omega\right)(t) = s^{-\upsilon}\left(1 + s^{-\nu}\right)^{-\phi}(s\Omega(s) - \Omega(0)).$$

The properties of the GFD in the nonsingular Prabhakar kernel are given as [21, 38, 41]:

$$\left({}_{\alpha}^{RL}\mathrm{D}_{\tau}^{(\nu)}(\Theta_1+\Theta_2)\right)(\tau)=\left({}_{\alpha}^{RL}\mathrm{D}_{\tau}^{(\nu)}\Theta_1\right)(\tau)+\left({}_{\alpha}^{RL}\mathrm{D}_{\tau}^{(\nu)}\Theta_2\right)(\tau),$$

$$ {}_{a}^{RL}\mathrm{D}_{\tau}^{(\nu)}1=\tau^{\upsilon-1}E_{\nu,\upsilon}^{\phi}(-\tau^{\nu}),$$

$$\Big|_{\alpha}^{LC}D_{\tau}^{(\nu)}1=0,$$

$$\left({}_{\alpha}^{LC}\mathrm{D}_{\tau}^{(\nu)}(\Theta_1+\Theta_2)\right)(\tau)=\left({}_{\alpha}^{LC}\mathrm{D}_{\tau}^{(\nu)}\Theta_1\right)(\tau)+\left({}_{\alpha}^{LC}\mathrm{D}_{\tau}^{(\nu)}\Theta_2\right)(\tau).$$

3 The Rheological Models with GFCs Involving the Nonsingular Kernels

3.1 *Complex Phenomena in Viscoelasticity*

The stress–strain-time relation with the positive-parametric Mittag–Leffler function can be written as

$$\sigma_{\nu}(\tau)=\mathrm{M}E_{\nu}(-\tau^{\nu})\varepsilon_{\nu}(0). \tag{45}$$

where $\sigma_{\nu}(\tau)$ is stress, $\varepsilon_{\nu}(0)$ is the initial strain, τ is time and M is the material constant.

There are

$$E_{\nu}(-\tau^{\nu})\propto\tau^{-\nu}, \tag{46}$$

$$E_{\nu,1}(-\tau^{\nu})\propto\tau^{-\nu}, \tag{47}$$

$$E_{\nu,1}^{1}(-\tau^{\nu})\propto\tau^{-\nu}, \tag{48}$$

which, after taking the Laplace transform, leads to

$$\sigma_{\nu}(s)=\mathrm{M}\varepsilon_{\nu}(0)s^{-1}\left(1+s^{-\nu}\right)^{-1}\propto\mathrm{M}\varepsilon_{\nu}(0)\Gamma(1-\nu)s^{\nu}. \tag{49}$$

The phenomena in rheological behaviors are called as the Nutting behaviors in the real materials.

In another hand, there may exist the stress–strain-time relation with the positive-parametric Mittag–Leffler function can be written as

$$\sigma_{\nu}(\tau)=\mathrm{M}E_{\nu}\left(-\tau^{-\nu}\right)\varepsilon_{\nu}(0). \tag{50}$$

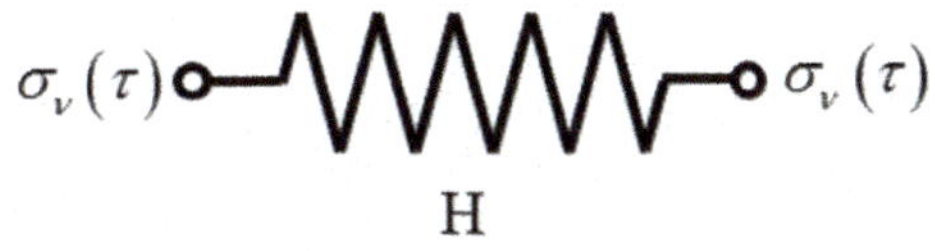

Fig. 1 The spring element

where $\sigma_\nu(\tau)$ is stress, $\varepsilon_\nu(0)$ is the initial strain, τ is time and M is the material constant.

There are

$$E_\nu\left(-\tau^{-\nu}\right) \propto \tau^\nu, \tag{51}$$

$$E_{\nu,1}\left(-\tau^{-\nu}\right) \propto \tau^\nu, \tag{52}$$

$$E_{\nu,1}^1\left(-\tau^{-\nu}\right) \propto \tau^\nu, \tag{53}$$

The phenomena in rheological behaviors are called as the anomalous Nutting behaviors in the real materials.

3.2 *The Viscoelastic Elements with GFDs*

3.2.1 The Spring Element

Model 1

As shown in Fig. 1, the spring element follows the Hooke's law given as [20, 21]

$$\sigma_\nu(\tau) = \mathrm{H}\varepsilon_\nu(\tau), \tag{54}$$

where H is the Young's modulus of the material.

3.2.2 The Viscoelastic Elements

As shown in Fig. 2, the viscoelastic elements with the FD and GFDs were presented to describe the viscoelastic behaviors in the real materials.

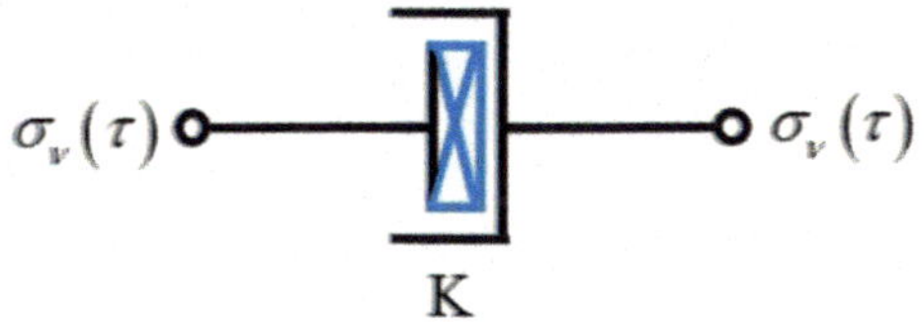

Fig. 2 The viscoelastic element

Model 2

The viscoelastic element with the FD in the singular power-law kernel is given as [20, 21]:

$$\sigma_{\nu}(\tau) = \mathrm{K}\left({}_{0}^{LC}\mathrm{D}_{t}^{(\nu)}\varepsilon_{\nu}\right)(\tau). \tag{55}$$

where K is the coefficient of viscosity.

Model 3

The viscoelastic element with the GFD in the singular power-law kernel is represented in the form:

$$\sigma_{\nu}(\tau) = \mathrm{K}\left({}_{0}^{LC}\mathrm{D}_{t}^{(\nu)}\varepsilon_{\nu}\right)(\tau). \tag{56}$$

where K is the coefficient of viscosity.

Model 4

The viscoelastic element with the GFD in the kernel of the Mittag–Leffler function is can be expressed as:

$$\sigma_{\nu}(\tau) = \mathrm{K}\left({}_{0}^{LC}\mathrm{D}_{t}^{(\nu)}\varepsilon_{\nu}\right)(\tau). \tag{57}$$

where K is the coefficient of viscosity.

Model 5

The viscoelastic element with the GFD in the kernel of the Wiman functions is represented as:

$$\sigma_{\nu}(\tau) = \mathrm{K}\left({}_{0}^{LC}\mathrm{D}_{t}^{(\nu)}\varepsilon_{\nu}\right)(\tau). \tag{58}$$

where K is the coefficient of viscosity.

Model 6

The viscoelastic element with the GFD in the kernel of the Prabhakar functions is represented in the form:

$$\sigma_{\nu}(\tau) = \mathrm{K}\left({}_{0}^{LC}\mathrm{D}_{t}^{(\nu)}\varepsilon_{\nu}\right)(\tau). \tag{59}$$

where K is the coefficient of viscosity.

The creep and relaxation representations are given through the equations of the Volterra type:

$$\varepsilon_\nu(\tau) = \sigma_\nu(0)J_\nu(\tau) + \int\limits_0^\tau J_\nu(\tau - t)\left({}_0^{CT}\mathrm{D}_t^{(\nu)}\sigma_\nu\right)(t)dt \tag{60}$$

and

$$\sigma_\nu(\tau) = \varepsilon_\nu(0)G_\nu(\tau) + \int\limits_0^\tau G_\nu(\tau - t)\left({}_0^{CT}\mathrm{D}_t^{(\nu)}\varepsilon_\nu\right)(t)dt, \tag{61}$$

where the creep compliance and relaxation modulus are given by: $J_\nu(\tau) = \varepsilon_\nu(\tau)/\sigma_\nu(0)$ and $G_\nu(\tau) = \sigma_\nu(\tau)/\varepsilon_\nu(0)$, respectively.

3.3 *The Maxwell Models with GFDs*

As shown in Fig. 3, the Maxwell models with the GFDs and FD consists of a Hookean element and a general fractional-order Newtonian element in series.

The constitutive equation of the Maxwell model with GFDs can be written as

$$\left({}_0^{CT}\mathrm{D}_\tau^{(\nu)}\varepsilon_\nu\right)(\tau) = \frac{\sigma_\nu(\tau)}{\mathrm{K}} + \frac{1}{\mathrm{H}}\left({}_0^{CT}\mathrm{D}_\tau^{(\nu)}\sigma_\nu\right)(\tau).$$

Case 1

The creep compliance of the Maxwell model with the FD in the singular power-law kernel can be written as [20, 21]

$$J_\nu(\tau) = \frac{1}{\mathrm{K}}\frac{\tau^\nu}{\Gamma(1+\nu)} + \frac{2}{\mathrm{H}},$$

and the relaxation modulus of the Maxwell model with the FD in the singular power-law kernel is given as [20, 21]

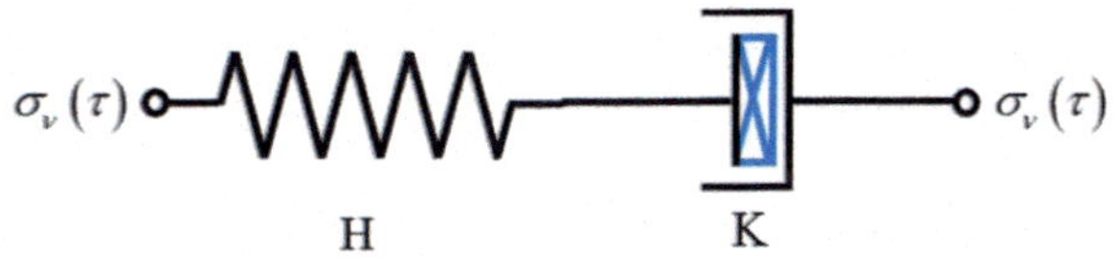

Fig. 3 The Maxwell model via FD and GFDs

$$G_\nu(\tau) = 2\mathrm{K}\tau^{-\nu} E_\nu\left(-\frac{\mathrm{K}}{\mathrm{H}}t^\nu\right).$$

Case 2

The creep compliance of the Maxwell model with the GFD in the nonsingular power-law kernel is

$$J_\nu(\tau) = \frac{1}{\mathrm{K}}\frac{\tau^{-\nu}}{\Gamma(1-\nu)} + \frac{2}{\mathrm{H}},$$

and the relaxation modulus of the Maxwell model with the GFD in the nonsingular power-law kernel can be given as

$$G_\nu(\tau) = 2\mathrm{K}\tau^{\nu} E_{\nu,1+\nu}\left(-\frac{\mathrm{K}}{\mathrm{H}}t^\nu\right).$$

Case 3

The creep compliance of the Maxwell model with GFD in the kernel of the Mittag–Leffler function can be written as

$$J_\nu(\tau) = \frac{1}{\mathrm{K}} + \frac{2}{\mathrm{H}} + \frac{1}{\mathrm{K}}\frac{\tau^{\nu}}{\Gamma(1+\nu)},$$

and the relaxation modulus of the Maxwell model with general fractional derivative in the kernel of the Mittag–Leffler function becomes

$$G_\nu(\tau) = \frac{\mathrm{K}}{\mathrm{H}+\mathrm{K}} E_\nu\left(-\frac{\mathrm{H}}{\mathrm{H}+\mathrm{K}}\tau^\nu\right).$$

Case 4

The creep compliance of the Maxwell model with general fractional derivative in the kernel of the Wiman function can be represented in the form:

$$J_\nu(\tau) = \frac{1}{\mathrm{K}}\left(\frac{\tau^{1-\upsilon}}{\Gamma(2-\upsilon)} + \frac{\tau^{1-\upsilon+\nu}}{\Gamma(2-\upsilon+\nu)} + \frac{\mathrm{K}}{\mathrm{H}}\right),$$

and the relaxation modulus of the Maxwell model with the GFD in the kernel of the Wiman function is

$$G_\nu(\tau) = \mathrm{K}E_{\nu,\nu}\left(-\left(\frac{\mathrm{K}}{\mathrm{H}}+1\right)\tau^\nu\right).$$

Case 5

The creep compliance of the Maxwell model with the GFD in the kernel of the Prabhakar function can be expressed as

$$J_{\upsilon}(\tau)=\frac{1}{\mathrm{K}}\left(\tau^{1-\upsilon}E_{\upsilon,\upsilon}^{-\phi}(-\tau^{\upsilon})+\frac{2}{\mathrm{H}}\right),$$

and the relaxation modulus of the Maxwell model with the GFD in the kernel of the Prabhakar function is

$$G_{\upsilon}(\tau)=2\mathrm{K}\sum_{n=0}^{\infty}\left(-\frac{\mathrm{K}}{\mathrm{H}}\right)^{n}\tau^{(n-1)(2-\upsilon)}E_{\upsilon,(n-1)(2-\upsilon)+1}^{(1-n)\phi}(-\tau^{\upsilon}).$$

3.4 The Voigt Models with GFDs

As shown in Fig. 4, the Voigt models with the GFDs and FD consists of a Hookean element and a general fractional-order Newtonian element in parallel.

The constitutive equation of the Voigt model can be written as

$$\sigma_{\upsilon}(\tau)=\mathrm{H}\varepsilon_{\upsilon}(\tau)+\mathrm{K}\left({}_{0}^{RLT}\mathrm{D}_{\tau}^{(\upsilon)}\varepsilon_{\upsilon}\right)(\tau).$$

Case 1

The creep compliance of the Voigt model with the FD in the singular power-law kernel can be written as [20, 21]

$$J_{\upsilon}(\tau)=\frac{1}{\mathrm{H}}\left(1-E_{\upsilon}\left(-\frac{\mathrm{K}}{\mathrm{H}}\tau^{\upsilon}\right)\right),$$

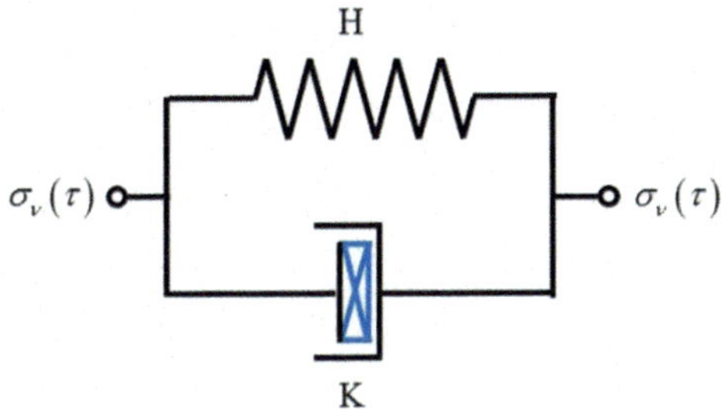

Fig. 4 The Voigt model via FD and GFDs

and the relaxation modulus of the Voigt model with the FD in the singular power-law kernel is given as [20, 21]

$$G_{\nu}(\tau) = \mathrm{H} + \mathrm{K}\frac{\tau^{-\nu}}{\Gamma(1-\nu)}.$$

Case 2

The creep compliance of the Voigt model with the GFD in the nonsingular power-law kernel is represented as

$$J_{\nu}(\tau) = \frac{1}{\mathrm{H}} E_{\nu}\left(-\frac{\mathrm{K}}{\mathrm{H}}\tau^{\nu}\right),$$

and the relaxation modulus of the Voigt model with the GFD in the nonsingular power-law kernel can be given as

$$G_{\nu}(\tau) = \mathrm{H} + \mathrm{K}\frac{\tau^{\nu}}{\Gamma(1+\nu)}.$$

Case 3

The creep compliance of the Voigt model with the GFD in the kernel of the Mittag–Leffler function is

$$J_{\nu}(\tau) = \frac{1}{\mathrm{H}+\mathrm{K}}\left(E_{\nu}\left(-\frac{\mathrm{H}}{\mathrm{H}+\mathrm{K}}\tau^{\nu}\right) + E_{\nu,\nu+1}\left(-\frac{\mathrm{H}}{\mathrm{H}+\mathrm{K}}\tau^{\nu}\right)\right),$$

and the relaxation modulus of the Voigt model with the GFD in the kernel of the Mittag–Leffler function is given as

$$G_{\nu}(\tau) = \mathrm{H} + \mathrm{K}E_{\nu}(-\tau^{\nu}).$$

Case 4

The creep compliance of the Voigt model with the GFD in the kernel of the Wiman function is expressed by

$$J_{\nu}(\tau) = \frac{\tau^{1-\upsilon}}{\Gamma(2-\upsilon)} + \frac{\tau^{1+\nu-\upsilon}}{\Gamma(2+\nu-\upsilon)} + \frac{2\mathrm{K}}{\mathrm{H}},$$

and the relaxation modulus of the Voigt model with the GFD in the kernel of the Wiman function can be written as

$$G_\nu(\tau) = \frac{2\mathrm{H}}{\mathrm{K}} E_{\nu,\nu}\left(-\left(\frac{\mathrm{K}}{\mathrm{H}}+1\right)\tau^\nu\right).$$

Case 5

The creep compliance of the Voigt model with the GFD in the kernel of the Prabhakar function is

$$J_\nu(\tau) = \frac{1}{\mathrm{H}}\left(\sum_{n=0}^{\infty}\left(-\frac{\mathrm{K}}{\mathrm{H}}\right)^{-n} \tau^{n(1-\upsilon)} E_{\nu,n(1-\upsilon)+1}^{-n\phi}(-\tau^\nu)\right.$$
$$\left.+\frac{\mathrm{K}}{\mathrm{H}}\sum_{n=0}^{\infty}\left(-\frac{\mathrm{K}}{\mathrm{H}}\right)^{n} \tau^{(n-1)(2-\upsilon)} E_{\nu,(n-1)(2-\upsilon)+1}^{(1-n)\phi}(-\tau^\nu)\right),$$

and the relaxation modulus of the Voigt model with the GFD in the kernel of the Prabhakar function is given as

$$G_\nu(\tau) = \mathrm{H} + \mathrm{K}\tau^{\upsilon-1} E_{\nu,\upsilon}^{\phi}(-\tau^\nu).$$

For more details of the applications of the GFC operators to the viscoelastic behaviors, see [20, 21].

4 Conclusion

In the present work, we investigated the basic formulations of the FC and GFC operators with the special functions with the power law. The Laplace transforms of the GFDs and GFIs formulations were discussed in detail. The anomalous Nutting behaviors in the real materials can be proposed for the first time. The applications of the GFC operators to the viscoelastic behaviors can be represented in the use of the complexity of the real materials. The Maxwell and Voigt models with the GFDs in the nonsingular kernels were obtained with the help of the Laplace transforms of the special functions. The results can be explained the complex phenomenon in the mining-rock materials.

Acknowledgements This work is supported by the State Key Research Development Program of the People's Republic of China (Grant No.2016YFC0600705), the Natural Science Foundation of China (Grant No.51323004), and the Priority Academic Program Development of Jiangsu Higher Education Institutions (PAPD2014).

References

1. Samko, S., Kilbas, A., Marichev, O.: Fractional Integrals and Derivatives: Theory and Applications. Gordon and Breach Science Publishers, Amsterdam (1993)
2. Kilbas, A.A., Srivastava, H.M., Trujillo, J.J.: Theory and Applications of Fractional Differential Equations. Academic (2006)
3. Machado, J.T., Kiryakova, V., Mainardi, F.: Recent history of fractional calculus. Commun. Nonlinear Sci. Numer. Simul. **16**(3), 1140–1153 (2011)
4. Liouville, J.: Mémoire sur quelques questions de géométrie et de mécanique, et sur un nouveau genre de calcul pour résoudre ces questions. J. de l'Ecole Polytechnique **13**, 1–69 (1832)
5. Riemann, B.: Versuch einer allgemeinen Auffassung der Integration und Differentiation, 14 Janvier. Bernhard Riemann's Gesammelte Mathematische Werke **1892**, 353–362 (1847)
6. Sonine, N.: Sur la différentiation à indice quelconque.Matematicheskii Sbornik **6**(1), 1–38 (1872)
7. Caputo, M.: Linear models of dissipation whose Q is almost frequency independent—II. Geophys. J. Int. **13**(5), 529–539 (1967)
8. Smit, W., De Vries, H.: Rheological models containing fractional derivatives. Rheol. Acta **9**(4), 525–534 (1970)
9. Dzhrbashyan, M., Nersesyan, A.: Drobnye proizvodnye i zadacha cauchy dlya di_erencial.nykh uravneniy drobnogo poryadka. Izvestiya Akademii Nauk Armyanskoj SSR, ser. Matematika **3**(1), 3–29 (1968)
10. Kaczorek, T., Rogowski, K.: Fractional Linear Systems and Electrical Circuits. Springer International Publishing, Switzerland (2015)
11. Vinagre, B.M., Podlubny, I., Hernandez, A., Feliu, V.: Some approximations of fractional order operators used in control theory and applications. Fractional Calculus Appl. Anal. **3**(3), 231–248 (2000)
12. Richard, H.: Fractional Calculus: An Introduction for Physicists. World Scientific (2014)
13. Drapaca, C.S., Sivaloganathan, S.: A fractional model of continuum mechanics. J. Elast. **107**(2), 105–123 (2012)
14. Blasiak, S.: Time-fractional heat transfer equations in modeling of the non-contacting face seals. Int. J. Heat Mass Transf. **100**, 79–88 (2016)
15. Machado, J.T., Mata, M.E.: Pseudo phase plane and fractional calculus modeling of western global economic downturn. Commun. Nonlinear Sci. Numer. Simul. **22**(1), 396–406 (2015)
16. Scalas, E., Gorenflo, R., Mainardi, F.: Fractional calculus and continuous-time finance. Phys. A **284**(1), 376–384 (2000)
17. Rivero, M., Trujillo, J.J., Vázquez, L., Velasco, M.P.: Fractional dynamics of populations. Appl. Math. Comput. **218**(3), 1089–1095 (2011)
18. Magin, R.L.: Fractional calculus models of complex dynamics in biological tissues. Comput. Math. Appl. **59**(5), 1586–1593 (2010)
19. Podlubny, I.: Fractional Differential Equations. Academic, San Diego (1998)
20. Mainardi, F.: Fractional Calculus and Waves in Linear Viscoelasticity: An Introduction to Mathematical Models. World Scientific (2010)
21. Yang, X.J., Gao, F., Ju, Y.: General Fractional Derivatives with Applications in Viscoelasticity. Academic (2020)
22. Gemant, A.: A method of analyzing experimental results obtained from elastiviscous bodies. Physics **7**, 311–317 (1936)
23. Blair, G.S.: The role of psychophysics in rheology. J. Colloid Sci. **2**(1), 21–32 (1947)
24. Gerasimov, A.N.: A generalization of linear laws of deformation and its application to problems of internal friction. Akad. Nauk SSSR. Prikl. Mat. Meh. **12**, 251–260 (1948). (Russian)
25. Caputo, M., Mainardi, F.: Linear models of dissipation in anelastic solids. La Rivista del Nuovo Cimento **1**(2), 161–198 (1971)
26. Rossikhin, Y.A., Shitikova, M.V.:. Application of fractional calculus for dynamic problems of solid mechanics: novel trends and recent results. Appl. Mech. Rev. **63**(1), 010801 (2010)

27. Bagley, R.: On the equivalence of the Riemann-Liouville and the Caputo fractional order derivatives in modeling of linear viscoelastic materials. Fractional Calculus and Applied Analysis **10**(2), 123–126 (2007)
28. Gao, F., Yang, X.J.: Fractional Maxwell fluid with fractional derivative without singular kernel. Therm. Sci. **20**, 871–877 (2016)
29. Yang, X.J., Gao, F., Srivastava, H.M.: New rheological models within local fractional derivative. Romanian Rep. Phys. **69**(3), 113 (2017)
30. Yang, X.J.: New general fractional-order rheological models within kernels of Mittag-Leffler functions. Rom. J. Phys. **69**(4), 115 (2017)
31. Rabotnov, Y.: Equilibrium of an elastic medium with after effect (in Russian). Prikladnaya Matematika i Mekhanika **12**(1), 53–62 (1948)
32. Stiassnie, M.: On the application of fractional calculus for the formulation of viscoelastic models. Appl. Math. Model. **3**(4), 300–302 (1979)
33. Koeller, R.C.: A theory relating creep and relaxation for linear materials with memory. J. Appl. Mech. **77**(3), 031008 (2010)
34. Bagley, R.L., Torvik, P.J.: On the fractional calculus model of viscoelastic behavior. J. Rheol. **30**(1), 133–155 (1986)
35. Heymans, N., Bauwens, J.C.: Fractal rheological models and fractional differential equations for viscoelastic behavior. Rheologica Acta **33**(3), 210–219 (1994)
36. Schiessel, H., Metzler, R., Blumen, A., Nonnenmacher, T.F.: Generalized viscoelastic models: their fractional equations with solutions. J. Phys. A **28**(23), 6567 (1995)
37. Nutting, P.G.: A new general law of deformation. J. Franklin Inst. **191**(5), 679–685 (1921)
38. Yang, X.-J.: Theoretical studies on general fractional-order viscoelasticity, Ph.D. Thesis, China University of Mining and Technology, Xuzhou, China (2017)
39. Yang, X.-J.: General fractional calculus: a new prospective on viscoelasticity. In: International Workshop on Theory and Applications of Fractional Partial Differential Equations. Qingdao, China (2018)
40. Yang, X.J.: General Fractional Derivatives: Theory, Methods and Applications. CRC Press (2019)
41. Yang, X.J., Machado, J.T., Baleanu, D.: Anomalous diffusion models with general fractional derivatives within the kernels of the extended Mittag-Leffler type functions. Romanian Rep. Phys **69**(4), 115 (2017)
42. Kilbas, A.A., Saigo, M., Saxena, R.K.: Generalized Mittag-Leffler function and generalized fractional calculus operators. Integr. Transf. Special Funct. **15**(1), 31-49 (2004)
43. Atangana, A., Baleanu, D.: New fractional derivatives with nonlocal and non-singular kernel: Theory and application to heat transfer model. Thermal Sci **20**(2), 763–769 (2016)
44. Kiryakova, V.S.: Generalized Fractional Calculus and Applications. CRC Press (1993)
45. Kochubei, A.N.: General fractional calculus, evolution equations, and renewal processes. Integr. Eqn. Oper. Theory **71**(4), 583–600 (2011)
46. Mittag-Leffler, G.M., Sur La Nouvelle Fonction $E_\alpha(x)$. Comptes Rendus de l'Acad′emie des Sciences, **137**, 554–558 (1903)
47. Wiman, A.: über Den Fundamental Satz in Der Theorie Der Funktionen $E_\alpha(x)$. Acta Math. **29**, 217–234 (1905)
48. Prabhakar, T.R.: A singular integral equation with a generalized mittag leffler function in the kernel. Yokohama Math. J. **19**(1), 7–15 (1971)
49. Gorenflo, R., Kilbas, A.A., Mainardi, F., Rogosin, S.V.: Mittag-Leffler Functions, Related Topics and Applications, vol. 2. Springer, Berlin (2014)
50. Hille, E., Tamarldn, J.D.: On the theory of linear integral equations. Ann. Math. **31**, 479–528 (1930)
51. Hille, E.: Notes on linear transformations. II. Analyticity of semi-groups. Ann. Math. **40**(1), 1–47 (1939)

Group Dynamical Systems on C^*-Algebras Generated by Countable Infinitely Many Semicircular Elements

Ilwoo Cho

Abstract In this paper, starting from a C^*-probability space $\mathfrak{X}_\varphi$ generated by mutually free, countable-infinitely many semicircular elements $\{s_n\}_{n=1}^\infty$, the free distributional data on $\mathfrak{X}_\varphi$ are characterized by joint free moments of $\{s_n\}_{n=1}^\infty$; and then, a certain group λ acting on $\mathfrak{X}_\varphi$ is constructed-and-studied under a group dynamical system Γ of λ. From the dynamics, the crossed product C^*-algebra $\mathbb{X}\Gamma$ is constructed, and the free probability on $\mathbb{X}\Gamma$ is considered in terms of that on $\mathfrak{X}_\varphi$. In particular, the free-distributional data of generating operators of $\mathbb{X}\Gamma$ are studied, and they illustrate how semicircularity works under our group-action.

Keywords Free probability · Semicircular elements · Free-isomorphisms · Groups · Group dynamical systems · Crossed product algebras

1991 Mathematics Subject Classification 46L10, 46L54, 47L55

1 Introduction

The main purposes of this paper are (i) to consider a *C^*-probability space* $\mathfrak{X}_\varphi = (\mathfrak{X},\ \varphi)$, where $\mathfrak{X}$ is a C^*-algebra generated by a set $X = \{s_n\}_{n=1}^\infty$ of mutually free, $|\mathbb{N}|$ -many *semicircular elements* x_n's, (ii) to characterize the free-distributional data on $\mathfrak{X}_\varphi$, (iii) to define a *group* λ acting on $\mathfrak{X}_\varphi$, preserving the free probability on $\mathfrak{X}_\varphi$, and the corresponding *group C^*-probability space* $(\Lambda,\ \tau)$ of the group C^*-algebra Λ of λ, and the *canonical trace* τ *on* Λ, (iv) to investigate a *group dynamical system*,

$$\Gamma = (\lambda,\ \Lambda \otimes_{\mathbb{C}} \mathfrak{X},\ \alpha),$$

I. Cho (✉)
St. Ambrose University, Department of Mathematics and Statistics, 421 Ambrose Hall, 518 W. Locust St., Davenport, Iowa 52803, USA
e-mail: choilwoo@sau.edu

J. Singh et al. (eds.), *Methods of Mathematical Modelling and Computation for Complex Systems*, Studies in Systems, Decision and Control 373,
https://doi.org/10.1007/978-3-030-77169-0_7

where $\otimes_{\mathbb{C}}$ is the *tensor product of* C^**-algebras*, (v) to study the corresponding *crossed product* C^**-algebra*,

$$\mathbb{X}\Gamma = (\Lambda \otimes_{\mathbb{C}} \mathfrak{X}) \times_{\alpha} \lambda,$$

generated by the dynamical system Γ of (iv), and (vi) to study the free probability on $\mathbb{X}\Gamma$ by investigating how a such dynamical system affects the semicircularity on $\mathfrak{X}_{\varphi}$.

1.1 Background

Not only in operator algebra theory, but also in statistical quantum physics, *semicircular elements* play major roles (e.g., [1, 2, 5–8, 10–12, 20, 21, 29, 30]). The (classical, or free) distributions of semicircular elements are well-known in functional analysis; they are called *the semicircular law*. In particular, *operators* satisfying the semicircular law have been studied, and well-characterized in the language of free probability theory (e.g., [1, 17, 18, 21, 28–30]).

As one can see in the (*free*) *central limit theorem*(*s*), e.g., see [2, 17, 19, 28–30], the semicircular law is roughly understood to be the noncommutative-algebraic version of the classical *Gaussian distribution*. In combinatorial approaches (e.g., [17, 22, 23]), the *free distributions* of semicircular elements are universally characterized by the *Catalan numbers* c_n,

$$c_n = \frac{1}{n+1}\binom{2n}{n} = \frac{(2n)!}{n!(n+1)!}, \tag{1}$$

for all $n \in \mathbb{N}_0 = \mathbb{N} \cup \{0\}$. i.e., the semicircular law is characterized by the *free-moment sequence*,

$$\left(\omega_n c_{\frac{n}{2}}\right)_{n=1}^{\infty} = (0,\ c_1,\ 0,\ c_2,\ 0,\ c_3,\ ...)\,, \tag{2}$$

with

$$\omega_n = \begin{cases} 1 & \text{if } n \text{ is even} \\ 0 & \text{if } n \text{ is odd,} \end{cases}$$

for all $n \in \mathbb{N}$, where $\{c_k\}_{k=1}^{\infty}$ in (2) are the Catalan numbers (1).

1.2 Motivation

Recently, it is shown that, from the analysis on p-adic number fields, semicircular elements are canonically constructed (e.g., [5, 12]). It provides other connections between operator algebra and quantum physics (e.g., [26, 27]). Motivated by [5, 12], semicircular elements are well-generated, as *Banach-space operators* (e.g., [13, 14]), acting on a *C^*-algebra* containing $|\mathbb{Z}|$-many *orthogonal projections* (e.g., [6–8, 10, 11]) different from earlier works.

Independently, characterizations, and estimations of *joint free distributions* of mutually free, multi semicircular elements are introduced in [9] (See Sects. 3 and 4 below). These results would characterize the free-distributional data on our C^*-probability space $\mathfrak{X}_{\varphi}$.

1.3 Overview

The main results of this paper show that: (I) the free probability on $\mathfrak{X}_{\varphi}$ is characterized by the joint free moments of generating, mutually free, $|\mathbb{N}|$-many semicircular elements, in Sects. 3 and 4; (II) the C^*-algebra $\mathfrak{X}$ is not only $*$-isomorphic to the C^*-algebra $\mathfrak{S}$ generated by mutually free, $|\mathbb{Z}|$-many semicircular elements $\{s_n\}_{n\in\mathbb{Z}}$, but also, *free-isomorphic to* $\mathfrak{S}$ in the sense that the free probability on $\mathfrak{X}$ is preserved to be that on $\mathfrak{S}$, in Sect. 5; (III) there are well-defined $*$-isomorphisms on $\mathfrak{S}$ (and hence, on $\mathfrak{X}$), preserving the free probability on $\mathfrak{S}$, and these $*$-isomorphisms generate a well-defined *group* λ and the corresponding *group C^*-algebra* Λ acting on $\mathfrak{S}$, in Sect. 6; (IV) the group λ of (III) induces a well-defined *group C^*-dynamical system* Γ, and the corresponding *crossed product C^*-algebra* $\mathbb{X}\Gamma$, in Sect. 7; and (V) the free probability on $\mathbb{X}\Gamma$ is considered, and the free-distributional data of the generating operators of $\mathbb{X}\Gamma$ are fully characterized in Sect. 8.

2 Preliminaries

In this section, we briefly introduce concepts and notations used in text.

2.1 Free Probability

For more about free probability, e.g., see [3, 15–17, 28–30]. *Free probability* is a noncommutative operator-algebraic version of classical *measure theory* and *statistical analysis* (implying probability theory and statistics). Free probability is not only a major branch of operator algebra theory (e.g., [9, 17, 20–22, 28, 29]), but also an interesting application in many related fields (e.g., [4–8, 10–12, 24, 25]).

In this paper, we use combinatorial approach of Speicher (e.g., [17, 22, 23]). Without introducing detailed definitions, or combinatorial backgrounds, the (joint) free moments and (joint) free cumulants are considered, and the (free-probabilistic) *free product* is used in text without precise introduction.

As usual in free probability theory, a mathematical pair (B, ψ) would be called a *free $*$-probability space*, if B is a (topological, or pure-algebraic noncommutative) $*$-algebra (over $\mathbb{C}$), and ψ is a (bounded, resp., unbounded) linear function on B.

2.2 Semicircularity

Let (A, φ) be a *topological $*$-probability space* (C^*-probability space, or W^*-probability space, or Banach $*$ -probability space, etc.), consisting of a topological $*$-algebra A (C^*-algebra, resp., W^*-algebra, resp., Banach $*$-algebra, etc.), and a bounded *linear functional* φ on A. Operators $a \in A$ are called *free random variables*, if one regards a as elements of (A, φ).

As in operator theory (e.g., [14]), a free random variable $a \in (A, \varphi)$ is said to be *self-adjoint*, if a is self-adjoint in A as an operator, i.e., $a^* = a$, where a^* is the *adjoint of* a. Note that the free distribution of a self-adjoint free random variable a is characterized by

$$\text{the free-moment sequence } \left(\varphi(a^n)\right)_{n=1}^{\infty}, \tag{3}$$

and

$$\text{the free-cumulant sequence } (k_n(a, \ ..., a))_{n=1}^{\infty},$$

by [17, 21], where $k_{\bullet}(.)$ is the *free cumulant on A in terms of φ*, under the *Möbius inversion* of [17, 21].

Definition 1 A self-adjoint free random variable $x \in (A, \varphi)$ is semicircular, if

$$\varphi(x^n) = \omega_n c_{\frac{n}{2}}, \text{ for all } n \in \mathbb{N}, \tag{4}$$

where ω_n are in the sense of (2), and c_k are the k-th Catalan numbers (1) for all $k \in \mathbb{N}_0$.

By the Möbius inversion, a self-adjoint free random variable x is semicircular in (A, φ), if and only if

$$k_n \left(x, \ ..., x\right) = \delta_{n,2} \tag{5}$$

for all $n \in \mathbb{N}$, where δ is the Kronecker delta.

Since free-moment sequence $(\varphi(x^n))_{n=1}^{\infty}$, and the free-cumulant sequence $(k_n(x, \ldots, x))_{n=1}^{\infty}$ provide equivalent free-distributional data of x in (A, φ) (e.g., [17, 21, 22]), one can use the definition (4) and the characterization (5) alternatively as the semicircularity, by (3).

i.e., *the semicircular law* is characterized by the free-moment sequence,

$$(0,\ c_1,\ 0,\ c_2,\ 0,\ c_3,\ 0,\ c_4,\ ...), \tag{6}$$

or, by the free-cumulant sequence,

$$(0, 1, 0, 0, 0, 0, \ldots), \tag{7}$$

by (4) and (5), respectively.

By definition, even though

$$x_l \in (A_l, \varphi_l), \text{ for } l = 1, 2,$$

are distinct semicircular elements in (possibly different) topological $*$-probability spaces, their free distributions are identical, by (6) and (7).

3 Catalan Numbers

In this section, we introduce some results known in [9], used in our later works. For $k \in \mathbb{N}_0$, let c_k be the k-th Catalan number,

$$c_k = \frac{1}{k+1}\binom{2k}{k} = \frac{(2k)!}{k!(k+1)!}.$$

Lemma 1 *Let $k_1 > k_2$ in $\mathbb{N}_0$. Then there exists a quantity $\beta_{k_1>k_2}$ in the set $\mathbb{R}^+$ of all positive real numbers, such that*

$$c_{k_1} = \beta_{k_1>k_2} c_{k_2}. \tag{8}$$

In particular,

$$\beta_{k_1>k_2} = 2^{k_1-k_2}\left(\frac{2k_2+1}{k_1+1}\right)\left(\prod_{l=1}^{k_1-k_2-1}\left(2 - \frac{1}{(k_1+1)-l}\right)\right) \tag{9}$$

in $\mathbb{R}^+$, with axiomatization:

$$\prod_{l=1}^{0}\left(2 - \frac{1}{(k_1+1)-l}\right) = 1.$$

Proof If $k_1 > k_2$ in $\mathbb{N}_0$, then

$$\begin{aligned}\frac{c_{k_1}}{c_{k_2}} &= \left(\frac{(2k_1)!}{(k_1)!(k_1+1)!}\right)\left(\frac{(k_2)!(k_2+1)!}{(2k_2)!}\right) \\ &= \left(\frac{(2k_1)!}{(2k_2)!}\right)\left(\frac{(k_2)!}{(k_1)!}\right)\left(\frac{(k_2+1)!}{(k_1+1)!}\right) \\ &= \frac{(2(k_1)2(k_1-1)2(k_1-2)\dots 2(k_2+1))\,((2k_1-1)(2k_1-3)\dots(2k_2+1))}{(k_1+1)(k_2+1)\,(k_1(k_1-1)\dots(k_2+2))^2} \\ &= 2^{k_1-k_2}\left(\frac{1}{k_1+1}\right)\left(\frac{2k_1-1}{k_1}\right)\left(\frac{2k_1-3}{k_1-1}\right)\cdots\left(\frac{2k_2+3}{k_2+2}\right)(2k_2+1)\,.\end{aligned}$$

Therefore, one obtains that

$$\frac{c_{k_1}}{c_{k_2}} = \beta_{k_1>k_2} \Longleftrightarrow c_{k_1} = \beta_{k_1>k_2} c_{k_2},$$

where $\beta_{k_1>k_2}$ is in the sense of (9). Therefore, the relation (8) holds. ∎

By (8), it is not hard to check that if $k_1 > k_2$ in $\mathbb{N}_0$, then

$$c_{k_1} c_{k_2} = c_{k_2} c_{k_1} = \beta_{k_1>k_2} c_{k_2}, \tag{10}$$

where $\beta_{k_1>k_2}$ is in the sense of (9).

Theorem 1 *Let $k_1 > k_2 > \cdots > k_N$ in $\mathbb{N}_0$, for some $N \in \mathbb{N} \setminus \{1\}$, and take k_l-th Catalan numbers c_{k_l}, for all $l = 1, \dots, N$. Now, take n_l-many c_{k_l}'s, for all $l = 1, \dots, N$, and hence, choose $\left(s = \sum_{l=1}^{N} n_l\right)$-many total Catalan numbers with repetition. For convenience, let's denote these totally s-many Catalan numbers by*

$$c_{j_1}, c_{j_2}, \dots, c_{j_s}.$$

Then, for every permutation α of the symmetric group S_X over $X = \{j_1, \dots, j_s\}$, we have that

$$\prod_{l=1}^{s} c_{\alpha(j_l)} = \left(\prod_{i=1}^{N-1} \beta_{k_i>k_{i+1}}^{\Sigma_{l=1}^{i} n_l}\right)\left(c_{k_N}^{\Sigma_{l=1}^{N} n_l}\right), \tag{11}$$

with

$$\beta_{k_i>k_{i+1}} = 2^{k_i-k_{i+1}}\left(\frac{2k_{i+1}+1}{k_i+1}\right)\left(\prod_{l=1}^{k_i-k_{i+1}-1}\left(2-\frac{1}{(k_i+1)-l}\right)\right),$$

for all $i = 1, \dots, N-1$.

Proof Under hypothesis, for every permutation α of the symmetric group S_X over $X = \{j_1, \dots, j_s\}$, one has that

$$\prod_{l=1}^{s} c_{\alpha(j_l)} = \prod_{l=1}^{N} c_{k_l}^{n_l} = c_{k_1}^{n_1} c_{k_2}^{n_2} ... c_{k_N}^{n_N}$$
$$= \left(\beta_{k_1>k_2} c_{k_2}\right)^{n_1} c_{k_2}^{n_2} c_{k_3}^{n_3} \cdots c_{k_N}^{n_N}$$

by (8)

$$= \beta_{k_1>k_2}^{n_1} c_{k_2}^{n_1+n_2} c_{k_3}^{n_3} \cdots c_{k_N}^{n_N}$$
$$= \beta_{k_1>k_2}^{n_1} \left(\beta_{k_2>k_3}^{n_1+n_2} c_{k_3}^{n_1+n_2}\right) c_{k_3}^{n_3} c_{k_4}^{n_4} \cdots c_{k_N}^{n_N}$$

by (8)

$$= \cdots$$
$$= \left(\beta_{k_1>k_2}^{n_1} \beta_{k_2>k_3}^{n_1+n_2} \cdots \beta_{k_{N-1}>k_N}^{n_1+n_2+\cdots+n_{N-1}}\right) c_{k_N}^{n_1+n_2+\cdots+n_{N-1}+n_N},$$

where $\beta_{k_i>k_{i+1}}$ are in the sense of (9), for all $i = 1, \ldots, N-1$. So, the relation (11) holds. ■

4 Free Distributions of Multi Semicircular Elements

In this section, we consider free distributions of mutually free, multi semicircular elements in a C^*-probability space (A, φ). Without loss of generality, readers can regard (A, φ) as an any topological $*$-probability space (W^*-probability space, or a Banach $*$-probability space, etc.).

Let (A, φ) be a fixed C^*-probability space, and suppose there are N-many semicircular elements $x_1, \ldots, x_N$ in (A, φ), for $N \in \mathbb{N}$. Assume further that they are free from each other in (A, φ). By the self-adjointness of $x_1, \ldots, x_N$ in A, the free distribution, say

$$\rho \overset{denote}{=} \rho_{x_1,\ldots,x_N}, \tag{12}$$

of them are characterized by the*joint free-moments*,

$$\bigcup_{n=1}^{\infty} \left(\bigcup_{(i_1,\ldots,i_n)\in\{1,\ldots,N\}^n} \left\{\varphi\left(x_{i_1} x_{i_2} ... x_{i_n}\right)\right\}\right) \tag{12'}$$

(e.g., [17, 22, 23]). More precisely, the free distribution ρ of (12), is characterized by the free-moments

$$\bigcup_{l=1}^{N} \left\{\varphi(x_l^n)\right\}_{n=1}^{\infty}, \tag{13}$$

and the "mixed" free-moments,

$$\bigcup_{s=2}^{\infty}\left\{\varphi\left(x_{i_1}^{n_1}x_{i_2}^{n_2}...x_{i_s}^{n_s}\right)\middle|\begin{array}{l}(i_1,\ldots,i_s)\in\{1,\ldots,N\}^s\\ \text{are mixed in }\{1,\ldots,N\}\end{array}\right\}, \tag{14}$$

by (12)′.

With help of the results of Sect. 3, we characterize the free distribution ρ of (12), by considering (13) and (14).

4.1 Free-Distributional Data (13) of ρ

Let $\rho = \rho_{x_1,\ldots,x_N}$ be the free distribution (12) of fixed N-many mutually free semicircular elements x_1, ..., x_N of (A, φ), characterized by the free-distributional data (12)′. The free-distributional data (13) are determined by the semicircularity (4), or (5), under the universality (6), respectively, (7).

Corollary 1 *The free-distributional data (13) of the free distribution ρ of (12) are characterized by the computations,*

$$\varphi\left(x_l^n\right) = \omega_n c_{\frac{n}{2}}, \textit{for all } n \in \mathbb{N}, \tag{15}$$

for all $l = 1$, ..., N.

Proof The formula (15) is proven by (4) and (6), since x_1, ..., x_N are semicircular. ■

4.2 Free-Distributional Data (14) of ρ

Throughout this section, for any $s \in \mathbb{N} \setminus \{1\}$, we fix an s-tuple I_s,

$$I_s \overset{denote}{=} (i_1, \ldots, i_s) \in \{1, \ldots, N\}^s, \tag{16}$$

which is mixed in $\{1,\ldots, N\}$, in the sense that there exists at least one entry i_{k_0} of I_s, satisfying $i_{k_0} \neq i_k$, for some $k \neq k_0$ in $\{1, \ldots, s\}$.

For example,

$$I_8 = (1, 1, 3, 2, 4, 2, 2, 1),$$

is a mixed 8-tuple in $\{1, 2, 3, 4, 5\}^8$.

From the sequence I_s of (16), define a set,

$$[I_s] = \{i_1, i_2, \ldots, i_s\}, \tag{17}$$

without considering repetition. For instance, if I_8 is as above, then

$$[I_8] = \{i_1, i_2, \ldots, i_8\},$$

with

$$i_1 = i_2 = i_8 = 1,$$
$$i_4 = i_6 = i_7 = 2,$$
$$i_3 = 3, \text{ and } i_5 = 4.$$

without considering repetition; for example, we regard all 1's in I_8 as different elements i_1, i_2 and i_8 in $[I_8]$.

Then from the set $[I_s]$ of (17), one can define a "noncrossing" partition $\pi_{(I_s)}$ in the noncrossing-partition lattice $NC\ ([I_s])$ (e.g., [17, 21, 22]), such that (i)

$$\forall V = \left(i_{j_1},\ i_{j_2},\ ...,\ i_{j_{|V|}}\right) \in \pi_{(I_s)}, \tag{18}$$

$\Longleftrightarrow$

$$\exists k \in \{1, \ldots, N\}, \text{ s.t., } i_{j_1} = i_{j_2} = \cdots = i_{j_{|V|}} = k,$$

(ii) such a partition $\pi_{(I_s)}$ of (i) is "maximal" satisfying (18), under the partial ordering on $NC\ ([I_s])$ (e.g., see [17, 21, 22]), and (iii) the first block of $\pi_{(I_s)}$ must be the maximal block starting from the first entry i_1 of I_s.

For example, if I_8 and $[I_8]$ are as above, then there exists a noncrossing partition

$$\pi_{(I_8)} = \{(i_1,\ i_2,\ i_8),\ (i_3),\ (i_4,\ i_6,\ i_7),\ (i_5)\}$$
$$= \{(1,\ 1,\ 1),\ (3),\ (2,\ 2,\ 2),\ (4)\},$$

in $NC([I_8])$, satisfying the conditions (i), (ii) and (iii).

Example 1 Let $I_{10} = (1,\ 1,\ 2,\ 2,\ 1,\ 1,\ 2,\ 2,\ 3,\ 3)$. Then the noncrossing partition $\pi_{(I_{10})}$ of (18) is

$$\pi_{(I_{10})} = \{(i_1, i_2, i_5, i_6), (i_3, i_4), (i_7, i_8), (i_9, i_{10})\},$$

where

$$i_1 = i_2 = i_5 = i_6 = 1,$$
$$i_3 = i_4 = i_7 = i_8 = 2,$$

and

$$i_9 = i_{10} = 3.$$

Note that we cannot have the block $(i_3,\ i_4,\ i_7,\ i_8)$ in $\pi_{(I_{10})}$ because of a crossing with $(i_1,\ i_2,\ i_5,\ i_6)$.

Now, suppose $\pi_{(I_s)} \in NC\ ([I_s])$ is the noncrossing partition (18), and let

$$\pi_{(I_s)} = \{U_1, \ldots, U_t\},$$

where $t \leq s$ and $U_k \in \pi_{(I_s)}$ are the blocks of (ii), satisfying the conditions (i) and (iii), for $k = 1, \ldots, t$. For the above $\pi_{(I_8)}$,

$$\pi_{(I_8)} = \{U_1, U_2, U_3, U_4\},$$

with

$$U_1 = \{i_1, i_2, i_8\}, U_2 = \{i_3\},$$
$$U_3 = \{i_4, i_6, i_7\}, U_4 = \{i_5\}.$$

Then the partition $\pi_{(I_s)}$ is regarded as the joint partition,

$$\pi_{(I_s)} = 1_{|U_1|} \vee 1_{|U_2|} \vee \ldots \vee 1_{|U_t|}, \tag{19}$$

where $1_{|U_k|}$ are the maximal elements, the one-block partitions $\{U_k\}$, of $NC\,(U_k)$, for all $k = 1, \ldots, t$.

Let I_s be in the sense of (16), and let $x_{i_1}, \ldots, x_{i_s}$ be the corresponding semicircular elements of (A, φ) induced by I_s, without considering repetition in the set $\{x_1, \ldots, x_N\}$. Define a free random variable $X[I_s]$ by

$$X[I_s] \overset{def}{=} \prod_{l=1}^{s} x_{i_l} \in (A, \varphi). \tag{20}$$

If $X[I_s]$ is in the sense of (20), then

$$\varphi\,(X[I_s]) = \sum_{\pi \in NC([I_s])} k_\pi$$

by the Möbius inversion of [17], where

$$k_\pi = \prod_{V \in \pi} k_V$$

with

$$k_V = k_{|V|}\left(x_{i_{k_1}}, \ldots, x_{i_{k_{|V|}}}\right),$$

whenever $V = (i_{k_1}, \ldots, i_{k_{|V|}})$, where $|V|$ is the cardinality of V in π, and hence, it goes to

$$= \sum_{\pi \in NC([I_s]),\ \pi \leq \pi_{(I_s)}} k_\pi$$

by the mutual-freeness of $x_1, \ldots, x_N$ in (A, φ) (e.g., [17, 22])

$$= \sum_{(\theta_1, \ldots, \theta_t) \in NC(U_1) \times \cdots \times NC(U_t)} k_{\theta_1 \vee \cdots \vee \theta_t}$$

by (19)

$$= \sum_{(\theta_1,\ldots,\theta_t)\in NC_2(U_1)\times\cdots\times NC_2(U_t)} k_{\theta_1\vee\cdots\vee\theta_t}$$

$$= \sum_{(\theta_1,\ldots,\theta_t)\in NC_2(U_1)\times\cdots\times NC_2(U_t)} \left(\prod_{l=1}^{t} k_{\theta_l}\right), \tag{21}$$

by the semicircularity (5) of $x_{i_1}, \ldots, x_{i_s}$ in (A, φ), where $NC_2(X)$ is the subset of the noncrossing-partition lattice $NC(X)$,

$$NC_2(X) = \{\pi \in NC(X) : \forall V \in \pi, |V| = 2\}, \tag{22}$$

over countable finite sets X.

By (21) and (22), it is not difficult to check that, if there exists at least one $k_0 \in \{1, \ldots, t\}$, such that $\left|U_{k_0}\right|$ is odd in $\mathbb{N}$, or, equivalently, if n is odd, then

$$\varphi(X[I_s]) = 0,$$

where $X[I_s]$ is the free random variable (20) of (A, φ).

So, the formula (21) is non-zero, only if

$$|U_k| \in 2\mathbb{N}, \text{ for all } k = 1, \ldots, t, \tag{23}$$

where $2\mathbb{N} = \{2n : n \in \mathbb{N}\}$.

Moreover, if the condition (23) is satisfied, then the summands $k_{\theta_1\vee\cdots\vee\theta_t}$ of (21) satisfy that

$$k_{\theta_1\vee\cdots\vee\theta_t} = \prod_{V\in\theta_1\vee\cdots\vee\theta_t} k_V = \prod_{V\in\theta_1\vee\cdots\vee\theta_t} \left(\prod_{i=1}^{t} 1^{\#(\theta_i)}\right) = 1, \tag{24}$$

by the semicircularity (5), where $\#(\theta_i)$ are the number of blocks of θ_i, for all $i = 1, \ldots, t$. Therefore, if the condition (23) holds, then

$$\begin{aligned} \varphi(X[I_s]) &= \sum_{(\theta_1,\ldots,\theta_t)\in NC_2([U_1])\times\cdots\times NC_2([U_t])} 1 \\ &= |NC_2(U_1) \times \cdots \times NC_2(U_t)|, \end{aligned} \tag{25}$$

by (21) and (24), where $|Y|$ mean the cardinalities of sets Y.

Lemma 2 *Let I_s be an s-tuple (16), and let $X[I_s] = \prod_{l=1}^{s} x_{i_l}$ be the corresponding free random variable (20) of (A, φ). If*

$$\pi_{(I_s)} = 1_{|U_1|} \vee \cdots \vee 1_{|U_t|},$$

in the sense of (18) and (19), then

$$\varphi(X[I_s]) = \begin{cases} \prod_{i=1}^{t} c_{\frac{|U_i|}{2}} & \textit{if } |U_k| \in 2\mathbb{N}, \\ & \textit{for all } k = 1, \ldots, t \\ 0 & \textit{otherwise.} \end{cases} \tag{26}$$

Proof Under hypothesis,

$$\varphi(X[I_s]) = \begin{cases} |NC_2(U_1) \times \cdots \times NC_2(U_t)| & \text{if } |U_k| \in 2\mathbb{N}, \\ & \text{for all } k = 1, \ldots, t \\ 0 & \text{otherwise,} \end{cases}$$

by (16).

Note that if the condition (23) is satisfied, then

$$|NC_2(U_k)| = \left| NC\left(\frac{|U_k|}{2}\right) \right|, \tag{27}$$

for all $k = 1, \ldots, t$ (e.g., see [17, 22, 23]).

So, the formula (25) goes to

$$\varphi(X[I_s]) = \begin{cases} \left|NC\left(\frac{U_1}{2}\right) \times \cdots \times NC\left(\frac{U_t}{2}\right)\right| & \text{if } |U_k| \in 2\mathbb{N}, \\ & \text{for all } k = 1, \ldots, t \\ 0 & \text{otherwise,} \end{cases}$$

by (27)

$$= \begin{cases} \prod_{l=1}^{t} c_{\frac{|U_l|}{2}} & \text{if } |U_l| \in 2\mathbb{N}, \\ & \text{for all } l = 1, \ldots, t \\ 0 & \text{otherwise,} \end{cases} \tag{28}$$

because $|NC(X)| = c_{|X|}$, for all finite sets X (e.g., [17, 22, 23]). Therefore, the formula (26) holds by (28). ■

By the above combinatorial result, one obtains the following analytic result, characterizing (14).

Theorem 2 *Let I_s be in the sense of (16), and $X[I_s]$, the corresponding free random variable (20) of (A, φ), and assume that*

$$\pi_{(I_s)} = \{U_1, \ldots, U_t\},$$

satisfies (18) and (19). Also, let

$$k'_l = \frac{|U_l|}{2} \in \mathbb{R}^+, \textit{for all } l = 1, \ldots, t,$$

and suppose there exist "mutually-distinct" $k'_{j_1}, \ldots, k'_{j_r}$ *among* $\{k'_1, \ldots, k'_t\}$, *where* $r \leq t$, *and assume that there are* n_{j_l}*-many such* k_{j_l}*'s, for* $l = 1, \ldots, r$, *satisfying*

$$\sum_{l=1}^{r} n_{j_l}(2k_{j_l}) = s.$$

Assume further that there is a permutation α *of the symmetric group* S_X *over* $X = \{k'_{j_1}, \ldots, k'_{j_r}\}$, *satisfying*

$$k_{j_1} > k_{j_2} > \cdots > k_{j_r} \textit{in } \mathbb{R}^+,$$

where

$$k_{j_l} = \alpha\left(k'_{j_l}\right) \in \mathbb{R}^+, \textit{for all } l = 1, \ldots, r,$$

Then

$$\varphi(X[I_s]) = \begin{cases} \left(\prod_{i=1}^{r-1} \beta_{k_{j_i} > k_{j_{i+1}}}^{\Sigma_{l=1}^{i} n_{j_l}}\right)\left(c_{k_{j_r}}^{\Sigma_{l=1}^{N} n_{j_l}}\right) & \textit{if } k_{j_1}, \ldots, k_{j_r} \in \mathbb{N} \\ 0 & \textit{otherwise} \end{cases} \tag{29}$$

with

$$\beta_{k_{j_i} > k_{j_{i+1}}} = 2^{k_{j_i} - k_{j_{i+1}}} \left(\frac{2k_{j_{i+1}} + 1}{k_{j_i} + 1}\right)\left(\prod_{l=1}^{k_{j_i} - k_{j_{i+1}} - 1} \left(2 - \frac{1}{(k_{j_i} + 1) - l}\right)\right),$$

for all $i = 1, \ldots, r - 1$.

Proof Under hypothesis, one has

$$\varphi(X[I_s]) = \begin{cases} \prod_{l=1}^{t} c_{k'_l} & \text{if } k'_l \in \mathbb{N}, \\ & \text{for all } l = 1, \ldots, t \\ 0 & \text{otherwise} \end{cases}$$

by (26)

$$= \begin{cases} \prod_{l=1}^{r} c_{k_{j_l}}^{n_{j_l}} & \text{if } k_{j_l} = \alpha\left(k'_{j_l}\right) \in \mathbb{N}, \\ & \text{for all } l = 1, \ldots, r \\ 0 & \text{otherwise} \end{cases}$$

by assumptions

$$= \begin{cases} \left(\prod_{i=1}^{r-1} \beta_{k_{j_i} > k_{j_{i+1}}}^{\Sigma_{l=1}^{i} n_{j_l}}\right) \left(c_{k_{j_r}}^{\Sigma_{l=1}^{r} n_{j_l}}\right) & \text{respectively,} \\ 0 \end{cases}$$

by (11), where

$$\beta_{k_{j_i} > k_{j_{i+1}}} = 2^{k_{j_i} - k_{j_{i+1}}} \left(\frac{2k_{j_{i+1}} + 1}{k_{j_i} + 1}\right) \left(\prod_{l=1}^{k_{j_i} - k_{j_{i+1}} - 1} \left(2 - \frac{1}{(k_{j_i} + 1) - l}\right)\right),$$

for all $i = 1, \ldots, r - 1$. Therefore, the free-distributional data (29) holds. ■

The above formula (29) characterizes the free-distributional data (14) of our free distribution $\rho = \rho_{x_1, \ldots, x_s}$ of (12 in (A, φ).

Example 2 Let x_1, x_2, x_3, x_4 be fixed mutually free semicircular elements of (A, φ), and let

$$W = x_1^2 x_2^4 x_1^2 x_3^2 \in (A, \varphi)$$

induced by $\{x_1, x_2, x_3, x_4\}$. Then one can take

$$I_W = (1,\ 1,\ 2,\ 2,\ 2,\ 2,\ 1,\ 1,\ 3,\ 3) \overset{let}{=} (i_1, \ldots, i_{10}),$$

and

$$\pi_{(I_W)} = \{(i_1, i_2, i_7, i_8), (i_3, i_4, i_5, i_6), (i_9, i_{10})\},$$

with $U_1 = \{i_1, i_2, i_7, i_8\} = \{1, 1, 1, 1\}$, having $k'_1 = \frac{|U_1|}{2} = 2$,
$U_2 = \{i_3, i_4, i_5, i_6\} = \{2, 2, 2, 2\}$, having $k'_2 = \frac{|U_2|}{2} = 2$,
and $U_3 = \{i_7, i_8\} = \{3, 3\}$, having $k'_3 = \frac{|U_3|}{2} = 1$.

Therefore, one can take

$$k_1 = 2 > 1 = k_2, \text{ and hence, } r = 2,$$

and

$$n_1 = 2, \text{ and } n_2 = 1,$$

since there are two 2's (represented by k_1' and k_2'), and there is only one 1 (represented by k_3').

Therefore, we have

$$\begin{aligned}\varphi(W) &= \left(\prod_{i=1}^{2-1} \beta_{k_i > k_{i+1}}^{\Sigma_{l=1}^{i} n_l}\right)\left(c_{k_r}^{\Sigma_{l=1}^{N} n_l}\right) = \beta_{k_1>k_2}^{n_1} c_{k_2}^{n_1+n_2} \\ &= \left(\beta_{2>1}^{2}\right) c_1^{2+1} \\ &= \left(2^{2-1}\left(\tfrac{2\cdot 1+1}{2+1}\right)\left(2-\tfrac{1}{(k_1+1)-l}\right)^0\right)^2 \left(\tfrac{2!}{1!2!}\right)^3 \\ &= 4,\end{aligned}$$

by (29).

4.3 *The Joint Free Distribution* $\rho_{x_1,\ldots,x_N}$

By Sects. 4.1 and 4.2, the free distribution $\rho = \rho_{x_1,\ldots,x_N}$ of (12) is determined by the free-distributional data of (15), characterizing the free moments (13), and those of (29), characterizing mixed free moments (14) of the mutually free, multi semicircular elements $x_1, \ldots, x_N$.

5 A C^*-Probability Space $(\mathfrak{X}, \varphi)$ Generated by $|\mathbb{N}|$-Many Semicircular Elements

In this section, we consider a structure theorem of a C^*-algebra $\mathfrak{X}$ generated by mutually free, $|\mathbb{N}|$-many semicircular elements $\{x_n\}_{n=1}^{\infty}$. Let (A, φ) be a C^* -probability space. Assume that it contains a family $X = \{s_n\}_{n=1}^{\infty}$ of mutually free semicircular elements s_n's. i.e., by [17, 22, 23], all mixed free cumulants of $\{s_n\}_{n=1}^{\infty}$ vanish with respect to φ. For notational convenience, we re-index the family

$$\{s_n\}_{n=1}^{\infty} = \{s_1, s_2, s_3, \ldots\}$$

to

$$\{x_n\}_{n=0}^{\infty} = \{x_0, x_1, x_2, \ldots\},$$

without loss of generality, by assigning s_n to x_{n-1}, for all $n \in \mathbb{N}$.

5.1 Free-Homomorphisms

Let (A_1, φ_1) and (A_2, φ_2) be C^*-probability spaces. The C^*-probability space (A_1, φ_1) is said to be *free-homomorphic* to the C^*-probability space (A_2, φ_2), if there exists a $*$-homomorphism Ω: $A_1 \to A_2$, such that

$$\varphi_2 (\Omega(a)) = \varphi_1(a), \text{ for all } a \in (A_1, \varphi_1).$$

In such a case, the $*$-homomorphism Ω is called a *free-homomorphism from* (A_1, φ_1) *to* (A_2, φ_2). We write this free-homomorphic relation by

$$(A_1, \varphi_1) \overset{\text{free-homo}}{\subseteq} (A_2 \varphi_2). \tag{30}$$

Definition 2 Suppose $(A_1, \varphi_1) \overset{\text{free-homo}}{\subseteq} (A_2, \varphi_2)$ in the sense of (30), via a free-homomorphism $\Omega : A_1 \to A_2$. If this free-homomorphism Ω is a $*$-isomorphism from A_1 onto A_2, then it is called a free-isomorphism. In this case, (A_1, φ_1) is said to be free-isomorphic to (A_2, φ_2), or (A_1, φ_1) and (A_2, φ_2) are free-isomorphic. This free-isomorphic relation is denoted by

$$(A_1, \varphi_1) \overset{\text{free-iso}}{=} (A_2, \varphi_2). \tag{31}$$

By the definitions (30) and (31), if two C^*-probability spaces are free-isomorphic, then they are regarded as the same C^*-probability space.

5.2 A C^*-Probability Space $(\mathfrak{X}, \varphi \mid_{\mathfrak{X}})$

Let (A, φ) be a fixed C^*-probability space containing a family $X = \{x_n\}_{n=0}^{\infty}$, consisting of mutually free semicircular elements x_n's, for $n \in \mathbb{N}_0$. Construct the C^*-subalgebra $\mathfrak{X} = C^*(X)$ of A, generated by the family X, where $C^*(Y)$ are the C^*-subalgebras of A generated by subsets,

$$Y \cup Y^* \cup \{1_A\} \text{of } A, \text{ with } Y^* = \{y^* \in A : y \in A\}.$$

Then one can obtain a canonical C^*-probabilistic sub-structure,

$$\mathfrak{X}_{\varphi} \overset{denote}{=} (\mathfrak{X}, \varphi = \varphi \mid_{\mathfrak{X}}), \tag{32}$$

in (A, φ).

Now, let (B, ψ) be a C^*-probability space, containing a family $S = \{s_n\}_{n \in \mathbb{Z}}$ of mutually free $|\mathbb{Z}|$ -many semicircular elements, and let

$$\mathfrak{S}_{\psi} \overset{denote}{=} (\mathfrak{S},\ \psi = \psi \mid_{\mathfrak{S}}) \tag{33}$$

be the corresponding C^*-probabilistic sub-structure of $(B,\ \psi)$, as in (32), where $\mathfrak{S} = C^*(S)$ is the C^*-subalgebra of B generated by the family S.

Remark 1 Such a topological $*$-probability space $\mathfrak{S}$ of (33) is well-determined naturally (e.g., [5, 12]), or artificially (e.g., [6–8, 10, 11]).

Consider the following structure theorem of the C^*-algebra $\mathfrak{X}$ in $(A,\ \varphi)$.

Proposition 1 *Let $\mathfrak{X}$ be the C^*-subalgebra $C^*(X)$ in $(A,\ \varphi)$. Then*

$$\mathfrak{X} \overset{*\text{-}iso}{=} \overline{\underset{n\in\mathbb{N}_0}{\star} C^*(\{x_n\})} \overset{*\text{-}iso}{=} \overline{C^*\left(\underset{n\in\mathbb{N}_0}{\star} \{x_n\}\right)}, \tag{34}$$

in $(A,\ \varphi)$, where "$\overset{*\text{-}iso}{=}$" means "being $*$-isomorphic."

Here, the free product $(\star)$ in the first $*$-isomorphic relation of (34) is the free-probability-theoretic free product of [17, 22, 28, 30], and the free product $(\star)$ in the second $*$-isomorphic relation of (34) is the pure-algebraic free product generating noncommutative free words in $X = \underset{n\in\mathbb{N}_0}{\cup} \{x_n\}$.

Proof Let $\mathfrak{X} = C^*(X)$ be the C^*-subalgebra of A. Then, by the assumption that X is a free family consisting of mutually free semicircular elements $\{x_n\}_{n\in\mathbb{N}_0}$ in $(A,\ \varphi)$, one can get that

$$\mathfrak{X} \overset{def}{=} C^*(X) = \overline{C^*\left(\{x_n\}_{n\in\mathbb{N}_0}\right)} \overset{*\text{-iso}}{=} \underset{n\in\mathbb{N}_0}{\star} \overline{C^*(\{x_n\})}. \tag{35}$$

Therefore, the first $*$-isomorphic relation of (34) holds by (35). i.e., all elements T of $\mathfrak{X}$ are the limits of linear combinations of free "reduced" words (under operator multiplication on A) in $\mathfrak{X}$ by [17, 22, 23, 30].

So, if we consider all noncommutative free words in the family $X = \{x_n\}_{n\in\mathbb{N}_0}$, then they have their unique operator forms in $\mathfrak{X}$, which are the free reduced words up to operator multiplication inherited from that on A, by (35). It shows that the second $*$-isomorphic relation of (34) holds, too. ■

The above proposition provides a structure theorem of $\mathfrak{X}$ in $(A,\ \varphi)$. So, by (34), one can understand the C^*-probability space $\mathfrak{X}_{\varphi}$ of (32) as an independent free-probabilistic structure,

$$\mathfrak{X}_{\varphi} = \left(\underset{n\in\mathbb{N}_0}{\star} C^*(\{x_n\}),\ \underset{n\in\mathbb{N}_0}{\star} \varphi \mid_{C^*(\{x_n\})}\right). \tag{36}$$

From below, we regards $\mathfrak{X}_{\varphi}$ of (32) and (36) as identical free-probabilistic objects.

Under similar arguments, one can get the following structure theorem of the C^*-probability space $\mathfrak{S}_{\psi}$ of (33)

Proposition 2 *Let $\mathfrak{S}_{\psi}$ be the C^*-probability space (33) in (B, ψ). Then*

$$\mathfrak{S}_{\psi} = \left(\underset{j \in \mathbb{Z}}{\star} C^* \left(\{s_j\}\right), \ \underset{j \in \mathbb{Z}}{\star} \psi \mid_{C^*(\{s_j\})} \right). \tag{37}$$

Proof Similar to the proof of (34), the C^*-subalgebra $\mathfrak{S} = C^*(S)$ of B satisfies that

$$\mathfrak{S} \overset{*\text{-iso}}{=} \overline{\underset{j \in \mathbb{Z}}{\star} C^*(\{s_j\})} \overset{*\text{-iso}}{=} \overline{C^* \left(\underset{j \in \mathbb{Z}}{\star} \{s_j\} \right)},$$

in (B, ψ), where $C^*(Z)$, here, mean the C^*-subalgebra of B generated by the subsets $Z \cup Z^*$ of B. Therefore, like in (34), the free-probabilistic structure (37) for the C^*-probability space $\mathfrak{S}_{\psi}$ of (33) holds. ■

To find a free-isomorphic relation between $\mathfrak{X}_{\varphi}$ and $\mathfrak{S}_{\psi}$, we regard them as the free-product C^*-probability spaces (36) and (37), respectively.

First, let's define a bijective function $g : \mathbb{N}_0 \to \mathbb{Z}$. To do that we partition $\mathbb{N}_0$ and $\mathbb{Z}$ as follows:

$$\mathbb{N}_0 = \{0\} \sqcup (2\mathbb{N}) \sqcup (2\mathbb{N} - 1), \tag{38}$$

and (38)

$$\mathbb{Z} = (-\mathbb{N}) \sqcup \{0\} \sqcup \mathbb{N},$$

where $\sqcup$ is the disjoint union, and

$$2\mathbb{N} = \{2n : n \in \mathbb{N}\}, 2\mathbb{N} - 1 = \{2n - 1 : n \in \mathbb{N}\},$$

and

$$-\mathbb{N} = \{-n : n \in \mathbb{N}\}.$$

From the partition (38), define a function $g : \mathbb{N}_0 \to \mathbb{Z}$ by

$$g(n) = \begin{cases} 0 & \text{if } n = 0 \\ \left(\frac{n+1}{2}\right) & \text{if } n \in 2\mathbb{N} - 1 \\ -\frac{n}{2} & \text{if } n \in 2\mathbb{N}, \end{cases} \tag{39}$$

in $\mathbb{Z}$, for all $n \in \mathbb{N}_0$. For instance,

$$g(0) = 0, g(1) = 1, g(2) = -1,$$

and

$$g(3) = 2, g(4) = -2,$$

etc.

Then the function g of (39) is a well-defined bijection. By this bijection g, one can construct a bijection

$$G : X \to S, \tag{40}$$

by (40)

$$G(x_n) = s_{g(n)}, \text{ for all } n \in \mathbb{N}_0.$$

where X and S are the generating free families of $\mathfrak{X}$ of (32), and $\mathfrak{S}$ of (33), respectively.

Since g is a bijection from $\mathbb{N}_0$ onto $\mathbb{Z}$, the function G of (40) is a bijection from the generator set X of $\mathfrak{X}$ onto the generator set S of $\mathfrak{S}$. Therefore, one can define the corresponding "multiplicative" linear transformation,

$$\Psi : \mathfrak{X} \to \mathfrak{S}, \tag{41}$$

satisfying

$$\Psi(x_n) = G(x_n) = s_{g(n)} \in S, \forall x_n \in X,$$

and $\Psi(1_A) = 1_B$, where G is in the bijection (40).

More precisely, for an alternating N-tuple $(n_1, ..., n_N) \in \mathbb{N}_0^N$, satisfying

$$n_1 \neq n_2, n_2 \neq n_3, \dots, n_{N-1} \neq n_N \text{in } \mathbb{N},$$

if one has a free reduced word $T = \prod_{l=1}^{N} x_{n_l}^{k_l} \in \mathfrak{X}_\varphi$, where $x_{n_1}, ..., x_{n_N} \in X$, for $k_1, ..., k_N \in \mathbb{N}$, for $N \in \mathbb{N}$, then

$$\Psi(T) = \Psi\left(\prod_{l=1}^{N} x_{n_l}^{k_l}\right) = \prod_{l=1}^{N} \Psi\left(x_{n_l}^{k_l}\right)$$

by the multiplicativity of Ψ

$$= \prod_{l=1}^{N} \left(\Psi(x_{n_l})\right)^{k_l}$$

by the multiplicativity of Ψ

$$= \prod_{l=1}^{N} s_{g(n_l)}^{k_l}, \tag{42}$$

in $\mathfrak{S}_{\psi}$, by (41).

Since $(n_1, \ldots, n_N)$ is an alternating N-tuple of $\mathbb{N}_0^N$, the N-tuple

$$(g(n_1), \ldots, g(n_N)) \in \mathbb{Z}^N$$

is an alternating N-tuple in $\mathbb{Z}$, too, by the bijectivity (39) and (40).

Thus, the formula (42) shows that the images $\Psi(T) \in \mathfrak{S}_{\psi}$ of all free reduced words $T \in \mathfrak{X}_{\varphi}$ with their lengths-N become free reduced words with the same lengths-N. i.e., the multiplicative linear transformation Ψ of (41) preserves the free structures of $\mathfrak{X}_{\varphi}$ to those of $\mathfrak{S}_{\psi}$. Also, it is not hard to check that the bijectivity of the function G of (40) guarantees the bijectivity and boundedness of Ψ by (34) and (37), because of the freeness-preserving property (42).

Lemma 3 *Let $\Psi : \mathfrak{X} \to \mathfrak{S}$ be the multiplicative linear transformation (41). Then it is a $*$-isomorphism. i.e.,*

$$\mathfrak{X} \overset{*\text{-}iso}{=} \mathfrak{S}. \tag{43}$$

Proof By the discussions in the very above paragraphs, the morphism Ψ is a bijective, bounded, freeness-preserving, multiplicative linear transformation. Remark again that all elements of $\mathfrak{X}$ (or, of $\mathfrak{S}$) are the limits of linear combinations of free reduced words in X (resp., in S) by (34) (resp., by (37)).

Observe that, for any $x_n \in X \subset \mathfrak{X}$, and $t \in \mathbb{C}$,

$$\Psi\left((tx_n)^*\right) = \Psi\left(\bar{t}\, x_n\right)$$

since $x_n^* = x_n$, by the semicircularity on the generator set X of $\mathfrak{X}$

$$= \bar{t}\Psi(x_n) = \bar{t}s_{g(n)} = \bar{t}s_{g(n)}^*$$

since $s_{g(n)}^* = s_{g(n)}$, by the semicircularity on the generator set S of $\mathfrak{S}$

$$= \left(ts_{g(n)}\right)^* = \left(\Psi(tx_n)\right)^*,$$

in $\mathfrak{S}$.

Therefore, by (34), (37), and (42),

$$\Psi\left(T^*\right) = \left(\Psi(T)\right)^* \text{ in} \mathfrak{S}, \tag{44}$$

for all $T \in \mathfrak{X}$.

Therefore, the morphism Ψ of (41) is a $*$-isomorphism from $\mathfrak{X}$ onto $\mathfrak{S}$ by (44). Equivalently, the $*$-isomorphic relation (43) holds. ■

By the above lemma, we obtain the following free-isomorphic relation.

Theorem 3 *Let $\mathfrak{X}_\varphi$ and $\mathfrak{S}_\psi$ be the C^*-probability spaces (36) and (37), respectively. If $\varphi(1_A) = \psi(1_B)$ in $\mathbb{C}$, then*

$$\mathfrak{X}_\varphi \overset{\text{free-iso}}{=} \mathfrak{S}_\psi. \tag{45}$$

Proof By (43), there exists a $*$-isomorphism Ψ of (41) from $\mathfrak{X}$ onto $\mathfrak{S}$. By the assumption that $\varphi(1_A) = \psi(1_B)$, one has

$$\psi(1_B) = \psi\left(\Psi(1_A)\right) = \varphi(1_A)\text{in}\mathbb{C}.$$

So, by (34) and (37), it suffices to show that the $*$-isomorphism Ψ preserves the free distributions of generators of $\mathfrak{X}_\varphi$ to those of generators of $\mathfrak{S}_\psi$.

Let $x_n \in X \subset \mathfrak{X}_\varphi$. Then

$$\psi\left((\Psi(x_n))^k\right) = \psi\left(s_{g(n)}^k\right) = \left(\omega_k c_{\frac{k}{2}}\right) = \varphi\left(x_n^k\right),$$

for all $k \in \mathbb{N}$, by the semicircularity (4 of $X \cup S$.

It shows that Ψ preserves the free probability on $\mathfrak{X}_\varphi$ to that on $\mathfrak{S}_\psi$, by (15) and (29). i.e., it is a free-isomorphism. Therefore, two C^*-probability space $\mathfrak{X}_\varphi$ and $\mathfrak{S}_\psi$ are free-isomorphic. ■

The above free-isomorphic relation (45) illustrates that the study of free probability on $\mathfrak{X}_\varphi$ is to study that on $\mathfrak{S}_\psi$. So, one can use the known results from [6–8, 11].

Assumption and Notation From below, we will identify $\mathfrak{X}_\varphi$ and $\mathfrak{S}_\psi$ as the same C^*-probability space, and denote it by $\mathfrak{X}_\varphi$. □

5.3 *Free-Distributional Data on $\mathfrak{X}_\varphi$*

Let $\mathfrak{X}_\varphi = (\mathfrak{X}, \varphi)$ be the C^*-probability space (37) identified with (36), with its free-generator set $X = \{x_j\}_{j \in \mathbb{Z}}$ of mutually free, $|\mathbb{Z}|$-many semicircular elements x_j's (by (45)).

Theorem 4 *Let $I_s = (i_1, ..., i_s)$ be an arbitrary s-tuple in $\mathbb{Z}^s$, for $s \in \mathbb{N}$, like in (16), and let $\pi_{(I_s)} \in NC(\{i_1, \ldots, i_s\})$ be in the sense of (19) for I_s. If $X[I_s]$ be a free random variable of $\mathfrak{X}_\varphi$, in the sense of (20). If $\varphi(X[I_s]) \neq 0$, then there exist $r \leq s$ in $\mathbb{N}$, and $i_{l_1}, ..., i_{l_r} \in \{i_1, ..., i_s\}$, such that*

$$k_{l_1} > k_{l_2} > \cdots > k_{l_r} in\ \mathbb{N}_0,$$

and

$$n_1, n_2, \ldots, n_r \in \mathbb{N},$$

such that

$$\sum_{l=1}^{r} n_l\ (2k_l) = s\ in\ \mathbb{N},$$

and

$$\varphi\,(X[I_s]) = \left(\prod_{i=1}^{r-1} \beta_{k_{l_i} > k_{l_{i+1}}}^{\Sigma_{l=1}^{i} n_l}\right)\left(c_{k_r}^{\Sigma_{l=1}^{r} n_l}\right), \tag{46}$$

with

$$\beta_{k_{l_i} > k_{l_{i+1}}} = 2^{k_{l_i} - k_{l_{i+1}}} \left(\frac{2k_{l_{i+1}} + 1}{k_{l_i} + 1}\right)\left(\prod_{u=1}^{k_{l_i} - k_{l_{i+1}} - 1} \left(2 - \frac{1}{(k_{l_i} + 1) - u}\right)\right),$$

for all $i = 1, \ldots, r - 1$.

Proof The free-distributional data (46) on $\mathfrak{X}_\varphi$ are obtained by (15), (29) and (45). In particular, if $i_1 = i_2 = \ldots = i_s$ in $\mathbb{Z}$, then the free-distributional data (46) is determined by (15); meanwhile, if there exists at least one $i_{k_0} \in \{i_1, \ldots, i_s\}$, such that $i_{k_0} \neq i_k$ in $\mathbb{Z}$, for some $k \in \{1, \ldots, s\}$, then the free-distributional data (46) is characterized by (29). ■

The above theorem characterizes the general free-distributional data on $\mathfrak{X}_\varphi$, because all elements of $\mathfrak{X}_\varphi$ are the limits of linear combinations of free reduced words in $X = \{x_j\}_{j \in \mathbb{Z}}$, by (36), (37) and (45).

6 Free-Isomorphisms on $\mathfrak{X}_\varphi$

Throughout this section, we let $\mathfrak{X}_\varphi$ be the C^* -probability space (37) generated by a *free semicircular family* $X = \{x_j\}_{j \in \mathbb{Z}}$ in the sense that: X is a free family whose mutually free elements are all semicircular.

6.1 *Shifts on* $\mathbb{Z}$

Define bijections h_+ and h_- on the set $\mathbb{Z}$ of all integers by

$$h_+(j) = j + 1, \text{ and } h_-(j) = j = 1, \tag{47}$$

for all $j \in \mathbb{Z}$.

Remark 2 Remark that, by (47), there are bijections h'_+ and h'_-,

$$h'_\pm = g^{-1} \circ h_\pm \circ g \text{on } \mathbb{N}_0, \tag{47'}$$

where g is the bijection (39) from $\mathbb{N}_0$ onto $\mathbb{Z}$. Thus, the existence of the bijections $h_\pm$ of (47 on $\mathbb{Z}$ guarantees the existence of bijections $h'_\pm$ on $\mathbb{N}_0$. From below, we focus on $h_\pm$ of (47).

Then one can define the bijections $h_\pm^{(n)}$ on $\mathbb{Z}$, by

$$h_\pm^{(n)} \overset{def}{=} \underbrace{h_\pm \circ h_\pm \circ \; .... \; \circ h_\pm}_{n\text{-times}}, \tag{48}$$

for all $n \in \mathbb{N}_0$, with axiomatization:

$$h_\pm^{(0)} = \text{the identity map } id_{\mathbb{Z}} \text{on } \mathbb{Z},$$

where ($\circ$) is the functional composition.

By (47) and (48), it is easy to check that

$$h_\pm^{(n)}(j) = j \pm n, \text{ for all } n \in \mathbb{N}_0.$$

Definition 3 We call the bijections $h_\pm^{(n)}$ of (48), the n-($\pm$)-shifts on $\mathbb{Z}$, for all $n \in \mathbb{N}_0$.

6.2 *Integer Shifts on* $\mathfrak{X}_\varphi$

Let $h_\pm^{(n)}$ be the n-($\pm$)-shifts (48) on $\mathbb{Z}$. In this section, we construct $*$-isomorphisms on $\mathfrak{X}_\varphi$ from the shifts $h_\pm^{(n)}$. For convenience, let

$$\mathbb{N}_0^\pm \overset{denote}{=} \{\pm\} \times \mathbb{N}_0.$$

For $(e, k) \in \mathbb{N}_0^\pm$, define a multiplicative linear transformation λ_e^k acting on $\mathfrak{X}_\varphi$ by the morphism satisfying

$$\lambda_e^k\left(x_j\right) = x_{jek}, \text{ for all } x_j \in X \subset \mathfrak{X}_\varphi, \tag{49}$$

where

$$jek = \begin{cases} j+k & \text{if } e=+ \\ j-k & \text{if } e=-, \end{cases}$$

in $\mathbb{Z}$.

By the multiplicativity (49) of the morphism λ_e^k, if $T = \prod_{l=1}^{N} x_{j_l}^{n_l}$ is a free reduced words of $\mathfrak{X}_\varphi$ with its length-N, then

$$\lambda_e^k(T) = \prod_{l=1}^{N} \lambda_e^k\left(x_{j_l}\right)^{n_l} = \prod_{l=1}^{N} x_{j_l ek}^{n_l}, \tag{50}$$

in $\mathfrak{X}_\varphi$.

Remark that, if $(j_1, ..., j_N) \in \mathbb{Z}^N$ is alternating, then

$$(j_1 ek, \ ..., \ j_N ek) \in \mathbb{Z}^N$$

is alternating, too, by the bijectivity of $h_\pm^{(k)}$, for all $k \in \mathbb{N}_0$.

So, the formula (50) demonstrates that λ_e^k assign free reduced words to free reduced words with same lengths in $\mathfrak{X}_\varphi$, for all $(e, k) \in \mathbb{N}_0^\pm$.

Note that, for any $t \in \mathbb{C}$, and $x_j \in X \subset \mathfrak{X}$,

$$\lambda_e^k\left((tx_j)^*\right) = \bar{t}x_{jek} = \bar{t}x_{jek}^* = \left(\lambda_e^k(tx_j)\right)^*,$$

implying that

$$\lambda_e^k\left(T^*\right) = \left(\lambda_e^k(T)\right)^*, \text{ for all } T \in \mathfrak{X}_\varphi, \tag{51}$$

in $\mathfrak{X}_\varphi$, by (37).

Theorem 5 *Let λ_e^k be a multiplicative linear transformation (49) on $\mathfrak{X}_\varphi$, for $(e, k) \in \mathbb{N}_0^\pm$. Then it is a free-isomorphism on $\mathfrak{X}_\varphi$.*

Proof By (51), the morphism λ_e^k of (49) is a well-defined $*$-homomorphism on $\mathfrak{X}_\varphi$. And, by the bijectivity of the k-(e)-shift $h_e^{(k)}$ on $\mathbb{Z}$, the restriction $\lambda_e^k\ |_X$ is a bijection on the free-generator set X of $\mathfrak{X}_\varphi$. Thus, by (50) and (37), it is bijective, and hence, it is a $*$-isomorphism on $\mathfrak{X}_\varphi$.

Observe now that

$$\varphi\left(\left(\lambda_e^k(x_j)\right)^n\right) = \varphi\left(x_{jek}^n\right) = \omega_n c_{\frac{n}{2}} = \varphi\left(x_j^n\right), \tag{52}$$

for all $n \in \mathbb{N}$, for all $x_j \in X$.

Therefore, for all s-tuple $I_s \in \mathbb{Z}^s$,

$$\varphi\left(X[I_s]\right) = \varphi\left(\lambda_e^k\left(X[I_s]\right)\right) \text{ in } \mathfrak{X}_\varphi,$$

by (52) and (46), where $X[I_s]$ are in the sense of (46). It guarantees that

$$\varphi\left(T\right) = \varphi\left(\lambda_e^k(T)\right), \text{ for all } T \in \mathfrak{X}_\varphi,$$

in $\mathfrak{X}_\varphi$, by (37). Therefore, this $*$-isomorphism λ_e^k preserves the free probability on $\mathfrak{X}_\varphi$ to that on $\mathfrak{X}_\varphi$, i.e., it is a free-isomorphism. ∎

Let $Aut\left(\mathfrak{X}_\varphi\right)$ be the *automorphism group of* $\mathfrak{X}_\varphi$,

$$Aut\left(\mathfrak{X}_\varphi\right) \overset{def}{=} \left(\left\{\alpha \,\middle|\, \begin{array}{c}\alpha \text{ is a}\\ *\text{-isomorphism}\\ \text{on } \mathfrak{X}_\varphi\end{array}\right\}, \cdot\right),$$

where $(\cdot)$ is the product (or composition) on $*$-isomorphisms.

Define now a subset λ of $Aut\left(\mathfrak{X}_\varphi\right)$ by

$$\lambda = \{\lambda_e^k : (e, k) \in \mathbb{N}_0^{\pm}\}, \tag{53}$$

where λ_e^k are the free-isomorphisms (49) on $\mathfrak{X}_\varphi$.

Theorem 6 *The subset λ of (53) is an abelian subgroup of $Aut(\mathfrak{X}_\varphi)$.*

Proof Let $\lambda_{e_1}^{k_1}, \lambda_{e_2}^{k_2} \in \lambda$. Then

$$\lambda_{e_1}^{k_1}\lambda_{e_2}^{k_2} = \lambda_{sgn(e_1k_1e_2k_2)}^{|e_1k_1e_2k_2|} \text{ in } \lambda,$$

where sgn is the sign map on $\mathbb{Z}$,

$$sgn(j) = \begin{cases} + & \text{if } j \geq 0 \\ - & \text{if } j < 0, \end{cases}$$

for all $j \in \mathbb{Z}$, and $|.|$ is the absolute value on $\mathbb{Z}$. Therefore, the algebraic structure $(\lambda, \cdot)$ is well-defined in $Aut\left(\mathfrak{X}_\varphi\right)$.

One can check that

$$\begin{aligned}\left(\lambda_{e_1}^{k_1}\lambda_{e_2}^{k_2}\right)\lambda_{e_3}^{k_3} &= \lambda_{sgn(e_1k_1e_2k_2)}^{|e_1k_1e_2k_2|}\lambda_{e_3}^{k_3} \\ &= \lambda_{sgn(e_1k_1e_2k_2e_3k_3)}^{||e_1k_1e_2k_2|e_3k_3|} = \lambda_{sgn(e_1k_1e_2k_2e_3k_3)}^{|e_1k_1|e_2k_2e_3k_3||} \\ &= \lambda_{e_1}^{k_1}\left(\lambda_{e_2}^{k_2}\lambda_{e_3}^{k_3}\right),\end{aligned}$$

in λ. Observe now that the subset λ contains

$$\lambda_e^0 = 1_{\mathfrak{X}_\varphi}, \text{ the identity map on } \mathfrak{X},$$

by (47) and (49), satisfying

$$1_{\mathfrak{X}_\varphi} \cdot \lambda_e^k = \lambda_e^k = \lambda_e^k \cdot 1_{\mathfrak{X}_\varphi} \text{ on } \mathfrak{X}_\varphi,$$

for all $(e, k) \in \mathbb{N}_0^{\pm}$.

Note that, for any $(e, k) \in \mathbb{N}_0^{\pm}$,

$$\lambda_e^k \lambda_{-e}^k = \lambda_{sgn(0)}^0 = 1_{\mathfrak{X}_\varphi} = \lambda_{-e}^k \lambda_e^k, \text{ in } \lambda.$$

It shows that every element $\lambda_e^k \in \lambda$ has its unique $(\cdot)$-inverse λ_{-e}^k, i.e.,

$$\left(\lambda_e^k\right)^{-1} = \lambda_{-e}^k, \text{ in } \lambda,$$

where y^{-1} mean the group-inverses of y. So, the pair $(\lambda, \cdot)$ is a subgroup of $Aut(\mathfrak{X}_\varphi)$.

It is not difficult to check that

$$\lambda_{e_1}^{k_1} \lambda_{e_2}^{k_2} = \lambda_{sgn(e_1k_1e_2k_2)}^{|e_1k_1e_2k_2|} = \lambda_{sgn(e_2k_2e_1k_1)}^{|e_2k_2e_1k_1|} = \lambda_{e_2}^{k_2} \lambda_{e_1}^{k_1},$$

in λ. Therefore, the subgroup $(\lambda, \cdot)$ is commutative in $Aut\left(\mathfrak{X}_\varphi\right)$. ■

The above theorem shows that the set λ of (53) is an abelian group. More precisely, we have the following structure theorem of λ.

Theorem 7 *Let* $\lambda \overset{denote}{=} (\lambda, \cdot)$ *be the abelian subgroup of* $Aut(\mathfrak{X}_\varphi)$. *Then*

$$\lambda \overset{Group}{=} (\mathbb{Z}, +), \tag{54}$$

where "$\overset{Group}{=}$" *means "being group-isomorphic."*

Proof Let λ be the subgroup (53) of $Aut(\mathfrak{X}_\varphi)$. Define a map $\Phi : \mathbb{Z} \to \lambda$ by

$$\Phi(j) = \lambda_{sgn(j)}^{|j|}, \text{ including} \Phi(0) = 1_{\mathfrak{X}_\varphi}.$$

Then it is a well-defined bijection from $\mathbb{Z}$ onto λ, because

$$j \longmapsto (sgn(j), \ |j|)$$

is bijective from $\mathbb{Z}$ onto $\mathbb{N}_0^{\pm}$; and it satisfies that

$$\Phi(j_1 + j_2) = \lambda_{sgn(j_1+j_2)}^{|j_1+j_2|} = \lambda_{sgn(j_1)}^{|j_1|} \lambda_{sgn(j_2)}^{|j_2|} = \Phi(j_1)\Phi(j_2),$$

in λ, for all $j_1, j_2 \in \mathbb{Z}$.

Therefore, the group-isomorphic relation (54) holds. ■

The above theorem characterizes the abelian group λ as an infinite cyclic abelian group in $Aut(\mathfrak{X}_\varphi)$.

By definition, there is a natural action θ of the group λ acting on our C^*-probability space $\mathfrak{X}_{\varphi}$;

$$\theta\left(\lambda_e^k\right)(T) = \lambda_e^k(T), \text{ for all } T \in \mathfrak{X}_{\varphi}. \tag{55}$$

Definition 4 The abelian group λ of (53), acting on $\mathfrak{X}_{\varphi}$ via the natural group-action θ of (55), is called the integer-shift group on $\mathfrak{X}_{\varphi}$.

6.3 *Free-Isomorphic Relations on* $\mathfrak{X}_{\varphi}$

In this section, we consider how the integer-shift group λ of (53) affects the free probability on $\mathfrak{X}_{\varphi}$, under the group-action θ of (55).

Theorem 8 *Let λ be the integer-shift group, and let θ be the group-action (55) of λ acting on $\mathfrak{X}_{\varphi}$. Then the free probability on $\mathfrak{X}_{\varphi}$ is preserved by θ. i.e.,*

$$\varphi\left(\theta\left(\lambda_e^k\right)(T)\right) = \varphi(T), \textit{for all } T \in \mathfrak{X}_{\varphi}, \tag{56}$$

for all $\lambda_e^k \in \lambda$.

Proof By (55), for any $T \in \mathfrak{X}_{\varphi}$,

$$\theta\left(\lambda_e^k\right)(T) = \lambda_e^k(T) \text{in } \mathfrak{X}_{\varphi}.$$

And all integer-shifts $\lambda_e^k \in \lambda$ are free-isomorphisms on $\mathfrak{X}_{\varphi}$, by (52). Therefore, for any $T \in \mathfrak{X}_{\varphi}$,

$$\varphi\left(\theta\left(\lambda_e^k\right)(T)\right) = \varphi\left(\lambda_e^k(T)\right) = \varphi(T).$$

Therefore, the action θ of λ preserves the free probability on $\mathfrak{X}_{\varphi}$. ■

The above theorem characterizes how the integer-shift group λ preserves the free probability on $\mathfrak{X}_{\varphi}$.

Notation From below, we denote the images $\theta\left(\lambda_e^k\right)(T) \in \mathfrak{X}_{\varphi}$ of $T \in \mathfrak{X}_{\varphi}$ simply by $\lambda_e^k(T)$, for all $\lambda_e^k \in \lambda$. □

7 A Group Dynamical System $(\lambda, \Lambda\mathfrak{X}, \alpha)$

Let $\lambda \subset Aut(\mathfrak{X}_{\varphi})$ be the integer-shift group (53) acting on the C^*-algebra $\mathfrak{X}_{\varphi}$ generated by the free family $\{x_j\}_{j\in\mathbb{Z}}$ of semicircular elements x_j's (via the action θ of (55)).

Remark that, by (54), the integer-shift group λ is a discrete group. From this discrete group λ, define the Hilbert space,

$$H_{\lambda} \overset{def}{=} l^{2}(\lambda), \tag{57}$$

the l^2-space generated by all non-trivial group elements $\lambda_e^k \neq 1_{\mathfrak{X}_\varphi} \in \lambda$. i.e., if h is a vector of H_λ, then

$$h = \sum_{g \in \lambda} t_g \xi_g, \text{ with } t_g \in \mathbb{C},$$

satisfying

$$< h, h >_2 = \sum_{g \in \lambda} |t_g|^2 < \infty,$$

where $|.|$ is the modulus on $\mathbb{C}$, and where $<, >_2$ is the inner product on H_λ,

$$\left\langle \sum_{g \in \lambda} t_g \xi_g, \sum_{i \in \lambda} s_i \xi_i \right\rangle_2 \overset{def}{=} \sum_{g \in \lambda} t_g \overline{s_g}.$$

Then this separable Hilbert space H_λ has its *orthonormal basis*,

$$\mathcal{B} = \{\xi_g : g \in \lambda^{\times} = \lambda \setminus \{1_{\mathfrak{X}_\varphi}\}\},$$

satisfying

$$\langle \xi_{g_1}, \xi_{g_2} \rangle_2 = \delta_{g_1, g_2}, \forall g_1, g_2 \in \lambda^{\times},$$

and

$$\|\xi_g\|_2 = \sqrt{\langle \xi_g, \xi_g \rangle_2} = 1, \forall g \in \lambda^{\times}.$$

Also, if we consider the subset

$$\mathcal{B}_1 = \mathcal{B} \cup \{\xi_{1_{\mathfrak{X}_\varphi}}\} \text{of } H_\lambda,$$

then

$$\xi_{g_1} \xi_{g_2} = \xi_{g_1 g_2} \text{in } \mathcal{B}_1, \text{ for all } g_1, g_2 \in \lambda,$$

implying that

$$\xi_g \xi_{g^{-1}} = \xi_{gg^{-1}} = \xi_{1_{\mathfrak{X}_\varphi}} = \xi_{g^{-1}g} = \xi_{g^{-1}} \xi_g,$$

in $\mathcal{B}_1$, for all $g \in \lambda$.

This l^2-Hilbert space H_λ of (57) is called the *group Hilbert space of* λ. Consider the operator algebra $B(H_\lambda)$, consisting of all bounded operators on H_λ. By definition, group elements g of λ act as multiplication operators M_g on H_λ, i.e.,

$$M_g(h) = \xi_g h, \text{ for all } h \in H_\lambda. \tag{58}$$

i.e., on the group Hilbert space H_λ, there is the group-action M of λ such that

$$M(g) = M_g \in B(H_\lambda), \text{ for all } g \in \lambda, \tag{59}$$

by (58). Equivalently, the pair (H_λ, m) forms a *Hilbert-space representation of* λ, called the *group representation of* λ.

Definition 5 Let (H_λ, M) be the group representation of the integer-shift group λ, consisting of the group Hilbert space H_λ of (57), and the group-action M of (59). Define the C^*-subalgebra Λ of the operator algebra $B(H_\lambda)$ by the C^*-algebra

$$\Lambda \overset{def}{=} C_M^*(\lambda) = C^*(M(\lambda)) \tag{60}$$

in $B(H_\lambda)$. Then the C^*-algebra Λ is called the group C^*-algebra of λ induced by (H_λ, M).

By definition, every element T of Λ has its expression,

$$T = \sum_{g \in \lambda} t_g M_g, \text{ with } t_g \in \mathbb{C},$$

where $\sum$ is the infinite (or, the limit of finite) sum(s) under the operator–norm-topology inherited from that for $B(H_\lambda)$. Note that the C^*-algebra Λ of (60) is equipped with the *canonical trace* τ *on* Λ, defined by

$$\tau\left(\sum_{g \in \lambda} t_g M_g\right) = t_{1_{\mathfrak{X}_\varphi}}. \tag{61}$$

Indeed, the linear functional τ on the group C^*-algebra Λ is a well-defined trace satisfying

$$\tau(a_1 a_2) = \tau(a_2 a_1), \text{ for all } a_1, a_2 \in \Lambda,$$

by (61).

Naturally, the pair (Λ, τ) forms a "unital" C^* -probability space satisfying

$$\tau(1_\Lambda) = \tau\left(M_{1_{\mathfrak{X}_\varphi}}\right) = 1,$$

where $1_\Lambda = M_{1_{\mathfrak{X}_\varphi}}$ is the unity of Λ.

Definition 6 The unital C^*-probability space (Λ, τ) is called the canonical group C^*-probability space of λ.

Let $\mathfrak{X}$ be the given C^*-algebra generated by the free semicircular family $\{x_j\}_{j \in \mathbb{Z}}$. Remark that, since $\mathfrak{X}$ is a C^*-algebra, there exists a well-determined *GNS-representation* $(H_{\mathfrak{X}}, \beta)$ of $\mathfrak{X}$, by Gelfand-Naimak-Segal, where $H_{\mathfrak{X}}$ is a Hilbert space where $\mathfrak{X}$ acts, and β is the corresponding GNS-representation (or algebra-action) of $\mathfrak{X}$ acting on $H_{\mathfrak{X}}$. From below, let's fix the *GNS-Hilbert space* $H_{\mathfrak{X}}$ satisfying $\mathfrak{X} \subset B(H_{\mathfrak{X}})$, where $B(H_{\mathfrak{X}})$ is the operator algebra consisting of all bounded operators on $H_{\mathfrak{X}}$.

Remark 3 Recall that our C^*-probability space $\mathfrak{X}_\varphi$ is originally introduced to be a C^*-probabilistic sub-structure of a C^*-probability space (A, φ). This C^*-algebra A is understood as a C^*-subalgebra of an operator algebra $B(\mathcal{H})$, where $\mathcal{H}$ is a Hilbert space that A acts on. So, the GNS-Hilbert space $H_{\mathfrak{X}}$ is Hilbert-space-homomorphic to $\mathcal{H}$, as a subspace. However, since we understand $\mathfrak{X}_\varphi$ as an independent C^*-probabilistic structure (37), we here focus on the GNS-Hilbert space $H_{\mathfrak{X}}$.

For the group Hilbert space H_λ of our integer-shift group λ and the GNS-Hilbert space $H_{\mathfrak{X}}$ of the C^* -probability space $\mathfrak{X}_\varphi$, define the *tensor product Hilbert space* H by

$$H \overset{def}{=} H_\lambda \otimes H_{\mathfrak{X}}, \tag{62}$$

where $\otimes$ is the tensor product of Hilbert spaces.

Then the *tensor product C^*-algebras*

$$\Lambda\mathfrak{X} \overset{denote}{=} \Lambda \otimes_{\mathbb{C}} \mathfrak{X} \tag{63}$$

acts on the Hilbert space H of (62), where $\otimes_{\mathbb{C}}$ is the tensor product of C^*-algebras. i.e., the C^*-algebra $\Lambda\mathfrak{X}$ of (63) is a C^*-subalgebra of the operator algebra $B(H)$.

7.1 A Group C^*-Dynamical System $\Gamma = (\lambda, \Lambda\mathfrak{X}, \alpha)$

Let $M_g \in \Lambda$ be multiplication operators (58) on the group Hilbert space H_λ of (57) with their symbols $g \in \lambda$.

Notation and Assumption 1 (in short, **NA** 1) From below, if there are no confusion, then we denote $M_{\lambda_e^k} \in \Lambda$ simply by λ_e^k, for all $\lambda_e^k \in \lambda \subset \Lambda$. Under this simplified notation, all elements Y of Λ are expressed by

$$Y = \sum_{g \in \lambda} t_g M_g \overset{denote}{=} \sum_{g \in \lambda} t_g g, \text{ in } \Lambda, \tag{64}$$

by (60). □

Define now a group-action α of λ acting on the tensor product C^*-algebra $\Lambda\mathfrak{X}$ of (63) by a morphism satisfying

$$\alpha\left(\lambda_e^k\right)\left(\lambda_f^l \otimes T\right) = \lambda_e^k \lambda_f^l \otimes T, \tag{65}$$

for all $\lambda_e^k \in \lambda$, and for all $\lambda_f^l \otimes T \in \Lambda\mathfrak{X}$, with $\lambda_f^l \in \lambda \subset \Lambda$, and $T \in \mathfrak{X}_\varphi$.

Observe now that

$$\alpha\left(\lambda_{e_1}^{k_1}\lambda_{e_2}^{k_2}\right)\left(\lambda_f^l \otimes T\right) = \alpha\left(\lambda_{sig(e_1k_1e_2k_2)}^{|e_1k_1e_2k_2|}\right)\left(\lambda_f^l \otimes T\right)$$

by (53)

$$= \lambda_{sgn(e_1k_1e_2k_2)}^{|e_1k_1e_2k_2|}\lambda_f^l \otimes T$$

by (65)

$$\begin{aligned}
&= \lambda_{e_1}^{k_1}\lambda_{e_2}^{k_2}\lambda_f^l \otimes T = \lambda_{e_1}^{k_1}\left(\lambda_{e_2}^{k_2}\lambda_f^l\right) \otimes T \\
&= \alpha\left(\lambda_{e_1}^{k_1}\right)\left(\lambda_{e_2}^{k_2}\lambda_f^l \otimes T\right) \\
&= \alpha\left(\lambda_{e_1}^{k_1}\right)\left(\alpha\left(\lambda_{e_2}^{k_2}\right)\left(\lambda_f^l \otimes T\right)\right) \\
&= \left(\alpha\left(\lambda_{e_1}^{k_1}\right)\alpha\left(\lambda_{e_2}^{k_2}\right)\right)\left(\lambda_f^l \otimes T\right),
\end{aligned}$$

for all $\lambda_f^l \otimes T \in \Lambda$, with $\lambda_f^l \in \lambda \subset \Lambda$, and $T \in \mathfrak{X}_\varphi$. So, by (60) and (63),

$$\alpha\left(\lambda_{e_1}^{k_1}\lambda_{e_2}^{k_2}\right) = \left(\alpha\left(\lambda_{e_1}^{k_1}\right)\right)\left(\alpha\left(\lambda_{e_2}^{k_2}\right)\right) \text{ on } \Lambda\mathfrak{X}, \tag{66}$$

for $\lambda_{e_l}^{k_l} \in \lambda$, for all $l = 1, 2$.

Also, one has

$$\begin{aligned}
\alpha\left(\left(\lambda_e^k\right)^{-1}\right)\left(\lambda_f^l \otimes T\right) &= \alpha\left(\lambda_{-e}^k\right)\left(\lambda_f^l \otimes T\right) \\
&= \lambda_{-e}^k\lambda_f^l \otimes T = \left(\alpha\left(\lambda_e^k\right)\right)^*\left(\lambda_f^l \otimes T\right),
\end{aligned}$$

on the Hilbert space H of (61), for all $\lambda_e^k \in \lambda$ and $\lambda_f^l \otimes T \in \Lambda\mathfrak{X}$, implying that

$$\alpha\left(\left(\lambda_e^k\right)^{-1}\right) = \alpha\left(\lambda_e^k\right)^*, \forall \lambda_e^k \in \lambda. \tag{67}$$

Therefore, by (66) and (67), the morphism α of (65) is indeed a well-defined group-action of λ acting on $\Lambda\mathfrak{X}$.

Proposition 3 *Let $\Gamma = (\lambda, \Lambda\mathfrak{X}, \alpha)$ be a mathematical triple consisting of the integer-shift group λ, the tensor product C^*-algebra $\Lambda\mathfrak{X}$ of (63), and the group-action α of (65). Then Γ is a well-defined group C^*-dynamical system.*

Proof It suffices to show that α is a well-defined group-action of λ acting on $\Lambda\mathfrak{X}$, but, by (66) and (67), it is. ■

So, we obtain a well-defined group C^*-dynamical system,

$$\Gamma = (\lambda, \Lambda\mathfrak{X}, \alpha) \tag{68}$$

of the integer-shift group λ acting on $\Lambda\mathfrak{X}$ via the group-action α.

Let

$$G = \sum_{\lambda_e^k \in \lambda} t_{\lambda_e^k} \lambda_e^k \in \Lambda, \text{ and } T = \prod_{l=1}^{N} x_{j_l}^{n_l} \in \mathfrak{X}_\varphi,$$

under **NA** 1 (See (64)), where the N-tuple $(j_1, ..., j_N)$ is alternating in $\mathbb{Z}$, and $n_1, \ldots, n_N \in \mathbb{N}$, and let

$$Y = G \otimes T \in \Lambda\mathfrak{X}. \tag{69}$$

By assumption, the tensor factor T of the operator Y of (69) is a free reduced word of $\mathfrak{X}_\varphi$ with its length-N, by (34), (37) and (45).

For any $\lambda_{e_0}^{k_0} \in \lambda$,

$$\alpha\left(\lambda_{e_0}^{k_0}\right)(Y) = \lambda_{e_0}^{k_0} G \otimes T = \left(\sum_{\lambda_e^k \in \lambda} t_e^k \lambda_{e_0}^{k_0} \lambda_e^k\right) \otimes T$$

by (65), where $t_e^k \overset{denote}{=} t_{\lambda_e^k}$ in $\mathbb{C}$

$$= \left(\sum_{\lambda_e^k \in \lambda} t_e^k \lambda_{sgn(e_0 k_0 ek)}^{|e_0 k_0 ek|}\right) \otimes T$$

$$= \sum_{\lambda_e^k \in \lambda} t_e^k \left(\left(\lambda_{sgn(e_0 k_0 ek)}^{|e_0 k_0 ek|}\right) \otimes T\right)$$

$$= \sum_{\lambda_e^k \in \lambda} t_e^k \left(\left(\lambda_{sgn(e_0 k_0 ek)}^{|e_0 k_0 ek|}\right) \otimes \left(\prod_{l=1}^{N} x_{j_l}^{n_l}\right)\right), \tag{70}$$

in $\Lambda\mathfrak{X}$, whenever $Y \in \Lambda\mathfrak{X}$ is in the sense of (69).

The above formula (70) shows that, to consider the images of the action α of λ in $\Lambda\mathfrak{X}$, it is sufficient to consider the images of $\lambda_e^k \otimes T \in \Lambda\mathfrak{X}$, where T are the free reduced words.

For the dynamical system Γ of (68), one can define the corresponding *crossed product C^*-algebra*

$$\mathbb{X}\Gamma \overset{def}{=} \Lambda\mathfrak{X} \times_\alpha \lambda, \tag{71}$$

by the C^*-algebra generated by $\Lambda\mathfrak{X}$ and $\{\alpha(g)\}_{g \in \lambda}$, satisfying the α- *relation*:

$$\left(W_1,\ \lambda_{e_1}^{k_1}\right)\left(W_2,\ \lambda_{e_2}^{k_2}\right) = \left(\left(\alpha\left(\lambda_{e_1}^{k_2}\lambda_{e_2}^{k_2}\right) W_1\right) W_2,\ \lambda_{e_1}^{k_1}\lambda_{e_2}^{k_2}\right), \tag{72}$$

and

$$\left(W,\ \lambda_e^k\right)^* = \left(\alpha\left(\lambda_{-e}^k\right)\left(W^*\right),\ \lambda_{-e}^k\right),$$

under NA 1, for all (e_1, k_1), (e_2, k_2), $(e, k) \in \mathbb{N}_0^{\pm}$, and $W_1, W_2, W \in \Lambda\mathfrak{X}$.

Definition 7 Let $\mathbb{X}\Gamma = \Lambda\mathfrak{X} \times_\alpha \lambda$ be the crossed product C^*-algebra (71) under the α-relation (72), induced by the group C^*-dynamical system $\Gamma = (\lambda, \Lambda\mathfrak{X}, \alpha)$ of (68). Then it is called the Γ-dynamical (C^*-)algebra of the integer-shift group λ (on $\mathfrak{X}_\varphi$).

7.2 The Γ-Dynamical Algebra $\mathbb{X}\Gamma$ of λ

Let $\mathbb{X}\Gamma = \Lambda\mathfrak{X} \times_\alpha \lambda$ be the Γ-dynamical algebra (71) of the integer-shift group λ induced by the group C^*-dynamical system Γ of (7.1.5). Now, we construct the tensor product C^*-algebra

$$\Lambda^2\mathfrak{X} \overset{def}{=} \Lambda\mathfrak{X} \otimes_{\mathbb{C}} \Lambda. \tag{73}$$

As a C^*-subalgebra of $\Lambda^2\mathfrak{X}$ of (73), define the "*conditional*" *tensor product C^*-algebra*

$$\mathbb{X}\Gamma' \overset{def}{=} \Lambda\mathfrak{X} \otimes_\alpha \Lambda, \tag{74}$$

by the C^*-algebra satisfying the *α-conditions*:

$$\left(W_1 \otimes_\alpha \lambda_{e_1}^{k_1}\right)\left(W_2 \otimes_\alpha \lambda_{e_2}^{k_2}\right) = \left(\alpha\left(\lambda_{e_1}^{k_1}\lambda_{e_2}^{k_2}\right)(W_1)\right) W_2 \otimes_\alpha \lambda_{e_1}^{k_1}\lambda_{e_2}^{k_2}, \tag{75}$$

and

$$\left(W \otimes_\alpha \lambda_e^k\right)^* = \left(\alpha\left(\lambda_{-e}^k\right)(W)\right) \otimes_\alpha \lambda_{-e}^k,$$

under **NA** 1, for all $W_1, W_2, W \in \Lambda\mathfrak{X}$, and $\lambda_{e_1}^{k_1}, \lambda_{e_2}^{k_2}, \lambda_e^k \in \lambda \subset \Lambda$.

Theorem 9 *Let $\mathbb{X}\Gamma$ be the Γ-dynamical algebra of the integer-shift group λ, and let $\mathbb{X}\Gamma'$ be the conditional tensor product C^*-algebra (74) satisfying the α-condition (75), as a C^*-subalgebra of $\Lambda^2\mathfrak{X}$ in the sense of (73). Then*

$$\mathbb{X}\Gamma \overset{*\text{-}iso}{=} \mathbb{X}\Gamma'. \tag{76}$$

Proof First, note that, by the definition (74), the conditional tensor product C^*-algebra $\mathbb{X}\Gamma'$ is generated by the generators,

$$\left(\lambda_{e_1}^{k_1} \otimes x_j\right) \otimes \lambda_{e_2}^{k_2},$$

for $(e_1, k_1), (e_2, k_2) \in \mathbb{N}_0^{\pm}$, and $x_j \in X$.

Define a linear transformation $\Theta : \mathbb{X}\Gamma \to \mathbb{X}\Gamma'$ by the generator-preserving morphism,

$$\Theta\left(\left(\lambda_{e_1}^{k_1} \otimes x_j,\ \lambda_{e_2}^{k_2}\right)\right) = \left(\lambda_{e_1}^{k_1} \otimes x_j\right) \otimes_\alpha \lambda_{e_2}^{k_2}. \tag{77}$$

It is not difficult to check that the morphism Θ is injective and bounded. Moreover, for any

$$T' = \left(\lambda_e^k \otimes x_j\right) \otimes_\alpha \left(\sum_{g \in \lambda} t_g g\right) = \sum_{g \in \lambda} t_g \left(\left(\lambda_e^k \otimes x_j\right) \otimes_\alpha g\right)$$

in $\mathbb{X}\Gamma'$, there exists a unique

$$T = \sum_{g \in \lambda} t_g \left(\lambda_e^k \otimes x_j,\ g\right) \text{ in } \mathbb{X}\Gamma,$$

such that $\Theta(T) = T'$, by the linearity of Θ, implying that Θ is surjective, too.

So, the linear morphism Θ of (77) is a bounded, bijective linear transformation from $\mathbb{X}\Gamma$ onto $\mathbb{X}\Gamma'$. Consider now that

$$\Theta\left(\left(\lambda_{e_1}^{k_1} \otimes x_j,\ \lambda_{e_2}^{k_2}\right)\left(\lambda_{e_3}^{k_3} \otimes x_i,\ \lambda_{e_4}^{k_4}\right)\right)$$
$$= \Theta\left(\lambda_{e_2}^{k_2}\lambda_{e_4}^{k_4}\lambda_{e_1}^{k_1}\lambda_{e_3}^{k_3} \otimes x_j x_i,\ \lambda_{e_2}^{k_2}\lambda_{e_4}^{k_4}\right)$$

by (72)

$$= \left(\lambda_{e_2}^{k_2}\lambda_{e_4}^{k_4}\lambda_{e_1}^{k_1}\lambda_{e_3}^{k_3} \otimes x_j x_i\right) \otimes_\alpha \lambda_{e_2}^{k_2}\lambda_{e_4}^{k_4}$$

by (77)

$$= \left(\left(\lambda_{e_1}^{k_1} \otimes x_j\right) \otimes_\alpha \lambda_{e_2}^{k_2}\right)\left(\left(\lambda_{e_3}^{k_3} \otimes x_i\right) \otimes_\alpha \lambda_{e_4}^{k_4}\right)$$

by (75)

$$= \Theta\left(\lambda_{e_1}^{k_1} \otimes x_j,\ \lambda_{e_2}^{k_2}\right)\Theta\left(\lambda_{e_3}^{k_3} \otimes x_i,\ \lambda_{e_4}^{k_4}\right). \tag{78}$$

By (78), the morphism Θ is multiplicative from $\mathbb{X}\Gamma$ to $\mathbb{X}\Gamma'$, i.e.,

$$\Theta(T_1 T_2) = \Theta(T_1)\Theta(T_2) \text{ in } \mathbb{X}\Gamma', \tag{78'}$$

for all $T_1, T_2 \in \mathbb{X}\Gamma$.

Furthermore, one can have that

$$\Theta\left(\left(\lambda_{e_1}^{k_1} \otimes x_j,\ \lambda_{e_2}^{k_2}\right)^*\right) = \Theta\left(\left(\lambda_{-e_2}^{k_2}\lambda_{e_1}^{k_1} \otimes x_j\right),\ \lambda_{-e_2}^{k_2}\right)$$

by (72)

$$= \left(\lambda_{-e_2}^{k_2}\lambda_{e_1}^{k_1} \otimes x_j\right) \otimes_\alpha \lambda_{-e_2}^{k_2} = \left((\lambda_{e_1}^{k_1} \otimes x_j) \otimes_\alpha \lambda_{e_2}^{k_2}\right)^*$$

by (75)

$$= \Theta\left(\left(\lambda_{e_1}^{k_1} \otimes x_j,\ \lambda_{e_2}^{k_2}\right)\right)^*. \tag{79}$$

By (79), the morphism Θ is adjoint-preserving in the sense that

$$\Theta\left(T^*\right) = (\Theta(T))^* \text{ in } \mathbb{X}\Gamma', \tag{79'}$$

for all $T \in \mathbb{X}\Gamma$.

By (78)′, and (79)′, the morphism Θ is a $*$-homomorphism.

Therefore, Θ is a well-defined $*$-isomorphism from $\mathbb{X}\Gamma$ onto $\mathbb{X}\Gamma'$, equivalently, the C^*-algebras $\mathbb{X}\Gamma$ and $\mathbb{X}\Gamma'$ are $*$-isomorphic. ■

By the isomorphism theorem (76), the C^*-algebras $\mathbb{X}\Gamma$ and $\mathbb{X}\Gamma'$ are $*$-isomorphic C^*-algebras. So, from below, one can regard our Γ-dynamical algebra $\mathbb{X}\Gamma$ as the conditional tensor product C^*-algebra $\mathbb{X}\Gamma'$ of (74) satisfying the α-condition (75).

Notation and Assumption 2 (in short, **NA** 2) In the following text, we understand $\mathbb{X}\Gamma$ as $\mathbb{X}\Gamma'$ (if needed), and we denote $\mathbb{X}\Gamma'$ by $\mathbb{X}\Gamma$ as an identical C^*-algebra. □

8 Free Probability on $\mathbb{X}\Gamma$

Let $\Gamma = (\lambda, \Lambda\mathfrak{X}, \alpha)$ be the group C^*-dynamical system (68) of the integer-shift group λ acting on the C^*-algebra $\Lambda\mathfrak{X}$ of (63) via the group-action α of (65), and $\mathbb{X}\Gamma = \Lambda\mathfrak{X} \times_\alpha \lambda$ be the Γ-dynamical algebra (71) of λ satisfying the α-relation (72), which is identified with the conditional tensor product C^*-algebra $\Lambda\mathfrak{X} \otimes_\alpha \Lambda$ of (74), by (76).

8.1 Free Probability on $\Lambda\mathfrak{X}_\varphi = (\Lambda\mathfrak{X},\ \tau_\varphi)$

Let Λ be the group C^*-algebra of λ. Then the tensor product C^*-algebra $\Lambda\mathfrak{X}$ of (63) induces the C^* -probabilistic structure with a linear functional τ_φ on $\Lambda\mathfrak{X}$ satisfying that

$$\tau_\varphi(G \otimes T) = \varphi(G(T)) \tag{80}$$

for all $G \in \Lambda$, $T \in \mathfrak{X}$.

By (80), each generating operator $\lambda_e^k \otimes x_j^n$ of $\Lambda\mathfrak{X}$, for $\lambda_e^k \in \lambda \subset \Lambda$, $x_j \in X \subset \mathfrak{X}_\varphi$, and $n \in \mathbb{N}$, satisfies that

$$\tau_\varphi\left(\lambda_e^k \otimes x_j^n\right) = \varphi\left(\lambda_e^k(x_j^n)\right) \\ = \varphi\left(x_{jek}^n\right) = \omega_n c_{\frac{n}{2}}, \tag{81}$$

by (46).

By (80) and (81), we obtain a well-defined C^*-probability space,

$$\Lambda\mathfrak{X}_\varphi \overset{denote}{=} \left(\Lambda\mathfrak{X},\ \tau_\varphi\right). \tag{82}$$

Now, let

$$u_j^{(e,k)} \overset{def}{=} \lambda_e^k \otimes x_j, \text{ for } (e,k) \in \mathbb{N}_0^{\pm},\ j \in \mathbb{Z}, \tag{83}$$

be a generating free random variable of the C^*-probability space $\Lambda\mathfrak{X}_\varphi$ of (82). Then

$$\left(u_j^{(e,k)}\right)^n = \left(\lambda_e^k\right)^n \otimes x_j^n$$

$$= \lambda_{sgn(n(ek))}^{|n(ek)|} \otimes x_j^n = \lambda_e^{nk} \otimes x_j^n,$$

because

$$sgn\,(n(ek)) = sgn\,(ekek...ek) = sgn(e),$$

and

$$|n(ek)| = |ekek...ek| = |nk| = nk,$$

in $\mathbb{Z}$. i.e.,

$$\left(u_j^{(e,k)}\right)^n = \left(\lambda_e^k \otimes x_j\right)^n = \lambda_e^{nk} \otimes x_j^n, \tag{84}$$

in $\Lambda\mathfrak{X}_\varphi$, for all $n \in \mathbb{N}$.

Theorem 10 *Let $u_j^{(e,k)} \in \Lambda\mathfrak{X}_\varphi$ be in the sense of (83). Then the free distribution of $u_j^{(e,k)}$ is characterized by the following joint free moments of $u_j^{(e,k)}$ and $u_j^{(-e,k)}$;*

$$\tau_\varphi\left(\left(u_j^{(e,k)}\right)^n\right) = \tau_\varphi\left(\left(u_j^{(-e,k)}\right)^n\right) = \omega_n c_{\frac{n}{2}}, \tag{85}$$

and

$$\tau_{\varphi}\left(\prod_{l=1}^{n} u_{j}^{(e_l,k_l)}\right) = \varphi\left(x_{jfl}^{n}\right) = \omega_n c_{\frac{n}{2}},$$

with

$$f = sgn\left(\sum_{l=1}^{n} e_l k_l\right), \text{ and } l = \left|\sum_{l=1}^{n} e_l k_l\right|,$$

for all $n \in \mathbb{N}$, *where* $(e_l, k_l) \in \{(e, k), (-e, k)\}$, *for* $l = 1, \ldots, n$.

Proof Let $u_j^{(e,k)}$ be a generating free random variable (83) of $\Lambda\mathfrak{X}_{\varphi}$. Then

$$\left(u_j^{(e,k)}\right)^{*} = \left(\lambda_e^k\right)^{*} \otimes x_j^{*} = \lambda_{-e}^{k} \otimes x_j = u_j^{(-e,k)},$$

in $\Lambda\mathfrak{X}_{\varphi}$.

Since $u_j^{(e,k)}$ is not self-adjoint in $\Lambda\mathfrak{X}$, in general, the free distribution of it is characterized by the "joint" free moments of $u_j^{(e,k)}$, those of $u_j^{(-e,k)}$, and their joinr free moments.

First of all, one can obtain that

$$\tau_{\varphi}\left(\left(u_j^{(e,k)}\right)^{n}\right) = \tau_{\varphi}\left(\lambda_e^{nk} \otimes x_j^n\right) = \varphi\left(\lambda_e^{nk}(x_j^n)\right)$$

$$= \varphi\left(x_{je(nk)}^{n}\right) = \omega_n c_{\frac{n}{2}},$$

for all $n \in \mathbb{N}$, by (84) and the semicircularity of $x_{je(nk)} \in X$ in $\mathfrak{X}_{\varphi}$. Similarly,

$$\tau_{\varphi}\left(\left(u_j^{(-e,k)}\right)^{n}\right) = \tau_{\varphi}\left(\lambda_{-e}^{nk} \otimes x_j^n\right) = \varphi\left(\lambda_{-e}^{nk}(x_j^n)\right)$$

$$= \varphi\left(x_{j-e(nk)}^{n}\right) = \omega_n c_{\frac{n}{2}},$$

for all $n \in \mathbb{N}$, by the semicircularity of $x_{j-e(nk)} \in X$ in $\mathfrak{X}_{\varphi}$.

Now, consider that, for $(e_l, k_l) \in \{(e, k), (-e, k)\}$, for $l = 1, \ldots, n$,

$$U = \prod_{l=1}^{n} u_j^{(e_l,k_l)} = \left(\prod_{l=1}^{n} \lambda_{e_l}^{k_l}\right) \otimes x_j^n = \lambda_{sgn(\Sigma_{l=1}^{n} e_l k_l)}^{|\Sigma_{l=1}^{n} e_l k_l|} \otimes x_j^n$$

in $\Lambda\mathfrak{X}_{\varphi}$, and hence,

$$\tau_{\varphi}(U) = \varphi\left(\lambda_f^l(x_j^n)\right) = \varphi\left(x_{jfl}^{n}\right) = \omega_n c_{\frac{n}{2}},$$

where

$$f = sgn\left(\sum_{l=1}^{n} e_l k_l\right), \text{ and } l = \left|\sum_{l=1}^{n} e_l k_l\right|.$$

Therefore, the free-distributional data (85) holds on $\Lambda\mathfrak{X}_{\varphi}$. ■

The above theorem characterizes the free-distributional data of generating free random variables of $\Lambda\mathfrak{X}_{\varphi}$, with help of (46). It shows that the free distributions of such free random variables are depending on the semicircular law, by (4) and (85). As a special case, the following corollary is obtained.

Corollary 2 *Let* $u_j^{(e,0)} = 1_{\mathfrak{X}_{\varphi}} \otimes x_j$ *be a free random variable (83) of* $\Lambda\mathfrak{X}_{\varphi}$. *Then it is semicircular in* $\Lambda\mathfrak{X}_{\varphi}$. *i.e.,*

$$u_j^{(e,0)} \textit{ is semicircular in } \Lambda\mathfrak{X}_{\varphi}. \tag{86}$$

Proof Let $u_j^{(e,0)}$ be as above in $\Lambda\mathfrak{X}_{\varphi}$. Then it is self-adjoint in $\Lambda\mathfrak{X}$, since

$$\left(u_j^{(e,0)}\right)^* = 1_{\mathfrak{X}_{\varphi}}^* \otimes x_j^* = u_j^{(e,0)}.$$

So, the free distribution of it is characterized by the free moments,

$$\left(\tau_{\varphi}\left(\left(u_j^{(e,0)}\right)^n\right)\right)_{n=1}^{\infty},$$

and

$$\tau_{\varphi}\left(\left(u_j^{(e,0)}\right)^n\right) = \varphi\left(x_j^n\right) = \omega_n c_{\frac{n}{2}},$$

for all $n \in \mathbb{N}$, by (85). Thus, the statement (86) holds true. ■

It is interesting that there do exist free random variables s of a certain C^*-probability space $(\mathbf{X}, \tau_{\mathbf{X}})$, such that: (i) s is not self-adjoint in $\mathbf{X}$, and (ii)

$$\tau_{\mathbf{X}}(s^n) = \tau_{\mathbf{X}}\left((s^*)^n\right) = \tau_{\mathbf{X}}\left(\prod_{l=1}^{n} s_l\right) = \omega_n c_{\frac{n}{2}},$$

where

$$(s_1, \ ..., s_n) \in \{s, s^*\}^n,$$

for all $n \in \mathbb{N}$. i.e., even though s is not self-adjoint in $(\mathbf{X}, \tau_{\mathbf{X}})$, the free distribution of s (and that of s^*) is followed by the semicircular law. Indeed, the free-distributional data (85) guarantees the existence of such free random variables.

8.2 Free-Distributional Data on $\mathbb{X}\Gamma$

Let $\mathbb{X}\Gamma$ be the Γ-dynamical algebra (71) or (74) of the integer-shift group λ acting on $\Lambda\mathfrak{X}$ via the group-action α. For convenience, we let

$$\mathbb{X}\Gamma = \Lambda\mathfrak{X} \otimes_\alpha \Lambda,$$

by (76).

As in Sect. 8.1, since a tensor-factor $\Lambda\mathfrak{X}$ of $\mathbb{X}\Gamma$ has its free-probabilistic structure (82), and the other tensor-factor Λ has its canonical free-probabilistic structure (61), one can consider the free probability on our Γ-dynamical algebra $\mathbb{X}\Gamma$.

Define a linear functional τ_φ^0 on $\mathbb{X}\Gamma$ by the linear morphism satisfying that

$$\tau_\varphi^0 \left((G \otimes T) \otimes_\alpha K \right) = \tau_\varphi \left(\tau(K)(G \otimes T) \right), \tag{87}$$

for all $G \otimes T \in \Lambda\mathfrak{X}$ and $K \in \Lambda$, where τ is the canonical trace (61) on Λ, and τ_φ is the linear functional (80) on $\Lambda\mathfrak{X}$.

By (80) and (87), we have i

$$\tau_\varphi^0 \left((G \otimes T) \otimes_\alpha K \right) = \tau(K)\varphi\left(G(T)\right).$$

Definition 8 Let $\mathbb{X}\Gamma$ be the Γ-dynamical algebra of λ, and τ_φ^0 be the linear functional (87) on $\mathbb{X}\Gamma$. Then the C^*-probability space,

$$\mathbb{X}\Gamma_0 \overset{denote}{=} \left(\mathbb{X}\Gamma, \ \tau_\varphi^0 \right) \tag{88}$$

is called the Γ-dynamical C^*-probability space (induced by λ acting on $\mathfrak{X}_\varphi$).

Now, let

$$U_{j,(e,k)}^{(f,l)} = u_j^{(e,k)} \otimes_\alpha \lambda_f^l = \left(\lambda_e^k \otimes x_j \right) \otimes_\alpha \lambda_f^l, \tag{89}$$

for $(e, k), (f, l) \in \mathbb{N}_0^\pm$, and $j \in \mathbb{Z}$. Then, such an operator $U_{j,(e,k)}^{(f,l)}$ is a generating operator of $\mathbb{X}\Gamma$, by (76).

Observe that

$$\begin{aligned}
\left(U_{j,(e,k)}^{(f,l)} \right)^n &= \left((\lambda_e^k \otimes x_j) \otimes_\alpha \lambda_f^l \right)^n \\
&= \left((\lambda_e^k \otimes x_j) \otimes_\alpha \lambda_f^l \right)^2 \left((\lambda_e^k \otimes x_j) \otimes_\alpha \lambda_f^l \right)^{n-2} \\
&= \left(\left((\lambda_f^l)^2 \lambda_e^k \otimes x_j \right) \otimes_\alpha (\lambda_f^l)^2 \right) \left((\lambda_e^k \otimes x_j) \otimes_\alpha \lambda_f^l \right)^{n-2}
\end{aligned}$$

by (75) (or (72))

$$= \left(\left(\lambda_{sgn(2(fl))}^{|2(fl)|}\lambda_e^k \otimes x_j^2\right) \otimes_\alpha \lambda_{sgn(2fl)}^{|2(fl)|}\right) \left((\lambda_e^k \otimes x_j) \otimes_\alpha \lambda_f^l\right)^{n-2}$$
$$= \left(\left((\lambda_f^{2l}\lambda_e^k) \otimes x_j^2\right) \otimes_\alpha \lambda_f^{2l}\right) \left((\lambda_e^k \otimes x_j) \otimes_\alpha \lambda_f^l\right)^{n-2}$$

since $flfl = f(2l)$ in $\mathbb{Z}$, i.e.,

$$sgn\,(2(fl)) = f, \text{ and } |2(fl)| = |2l| = 2l,$$

for all $(f, l) \in \mathbb{N}_0^{\pm}$, and hence, it goes to

$$\begin{aligned}
&\left((\lambda_f^{2l}\lambda_e^k \otimes x_j^2) \otimes_\alpha \lambda_f^{2l}\right)\\
&\cdot \left((\lambda_e^k \otimes x_j) \otimes_\alpha \lambda_f^l\right) \left((\lambda_e^k \otimes x_j) \otimes_\alpha \lambda_f^l\right)^{n-3}\\
&\left((\lambda_f^{3l}\lambda_e^k \otimes x_j^3) \otimes_\alpha \lambda_f^{3l}\right) \left((\lambda_e^k \otimes x_j) \otimes_\alpha \lambda_f^l\right)^{n-3}\\
&= \cdots\\
&= \left((\lambda_f^{nl}\lambda_e^k) \otimes x_j^n\right) \otimes_\alpha \lambda_f^{nl}\\
&= \left(\lambda_{sgn(f(nl)ek)}^{|f(nl)ek|} \otimes x_j^n\right) \otimes_\alpha \lambda_f^{nl},
\end{aligned} \tag{90}$$

in $\mathbb{X}\Gamma$, for all $n \in \mathbb{N}$.

Therefore, by (87) and (90), we obtain that if

$$U_{j,(e,k)}^{(f,l)} = u_j^{(e,k)} \otimes_\alpha \lambda_f^l \in \mathbb{X}\Gamma_0$$

is a generating free random variable (89), then

$$\tau_\varphi^0\left(\left(U_{j,(e,k)}^{(f,l)}\right)^n\right) = \tau_\varphi^0\left(\left(\lambda_{sgn(f(nl)ek)}^{|f(nl)ek|} \otimes x_j^n\right) \otimes_\alpha \lambda_f^{nl}\right)$$

by (90)

$$= \tau\left(\lambda_f^{nl}\right)\varphi\left(\lambda_e^k\left(x_j^n\right)\right)$$

where

$$k = |f(nl)ek|\,, \text{ and } e = sgn\,(f(nl)ek)\,,$$

then it goes to

$$= \delta_{\lambda_f^{nl},\, 1_{\mathfrak{X}_\varphi}}\varphi\left(x_{jek}^n\right) = \delta_{\lambda_f^{nl},\, 1_{\mathfrak{X}_\varphi}}\left(\omega_n c_{\frac{n}{2}}\right), \tag{91}$$

by (61), for all $n \in \mathbb{N}$, where δ is the *Kronecker delta*.

Lemma 4 *Let $U_{j,(e,k)}^{(f,l)}$ be a generating free random variable (89) of the Γ-dynamical C^*-probability space $\mathbb{X}\Gamma_0$ of (88). Then*

$$\tau_{\varphi}^{0}\left(\left(U_{j,(e,k)}^{(f,l)}\right)^{n}\right) = \delta_{\lambda_f^l,\, 1_{\mathfrak{X}_\varphi}}\left(\omega_n c_{\frac{n}{2}}\right)$$
$$= \tau_{\varphi}^{0}\left(\left(\left(U_{j,(e,k)}^{(f,l)}\right)^{*}\right)^{n}\right), \tag{92}$$

for all $n \in \mathbb{N}$.

Proof The first equality of (92) is proven by (91). Indeed,

$$\delta_{\lambda_f^{nl},\, 1_{\mathfrak{X}_\varphi}} = \delta_{\lambda_f^l,\, 1_{\mathfrak{X}_\varphi}} \text{in } \{0, 1\},$$

by (54), i.e.,

$$\lambda_f^l \neq 1_{\mathfrak{X}_\varphi} \iff \lambda_f^{nl} \neq 1_{\mathfrak{X}_\varphi}, \text{ for all } n \in \mathbb{N},$$

in $\lambda \subset \Lambda$. Therefore, the formula (91) implies the first equality of (92).

Observe now that

$$\left(U_{j,(e,k)}^{(f,l)}\right)^{*} = \left((\lambda_e^k \otimes x_j) \otimes_\alpha \lambda_f^l\right)^{*}$$
$$= \left(\lambda_{-f}^l \lambda_e^k \otimes x_j\right) \otimes_\alpha \lambda_{-f}^l$$
$$= \left(\lambda_{sgn(ek-fl)}^{|ek-fl|} \otimes x_j\right) \otimes_\alpha \lambda_{-f}^l,$$

in $\mathbb{X}\Gamma$, by (75) (or (72)). So, one can get that

$$\left(\left(U_{j,(e,k)}^{(f,l)}\right)^{*}\right)^{n} = \left(\lambda_{-f}^{nl} \lambda_{sgn(ek-fl)}^{|ek-fl|} \otimes x_j^n\right) \otimes \lambda_{-f}^{nl},$$

in $\mathbb{X}\Gamma_0$, for all $n \in \mathbb{N}$, by (90). And hence,

$$\tau_{\varphi}^{0}\left(\left(\left(U_{j,(e,k)}^{(f,l0}\right)^{*}\right)^{n}\right) = \delta_{\lambda_{-f}^{nl},\, 1_{\mathfrak{X}_\varphi}}\left(\omega_n c_{\frac{n}{2}}\right), \tag{93}$$

for all $n \in \mathbb{N}$. But

$$\delta_{\lambda_{-f}^{nl},\, 1_{\mathfrak{X}_\varphi}} = \delta_{\lambda_f^{nl},\, 1_{\mathfrak{X}_\varphi}} = \delta_{\lambda_f^l,\, 1_{\mathfrak{X}_\varphi}} \text{in}\{0, 1\},$$

by (54). Indeed,

$$\lambda_f^l \neq 1_{\mathfrak{X}_\varphi} \iff \left(\lambda_f^l\right)^{*} \neq \left(1_{\mathfrak{X}_\varphi}\right)^{*} = 1_{\mathfrak{X}_\varphi}$$

in $\lambda \subset \Lambda$.

Therefore, the second equality of (92) holds by (93). ■

The above lemma provides the free-momental information of $U_{j,(e,k)}^{(f,l)}$, and those of the adjoint $\left(U_{j,(e,k)}^{(f,l)}\right)^{*}$ in $\mathbb{X}\Gamma_{0}$, by (92). Note that, by the proof of (92), one has that

$$\left(U_{j,(e,k)}^{(f,l)}\right)^{*} = U_{j,(sgn(ek-fl),\ |ek-fl|)}^{(-f,l)} \text{in } \mathbb{X}\Gamma_{0}, \tag{94}$$

in the sense of (89).

Let's now consider the mixed free moments of them, to fully characterize the joint free moments of $U_{j,(e,k)}^{(f,l)}$ in $\mathbb{X}\Gamma_{0}$.

Notation and Assumption 3 (in shift, **NA** 3) Let $U_{j,(e,k)}^{(f,l)} \in \mathbb{X}\Gamma_{0}$ be in the sense of (89).

Then

$$e^{*} \overset{denote}{=} sgn\,(ek - fl) \text{ in } \{\pm 1\},$$

and

$$k^{*} \overset{denote}{=} |ek - fl| \text{ in } \mathbb{N}_{0}.$$

Then the adjoint $\left(U_{j,(e,k)}^{(f,l)}\right)^{*}$ of (94) is identified with

$$\left(U_{j,(e,k)}^{(f,l)}\right)^{*} = U_{j,(e^{*},k^{*})}^{(-f,l)} \text{in } \mathbb{X}\Gamma_{0}.$$

Also, for convenience, let's denote

$$U_{1} = U_{j,(e,k)}^{(f,l)} \text{and } U_{2} = U_{j,(e^{*},k^{*})}^{(-f,l)}.$$

□

Let U_1 and U_2 be in the sense of **NA** 3 in $\mathbb{X}\Gamma_{0}$, and take a "mixed" N-tuple

$$J = (i_1, \ldots, i_N) \in \{1, 2\}^{N}, \text{ for } N \in \mathbb{N} \setminus \{1\},$$

i.e., there exists at least one i_{k_0}, for $k_0 \in \{1, \ldots, N\}$, such that $i_{k_0} \neq i_k$, for some $k \in \{1, \ldots, N\}$.

For the above mixed N-tuple J, construct a free random variable W_J,

$$W_{J} = \prod_{l=1}^{N} U_{i_l} \in \mathbb{X}\Gamma_{0}. \tag{95}$$

It is not hard to check that neither

$$J = (1, 2, 1, 2, \ldots, 1, 2),$$

nor

$$J = (2, 1, 2, 1, \ldots, 2, 1),$$

if and only if

$$\tau_{\varphi}^{0}(W_J) = 0,$$

by the very definition (87) of our linear functional τ_{φ}^{0}, because

$$\tau\left((\lambda_f^l \lambda_{-f}^l)....(\lambda_f^l \lambda_{-f}^l)\right) = \tau\left(1_{\mathfrak{X}_{\varphi}}\right) = 1$$
$$= \tau\left((\lambda_{-f}^l \lambda_{-f}^l)...(\lambda_{-f}^l \lambda_{-f}^l)\right),$$

and 0, otherwise, on Λ, by (61).

Lemma 5 *Let $J = (i_1, ..., i_N)$ be a mixed N-tuple of $\{1, 2\}$, and let W_J be the corresponding free random variable (95) of $\mathbb{X}\Gamma_0$ under* **NA** *3. Then*

$$\tau_{\varphi}^{0}(W_J) = 0,$$

if and only if neither

$$J = (1, 2, 1, 2, \ldots, 1, 2), \tag{96}$$

nor

$$J = (2, 1, 2, 1, \ldots, 2, 1).$$

Proof The proof is done by the discussions in the very above paragraph. ■

By (96), to consider the "non-zero" mixed free moments of U_1 and U_2 (under **NA** 3) is to compute $\tau_{\varphi}^{0}(W_J)$, where

$$J = (1,\ 2,\ 1,\ 2,\ ...,\ 1,\ 2)$$

or

$$J = (2, 1, 2, 1, \ldots, 2, 1)$$

is an N-tuples.

First, assume that

$$J = (1, 2, 1, 2, \ldots, 1, 2),$$

and W_J be the corresponding free random variable (95) of $\mathbb{X}\Gamma_0$. Then there exists $(e_J, k_J) \in \mathbb{N}_0^{\pm}$, such that

$$\begin{aligned} W_J &= U_1 U_2 U_1 U_2 ... U_1 U_2 \\ &= \left(\lambda_{e_J}^{k_J} \otimes x_j^N\right) \otimes_\alpha 1_{\mathfrak{X}_\varphi}, \end{aligned} \tag{97}$$

by the induction on the following computation (98) below:

$$\begin{aligned} U_1 U_2 &= \left(U_{j,(e,k)}^{(f,l)}\right)\left(U_{j,(e^*,k^*)}^{(-f,l)}\right) \\ &= \left(u_{j,(e,k)} \otimes_\alpha \lambda_f^l\right)\left(u_j^{(e^*,k^*)} \otimes_\alpha \lambda_{-f}^l\right) \\ &= \left(\left(\lambda_f^l \lambda_{-f}^l \lambda_e^k \otimes x_j\right)\left(\lambda_{e^*}^{k^*} \otimes x_j\right)\right) \otimes_\alpha \left(\lambda_f^l \lambda_{-f}^l\right) \\ &= \left(\left(1_{\mathfrak{X}_\varphi} \lambda_e^k \otimes x_j\right)\left(\lambda_{e^*}^{k^*} \otimes x_j\right)\right) \otimes_\alpha 1_{\mathfrak{X}_\varphi} \\ &= \left(\lambda_e^k \lambda_{e^*}^{k^*} \otimes x_j^2\right) \otimes_\alpha 1_{\mathfrak{X}_\varphi} \\ &= \left(\lambda_{sgn(eke^*k^*)}^{|eke^*k^*|} \otimes x_j^2\right) \otimes_\alpha 1_{\mathfrak{X}_\varphi}, \end{aligned} \tag{98}$$

with

$$eke^*k^* = ekek - fl = 2(ek) - fl,$$

in $\mathbb{Z}$.

By (97), one can realize that

$$\tau_\varphi^0\left(W_J\right) = \tau\left(1_{\mathfrak{X}_\varphi}\right)\varphi\left(x_{je_Jk_J}^N\right) = \omega_N c_{\frac{N}{2}} = c_{\frac{N}{2}}, \tag{99}$$

because N is the length $|J|$ of J, which is even in $\mathbb{N}$.

Similarly, if $J = (2, 1, 2, 1, \ldots, 2, 1)$, then

$$\tau_\varphi^0\left(W_J\right) = c_{\frac{N}{2}}. \tag{99$'$}$$

Lemma 6 *Suppose either*

$$J = (1, 2, 1, 2, \ldots, 1, 2),$$

or

$$J = (2, 1, 2, 1, \ldots, 2, 1),$$

*and let W_J be the corresponding free random variable (95) of $\mathbb{X}\Gamma_0$ under **NA** 3. Then*

$$\tau_{\varphi}^{0}\left(W_J\right) = c_{\frac{|J|}{2}}, \tag{100}$$

where $|J|$ is the length of the finite sequence J.

Proof The proof of the free-distributional data (100) is done by (99) and (99)′. ∎

By the above three lemmas, we obtain the following free-probabilistic information on $\mathbb{X}\Gamma_0$.

Theorem 11 *Let $U_1 = U_{j,(e,k)}^{(f,l)}$ be a generating free random variable of the Γ-dynamical C^*-probability space $\mathbb{X}\Gamma_0$, and let $U_2 = U_1^* \in \mathbb{X}\Gamma_0$. Then the free distribution of U_1 (or that of U_2) is characterized by the following joint free-moment computations:*

$$\tau_{\varphi}^{0}\left(U_1^n\right) = \delta_{\lambda_f^l,\, 1_{\mathfrak{X}_\varphi}}\left(\omega_n c_{\frac{n}{2}}\right) = \tau_{\varphi}^{0}\left(U_2^n\right), \tag{101}$$

for all $n \in \mathbb{N}$, and if J is a mixed N-tuple $(i_1, ..., i_N)$ of $\{1, 2\}$, then

$$\tau_{\varphi}^{0}\left(\prod_{l=1}^{N} U_{i_l}\right) = \begin{cases} c_{\frac{N}{2}} & \text{if either } J = (1, 2, 1, 2, \ldots, 1, 2), \\ & \quad \text{or } J = (2, 1, 2, 1, \ldots, 2, 1) \\ 0 & \text{otherwise,} \end{cases} \tag{102}$$

for all $N \in \mathbb{N} \setminus \{1\}$.

Proof Since a given generating free random variable U_1 is not self-adjoint in $\mathbb{X}\Gamma_0$, the free distribution of U_1 (or that of $U_1^* = U_2$) is characterized by the joint free moments of U_1 and U_2. But the free moments $\left(\tau_{\varphi}^{0}\left(U_i^n\right)\right)_{n=1}^{\infty}$ is obtained by (92), for all $i = 1, 2$; and the mixed free moments of U_1 and U_2 are computed in (96) and (100).

Therefore, the free-moment formula (101) is proven by (92), and the mixed-free-momental data (102) is shown by (96) and (100). ∎

The above theorem fully characterizes the free-distributional information of generating free random variables of our Γ-dynamical C^*-probability space $\mathbb{X}\Gamma_0$, by (101) and (102).

Also, the free distributions of the generating operators $U_{j,(e,k)}^{(f,l)}$ illustrates how the group-action α of λ affects the free probability on the C^*-probability space $\Lambda\mathfrak{X}_\varphi$, under dynamics.

References

1. Ahsanullah, M.: Some inferences on semicircular distribution. J. Stat. Theo. Appl. **15**(3), 207–213 (2016)
2. Bercovici, H., Voiculescu, D.: Superconvergence to the central limit and failure of the cramer theorem for free random variables. Probab. Theo. Related Fields **103**(2), 215–222 (1995)
3. Bozejko, M., Ejsmont, W., Hasebe, T.: Noncommutative Probability of Type D. Internat. J. Math. **28**(2), 1750010, 30 (2017)
4. Bozheuiko, M., Litvinov, E.V., Rodionova, I.V.: An Extended Anyon Fock Space and Non-commutative Meixner-Type Orthogonal Polynomials in the Infinite-Dimensional Case. Uspekhi Math. Nauk. **70**(5), 75–120 (2015)
5. Cho, I.: Semicircular Families in Free Product Banach $*$-Algebras Induced by p-Adic Number Fields over Primes p. Compl. Anal. Oper. Theo. **11**(3), 507–565 (2017)
6. Cho, I.: Semicircular-like laws and the semicircular law induced by orthogonal projections. Compl. Anal. Oper. Theo. (2018) To Appear
7. Cho, I.: Free stochastic integrals for weighted-semicircular motion induced by orthogonal projections. Monograph Ser. Appl. Math. Anal: Theo. Methods Appl. Published by Taylor & Fransis, (2019) To Appear
8. Cho, I.: Acting Semicircular Elements Induced by Orthogonal Projections on von Neumann Algebras. Mathematics **5**, 74 (2017). https://doi.org/10.3390/math5040074
9. Cho, I., Dong, J.: Numbers, catalan., distributions, free, of mutually free multi semicircular elements Submitted to J. Probab, Related Fields (2018)
10. Cho, I., Jorgensen, P.E.T.: Semicircular Elements Induced by Projections on Separable Hilbert Spaces, Monograph Ser., Operator Theory: Advances & Applications. Published by Birkhauser, Basel (2019) To Appear
11. Cho, I., Jorgensen, P.E.T.: Banach $*$-Algebras Generated by Semicircular Elements Induced by Certain Orthogonal Projections. Opuscula Math. **38**(4), 501–535 (2018)
12. Cho, I., Jorgensen, P.E.T.: Semicircular Elements Induced by p-Adic Number Fields. Opuscula Math. **35**(5), 665–703 (2017)
13. Connes, A.: Geometry, Noncommutative. Academic, San Diego, CA (1994). ISBN: 0-12-185860-X
14. Halmos, P.R.: Hilbert space problem books, grad. Texts Math. **19** (1982). ISBN: 978-0387906850. Published by Springer
15. Kaygorodov, I., Shestakov, I.: Free generic poisson fields and algebras. Commun. Alg. **46**(4) (2018). https://doi.org/10.1080/00927872.2017.1358269
16. Makar-Limanov, L., Shestakov, I.: Polynomials and poisson dependence in free poisson algebras and free poisson fields. J. Alg. **349**(1), 372–379 (2012)
17. Nica, A., Speicher, R.: Lectures on the Combinatorics of Free Probability, 1st edn., vol. 335. London Math. Soc. Lecture Note Ser. Cambridge Univ. Press (2006). ISBN-13:978-0521858526
18. Nourdin, I., Peccati, G., Speicher, R.: Multi-dimensional semicircular limits on the free wigner Chaos. Progr. Probab. **67**, 211–221 (2013)
19. Pata, V.: The central limit theorem for free additive convolution. J. Funct. Anal. **140**(2), 359–380 (1996)
20. Radulescu, F.: Random matrices, amalgamated free products and subfactors of the C^*-algebra of a free group of nonsingular index. Invent. Math. **115**, 347–389 (1994)
21. Radulescu, F.: Free group factors and hecke operators, notes taken. In: Ozawa, N (ed.) Proceedings 24th Conference in Oper. Theo., Theta Advanced Series in Math. (2014) Theta Foundation
22. R. Speicher, Combinatorial Theory of the Free Product with Amalgamation and Operator-Valued Free Probability Theory, Amer. Math. Soc. Mem., vol 132, no. 627, (1998)
23. Speicher, R.: A conceptual proof of a basic result in the combinatorial approach to freeness. Infinit. Dimention. Anal. Quant. Prob. Related Top. **3**, 213–222 (2000)
24. Speicher, R., Kemp, T.: Strong haagerup inequalities for free R-diagonal elements. J. Funct. Anal. **251**(1), 141–173 (2007)

25. Speicher, R., Haagerup, U.: Brown's spectrial distribution measure for R-diagonal elements in finite Von Neumann algebras. J. Funct. Anal. **176**(2), 331–367 (2000)
26. Vladimirov, V.S.: p-adic quantum mechanics. Commun. Math. Phy. **123**(4), 659–676 (1989)
27. Vladimirov, V.S., Volovich, I.V., Zelenov, E.I.: p-Adic Analysis and Mathematical Physics, Ser. Soviet & East European Math., vol 1. World Scientific (1994). ISBN: 978-981-02-0880-6
28. Voiculescu, D.: Aspects of free analysis. Jpn. J. Math. **3**(2), 163–183 (2008)
29. Voiculescu, D.: Free probability and the Von Neumann algebras of free groups. Rep. Math. Phys. **55**(1), 127–133 (2005)
30. Voiculescu, D., Dykemma, K., Nica, A.: Variables, Free Random., Monograph Series, C.R.M., vol 1. Published by Amer. Math, Soc (1992). ISBN-13: 978–0821811405

Lie Group Theory for Nonlinear Fractional K(m, n) Type Equation with Variable Coefficients

H. Jafari, N. Kadkhoda, and Dumitru Baleanu

Abstract We investigated the analytical solution of fractional order K(m, n) type equation with variable coefficient which is an extended type of KdV equations into a genuinely nonlinear dispersion regime. By using the Lie symmetry analysis, we obtain the Lie point symmetries for this type of time-fractional partial differential equations (PDE). Also we present the corresponding reduced fractional differential equations (FDEs) corresponding to the time-fractional K(m, n) type equation.

Keywords Fractional differential equation · Lie symmetry analysis method · Reduced equation · Fractional order k(m, n) type equation.

2010 MSC 31B10 · 44A10 · 26A33.

1 Introduction

Most problems in engineering, biology, applied mathematics and physics might be better modeled by using ordinary/partial differential equations with fractional (arbitrary) order. The method of group analysis for ordinary/partial differential equations, originally advocated by the Norwegian mathematician Sophus Lie during 1870s. The

H. Jafari (✉)
Department of Mathematics, University of Mazandaran, Babolsar, Iran

Department of Mathematical Sciences, University of South Africa, UNISA0003, Pretoria, South Africa

N. Kadkhoda
Department of Mathematics, Faculty of Basic Sciences, Bozorgmehr University of Qaenat, Qaenat, Iran
e-mail: kadkhoda@buqaen.ac.ir

D. Baleanu
Department of Mathematics, Faculty of Art and Sciences, Cankaya University, Ankara, Turkey
e-mail: dumitru@cankaya.edu.tr

Institute of Space Sciences, Magurele-Bucharest, Magurele, Romania

J. Singh et al. (eds.), *Methods of Mathematical Modelling and Computation for Complex Systems*, Studies in Systems, Decision and Control 373,
https://doi.org/10.1007/978-3-030-77169-0_8

tangent structural equations under transformation groups is the fundamental idea of symmetry analysis. Numerous methods developed to solve differential equations based on Lie symmetry analysis.

In last few decades, many researcher studied different class of the fractional partial differential equations (FPDEs). These equations arise in various branches of sciences such as physics, biology, viscoelastic materials, electrochemistry, signal processing, fluid mechanics [1, 14, 19, 26, 28, 33]. Integrals and derivatives are of any order in the fractional calculus [19, 26]. In the recent years, finding exact solutions of FDEs has gained much attention.

Many researchers have presented various techniques and methods for obtaining the numerical and analytical solutions of FDEs, such as separating variables method [7], variational iteration method [11], fractional complex transform [12], operational matrices [16], first integral method [21], and so on. In many years ago, there are many articles to obtain the analytical solutions of nonlinear PDE using Lie group theory. It is important to know, however, that few of them involve FDEs [5, 9, 15, 18, 20, 31, 32]. Furthermore, already few articles done in symmetries of variable coefficients FDEs such as [10, 22]. Our purpose is to study the time-fractional K(m, n) equation:

$$\frac{\partial^\alpha u}{\partial t^\alpha} + \zeta(u^m)_x + g(t)(u^n)_{xxx} = 0, \quad t > 0,\ 0 < \alpha \leq 1, \tag{1}$$

or equivalently

$$D_t^\alpha u + \zeta(u^m)_x + g(t)(u^n)_{xxx} = 0, \quad t > 0,\ 0 < \alpha \leq 1. \tag{2}$$

here m and $n \neq 0$ are arbitrary constants, $\zeta = \pm 1$ and $g(t)$ is an arbitrary nonvanishing function of the variable t. This equation for $\alpha = 1$ and also with constant coefficients for $0 < \alpha < 1$ has been discussed in [6, 30].

In the follow, we study the above equation with $m = 2, n = 3$. Equation (1) is called the K(m, n) equation, when $\zeta = g(t) = 1$. Rosenau introduced this equation in 1998 [27, 34] which is described the process of interpretation the role nonlinear dispersion in the formation of structures in liquid drops.

This article is organized as follows. In the next section, it is given the analysis of Lie symmetry group for a FPDE. Then in Sect. 3, using Lie group, the Lie point symmetries of Eq. (1) are obtained. In Sect. 4, we perform Lie group on the Eq. (1) for obtaining invariant solutions and reduced fractional ODEs. Conclusions are given in the Sect. 5.

2 Lie Symmetry Analysis Method for FPDEs

According to the importance of FPDEs in mathematics and physics, finding the exact solutions for these equations is very important. Although nonlinear FPDEs are difficult to solve, but many papers have been presented by scientists. Studying differential equations using the fundamental method of the Lie symmetries are interesting for many researchers. In the past century, many researchers have studied in the field of the Lie groups. Some of them are Baumann [2], Bluman [3], Ibragimov [13], Olver [24], Ovsiannikov [25]. In this section, finding infinitesimal functions of FPDEs are given. Let us consider the below form of FPDEs:

$$\mathcal{D}_t^\alpha u = F(x, t, u, u_{(1)}, \ldots), \quad \alpha > 0. \tag{3}$$

where D_t^α fractional derivative in the sense of Riemann-Liouville [15] and u is depend to x, t:

$$D_t^\alpha u = \begin{cases} \frac{\partial^m u}{\partial t^m}; & \alpha = m \in N, \\ \\ \frac{1}{\Gamma(m-\alpha)} \frac{\partial^m}{\partial t^m} \int_0^t \frac{u(\tau,x)}{(t-\tau)^{\alpha+1-m}} d\tau; & m-1 < \alpha < m, \ m \in N. \end{cases} \tag{4}$$

Similar discussion of PDEs[4, 24], we can write

$$\mathcal{D}_{\bar{t}}^\alpha \bar{u} = \mathcal{D}_t^\alpha u + \varepsilon[\eta_t^{(\alpha)}(t, x, u, u_{(1)}, u_{(\alpha)}, \ldots)] + o(\varepsilon^2). \tag{5}$$

In view of by the prolongation formula, for $\eta_t^{(\alpha)}$ we have [9]

$$\eta_t^{(\alpha)} = \mathcal{D}_t^\alpha(\eta) + \xi_x \mathcal{D}_t^\alpha(u_x) - \mathcal{D}_t^\alpha(\xi_x u_x) + \mathcal{D}_t^\alpha(D_t(\xi_t)u) - \mathcal{D}_t^{\alpha+1}(\xi_t u) + \xi_t D_t^{\alpha+1} u, \tag{6}$$

and the total derivative operator D_t is defined by

$$D_t = \frac{\partial}{\partial t} + u_t \frac{\partial}{\partial u} + u_{xt} \frac{\partial}{\partial u_x} + u_{tt} \frac{\partial}{\partial u_t} + u_{xxt} \frac{\partial}{\partial u_{xx}} + \cdots. \tag{7}$$

Simplifying (6) using the Leibnitz formula [29]

$$\mathcal{D}_t^\alpha[f(t)g(t)] = \sum_{n=0}^{\infty} \frac{(-1)^{n-1}\alpha\Gamma(n-\alpha)}{\Gamma(1-\alpha)\Gamma(n+1)} \mathcal{D}_t^{\alpha-n} f(t) D_t^n g(t), \quad \alpha > 0, \tag{8}$$

we can write [17]:

$$\begin{aligned} \eta_t^{(\alpha)} = {} & \frac{\partial^\alpha \eta}{\partial t^\alpha} + (\eta_u - \alpha D_t(\xi_t)) \frac{\partial^\alpha u}{\partial t^\alpha} - u \frac{\partial^\alpha \eta_u}{\partial t^\alpha} + \sum_{m=1}^{\infty} \left[\binom{\alpha}{m} \frac{\partial^m(\eta_u)}{\partial t^m} \right. \\ & \left. - \binom{\alpha}{m+1} D_t^{m+1}(\xi_t) \right] \mathcal{D}_t^{\alpha-m}(u) - \sum_{m=1}^{\infty} \binom{\alpha}{m} \mathcal{D}_t^{\alpha-m}(u_x) D_t^m(\xi_x). \end{aligned} \tag{9}$$

To obtain coefficients of X, we must have:

$$X^{(\alpha)}[\mathcal{D}_t^\alpha u - F(t, x, u, u_{(1)}, \ldots)]_{\mathcal{D}_t^\alpha u = F(t,x,u_{(1)},\ldots)} = 0, \tag{10}$$

where

$$X^{(\alpha)} = \xi_x(t,x,u)\frac{\partial}{\partial x} + \xi_t(t,x,u)\frac{\partial}{\partial t} + \eta(t,x,u)\frac{\partial}{\partial u} + \eta_i^{(1)}(t,x,u,u_{(1)})\frac{\partial}{\partial u_i} + \cdots$$
$$+ \eta^{(k)}_{i_1 i_2 \dots, i_k}(t,x,u,u_{(1)},\dots u_{(k)})\frac{\partial}{\partial u_{i_1 i_2 \dots, i_k}} + \eta_t^{(\alpha)}(t,x,u,\dots,u_{(\alpha),\dots})\frac{\partial}{\partial u_t^{(\alpha)}}. \quad (11)$$

Using these relations, we obtain the Lie symmetries.

3 Fractional Lie Symmetries for Time-Fractional K(m, n)

Here, we obtain the infinitesimal generator of the time-fractional K(m, n) equation

$$\frac{\partial^\alpha u}{\partial t^\alpha} + (u^m)_x + g(t)(u^n)_{xxx} = 0, \quad t > 0,\ 0 < \alpha \leq 1. \quad (12)$$

Theorem 1 *Lie symmetries for Eq.* (12)*, which those are solutions of determining equations depend on the selection of the function* $g(t)$ *for* $0 < \alpha < 1,\ m-1-n,\ 3m-n-2,\ -3n+m-1,\ \alpha(m+1)-m+1,\ \alpha(m+2)-2m+2 \neq 0$*, are*

Case 1: $g(t)$ *be an non-vanishing arbitrary function.*
In this case, the infinitesimal generator is given by

$$X_{1.1} = \frac{\partial}{\partial x}. \quad (13)$$

Proof The one-parameter Lie group of transformations in x, t, u with ε as the group parameter are given

$$t^* = t + \varepsilon\xi_t(t,x,u) + O(\varepsilon^2),$$
$$x^* = x + \varepsilon\xi_x(t,x,u) + O(\varepsilon^2),$$
$$u^* = u + \varepsilon\eta_u(t,x,u) + O(\varepsilon^2),$$

the Lie algebra of K(m, n) equation (Eq. (12)) is spanned by vector fields

$$X = \xi_x(t,x,u)\frac{\partial}{\partial x} + \xi_t(t,x,u)\frac{\partial}{\partial t} + \eta_u(x,t,u)\frac{\partial}{\partial u}, \quad (14)$$

where

$$\xi_x = \frac{dx^*}{d\varepsilon}|_{\varepsilon=0}, \quad \xi_t = \frac{dt^*}{d\varepsilon}|_{\varepsilon=0}, \quad \eta_u = \frac{du^*}{d\varepsilon}|_{\varepsilon=0}. \quad (15)$$

Applying the $X^{(\alpha)}$ to (12), leads

$$X^{(\alpha)}\left[\frac{\partial^\alpha u}{\partial t^\alpha}+(u^m)_x+g(t)(u^n)_{xxx}\right]_{\frac{\partial^\alpha u}{\partial t^\alpha}+(u^m)_x+g(t)(u^n)_{xxx}=0}=0. \tag{16}$$

Expanding the (16), we obtain the following overdetermined system of linear partial differential equations:

$$\begin{aligned}
&u^{n-3}\alpha\, g(t)\, n^2 \frac{\partial}{\partial t}\xi_t(x,t,u)+\eta_u(x,t,u)\, g(t)\, u^{n-4}n^3+\xi_t(x,t,u)\left(\frac{d}{dt}g(t)\right)u^{n-3}n^2\\
&-3\left(\frac{\partial}{\partial x}\xi_x(x,t,u)\right)g(t)\,u^{n-3}n^2+2\left(\frac{\partial}{\partial u}\eta_u(x,t,u)\right)u^{n-3}g(t)\,n^2-3\,u^{n-3}\alpha\, g(t)\, n\frac{\partial}{\partial t}\xi_t(x,t,u)\\
&-6\,\eta_u(x,t,u)\,g(t)\,u^{n-4}n^2-3\,\xi_t(x,t,u)\left(\frac{d}{dt}g(t)\right)u^{n-3}n+9\left(\frac{\partial}{\partial x}\xi_x(x,t,u)\right)g(t)\,u^{n-3}n\\
&-6\left(\frac{\partial}{\partial u}\eta_u(x,t,u)\right)u^{n-3}g(t)\,n+2\,u^{n-3}\left(\frac{\partial}{\partial t}\xi_t(x,t,u)\right)g(t)\,\alpha+3\,g(t)\left(\frac{\partial^2}{\partial u^2}\eta_u(x,t,u)\right)u^{n-2}n\\
&-6\,g(t)\left(\frac{\partial^2}{\partial x\partial u}\xi_x(x,t,u)\right)u^{n-2}n+11\,\eta_u(x,t,u)\,g(t)\,u^{n-4}n+2\,\xi_t(x,t,u)\left(\frac{d}{dt}g(t)\right)u^{n-3}\\
&-6\left(\frac{\partial}{\partial x}\xi_x(x,t,u)\right)g(t)\,u^{n-3}+4\left(\frac{\partial}{\partial u}\eta_u(x,t,u)\right)u^{n-3}g(t)-3\,g(t)\left(\frac{\partial^2}{\partial u^2}\eta_u(x,t,u)\right)u^{n-2}\\
&+6\,g(t)\left(\frac{\partial^2}{\partial x\partial u}\xi_x(x,t,u)\right)u^{n-2}+u^{n-1}\left(\frac{\partial^3}{\partial u^3}\eta_u(x,t,u)\right)g(t)-3\,u^{n-1}\left(\frac{\partial^3}{\partial x\partial u^2}\xi_x(x,t,u)\right)g(t)\\
&-6\,\eta_u(x,t,u)\,g(t)\,u^{n-4}=0,\\
&3\,u^{n-3}\left(\frac{\partial}{\partial u}\xi_x(x,t,u)\right)n^2-9\,u^{n-3}\left(\frac{\partial}{\partial u}\xi_x(x,t,u)\right)n+3\,u^{n-2}\left(\frac{\partial^2}{\partial u^2}\xi_x(x,t,u)\right)n\\
&+6\,u^{n-3}\frac{\partial}{\partial u}\xi_x(x,t,u)-3\,u^{n-2}\frac{\partial^2}{\partial u^2}\xi_x(x,t,u)+u^{n-1}\frac{\partial^3}{\partial u^3}\xi_x(x,t,u)=0,
\end{aligned}$$

$$\begin{aligned}
&\eta_u(x,t,u)\,g(t)\,u^{n-3}n^2+u^{n-2}\alpha\, g(t)\, n\frac{\partial}{\partial t}\xi_t(x,t,u)-3\,\eta_u(x,t,u)\,g(t)\,u^{n-3}n\\
&+\xi_t(x,t,u)\left(\frac{d}{dt}g(t)\right)u^{n-2}n-3\left(\frac{\partial}{\partial x}\xi_x(x,t,u)\right)g(t)\,u^{n-2}n+\left(\frac{\partial}{\partial u}\eta_u(x,t,u)\right)u^{n-2}g(t)\,n\\
&-u^{n-2}\left(\frac{\partial}{\partial t}\xi_t(x,t,u)\right)g(t)\,\alpha+2\,\eta_u(x,t,u)\,g(t)\,u^{n-3}-\xi_t(x,t,u)\left(\frac{d}{dt}g(t)\right)u^{n-2}\\
&+3\left(\frac{\partial}{\partial x}\xi_x(x,t,u)\right)g(t)\,u^{n-2}-\left(\frac{\partial}{\partial u}\eta_u(x,t,u)\right)u^{n-2}g(t)+u^{n-1}\left(\frac{\partial^2}{\partial u^2}\eta_u(x,t,u)\right)g(t)\\
&-3\,u^{n-1}\left(\frac{\partial^2}{\partial x\partial u}\xi_x(x,t,u)\right)g(t)=0,
\end{aligned}$$

$$3\left(\frac{\partial}{\partial x}\eta_u(x,t,u)\right)g(t)\,u^{n-3}n^3-9\left(\frac{\partial}{\partial x}\eta_u(x,t,u)\right)g(t)\,u^{n-3}n^2+6\left(\frac{\partial}{\partial x}\eta_u(x,t,u)\right)$$
$$g(t)\,u^{n-3}n+6\,g(t)\left(\frac{\partial^2}{\partial x\partial u}\eta_u(x,t,u)\right)u^{n-2}n^2-6\,g(t)\left(\frac{\partial^2}{\partial x\partial u}\eta_u(x,t,u)\right)u^{n-2}n$$
$$-3\,g(t)\left(\frac{\partial^2}{\partial x^2}\xi_x(x,t,u)\right)u^{n-2}n^2+3\,g(t)\left(\frac{\partial^2}{\partial x^2}\xi_x(x,t,u)\right)u^{n-2}n+3\,g(t)\,u^{n-1}n$$
$$\frac{\partial^3}{\partial x\partial u^2}\eta_u(x,t,u)-3\,g(t)\,u^{n-1}n\frac{\partial^3}{\partial x^2\partial u}\xi_x(x,t,u)-\left(\frac{\partial}{\partial u}\xi_x(x,t,u)\right)u^{m-1}m=0,$$

$$u^{n-3}\left(\frac{\partial}{\partial u}\xi_t(x,t,u)\right)\alpha\,n^2-3\,u^{n-3}\left(\frac{\partial}{\partial u}\xi_t(x,t,u)\right)\alpha\,n-3\,u^{n-3}\left(\frac{\partial}{\partial u}\xi_t(x,t,u)\right)n^2$$
$$+2\,u^{n-3}\left(\frac{\partial}{\partial u}\xi_t(x,t,u)\right)\alpha+9\,u^{n-3}\left(\frac{\partial}{\partial u}\xi_t(x,t,u)\right)n-3\,u^{n-2}\left(\frac{\partial^2}{\partial u^2}\xi_t(x,t,u)\right)n$$
$$-6\,u^{n-3}\frac{\partial}{\partial u}\xi_t(x,t,u)+3\,u^{n-2}\frac{\partial^2}{\partial u^2}\xi_t(x,t,u)-u^{n-1}\frac{\partial^3}{\partial u^3}\xi_t(x,t,u)=0,$$

$$u^{m-1}\alpha\,m\frac{\partial}{\partial t}\xi_t(x,t,u)+3\,g(t)\left(\frac{\partial^2}{\partial x^2}\eta_u(x,t,u)\right)u^{n-2}n^2-3\,g(t)\left(\frac{\partial^2}{\partial x^2}\eta_u(x,t,u)\right)u^{n-2}n$$
$$+3\,g(t)\,u^{n-1}n\frac{\partial^3}{\partial x^2\partial u}\eta_u(x,t,u)-g(t)\,u^{n-1}n\frac{\partial^3}{\partial x^3}\xi_x(x,t,u)+\eta_u(x,t,u)\,u^{m-2}m^2$$
$$-\eta_u(x,t,u)\,u^{m-2}m-\left(\frac{\partial}{\partial x}\xi_x(x,t,u)\right)u^{m-1}m=0,$$

$$u^{n-3}\left(\frac{\partial}{\partial x}\xi_t(x,t,u)\right)n^2-3\,u^{n-3}\left(\frac{\partial}{\partial x}\xi_t(x,t,u)\right)n+2\,u^{n-2}\left(\frac{\partial^2}{\partial x\partial u}\xi_t(x,t,u)\right)n$$
$$+2\,u^{n-3}\frac{\partial}{\partial x}\xi_t(x,t,u)-2\,u^{n-2}\frac{\partial^2}{\partial x\partial u}\xi_t(x,t,u)+u^{n-1}\frac{\partial^3}{\partial x\partial u^2}\xi_t(x,t,u)=0,$$

$$-\left(\frac{\partial}{\partial u}\xi_t(x,t,u)\right)u^{m-1}m+u^{m-1}\alpha\,m\frac{\partial}{\partial u}\xi_t(x,t,u)-3\,g(t)\left(\frac{\partial^2}{\partial x^2}\xi_t(x,t,u)\right)u^{n-2}n^2$$
$$+3\,g(t)\left(\frac{\partial^2}{\partial x^2}\xi_t(x,t,u)\right)u^{n-2}n-3\,g(t)\,u^{n-1}n\frac{\partial^3}{\partial x^2\partial u}\xi_t(x,t,u)=0,$$
$$u^{n-2}\left(\frac{\partial}{\partial u}\xi_t(x,t,u)\right)\alpha\,n-u^{n-2}\left(\frac{\partial}{\partial u}\xi_t(x,t,u)\right)\alpha-2\,u^{n-2}\left(\frac{\partial}{\partial u}\xi_t(x,t,u)\right)n$$
$$+2\,u^{n-2}\frac{\partial}{\partial u}\xi_t(x,t,u)-u^{n-1}\frac{\partial^2}{\partial u^2}\xi_t(x,t,u)=0,$$
$$\eta_u(x,t,u)\,g(t)\,u^{n-2}n+u^{n-1}\left(\frac{\partial}{\partial t}\xi_t(x,t,u)\right)g(t)\,\alpha-\eta_u(x,t,u)\,g(t)\,u^{n-2}$$
$$+u^{n-1}\left(\frac{\mathrm{d}}{\mathrm{d}t}g(t)\right)\xi_t(x,t,u)-3\,u^{n-1}\left(\frac{\partial}{\partial x}\xi_x(x,t,u)\right)g(t)=0,$$
$$u^{n-2}\left(\frac{\partial}{\partial x}\eta_u(x,t,u)\right)n-u^{n-2}\frac{\partial}{\partial x}\eta_u(x,t,u)+u^{n-1}\frac{\partial^2}{\partial x\partial u}\eta_u(x,t,u)$$

$$-u^{n-1}\frac{\partial^2}{\partial x^2}\xi_x(x,t,u)=0,\ \frac{\partial^3}{\partial u^3}\xi_t(x,t,u)=0,\ \frac{\partial^3}{\partial u\partial t^2}\xi_t(x,t,u)=0,$$

$$-\alpha\frac{\partial^3}{\partial t^3}\xi_t(x,t,u)+3\frac{\partial^3}{\partial u\partial t^2}\eta_u(x,t,u)+2\frac{\partial^3}{\partial t^3}\xi_t=0,\ \frac{\partial^3}{\partial u^2\partial t}\xi_t(x,t,u)=0,$$
$$2u^{n-2}\left(\frac{\partial}{\partial u}\xi_x(x,t,u)\right)n-2u^{n-2}\frac{\partial}{\partial u}\xi_x(x,t,u)+u^{n-1}\frac{\partial^2}{\partial u^2}\xi_x(x,t,u)=0,$$
$$u^{n-2}\left(\frac{\partial}{\partial x}\xi_t(x,t,u)\right)n-u^{n-2}\frac{\partial}{\partial x}\xi_t(x,t,u)+u^{n-1}\frac{\partial^2}{\partial x\partial u}\xi_t(x,t,u)=0,$$
$$g(t)n\left(u^{n-2}\left(\frac{\partial}{\partial x}\xi_t(x,t,u)\right)n-u^{n-2}\frac{\partial}{\partial x}\xi_t(x,t,u)+u^{n-1}\frac{\partial^2}{\partial x\partial u}\xi_t(x,t,u)\right)=0,$$

$$2u^{n-2}\left(\frac{\partial}{\partial u}\xi_t(x,t,u)\right)n-2u^{n-2}\frac{\partial}{\partial u}\xi_t(x,t,u)+u^{n-1}\frac{\partial^2}{\partial u^2}\xi_t(x,t,u)=0,$$
$$\alpha\frac{\partial^2}{\partial t^2}\xi_t(x,t,u)-2\frac{\partial^2}{\partial u\partial t}\eta_u(x,t,u)-\frac{\partial^2}{\partial t^2}\xi_t(x,t,u)=0,\ \frac{\partial^2}{\partial u\partial t}\xi_t(x,t,u)0,$$
$$g(t)u^{n-1}n\frac{\partial^3}{\partial x^3}\xi_t(x,t,u)+\left(\frac{\partial}{\partial x}\xi_t(x,t,u)\right)u^{m-1}m=0,\ g(t)u^{n-1}n\frac{\partial}{\partial u}\xi_t(x,t,u)=0,$$
$$\frac{\partial^2}{\partial u^2}\xi_x(x,t,u)=0,\ \frac{\partial^2}{\partial t^2}\xi_x(x,t,u)=0,\ \frac{\partial^2}{\partial u^2}\xi_t(x,t,u)=0,\ \frac{\partial^2}{\partial u\partial t}\xi_x(x,t,u)=0,$$

$$\frac{\partial^2}{\partial u\partial t}\xi_t(x,t,u)=0,\ \frac{\partial}{\partial u}\xi_x(x,t,u)=0,\ \frac{\partial}{\partial u}\xi_t(x,t,u)=0,$$
$$g(t)u^{n-1}n\frac{\partial^2}{\partial x^2}\xi_t(x,t,u)=0,$$
$$g(t)u^{n-1}n\frac{\partial}{\partial u}\xi_x(x,t,u)=0,\ g(t)u^{n-1}n\frac{\partial}{\partial u}\xi_x(x,t,u)=0,$$
$$g(t)u^{n-1}n\frac{\partial}{\partial x}\xi_t(x,t,u)=0,$$
$$g(t)u^{n-1}n\frac{\partial}{\partial u}\xi_t(x,t,u)=0,\ \frac{\partial}{\partial u}\xi_x(x,t,u)=0,$$
$$\frac{\partial}{\partial t}\xi_x(x,t,u)=0,\ \frac{\partial^2}{\partial u^2}\eta_u(x,t,u)=0.$$

Solving this obtained system using the Maple, we obtain

$$\xi_x(x,t,u)=C_1,\ \xi_t(x,t,u)=0,\ \eta_u(x,t,u)=0,$$

therefore, we have the following infinitesimal generator:

$$X_{1.1}=\frac{\partial}{\partial x}.$$

For mentioned g(t) as follows, we have additional symmetries.

Case 2: $g(t) = kt^b$.
For this case, we have the following infinitesimal generators:

$$X_{2.1} = \frac{\partial}{\partial x}, \quad X_{2.2} = (3mt - (n+2)t)\frac{\partial}{\partial t} + (-2\alpha u + bu)\frac{\partial}{\partial u} + (x\alpha(m-n) + xb(m-1))\frac{\partial}{\partial x}. \tag{17}$$

Case 3: $g(t) = k$.
In this case, the infinitesimal generators are as follows

$$X_{3.1} = \frac{\partial}{\partial x}, \quad X_{3.2} = (3mt - (n+2)t)\frac{\partial}{\partial t} - 2\alpha u\frac{\partial}{\partial u} + (x\alpha(m-n))\frac{\partial}{\partial x}. \tag{18}$$

Case 4: $g(t) = ke^{bt}$.
In this case, the infinitesimal generators are as follows

$$X_{4.1} = \frac{\partial}{\partial x}, \quad X_{4.2} = xb(m-1))\frac{\partial}{\partial x} + ub\frac{\partial}{\partial u}. \tag{19}$$

4 Fractional Lie Symmetries for Time-Fractional K(2, 3)

Now, we obtain the infinitesimal generator of the time-fractional K(2, 3) equation

$$\frac{\partial^\alpha u}{\partial t^\alpha} + (u^2)_x + g(t)(u^3)_{xxx} = 0, \quad t > 0,\ 0 < \alpha < 1. \tag{20}$$

Theorem 2 *Lie symmetries for Eq. (20), which those are solutions of determining equations depend on the selection of the function $g(t)$, are*

Case 1: $0 < \alpha < 1,\ \alpha \neq \frac{1}{2}, \frac{1}{3},\ k, b \neq 0.$

Case 1.1: $g(t)$ *be an non-vanishing arbitrary function.*
In this case, the infinitesimal generator is given by

$$X_{1.1} = \frac{\partial}{\partial x}. \tag{21}$$

For mentioned $g(t)$ as follows, we have additional symmetries.
Case 1.2: $g(t) = kt^b$.
For this case, we have

$$X_{1.2.1} = \frac{\partial}{\partial x}, \quad X_{1.2.2} = -t\frac{\partial}{\partial t} + (\alpha - b)x\frac{\partial}{\partial x} + (2\alpha - b)u\frac{\partial}{\partial u}. \tag{22}$$

Case 1.3: $g(t) = k$.
In this case, the infinitesimal generators are as follows

$$X_{1.3.1} = \frac{\partial}{\partial x}, \quad X_{1.3.2} = \alpha x\frac{\partial}{\partial x} - t\frac{\partial}{\partial t} + 2\alpha u\frac{\partial}{\partial u}. \tag{23}$$

Case 2: $\alpha = \frac{1}{2}$, $k, b \neq 0$.
For $\alpha = \frac{1}{2}$, *functions of* $g(t)$ *can be obtained as follows*

$$g(t) = ke^{bt}, \; kt^b, \; k.$$

Case 2.1: $g(t) = ke^{bt}$.
In this case, the infinitesimal generator is given by

$$X_{2.1} = \frac{\partial}{\partial x}. \tag{24}$$

Case 2.2: $g(t) = kt^b$.
The infinitesimal generators in this case are

$$X_{2.2.1} = \frac{\partial}{\partial x}, \quad X_{2.2.2} = (2b - 1)x\frac{\partial}{\partial x} + 2t\frac{\partial}{\partial t} + 2(b - 1)u\frac{\partial}{\partial u}. \tag{25}$$

Case 2.3: $g(t) = k$.
We obtain the infinitesimal generators as follows

$$X_{2.3.1} = \frac{\partial}{\partial x}, \quad X_{2.3.2} = x\frac{\partial}{\partial x} - 2t\frac{\partial}{\partial t} + 2u\frac{\partial}{\partial u}. \tag{26}$$

Case 3: $\alpha = \frac{1}{3}$, $k, b \neq 0$.
For $\alpha = \frac{1}{3}$, *functions of* $g(t)$ *can be obtained as follows*

$$g(t) = k(t - b)^{\frac{2}{3}}, \; k(t^2 - b)^{\frac{1}{3}}, \; ke^{bt}, \; kt^b, \; k.$$

Case 3.1: $g(t) = k(t - b)^{\frac{2}{3}}, \; k(t^2 - b)^{\frac{1}{3}}, \; ke^{bt}$.
In these cases, the infinitesimal generator is given by

$$X_{3.1} = \frac{\partial}{\partial x}. \tag{27}$$

Case 3.2: $g(t) = kt^b$.
The infinitesimal generators in this case are

$$X_{3.2.1} = \frac{\partial}{\partial x}, \quad X_{3.2.2} = (3b-1)x\frac{\partial}{\partial x} + 3t\frac{\partial}{\partial t} + (3b-2)u\frac{\partial}{\partial u}. \tag{28}$$

Case 3.3: $g(t) = k$.

In this case, we obtain the infinitesimal generators as follows

$$X_{3.3.1} = \frac{\partial}{\partial x}, \quad X_{3.3.2} = -3t\frac{\partial}{\partial t} + x\frac{\partial}{\partial x} + 2u\frac{\partial}{\partial u}. \tag{29}$$

Proof The one-parameter Lie group of transformations in x, t, u with ε as the group parameter are given

$$\begin{aligned} t^* &= t + \varepsilon\xi_t(t, x, u) + O(\varepsilon^2), \\ x^* &= x + \varepsilon\xi_x(t, x, u) + O(\varepsilon^2), \\ u^* &= u + \varepsilon\eta_u(t, x, u) + O(\varepsilon^2), \end{aligned}$$

the Lie algebra of K(2, 3) equation (Eq. (20)) is spanned by vector fields

$$X = \xi_x(t, x, u)\frac{\partial}{\partial x} + \xi_t(t, x, u)\frac{\partial}{\partial t} + \eta_u(x, t, u)\frac{\partial}{\partial u}, \tag{30}$$

where

$$\xi_x = \frac{dx^*}{d\varepsilon}|_{\varepsilon=0}, \quad \xi_t = \frac{dt^*}{d\varepsilon}|_{\varepsilon=0}, \quad \eta_u = \frac{du^*}{d\varepsilon}|_{\varepsilon=0}. \tag{31}$$

Applying the $X^{(\alpha)}$ to (20), leads

$$X^{(\alpha)}\left[\frac{\partial^\alpha u}{\partial t^\alpha} + (u^2)_x + g(t)(u^3)_{xxx}\right]_{\frac{\partial^\alpha u}{\partial t^\alpha} + (u^2)_x + g(t)(u^3)_{xxx} = 0} = 0. \tag{32}$$

Expanding the (32), and solving this obtained set using the Maple, we can distinguish all selections of the function $g(t)$. Finally, the Lie point symmetries for (20) can be obtained as follow.

- If $0 < \alpha < 1$, $\alpha \neq \frac{1}{2}, \frac{1}{3}$, $k, b \neq 0$, and $g(t)$ be an arbitrary nonvanishing function then we have:

$$\xi_x = c_1, \quad \xi_t = 0, \quad \eta_u = 0.$$

Thus, the infinitesimal generator is given by

$$X_1 = \frac{\partial}{\partial x}.$$

- If $0 < \alpha < 1$, $\alpha \neq \frac{1}{2}, \frac{1}{3}$, $k, b \neq 0$, and $g(t) = k$ then we have:

$$\xi_x = c_1 + c_2\alpha x, \quad \xi_t = -c_2 t, \quad \eta_u = 2c_2\alpha u.$$

Therefore, the infinitesimal generators are given by

$$X_1 = \frac{\partial}{\partial x}, \; X_2 = \alpha x\frac{\partial}{\partial x} - t\frac{\partial}{\partial t} + 2\alpha u\frac{\partial}{\partial u}.$$

- If $0 < \alpha < 1$, $\alpha \neq \frac{1}{2}, \frac{1}{3}$, $k, b \neq 0$, and $g(t) = kt^b$. Applying the $X^{(\alpha)}$ to (20) with $g(t) = kt^b$, leads

$$X^{(\alpha)}\left[\frac{\partial^\alpha u}{\partial t^\alpha} + (u^2)_x + kt^b(u^3)_{xxx}\right]_{\frac{\partial^\alpha u}{\partial t^\alpha}+(u^2)_x+kt^b(u^3)_{xxx}=0} = 0. \tag{33}$$

Expanding the (33), we obtain the following overdetermined system of linear partial differential equations:

$$\left(\frac{\partial^3}{\partial u^3}\eta_u(x,t,u)\right)t^b u^2 - 3\left(\frac{\partial^3}{\partial x\partial u^2}\xi_x(x,t,u)\right)t^b u^2 + 2t^{b-1}\xi_t(x,t,u)\,b$$
$$+6\left(\frac{\partial^2}{\partial u^2}\eta_u(x,t,u)\right)t^b u - 12\left(\frac{\partial^2}{\partial x\partial u}\xi_x(x,t,u)\right)t^b u + 2\left(\frac{\partial}{\partial t}\xi_t(x,t,u)\right)t^b\alpha$$
$$+4\left(\frac{\partial}{\partial u}\eta_u(x,t,u)\right)t^b - 6\left(\frac{\partial}{\partial x}\xi_x(x,t,u)\right)t^b = 0,$$

$$2t^{b-1}\xi_t(x,t,u)\,bu + \left(\frac{\partial^2}{\partial u^2}\eta_u(x,t,u)\right)t^b u^2 - 3\left(\frac{\partial^2}{\partial x\partial u}\xi_x(x,t,u)\right)t^b u^2$$
$$+2\left(\frac{\partial}{\partial t}\xi_t(x,t,u)\right)t^b\alpha u + 2\left(\frac{\partial}{\partial u}\eta_u(x,t,u)\right)t^b u - 6\left(\frac{\partial}{\partial x}\xi_x(x,t,u)\right)t^b u$$
$$+2\eta_u(x,t,u)t^b = 0, \quad \frac{\partial^2}{\partial u^2}\eta_u(x,t,u) = 0, \quad \frac{\partial}{\partial t}\xi_x(x,t,u) = 0,$$
$$-2\left(\frac{\partial}{\partial u}\xi_x(x,t,u)\right)u + 36kt^b u\frac{\partial^2}{\partial x\partial u}\eta_u(x,t,u) - 18kt^b u\frac{\partial^2}{\partial x^2}\xi_x(x,t,u)$$
$$+9kt^b u^2\frac{\partial^3}{\partial x\partial u^2}\eta_u(x,t,u) - 9kt^b u^2\frac{\partial^3}{\partial x^2\partial u}\xi_x(x,t,u) + 18\left(\frac{\partial}{\partial x}\eta_u(x,t,u)\right)kt^b = 0,$$
$$-\alpha\frac{\partial^3}{\partial t^3}\xi_t(x,t,u) + 3\frac{\partial^3}{\partial u\partial t^2}\eta_u(x,t,u) + 2\frac{\partial^3}{\partial t^3}\xi_t(x,t,u) = 0,$$
$$\left(\frac{\partial^3}{\partial u^3}\xi_t(x,t,u)\right)u^2 + 6\left(\frac{\partial^2}{\partial u^2}\xi_t(x,t,u)\right)u - 2\alpha\frac{\partial}{\partial u}\xi_t(x,t,u) + 6\frac{\partial}{\partial u}\xi_t(x,t,u) = 0,$$
$$2\eta_u(x,t,u) - 2\left(\frac{\partial}{\partial x}\xi_x(x,t,u)\right)u + 18kt^b u\frac{\partial^2}{\partial x^2}\eta_u(x,t,u) - 3kt^b u^2\frac{\partial^3}{\partial x^3}\xi_x(x,t,u)$$
$$+9kt^b u^2\frac{\partial^3}{\partial x^2\partial u}\eta_u(x,t,u) + 2\alpha u\frac{\partial}{\partial t}\xi_t(x,t,u) = 0,$$
$$\frac{9}{2}\left(\frac{\partial^3}{\partial x^2\partial u}\xi_t(x,t,u)\right)t^b ku + 9\left(\frac{\partial^2}{\partial x^2}\xi_t(x,t,u)\right)t^b k - \alpha\frac{\partial}{\partial u}\xi_t(x,t,u) + \frac{\partial}{\partial u}\xi_t(x,t,u) = 0,$$
$$\frac{\partial^3}{\partial u^3}\xi_t(x,t,u) = 0, \quad \frac{\partial^3}{\partial u^2\partial t}\xi_t(x,t,u) = 0, \quad \frac{\partial^3}{\partial u\partial t^2}\xi_t(x,t,u) = 0, \quad \frac{\partial^2}{\partial u^2}\xi_t(x,t,u) = 0,$$
$$t^{b-1}\xi_t(x,t,u)\,bu + \left(\frac{\partial}{\partial t}\xi_t(x,t,u)\right)t^b\alpha u - 3\left(\frac{\partial}{\partial x}\xi_x(x,t,u)\right)t^b u + 2\eta_u(x,t,u)t^b = 0,$$

$$\left(\frac{\partial^3}{\partial u^3}\xi_x(x,t,u)\right)u^2+6\left(\frac{\partial^2}{\partial u^2}\xi_x(x,t,u)\right)u+6\frac{\partial}{\partial u}\xi_x(x,t,u)=0,\ \frac{\partial}{\partial u}\xi_x(x,t,u)=0,$$
$$\left(\frac{\partial^3}{\partial x\partial u^2}\xi_t(x,t,u)\right)u^2+4\left(\frac{\partial^2}{\partial x\partial u}\xi_t(x,t,u)\right)u+2\frac{\partial}{\partial x}\xi_t(x,t,u)=0,\ \frac{\partial}{\partial x}\xi_t(x,t,u)=0,$$
$$\left(\frac{\partial^2}{\partial x\partial u}\eta_u(x,t,u)\right)u-\left(\frac{\partial^2}{\partial x^2}\xi_x(x,t,u)\right)u+2\frac{\partial}{\partial x}\eta_u(x,t,u)=0,\ \frac{\partial}{\partial u}\xi_t(x,t,u)=0,$$
$$\left(\frac{\partial^2}{\partial u^2}\xi_t(x,t,u)\right)u-2\alpha\frac{\partial}{\partial u}\xi_t(x,t,u)+4\frac{\partial}{\partial u}\xi_t(x,t,u)=0,\ \frac{\partial}{\partial u}\xi_x(x,t,u)=0,$$
$$\frac{\partial^2}{\partial u\partial t}\xi_t(x,t,u)=0,\ \frac{\partial}{\partial u}\xi_t(x,t,u)=0,\ \frac{\partial^2}{\partial u^2}\xi_x(x,t,u)=0,\ \frac{\partial^2}{\partial t^2}\xi_x(x,t,u)=0,$$
$$\left(\frac{\partial^2}{\partial u^2}\xi_x(x,t,u)\right)u+4\frac{\partial}{\partial u}\xi_x(x,t,u)=0,\ \left(\frac{\partial^2}{\partial x\partial u}\xi_t(x,t,u)\right)u+2\frac{\partial}{\partial x}\xi_t(x,t,u)=0,$$
$$\left(\frac{\partial^2}{\partial x\partial u}\xi_t(x,t,u)\right)u+2\frac{\partial}{\partial x}\xi_t(x,t,u)=0,\ \left(\frac{\partial^2}{\partial u^2}\xi_t(x,t,u)\right)u+4\frac{\partial}{\partial u}\xi_t(x,t,u)=0,$$
$$3\left(\frac{\partial^3}{\partial x^3}\xi_t(x,t,u)\right)t^bku+2\frac{\partial}{\partial x}\xi_t(x,t,u)=0,\ \frac{\partial}{\partial u}\xi_t(x,t,u)=0,\ \frac{\partial^2}{\partial u\partial t}\xi_x(x,t,u)=0,$$
$$\frac{\partial^2}{\partial u\partial t}\xi_t(x,t,u)=0,\ \frac{\partial}{\partial u}\xi_x(x,t,u)=0,\ \frac{\partial}{\partial u}\xi_t(x,t,u)=0,\ \frac{\partial^2}{\partial x^2}\xi_t(x,t,u)=0,$$
$$-\alpha\frac{\partial^2}{\partial t^2}\xi_t(x,t,u)+2\frac{\partial^2}{\partial u\partial t}\eta_u(x,t,u)+\frac{\partial^2}{\partial t^2}\xi_t(x,t,u)=0,\ \frac{\partial^2}{\partial u^2}\xi_t(x,t,u)=0.$$

Solving this obtained system using the Maple, we obtain:

$$\xi_x=c_1+c_2(\alpha-b)x,\quad \xi_t=-c_2t,\quad \eta_u=c_2(2\alpha-b)u.$$

So, the infinitesimal generators are

$$X_1=\frac{\partial}{\partial x},\ X_2=(\alpha-b)x\frac{\partial}{\partial x}-t\frac{\partial}{\partial t}+(2\alpha-b)u\frac{\partial}{\partial u}.$$

The proof for $\alpha=\frac{1}{2},\frac{1}{3}$ are similar. Therefore, proof is completed.

5 Reduced Equations and Invariant Solution of Eq. (20)

Our purpose for Eq. (20) is to reduce it the coordinates (x,t,u) using invariants (r,z) to a new coordinates[23].
Let us consider

$$X=\xi_t(t,x,u)\frac{\partial}{\partial t}+\xi_x(t,x,u)\frac{\partial}{\partial x}+\eta_u(t,x,u)\frac{\partial}{\partial u},$$

as a Lie point symmetry of the time-fractional K(2, 3) equation

$$\frac{\partial^\alpha u}{\partial t^\alpha}+(u^2)_x+g(t)(u^3)_{xxx}=0,\quad 0<\alpha<1,\ t>0.$$

We use two invariants $z = \psi(x, t)$ and $r = \varphi(x, t)$ which are linearly independent in the characteristic equations

$$\frac{dt}{\xi_t(t, x, u)} = \frac{dx}{\xi_x(t, x, u)} = \frac{du}{\eta_u(t, x, u)},$$

for obtaining the invariant solutions. After that, we assume one of those invariants is depend to another,

$$z = h(r), \tag{34}$$

then we solve (34) for u. Finally, substituting u in Eq. (20) for the unknown function h, a fractional ODE can be obtained. Now, we obtain corresponding reduced equations, invariants and group invariant solutions of equation (20) for different cases of $g(t)$ and α as follows.

Case1:

- Case 1.1: $0 < \alpha < 1, \alpha \neq \frac{1}{2}, \frac{1}{3}$ and $g(t)$ is a nonvanishing arbitrary function.
- Case 1.2: $\alpha = \frac{1}{2}, g(t) = \{k,\ kt^b,\ ke^{bt}\}$
- Case 1.3: $\alpha = \frac{1}{3}, g(t) = \{k,\ kt^b,\ ke^{bt},\ k(t-b)^{\frac{2}{3}},\ k(t^2-b)^{\frac{1}{3}}\}$

In these cases, according to the infinitesimal generator $X = \frac{\partial}{\partial x}$, the similarity variables using the method of characteristics are as follows:

$$z = u, \ \ r = t, \tag{35}$$

and a solution is

$$z = h(r) \Rightarrow u = h(t). \tag{36}$$

By substituting (36) into Eq. (20) we find the $h(r)$. Thus $h(r)$ must be satisfied:

$$\frac{d^{\alpha}h(t)}{dt^{\alpha}} = 0. \tag{37}$$

Then by solving the above equation by the Laplace transform[26], we have

$$h(t) = \frac{\kappa t^{\alpha-1}}{\Gamma(\alpha)}, \quad \kappa \text{ is a constant.} \tag{38}$$

Case2:

- Case 2.1: $\alpha \neq \frac{1}{2}, \frac{1}{3}, g(t) = kt^b$.
  In this case

$$X_{1.2.2} = -t\frac{\partial}{\partial t} + (\alpha - b)x\frac{\partial}{\partial x} + (2\alpha - b)u\frac{\partial}{\partial u}, \tag{39}$$

so the similarity variables for this Lie point symmetry using the method of characteristics are as follows:

$$r = tx^{\frac{1}{\alpha-b}}, \quad z = ux^{\frac{b-2\alpha}{\alpha-b}}, \tag{40}$$

because

$$-\frac{dt}{t} = \frac{dx}{(\alpha-b)x} \Rightarrow -\ln t + \ln r = \frac{1}{\alpha-b}\ln x \Rightarrow \frac{1}{\alpha-b}\ln x + \ln t = \ln r \Rightarrow r = tx^{\frac{1}{\alpha-b}},$$
$$\frac{dx}{(\alpha-b)x} = \frac{du}{(2\alpha-b)u} \Rightarrow \frac{2\alpha-b}{\alpha-b}\ln x + \ln z = \ln u \Rightarrow \ln z = \ln u + \ln(x^{\frac{b-2\alpha}{\alpha-b}}) \Rightarrow$$
$$z = ux^{\frac{b-2\alpha}{\alpha-b}}.$$

And a solution for Eq. (20) is

$$z = h(r) \Rightarrow u = x^{\frac{b-2\alpha}{b-\alpha}} h(tx^{\frac{1}{\alpha-b}}). \tag{41}$$

We substitute (41) into Eq. (20) to find the $f(r)$ and $f(r)$ must be satisfied in the fractional ODE as follows:

$$\begin{aligned}
&(b-\alpha)^3\frac{\partial^\alpha h}{\partial r^\alpha} - 2(b-\alpha)^2 r h(r)h'(r) + 18(-1+2b-5\alpha)kr^{2+b}h(r)h'(r)^2 \\
&-18kr^{3+b}h(r)h'(r)h''(r) + 3(2b^3 - 17b^2\alpha + 46b\alpha^2 - 40\alpha^3)kr^b h(r)^3 \\
&-6kr^{3+b}h'(r)^3 + (2b^3 - 8b^2\alpha + 10b\alpha^2 - 4\alpha^3)h(r)^2 - 3(11b^2 + 15\alpha + 74\alpha^2 \\
&-2b(29\alpha+3)+1)kr^{1+b}h'(r)h(r)^2 + 9(-1+2b-5\alpha)kr^{2+b}h''(r)h(r)^2 \\
&-3kr^{3+b}h'''(r)h(r)^2 = 0.
\end{aligned}$$

where $\alpha \neq \frac{1}{2}, \frac{1}{3}$.

- Case 2.2: $\alpha \neq \frac{1}{2}, \frac{1}{3}, g(t) = k$.
  For this case we have

$$X_{1.3.2} = \alpha x\frac{\partial}{\partial x} - t\frac{\partial}{\partial t} + 2\alpha u\frac{\partial}{\partial u}, \tag{42}$$

so the similarity variables for this Lie point symmetry using the method of characteristics are as follows:

$$r = tx^{\frac{1}{\alpha}}, \quad z = ux^{-2}, \tag{43}$$

because

$$\frac{dx}{\alpha x} = -\frac{dt}{t} \Rightarrow \frac{1}{\alpha}\ln x = -\ln t + \ln r \Rightarrow \ln r = \ln(tx^{\frac{1}{\alpha}}) \Rightarrow r = tx^{\frac{1}{\alpha}},$$
$$\frac{dx}{\alpha x} = \frac{du}{2\alpha u} \Rightarrow 2\ln x + \ln z = \ln u \Rightarrow \ln z = \ln u - \ln(x^2) = \ln u + \ln(x^{-2}) \Rightarrow z = ux^{-2}.$$

And a solution for (20) is

$$z = h(r) \Rightarrow u = x^2 h(tx^{\frac{1}{\alpha}}). \tag{44}$$

Again we substitute (44) into Eq. (20) to obtain the $f(r)$. So $h(r)$ must satisfy in the fractional ODE as follows:

$$\begin{aligned}
&\alpha^3 \frac{\partial^\alpha h}{\partial r^\alpha} + 2\alpha^2 r h(r) h'(r) + 18(5\alpha + 1) k r^2 h(r) h'(r)^2 \\
&+18kr^3 h(r) h'(r) h''(r) + 120k\alpha^3 h(r)^3 + 6kr^3 h'(r)^3 + 4\alpha^3 h(r)^2 \\
&+3(74\alpha^2 + 15\alpha + 1) krh(r)^2 h'(r) + 9(5\alpha + 1) kr^2 h(r)^2 h''(r) \\
&+3kr^3 h(r)^2 h'''(r) = 0.
\end{aligned}$$

where $\alpha \neq \frac{1}{2}, \frac{1}{3}$.

Case3:

- Case 3.1: $\alpha = \frac{1}{2}$, $g(t) = kt^b$.
  For this case we have

$$X_{2.2.2} = (2b - 1)x\frac{\partial}{\partial x} + 2t\frac{\partial}{\partial t} + 2(b - 1)u\frac{\partial}{\partial u}, \tag{45}$$

so the similarity variables for this Lie point symmetry using the method of characteristics are as follows:

$$r = tx^{\frac{2}{1-2b}}, \quad z = ux^{\frac{2b-2}{1-2b}}, \tag{46}$$

because

$$\frac{dx}{(2b-1)x} = \frac{dt}{2t} \Rightarrow \frac{2}{2b-1}\ln x + \ln r = \ln t \Rightarrow \ln r = \ln t - \frac{2}{2b-1}\ln x \Rightarrow \ln r = \ln(tx^{\frac{2}{1-2b}})$$
$$\Rightarrow r = tx^{\frac{2}{1-2b}},$$
$$\frac{dx}{(2b-1)x} = \frac{du}{2(b-1)u} \Rightarrow \frac{2(b-1)}{2b-1}\ln x + \ln z = \ln u \Rightarrow \ln z = \ln u - \frac{2(b-1)}{2b-1}\ln x$$
$$\Rightarrow z = ux^{\frac{2b-2}{1-2b}}.$$

And in view of (36), a solution for (20) is

$$u = x^{\frac{2-2b}{1-2b}} h\left(tx^{\frac{2}{1-2b}}\right). \tag{47}$$

We substitute (51) into Eq. (20) to obtain $h(r)$. After that $h(r)$ must be satisfied in the FDE as follows:

$$\frac{(2b-1)^3}{4}\frac{\partial^\alpha h}{\partial r^\alpha} - (1-2b)^2 r h(r)h'(r) + 18(4b-7)kr^{2+b}h(r)h'(r)^2$$
$$-36kr^{3+b}h(r)h'(r)h''(r) + 3(4b^3-17b^2+23b-10)kr^b h(r)^3$$
$$-12kr^{3+b}h'(r)^3 + (4b^3-8b^2+5b-1)h(r)^2 - 6(11b^2-35b+27)kr^{1+b}h'(r)h(r)^2$$
$$+9(4b-7)kr^{2+b}h''(r)h(r)^2 - 6kr^{3+b}h'''(r)h(r)^2 = 0.$$

where $\alpha = \frac{1}{2}$.

- Case 3.2: $\alpha = \frac{1}{2}$, $g(t) = k$.
  For this case we have

$$X_{2.3.2} = x\frac{\partial}{\partial x} - 2t\frac{\partial}{\partial t} + 2u\frac{\partial}{\partial u}, \tag{48}$$

Applying the $X^{(\frac{1}{2})}$ to (20) with $g(t) = k$ and $\alpha = \frac{1}{2}$, leads

$$X^{(\frac{1}{2})}\left[\frac{\partial^{\frac{1}{2}}u}{\partial t^{\frac{1}{2}}} + (u^2)_x + k(u^3)_{xxx}\right]_{\frac{\partial^{\frac{1}{2}}u}{\partial t^{\frac{1}{2}}}+(u^2)_x+k(u^3)_{xxx}=0} = 0. \tag{49}$$

Expanding the (49), we obtain the following overdetermined system of linear partial differential equations:

$$-\left(\frac{\partial^3}{\partial u^3}\eta_u(x,t,u)\right)u^2 + 3\left(\frac{\partial^3}{\partial x\partial u^2}\xi_x(x,t,u)\right)u^2 - 6\left(\frac{\partial^2}{\partial u^2}\eta_u(x,t,u)\right)u$$
$$+12\left(\frac{\partial^2}{\partial x\partial u}\xi_x(x,t,u)\right)u + 6\frac{\partial}{\partial x}\xi_x(x,t,u) - 4\frac{\partial}{\partial u}\eta_u(x,t,u) - \frac{\partial}{\partial t}\xi_t(x,t,u) = 0,$$
$$\frac{3}{8}\frac{\partial^3}{\partial t^3}\xi_t(x,t,u) + \frac{3}{4}\frac{\partial^3}{\partial u\partial t^2}\eta_u(x,t,u) = 0, \quad \frac{\partial^3}{\partial u^3}\xi_t(x,t,u) = 0,$$
$$18\left(\frac{\partial}{\partial x}\eta_u(x,t,u)\right)k - 2\left(\frac{\partial}{\partial u}\xi_x(x,t,u)\right)u + 36\,ku\frac{\partial^2}{\partial x\partial u}\eta_u(x,t,u)$$
$$-18\,ku\frac{\partial^2}{\partial x^2}\xi_x(x,t,u) + 9\,ku^2\frac{\partial^3}{\partial x\partial u^2}\eta_u(x,t,u) - 9\,ku^2\frac{\partial^3}{\partial x^2\partial u}\xi_x(x,t,u) = 0,$$
$$2\,\eta_u(x,t,u) - 2\left(\frac{\partial}{\partial x}\xi_x(x,t,u)\right)u + 18\,ku\frac{\partial^2}{\partial x^2}\eta_u(x,t,u) + 9\,ku^2\frac{\partial^3}{\partial x^2\partial u}\eta_u(x,t,u)$$
$$-3\,ku^2\frac{\partial^3}{\partial x^3}\xi_x(x,t,u) + \left(\frac{\partial}{\partial t}\xi_t(x,t,u)\right)u = 0, \quad \frac{\partial^3}{\partial u^2\partial t}\xi_t(x,t,u) = 0,$$

$$-\left(\frac{\partial^2}{\partial u^2}\eta_u(x,t,u)\right)u^2 + 3\left(\frac{\partial^2}{\partial x\partial u}\xi_x(x,t,u)\right)u^2 + 6\left(\frac{\partial}{\partial x}\xi_x(x,t,u)\right)u$$
$$-2\left(\frac{\partial}{\partial u}\eta_u(x,t,u)\right)u - \left(\frac{\partial}{\partial t}\xi_t(x,t,u)\right)u - 2\,\eta_u(x,t,u) = 0,$$

$$\left(\frac{\partial^3}{\partial u^3}\xi_t(x,t,u)\right)u^2+6\left(\frac{\partial^2}{\partial u^2}\xi_t(x,t,u)\right)u+5\frac{\partial}{\partial u}\xi_t(x,t,u)=0,$$

$$9\left(\frac{\partial^3}{\partial x^2\partial u}\xi_t(x,t,u)\right)ku+18\left(\frac{\partial^2}{\partial x^2}\xi_t(x,t,u)\right)k+\frac{\partial}{\partial u}\xi_t(x,t,u)=0,$$

$$-\frac{1}{4}\frac{\partial^2}{\partial t^2}\xi_t(x,t,u)-\frac{\partial^2}{\partial u\partial t}\eta_u(x,t,u)=0,\ \frac{\partial^3}{\partial u\partial t^2}\xi_t(x,t,u)=0,$$

$$\frac{\partial^2}{\partial u\partial t}\xi_t(x,t,u)=0,\ \frac{\partial^2}{\partial u^2}\xi_t(x,t,u)=0,\ \frac{\partial}{\partial u}\xi_t(x,t,u)=0,$$

$$\left(\frac{\partial^3}{\partial x\partial u^2}\xi_t(x,t,u)\right)u^2+4\left(\frac{\partial^2}{\partial x\partial u}\xi_t(x,t,u)\right)u+2\frac{\partial}{\partial x}\xi_t(x,t,u)=0,$$

$$\left(\frac{\partial^3}{\partial u^3}\xi_x(x,t,u)\right)u^2+6\left(\frac{\partial^2}{\partial u^2}\xi_x(x,t,u)\right)u+6\frac{\partial}{\partial u}\xi_x(x,t,u)=0,\ \frac{\partial}{\partial u}\xi_x(x,t,u)=0,$$

$$\left(\frac{\partial^2}{\partial x\partial u}\eta_u(x,t,u)\right)u-\left(\frac{\partial^2}{\partial x^2}\xi_x(x,t,u)\right)u+2\frac{\partial}{\partial x}\eta_u(x,t,u)=0,\ \frac{\partial}{\partial t}\xi_x(x,t,u)=0,$$

$$\left(\frac{\partial^2}{\partial u^2}\xi_t(x,t,u)\right)u+3\frac{\partial}{\partial u}\xi_t(x,t,u)=0,\ \frac{\partial}{\partial u}\xi_t(x,t,u)=0,\ \frac{\partial}{\partial u}\xi_t(x,t,u)=0,$$

$$6\left(\frac{\partial}{\partial x}\xi_x(x,t,u)\right)u-\left(\frac{\partial}{\partial t}\xi_t(x,t,u)\right)u-4\eta_u(x,t,u)=0,\ \frac{\partial}{\partial x}\xi_t(x,t,u)=0,$$

$$\frac{\partial^2}{\partial u^2}\xi_x(x,t,u)=0,\ \frac{\partial^2}{\partial t^2}\xi_x(x,t,u)=0,\ \frac{\partial^2}{\partial u^2}\xi_t(x,t,u)=0,\ \frac{\partial^2}{\partial x^2}\xi_t(x,t,u)=0,$$

$$3\left(\frac{\partial^3}{\partial x^3}\xi_t(x,t,u)\right)ku+2\frac{\partial}{\partial x}\xi_t(x,t,u)=0,\ \frac{\partial^2}{\partial u\partial t}\xi_x(x,t,u)=0,\ \frac{\partial}{\partial u}\xi_x(x,t,u)=0,$$

$$\frac{\partial^2}{\partial u\partial t}\xi_t(x,t,u)=0,\ \left(\frac{\partial^2}{\partial u^2}\xi_x(x,t,u)\right)u+4\frac{\partial}{\partial u}\xi_x(x,t,u)=0,\ \frac{\partial}{\partial u}\xi_x(x,t,u)=0,$$

$$\left(\frac{\partial^2}{\partial x\partial u}\xi_t(x,t,u)\right)u+2\frac{\partial}{\partial x}\xi_t(x,t,u)=0,\ \frac{\partial}{\partial u}\xi_t(x,t,u)=0,$$

$$\left(\frac{\partial^2}{\partial x\partial u}\xi_t(x,t,u)\right)u+2\frac{\partial}{\partial x}\xi_t(x,t,u)=0,\ \frac{\partial}{\partial u}\xi_x(x,t,u)=0,\ \frac{\partial^2}{\partial u^2}\eta_u(x,t,u)=0,$$

$$\left(\frac{\partial^2}{\partial u^2}\xi_t(x,t,u)\right)u+4\frac{\partial}{\partial u}\xi_t(x,t,u)=0,$$

Solving this obtained system, the infinitesimal generator can be obtained as follows:

$$X=x\frac{\partial}{\partial x}-2t\frac{\partial}{\partial t}+2u\frac{\partial}{\partial u}.$$

So the similarity variables for this Lie point symmetry using the method of characteristics are as follows:

$$r=tx^2,\quad z=ux^{-2}, \tag{50}$$

because

$$\frac{dx}{x} = -\frac{dt}{2t} \Rightarrow -2\ln x + \ln r = \ln t \Rightarrow \ln r = \ln(tx^2) \Rightarrow r = tx^2,$$
$$\frac{dx}{x} = \frac{du}{2u} \Rightarrow 2\ln x + \ln z = \ln u \Rightarrow \ln z = \ln u - 2\ln x \Rightarrow z = ux^{-2}.$$

And in view of (36), a solution for (20) is

$$u = x^2 h\left(tx^2\right). \tag{51}$$

To obtain $f(r)$, we substitute (51) into Eq. (20). Then $f(r)$ must satisfy in the fractional ODE as follows:

$$\frac{1}{4}\frac{\partial^\alpha f}{\partial r^\alpha} + 30kf(r)^3 + 12kr^3 f'(r)^3 + rf(r)f'(r) + 126kr^2 f(r)f'(r)^2$$
$$+36kr^3 f(r)f'(r)f''(r) + f(r)^2 + 162krf'(r)f(r)^2 + 63kr^2 f(r)^2 f''(r)$$
$$+6kr^3 f(r)^2 f'''(r) = 0.$$

where $\alpha = \frac{1}{2}$.
Case4:

- Case 4.1: $\alpha = \frac{1}{3}$, $g(t) = kt^b$.
  For this case we have

$$X_{3.2.2} = (3b-1)x\frac{\partial}{\partial x} + 3t\frac{\partial}{\partial t} + (3b-2)u\frac{\partial}{\partial u}, \tag{52}$$

so the similarity variables for this Lie point symmetry using the method of characteristics are as follows:

$$r = tx^{\frac{3}{1-3b}}, \quad z = ux^{\frac{3b-2}{1-3b}}, \tag{53}$$

because

$$\frac{dx}{(3b-1)x} = \frac{dt}{3t} \Rightarrow \frac{3}{3b-1}\ln x + \ln r = \ln t \Rightarrow \ln r = \ln t - \frac{3}{3b-1}\ln x \Rightarrow \ln r = \ln(tx^{\frac{3}{1-3b}})$$
$$\Rightarrow r = tx^{\frac{3}{1-3b}},$$
$$\frac{dx}{(3b-1)x} = \frac{du}{(3b-2)u} \Rightarrow \frac{3b-2}{3b-1}\ln x + \ln z = \ln u \Rightarrow \ln z = \ln u - \frac{3b-2}{3b-1}\ln x$$
$$\Rightarrow z = ux^{\frac{3b-2}{1-3b}}.$$

And a solution to our equation is

$$z = h(r) \Rightarrow u = x^{\frac{2-3b}{1-3b}} h\left(tx^{\frac{3}{1-3b}}\right). \tag{54}$$

We substitute (54) into Eq. (20) to determine the $h(r)$. Then $h(r)$ must satisfy in the fractional ODE as follows:

$$(3b-1)^3\frac{\partial^\alpha h}{\partial r^\alpha}-6(1-3b)^2 rh(r)h'(r)+324(3b-4)kr^{2+b}h(r)h'(r)^2$$
$$-486kr^{3+b}h(r)h'(r)h''(r)+3(54b^3-153b^2+138b-40)kr^b h(r)^3$$
$$-162kr^{3+b}h'(r)^3+(54b^3-72b^2+30b-4)h(r)^2-9(99b^2-228b+128)kr^{1+b}h'(r)h(r)^2$$
$$+162(3b-4)kr^{2+b}h''(r)h(r)^2-81kr^{3+b}h'''(r)h(r)^2=0.$$

where $\alpha=\frac{1}{3}$.

- Case 4.2: $\alpha=\frac{1}{3},\ g(t)=k.$
  For this case we have

$$X_{3.3.2}=x\frac{\partial}{\partial x}-3t\frac{\partial}{\partial t}+2u\frac{\partial}{\partial u}, \tag{55}$$

so the similarity variables for this Lie point symmetry using the method of characteristics are as follows:

$$r=tx^3,\quad z=ux^{-2}, \tag{56}$$

because

$$\frac{dx}{x}=-\frac{dt}{3t}\Rightarrow -3\ln x+\ln r=\ln t\Rightarrow \ln r=\ln t+3\ln x\Rightarrow \ln r=\ln(tx^3)$$
$$\Rightarrow r=tx^3,$$
$$\frac{dx}{x}=\frac{du}{2u}\Rightarrow 2\ln x+\ln z=\ln u\Rightarrow \ln z=\ln u-2\ln x\Rightarrow z=ux^{-2}.$$

And a solution to our equation is

$$z=h(r)\Rightarrow u=x^2\,h\left(tx^3\right). \tag{57}$$

We substitute (57) into Eq. (20) to determine the $h(r)$. Then $h(r)$ must satisfy in the fractional ODE as follows:

$$\frac{\partial^\alpha h}{\partial r^\alpha}+120kh(r)^3+162kr^3h'(r)^3+6rh(r)h'(r)+1296kr^2h(r)h'(r)^2$$
$$+486kr^3h(r)h'(r)h''(r)+4h(r)^2+1152krh'(r)h(r)^2+648kr^2h(r)^2h''(r)$$
$$+81kr^3h(r)^2\,h'''(r)=0.$$

where $\alpha=\frac{1}{3}$.

6 Conclusion

In the present study, we investigated the efficiency of the classical Lie symmetry group analysis to the fractional differential equations. The fractional Lie symmetries method is considered for application to the time-fractional K(m, n) equation with variable coefficients with Riemann-Liouville derivative. Also, we reduced this time-fractional equation into a nonlinear ODE of fractional order. For this propose, the Lie group method and the symmetry properties have been investigated for the governing equations. They have been used to reduce the given FPDE to a corresponding nonlinear ODE of fractional order which might be solved easily.

Conflict of interest The author declare that they have no conflict of interest.

Funding Not applicable.

Authors' contributions All authors have read and approved the final manuscript.

Acknowledgements The authors would like to extend great gratitude to the Editor and anonymous reviewers, whose insightful comments and constructive suggestions helped us to significantly improve the quality of this paper.

References

1. Baleanu, D., Diethelm, K., Scalas, E., Trujillo, J.J.: Fractional Calculus Models and Numerical Methods (Series on Complexity. Nonlinearity and Chaos, World Scientific) (2012)
2. Baumann, G.: Symmetry Analysis of Differential Equations with Mathematica. Springer, New York, Telos (2000)
3. Bluman, G.W., Anco, S.C.: Symmetry and Integtation Methods for Differential Equations. Springer, New York (2002)
4. Bluman, G.W., Kumei, S.: Symmetries and Differential Equations, Applied Mathematical Sciences, 81. Springer, New York (1989)
5. Buckwar, E., Luchko, Yu.: Invariance of a partial differential equation of fractional order under the Lie group of scaling transformations. J. Math. Anal. Appl. **227**, 81–97 (1998)
6. Charalambous, K., Vaneeva, O., Sophocleous, C.: Group classification of variable coefficient K(m, n) equations, geometry and symmetry in physics **33**, 79–90 (2014)
7. Chen, J., Liu, F., Anh, V.: Analytical solution for the time-fractional telegraph equation by the method of separating variables. J. Math. Anal. Appl. **338**(2), 1364–1377 (2008)
8. Dehghan, M., Manafian, J., Saadatmandi, A.: Solving nonlinear fractional partial differential equations using the homotopy analysis method. Numer. Meth. Part. D. E. **26**(2), 448–479 (2010)
9. Gazizov, R.K., Kasatkin, A.A., Lukashchuk, S.Y.: Symmetry properties of fractional diffusion equations. Phys. Scr. **T136** (2009)
10. Manoj, G., Singh K.: Symmetry classification and exact solutions of a variable coefficient space-time fractional potential burgers equation. Int. J. DiffER. Equ. **2016**, Article ID 4270724, 8 (2016). https://doi.org/10.1155/2016/4270724
11. He, J.H., Wu, G.C., Austin, F.: The variational iteration method which should be followed. Nonlinear Sci. Lett. A **1**, 1–30 (2010)
12. He, J.H.: A new fractal derivation. Therm. Sci. **15**, 145–147 (2011)

13. Ibragimov, N.H.: Handbook of Lie Group Analysis of Differential Equations, vol. 1, 2, 3. CRC Press, Boca Raton, Ann Arbor, London, Tokyo (1994, 1995, 1996)
14. Jafari, H., Kadkhoda, N., Azadi, M., Yaghobi, M.: Group classification of the time-fractional Kaup-Kupershmidt equation. Scientia Iranica **24**(1), 302–307 (2017)
15. Jafari, H., Kadkhoda, N., Baleanu, D.: Fractional Lie group method of the time-fractional Boussinesq equation. Nonlinear Dyn. **81**, 1569–1574 (2015). https://doi.org/10.1007/s11071-015-2091-4
16. Jafari, H.: Numerical solution of time-fractional klein-gordon equation by using the decomposition methods. ASME. J. Comput. Nonlinear Dyn. **11**(4), 041015 (2016)
17. Jefferson, G.F., Carminati, J.: FracSym: automated symbolic computation of Lie symmetries of fractional differential equations. Comput. Phys. Commun. **185**, 430–441 (2014)
18. Kasatkin, A.A.: Symmetry properties for systems of two ordinary fractional differential equations. Ufa Math. J. **4**, 65–75 (2012)
19. Kilbas, A.A., Srivastava, H.M., Trujillo, J.J.: Theory and Applications of Fractional Differential Equations. North-Holland Mathematics Studies, vol. 204. Elsevier Science B.V, Amsterdam, The Netherlands (2006)
20. Liu, H.Z.: Complete group classifications and symmetry reductions of the fractional fifth-order KdV types of equations. Stud. Appl. Math. **131**, 317–330 (2013)
21. Lu, B.: The first integral method for some time fractional differential equations. J. Math. Anal. Appl. **395**, 684–693 (2012)
22. Lukashchuk, SYu., Makunin, A.V.: Group classification of nonlinear time-fractional diffusion equation with a source term. Appl. Math. Comput. **257**, 335–343 (2015)
23. Nadjafikhah, M., Ahangari, F.: Symmetry reduction of two-dimensional damped Kuramoto-Sivashinsky equation. Commun. Theor. Phys. **56**, 211–217 (2011)
24. Olver, P.J.: Applications of Lie Groups to Differential Equations, Graduate Texts in Mathematics, 107,2nd edition. Springer, Berlin (1993)
25. Ovsyannikov, L.V.: Group Analysis of Differential Equations. Academic, New York (1982)
26. Podlubny, I.: Fractional differential equations: an introduction to fractional derivatives, fractional differential equations, to methods of their solution and some of their applications. In: Mathematics in Science and Engineering, vol. 198. Academic, San Diego, Calif, USA (1999)
27. Rosenau, P.: On a class of nonlinear dispersive-dissipative interaction. Physica D **123**, 1–4 (1998)
28. Samko, G., Kilbas, A.A., Marichev, O.I.: Fractional Integrals and Derivatives: Theory and Applications. Gordon and Breach, Yverdon (1993)
29. Wang, G.W., Liu, X.Q., Zhang, Y.Y.: Lie symmetry analysis to the time fractional generalized fith-order KdV equation. Commun. Nonlinear Sci. Numer. Simul. (2013). https://doi.org/10.1016/j.cnsns.2012.11.032
30. Wang, G.W., Hashemi, M.S.: Lie symmetry analysis and soliton solutions of time-fractionalK(m, n) equation. Pramana J. Phys. **88**(7), 1–7 (2017)
31. Gün Polat G, Özer T. new conservation laws, lagrangian forms, and exact solutions of modified emden equation. ASME. J. Comput. Nonlinear Dyn. **12**(4), 041001–041001-15 (2017). https://doi.org/10.1115/1.4035408.
32. Wu, G.C.: A fractional lie group method for anomalous diffusion equations. Commun. Frac. Calc. **1**, 27–31 (2010)
33. Wu, G.C., Baleanu, D.: Discrete fractional logistic map and its chaos. Nonlinear Dyn. **75**(1–2), 283–287 (2014)
34. Zilburg, A., Rosenau, P.: Early and late stages of K(m, n) compactons interaction. Phys. Lett. A **383**, 991–996 (2019)

Generalized Rayleigh Wave Propagation in Heterogeneous Substratum Over Homogeneous Half-Space Under Gravity

Pulak Patra, Asit Kumar Gupta, and Santimoy Kundu

Abstract The present work considering the propagation of Rayleigh waves in an incompressible heterogeneous medium with a general variation of rigidity; resting over another incompressible homogeneous half-space under the effect of gravity. Instead of using Whittaker's function, the expansion formula proposed by Newland's has been used to solve the equation of motion for better result in the incompressible half space. Newland's method gives better results for shallow depth. The velocity equations have been calculated and the results shown in the figures. The study may reveal the fact that, except for linear and quadratic variation of rigidity, the relation between the phase velocity of Rayleigh wave and the gravity being directly proportional to each other. The phase velocity of Rayleigh wave in absence of gravity is smaller than the presence of gravity in all cases except linear and quadratic variation in rigidity.

Keywords Rayleigh wave · Incompressible substratum · Biot's gravity parameter · Newland's method · Propagation of waves

1 Introduction

Study of surface waves are always interesting for researcher due to its practical applicability in various fields of Seismology, Geology etc. Basically in coastal belt area, the effect of surface waves shown its characteristics in different manner. In fact most of the time this unknown effect being one of the major reason for the

P. Patra (✉)
Department of Mathematics, Brainware Group of Institutions-SDET, Kolkata 700125, West Bengal, India

A. Kumar Gupta
Department of Physics, Asansol Engineering College, Asansol 713305, West Bengal, India

S. Kundu
Department of Applied Mathematics, IIT(ISM), Dhanbad 826004, Jharkhand, India

J. Singh et al. (eds.), *Methods of Mathematical Modelling and Computation for Complex Systems*, Studies in Systems, Decision and Control 373,
https://doi.org/10.1007/978-3-030-77169-0_9

destructiveness. Hence, the authors are mainly focused on the coastal security issues in this work.

If we had turned the page of history, we found that around 18th century a famous physicist Lord Rayleigh first observed such surface waves (viz.- Rayleigh Wave) during earthquake. Interestingly he was not just made an observations, but also analysed it in details and published some interesting facts in the form of an scientific article [23] which ends with a fundamental differences in the characteristics of half space and an infinite medium. Lamb [19] was the first who introduced the propagation of Rayleigh waves in a half-space and concluded that the deformations produced by it are much greater than other associated compressional and shear waves. Thus Rayleigh waves play a vital role in the transmission of seismic disturbances at the surface of the earth during an earthquakes. In this consequent it does not need to mention the importance of it to the researcher, seismologists or geologists. After Lamb [19] a good number of literatures being available at different books written by Achenbach [1] or Ewing [14] or Cowin et al. [10] or Akbarov [2]; related to the propagation of Rayleigh waves in homogeneous and non-homogeneous medium in presence of one or more superficial layer(s).Most of the study regarding Rayleigh waves are considered the asymptotic expansion of Whittaker functions [27], but the expansions are only valid for the large depths and unable to concludes the result properly in case of shallow depths. So on this note, authors are considering Newland's method [21] to study the present paper in incompressible heterogeneous medium with shear modulus $\mu = \mu_0(1 + bz)^m$. The effect of gravity on Rayleigh waves in a half space was considered by Dey et al. [13] and ended with some interesting results regarding the propagation of the wave; whereas Dey et al. [12] studies the propagation of the wave in incompressible medium and concluded with the variation of velocities during propagation at the said medium. In recent past Burlak et al. [9] study the nature of the waves in a two layered ocean-earth model and concluded its characteristics on such situation. Novikiva et al. [22] in his work investigates the nature of the wave in sedimentary layer and Abd-Alla et al. [4] was studied the influence of the initial stress and rotation of these wave fronts; interestingly both the mentioned work [4, 22] was done under the effect of gravity for its practicability. Kakar [18] in very recent made an observation on the effect of gravity and non-homogeneity on the Rayleigh waves in case of higher order elastic-viscoelastic half spaces and concluded with some interesting results. Akbarov et al. [3] studied the propagation of waves in layered media. Applicability of Rayleigh waves in different coastal areas are studied by Chen et al. [7], Hu et al. [15, 16] and Suzuki et al. [24] etc.

As it was already declared that; in this present paper the authors are considering the propagation of Rayleigh waves in an incompressible heterogeneous medium with general variation of rigidity lying over another incompressible homogeneous half space under the effect of gravity with general variation of rigidity and the authors are reveals the fact that the phase velocity of Rayleigh wave at the said medium under the assumed condition is proportional with the values of gravity except for the linear and quadratic variation of rigidity. In fact the authors are mainly focused on the applicability of the surface wave in coastal safety issues and the authors are

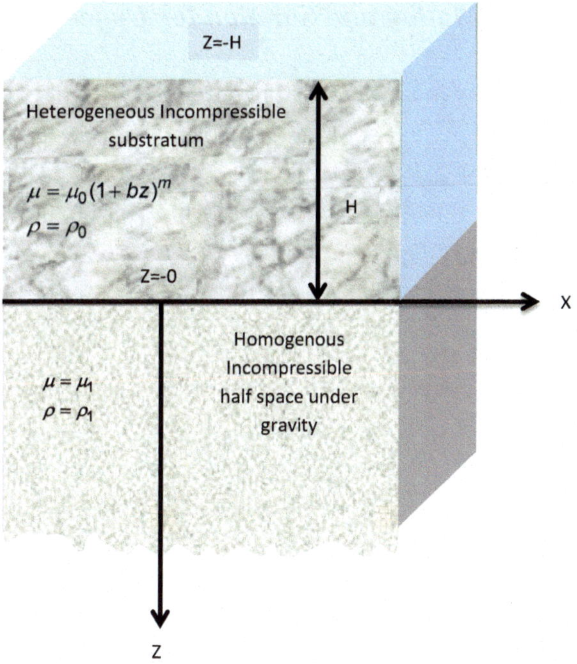

Fig. 1 Geometry of the problem

believed, it will be definitely help to develop a new safety model which might help to increase the safety of coastal areas.

2 Formulation of the Problem

Authors are considered an incompressible layer of thickness H with shear modulus of rigidity $\mu = \mu_0(1 + bz)^m$ and density $\rho = \rho_0$ lying over another incompressible half-space under the initial hydro static stress due to the effect of gravity field. Considering the rigidity and density of the half-space as μ_1 and ρ_1 respectively. Taken z as the vertical distance from the origin situated at the interface of the incompressible medium and the half-space, and the downward direction of z being taken as positive (Fig. 1).

Assuming that the wave propagates along x-axis with phase velocity c and wavelength of $2\pi/K$ and u, w be the displacement components along the x and z directions respectively at a point (x, y, z) at any time .For smooth calculation considering that apart from the factor $e^{ik(x-ct)}$, u and w are functions of z only.

2.1 Equations of Motion and Solution for Upper Layer

The equation of vibratory motion in two dimensions for the upper layer are

$$\frac{\partial}{\partial x}\left(\lambda\Delta + 2\mu\frac{\partial u}{\partial x}\right) + \frac{\partial}{\partial z}\left\{\mu\left(\frac{\partial u}{\partial z} + \frac{\partial w}{\partial x}\right)\right\} = \rho_0\frac{\partial^2 u}{\partial t^2} \tag{1}$$

$$\frac{\partial}{\partial x}\left\{\mu\left(\frac{\partial u}{\partial z} + \frac{\partial w}{\partial x}\right)\right\} + \frac{\partial}{\partial z}\left(\lambda\Delta + 2\mu\frac{\partial w}{\partial z}\right) = \rho_0\frac{\partial^2 w}{\partial t^2} \tag{2}$$

where, λ and μ are the Lame's Constant.

Since the medium is incompressible then the value of dilation (Δ) takes the value as

$$\begin{aligned}\Delta = \left(\frac{\partial u}{\partial x} + \frac{\partial w}{\partial z}\right) = 0 \\ \Rightarrow \nabla^2\phi = 0\end{aligned} \tag{3}$$

with the following relations

$$\begin{aligned}u &= \left(\frac{\partial \phi}{\partial x} + \frac{\partial \chi}{\partial z}\right) \\ w &= \left(\frac{\partial \phi}{\partial z} - \frac{\partial \chi}{\partial x}\right)\end{aligned} \tag{4}$$

where, ϕ is the scalar potential and χ is the vector potential.

Since the shear modulus for the upper layer being $\mu = \mu_0(1 + bz)^m$, with the incompressible condition $\Delta = 0$, and together with

$$\lim_{\lambda\to\infty,\Delta\to 0} \lambda\Delta \to -P_1$$

(where P_1 is the hydrostatic stress). The Eqs. (1) and (2) becomes

$$\begin{aligned}\frac{\partial}{\partial x}\Big[-P_1 + 2bm\mu_0\left(1 + bz\right)^{m-1} w + \rho_0 c^2 K^2\phi\Big] \\ + \frac{\partial}{\partial z}\Big[\mu(\nabla^2\chi) + \rho_0 c^2 K^2\chi\Big] = 0\end{aligned} \tag{5}$$

$$\begin{aligned}\frac{\partial}{\partial z}\Big[-P_1 + 2bm\mu_0\left(1 + bz\right)^{m-1} w + \rho_0 c^2 K^2\phi\Big] \\ - \frac{\partial}{\partial x}\Big[\mu(\nabla^2\chi) + \rho_0 c^2 K^2\chi\Big] = 0\end{aligned} \tag{6}$$

combining Eqs. (5) and (6) which are satisfies-

$$P_1 = 2\mu_0 bm(1+bz)^{m-1} w + \rho_0 c^2 K^2 \phi \tag{7}$$

and

$$\mu\left(\nabla^2 \chi\right) = -\rho_0 c^2 K^2 \chi \tag{8}$$

together with

$$\nabla^2 \phi = 0 \tag{9}$$

where $\nabla^2 \equiv \frac{\partial^2}{\partial x^2} + \frac{\partial^2}{\partial z^2}$.

Without ambiguity, considering the initial solutions as-
$\phi = \phi(z)\cos K(x - ct)$ and $\chi = \chi(z)\sin K(x - ct)$
then the Eqs. (8) and (9) becomes,respectively-

$$\frac{d^2\chi}{dz^2} + k^2\left[\frac{\rho_0 c^2}{\mu_0(1+bz)^m} - 1\right]\chi(z) = 0 \tag{10}$$

and

$$\frac{d^2\phi}{dz^2} - k^2\phi = 0 \tag{11}$$

Instead of solving the above equations by Whittaker functions; authors are attempting to solve it directly by help of Newland's method [21] as follows-
assuming $(1 + bz) = Z$, the Eq. (10) becomes-

$$\frac{d^2\chi}{dZ^2} - \left(\frac{K}{b}\right)^2\left(1 - \frac{\rho_0 c^2}{\mu_0 Z^m}\right)\chi = 0 \tag{12}$$

now, from Eq. (12) it can be expanding into the series solution in power of $\left(\frac{K}{b}\right)^2$ as-

$$\chi(Z) = \chi_0(Z) + \left(\frac{K}{b}\right)^2 \chi_1(Z) + \left(\frac{K}{b}\right)^4 \chi_2(Z) + \cdots + \left(\frac{K}{b}\right)^{2n} \chi_n(Z) + \cdots \tag{13}$$

using (Eq. 13) on (Eq. 12) it gives-

$$\left[\chi_0''(Z) + \left(\frac{K}{b}\right)^2 \chi_1''(Z) + \left(\frac{K}{b}\right)^4 \chi_2''(Z) + \cdots + \left(\frac{K}{b}\right)^{2n+2} \chi_{n+1}''(Z) + \cdots\right] = \left(\frac{K}{b}\right)^2 \left(1 - \frac{\rho_0 c^2}{\mu_0 Z^m}\right) \left[\chi_0(Z) + \left(\frac{K}{b}\right)^2 \chi_1(Z) + \left(\frac{K}{b}\right)^4 \chi_2(Z) + \cdots + \left(\frac{K}{b}\right)^{2n} \chi_n(Z) + \cdots\right] \tag{14}$$

From both sides of Eq. (14), which gives-

$$\begin{aligned} \chi_0''(Z) &= 0 \\ \chi_1''(Z) &= \left(1 - \frac{\rho_0 c^2}{\mu_0 Z^m}\right)\chi_0(Z) \\ \chi_2''(Z) &= \left(1 - \frac{\rho_0 c^2}{\mu_0 Z^m}\right)\chi_1(Z) \\ &\cdots\cdots\cdots\cdots \\ &\cdots\cdots\cdots\cdots \\ \chi_{n+1}''(Z) &= \left(1 - \frac{\rho_0 c^2}{\mu_0 Z^m}\right)\chi_n(Z) \end{aligned} \tag{15}$$

Thus it is quite clear that if Eq. (15) and the resulting series (Eq. 13) both converges and involves two arbitrary constants, it is a valid solution and interestingly most general solution of Eq. (10).

Thus for converging Eq. (13),it should be

$$\chi_n'(1) = \chi_n(1) = 0 \tag{16}$$

again from Eq. (15) it gives

$$\chi_0'(Z) = \frac{A_2}{b} \tag{17}$$

and

$$\chi_0(Z) = A_1 + \frac{A_2}{b}(Z - 1) \tag{18}$$

From the linearity of equation (Eq. 12) by using Eq. (13) with the help of equation (18) one may write

$$\chi(Z) = A_1\chi^{(1)} + A_2\chi^{(2)} \tag{19}$$

where,
$\chi^{(1)} = \left\{\chi_0^{(1)} + \left(\frac{K}{b}\right)^2 \chi_1^{(1)} + \left(\frac{K}{b}\right)^4 \chi_2^{(1)} + \cdots\right\}$
and
$\chi^{(2)} = \left\{\chi_0^{(2)} + \left(\frac{K}{b}\right)^2 \chi_1^{(2)} + \left(\frac{K}{b}\right)^4 \chi_2^{(2)} + \cdots\right\}$

Again considering,

$$\chi_0^{(1)} = 1$$
$$\chi_0^{(2)} = z = \frac{Z-1}{b} \tag{20}$$

Now from Eq. (15) the value of χ_1 may be obtained as
$\chi_1(Z) = \int_1^Z \left\{ \int_1^\zeta \left(1 - \frac{\rho_0 c^2}{\mu_0 t^m}\right) \chi_0(t)dt \right\} d\zeta$
to evaluate the above integral authors are considering different values of m as follows-

Case-I: When $m = 1$
$\chi_1(Z) = A_1\left[\frac{1}{2}(Z-1)^2 - \frac{\rho_0 c^2}{\mu_0}\left\{Z\log Z - (Z-1)\right\}\right] + \frac{A_2}{b}\left[\frac{1}{6}(Z-1)^3 + \frac{\rho_0 c^2}{\mu_0}\left\{Z\log Z + \frac{1}{2} - \frac{Z^2}{2}\right\}\right]$
and the solution of $\chi^{(1)}$ and $\chi^{(2)}$ becomes

$$\chi^{(1)} = 1 + \left(\frac{K}{b}\right)^2 \left[\frac{1}{2}(Z-1)^2 - \frac{\rho_0 c^2}{\mu_0}(Z\log Z - Z + 1)\right] \tag{21}$$

$$\chi^{(2)} = (Z-1) + \left(\frac{K}{b}\right)^2 \left[\frac{1}{6}(Z-1)^3 + \frac{\rho_0 c^2}{\mu_0}\left(Z\log Z - \frac{Z^2}{2} + \frac{1}{2}\right)\right] \tag{22}$$

Case-II: When $m = 2$
$\chi_1(Z) = A_1\left[\frac{1}{2}(Z-1)^2 + \frac{\rho_0 c^2}{\mu_0}\left\{\log Z - (Z-1)\right\}\right] + \frac{A_2}{b}\left[\frac{1}{6}(Z-1)^3 - \frac{\rho_0 c^2}{\mu_0}\right.$
$\left.\left\{(Z+1)\log Z + 2(1-Z)\right\}\right]$
and the solution of $\chi^{(1)}$ and $\chi^{(2)}$ becomes

$$\chi^{(1)} = 1 + \left(\frac{K}{b}\right)^2 \left[\frac{1}{2}(Z-1)^2 + \frac{\rho_0 c^2}{\mu_0}(\log Z - Z + 1)\right] \tag{23}$$

$$\chi^{(2)} = (Z-1) + \left(\frac{K}{b}\right)^2 \left[\frac{1}{6}(Z-1)^3 - \frac{\rho_0 c^2}{\mu_0} \left\{(Z+1)\log Z + 2(1-Z)\right\}\right] \tag{24}$$

Case-III: When $m = 3$
$\chi_1(Z) = A_1\left[\frac{1}{2}(Z-1)^2 + \frac{\rho_0 c^2}{\mu_0}\left\{1 - \frac{Z}{2} - \frac{1}{2Z}\right\}\right] + \frac{A_2}{b}\left[\frac{1}{6}(Z-1)^3 + \frac{\rho_0 c^2}{\mu_0}\left\{\log Z + \right.\right.$

$\frac{1}{2Z} - \frac{Z}{2}\Big\}\Big]$

and the solution of $\chi^{(1)}$ and $\chi^{(2)}$ becomes

$$\chi^{(1)} = 1 + \left(\frac{K}{b}\right)^2 \left[\frac{1}{2}(Z-1)^2 + \frac{\rho_0 c^2}{\mu_0}\left(1 - \frac{Z}{2} - \frac{1}{2Z}\right)\right] \tag{25}$$

$$\chi^{(2)} = (Z-1) + \left(\frac{K}{b}\right)^2 \left[\frac{1}{6}(Z-1)^3 + \frac{\rho_0 c^2}{\mu_0}\left\{\log Z + \frac{1}{2Z} - \frac{Z}{2}\right\}\right] \tag{26}$$

Case-IV: For any value of m other than $m \neq 1, m \neq 2$ and $m \neq 3$

$\chi_1(Z) = A_1\left[\frac{1}{2}(Z-1)^2 + \frac{\rho_0 c^2}{\mu_0}\frac{1}{(1-m)}\left\{(Z-1) + \frac{1}{(2-m)}\left(1 - \frac{1}{Z^{m-2}}\right)\right\}\right] + \frac{A_2}{b}\left[\frac{1}{6}(Z-1)^3 + \frac{\rho_0 c^2}{\mu_0}\frac{1}{(2-m)}\left\{(Z-1) + \frac{1}{(3-m)}\left(1 - \frac{1}{Z^{m-3}}\right)\right\} + \frac{\rho_0 c^2}{\mu_0}\frac{1}{(1-m)}\left\{(1-Z) + \frac{1}{(2-m)}\left(\frac{1}{Z^{m-2}} - 1\right)\right\}\right]$

and the solution of $\chi^{(1)}$ and $\chi^{(2)}$ becomes

$$\chi^{(1)} = 1 + \left(\frac{K}{b}\right)^2 \left[\frac{1}{2}(Z-1)^2 + \frac{\rho_0 c^2}{\mu_0}\frac{1}{(1-m)}\left\{(Z-1) + \frac{1}{(2-m)}\left(1 - \frac{1}{Z^{m-2}}\right)\right\}\right] \tag{27}$$

$$\chi^{(2)} = (Z-1) + \left(\frac{K}{b}\right)^2 \left[\frac{1}{6}(Z-1)^3 + \frac{\rho_0 c^2}{\mu_0}\frac{1}{(2-m)}\left\{(Z-1) + \frac{1}{(3-m)}\left(1 - \frac{1}{Z^{m-3}}\right)\right\} + \frac{\rho_0 c^2}{\mu_0}\frac{1}{(1-m)}\left\{(1-Z) + \frac{1}{(2-m)}\left(\frac{1}{Z^{m-2}} - 1\right)\right\}\right] \tag{28}$$

Then the solution for the upper layer can be considered as

$$\begin{aligned}\phi &= \left(P\cosh Kz + Q\sinh Kz\right)\cos K(x - ct)\\ \chi &= \left(A_1\chi^{(1)}(Z) + A_2\chi^{(2)}(Z)\right)\sin K(x - ct)\end{aligned} \tag{29}$$

where $\chi^{(1)}$ and $\chi^{(2)}$ have the values given by the Eq. (21) to Eq. (28) for different values of m.

So, The displacement components at the inhomogeneous upper layer become as-

$$u = \left[-K\left(P\cosh Kz + Q\sinh Kz\right) + \left(A_1\frac{\partial}{\partial z}\chi^{(1)}(Z) + A_2\frac{\partial}{\partial z}\chi^{(2)}(Z)\right)\right]\sin K(x-ct) \tag{30}$$

$$w = \left[K\left(P\sinh Kz + Q\cosh Kz\right) - K\left(A_1\chi^{(1)}(Z) + A_2\chi^{(2)}(Z)\right)\right]\cos K(x-ct) \tag{31}$$

and the stress components of the upper layer become as-

$$\sigma_{xz} = 2K\left[-\mu\frac{\partial\phi}{\partial z} + K\left(\mu - \frac{\varepsilon\mu_0}{2}\right)\chi\right]\sin K(x-ct) \tag{32}$$

$$\sigma_{zz} = 2\left[K^2\phi\left(\mu - \frac{\varepsilon\mu_0}{2}\right) - \mu_0 bm\left(1+bz\right)^{m-1}\frac{\partial\phi}{\partial z} + \mu_0 bm\left(1+bz\right)^{m-1}K\chi - \mu K\frac{\partial\chi}{\partial z}\right]\cos K(x-ct) \tag{33}$$

where $\varepsilon = \frac{\rho_0 c^2}{\mu_0}$

2.2 Equations of Motion and Solution for Half Space

Since the authors are considered that the half space being incompressible, so the equations of motion being

$$\begin{aligned}\frac{\partial s_{11}}{\partial x} + \frac{\partial s_{12}}{\partial z} + \rho_1\omega g - \frac{\partial s}{\partial z}e_{xz} &= \rho_1\frac{\partial^2 u}{\partial t^2}\\ \frac{\partial s_{12}}{\partial x} + \frac{\partial s_{22}}{\partial z} - \frac{\partial s}{\partial z}e_{zz} &= \rho_1\frac{\partial^2 w}{\partial t^2}\end{aligned} \tag{34}$$

where, $e_{xz} = \frac{1}{2}\left(\frac{\partial w}{\partial x} + \frac{\partial u}{\partial z}\right)$, $\omega = \frac{1}{2}\left(\frac{\partial w}{\partial x} - \frac{\partial u}{\partial z}\right)$ and $e_{zz} = \frac{\partial w}{\partial z}$

Then, from Eq. (34) it can be re-write as

$$\begin{aligned}\frac{\partial}{\partial x}(s_{11}+\rho_1 g w)+\frac{\partial s_{12}}{\partial z}&=\rho_1\frac{\partial^2 u}{\partial t^2}\\ \frac{\partial s_{12}}{\partial x}+\frac{\partial}{\partial z}\left(s_{22}+\rho_1 g w\right)&=\rho_1\frac{\partial^2 w}{\partial t^2}\end{aligned} \tag{35}$$

Now introducing the fictitious stresses as-

$$s_{11}+\rho_1 g w=s_{11}^{'},\qquad s_{22}+\rho_1 g w=s_{22}^{'},\qquad s_{12}=s_{12}^{'} \tag{36}$$

then the Eq. (35) may be written as-

$$\begin{aligned}\frac{\partial s_{11}^{'}}{\partial x}+\frac{\partial s_{12}^{'}}{\partial z}&=\rho_1\frac{\partial^2 u}{\partial t^2}\\ \frac{\partial s_{12}^{'}}{\partial x}+\frac{\partial s_{22}^{'}}{\partial z}&=\rho_1\frac{\partial^2 w}{\partial t^2}\end{aligned} \tag{37}$$

Now according to M. A. Biot [5] , the stress-strain relation for incompressible material were as follows-

$$s_{11}-s=2\mu_1 e_{xx},\quad s_{22}-s=2\mu_1 e_{zz},\qquad s_{12}=2\mu_1 e_{xz} \tag{38}$$

using the relation (Eq. 36) in (Eq. 38) , it can be changes into

$$\begin{aligned}s_{11}^{'}-s^{'}&=2\mu_1 e_{xx}\\ s_{22}^{'}-s^{'}&=2\mu_1 e_{zz}\\ s_{12}^{'}&=2\mu_1 e_{xz}\end{aligned} \tag{39}$$

and

$$s^{'}=\frac{1}{2}(s_{11}^{'}+s_{22}^{'})=s+\rho_1 g w \tag{40}$$

For changing the strain components in terms of displacement components, substituting the relation (Eq. 39) into the Eq. (37) and utilizing the incompressible condition (Eq. 3), which gives-

$$\begin{aligned}\frac{\partial s^{'}}{\partial x}+\mu_1\nabla^2 u&=\rho_1\frac{\partial^2 u}{\partial t^2}\\ \frac{\partial s^{'}}{\partial z}+\mu_1\nabla^2 w&=\rho_1\frac{\partial^2 w}{\partial t^2}\end{aligned} \tag{41}$$

The incompressibility condition is satisfied by-

$$u=-\frac{\partial\phi}{\partial z},\qquad w=\frac{\partial\phi}{\partial x} \tag{42}$$

Using relation (Eq. 42) in Eq. (41) and eliminating s', gives-

$$\mu_1\left[\frac{\partial^4\phi}{\partial x^4}+2\frac{\partial^4\phi}{\partial x^2\partial z^2}+\frac{\partial^4\phi}{\partial z^4}\right]=\rho_1\frac{\partial^2}{\partial t^2}\left[\frac{\partial^2\phi}{\partial x^2}+\frac{\partial^2\phi}{\partial z^2}\right] \tag{43}$$

Considering the solution as $\phi=\eta(z)\sin K(x-ct)$, from the Eq. (43) which gives-

$$\eta'''' - K^2\eta''\left(1+\beta^2\right)+K^4\eta\beta^2=0 \tag{44}$$

where, $\beta^2=\left(1-\frac{\rho_1c^2}{\mu_1}\right)$

Therefore the solution of equation (Eq. 44), where it is important that as $z\to\infty,\eta\to0$ can be written as
$\eta=\left(A_3e^{-Kz}+A_4e^{-\beta Kz}\right)$ and hence,

$$\phi=\left(A_3e^{-Kz}+A_4e^{-\beta Kz}\right)\sin K(x-ct) \tag{45}$$

Therefore,

$$u=K\left(A_3e^{-Kz}+\beta A_4e^{-\beta Kz}\right)\sin K(x-ct) \tag{46}$$

$$w=K\left(A_3e^{-Kz}+A_4e^{-\beta Kz}\right)\cos K(x-ct) \tag{47}$$

According to Biot [5] the stress components for the half space become-

$$\sigma_{xz}=-\mu_1K^2\left[2A_3e^{-Kz}+A_4\left(1+\beta^2\right)e^{-K\beta z}\right]\sin K(x-ct) \tag{48}$$

$$\sigma_{zz}=\mu_1K^2\left[A_3\left\{\left(1-\beta^2\right)-\frac{\rho_1g}{\mu_1K}-2\right\}e^{-Kz}-A_4\left(2\beta+\frac{\rho_1g}{\mu_1K}\right)e^{-K\beta z}\right]\cos K(x-ct) \tag{49}$$

3 Boundary Conditions

On the basis of geometry for the problem, authors are defined the boundary conditions into two sets as-
(i) the displacement components and stress components are continuous at $z=0$; i.e. u, w, σ_{xz} and σ_{zz} are continuous at $z=0$ or
$(u)_1=(u)_2$ at $z=0$

$(w)_1 = (w)_2$ at $z = 0$
$(\sigma_{xz})_1 = (\sigma_{xz})_2$ at $z = 0$
$(\sigma_{zz})_1 = (\sigma_{zz})_2$ at $z = 0$
and
(ii) the stress components should be vanish at $z = -H$; i.e. σ_{xz} and σ_{zz} are vanish at $z = -H$ or
$(\sigma_{xz})_1 = 0$ at $z = -H$
$(\sigma_{zz})_1 = 0$ at $z = -H$
Check (Fig. 1) for better understanding.

Using set (i) it gives-

$$-KP + A_2 b - A_3 K - A_4 K\beta = 0 \tag{50}$$

$$Q - A_1 - A_3 - A_4 = 0 \tag{51}$$

$$-Q + A_1\left(1 - \frac{\varepsilon}{2}\right) + A_3\frac{\mu_1}{\mu_0} + A_4\frac{\mu_1}{2\mu_0}\left(1 + \beta^2\right) = 0 \tag{52}$$

$$\begin{aligned} &P\left[K\left(1 - \frac{\varepsilon}{2}\right)\right] - Q(bm) + A_1(bm) - A_2 b \\ &-A_3\left[\frac{\mu_1 K}{2\mu_0}\left\{(1 - \beta^2) - \frac{\rho_1 g}{\mu_1 K} - 2\right\}\right] \\ &+A_4\left[\frac{\mu_1 K}{2\mu_0}\left(2\beta + \frac{\rho_1 g}{\mu_1 K}\right)\right] = 0 \end{aligned} \tag{53}$$

and using set (ii) it gives-

$$\begin{aligned} &P\sinh(-KH) + Q\cosh(-KH) - \left(1 - \frac{\varepsilon\mu_0}{2\mu_2}\right) \\ &\left[A_1\chi^{(1)}(Z) + A_2\chi^{(2)}(Z)\right]_{z=-H} = 0 \end{aligned} \tag{54}$$

$$\begin{aligned} &P\left[K\cosh(Kz)\left(\mu - \frac{\varepsilon\mu_0}{2}\right) - \mu_0 bm(1+bz)^{m-1}\sinh(Kz)\right]_{z=-H} \\ &+Q\left[K\sinh(Kz)\left(\mu - \frac{\varepsilon\mu_0}{2}\right) - \mu_0 bm(1+bz)^{m-1}\cosh(Kz)\right]_{z=-H} \\ &+A_1\left[\mu_0 bm(1+bz)^{m-1}\chi^{(1)}(Z) - \mu\frac{\partial}{\partial z}\chi^{(1)}(Z)\right]_{z=-H} \\ &+A_2\left[\mu_0 bm(1+bz)^{m-1}\chi^{(2)}(Z) - \mu\frac{\partial}{\partial z}\chi^{(2)}(Z)\right]_{z=-H} = 0 \end{aligned} \tag{55}$$

where, $\mu_2 = \mu_0(1 - bH)^m$.

From the Eqs. (50), (51), (52) and (53) the relationship between P, Q, A_1, A_2 may be obtained as

$$\begin{aligned}\frac{1}{2}\varepsilon P &= a_1A_3 + b_1A_4\\ \frac{1}{2}\varepsilon Q &= a_2A_3 + b_2A_4\\ \frac{1}{2}\varepsilon A_1 &= a_3A_3 + b_3A_4\\ \frac{1}{2}\varepsilon A_2 &= a_4A_3 + b_4A_4\end{aligned} \tag{56}$$

where,
$a_1 = \frac{1}{K}\left\{\frac{mba_3}{\varepsilon/2} - \frac{mba_2}{\varepsilon/2} + \frac{\mu_1K}{2\mu_0} + \frac{\beta^2\mu_1K}{2\mu_0} + G\frac{\mu_1K}{2\mu_0} - K\right\}$,as $G = \frac{\rho_1 g}{\mu_1 K}$ is the Biot's gravity parameter
$a_2 = \frac{\varepsilon}{2} + a_3$
$a_3 = \frac{\mu_1}{\mu_0} - 1$
$a_4 = \frac{1}{b}\left(a_1K + \frac{K\varepsilon}{2}\right)$
$b_1 = \frac{1}{K}\left\{\frac{bmb_3}{\varepsilon/2} - \frac{bmb_2}{\varepsilon/2} + K\beta + \beta\frac{\mu_1K}{\mu_0} + G\frac{\mu_1K}{2\mu_0} - K\beta\varepsilon\right\}$
$b_2 = \frac{\varepsilon}{2} + b_3$
$b_3 = \frac{\mu_1}{2\mu_0} + \frac{\beta^2\mu_1}{2\mu_0} - 1$
$b_4 = \frac{1}{b}\left(b_1K + \frac{K\beta\varepsilon}{2}\right)$
Again, from the Eqs. (54) and (55) assuming-
$c_1 = \sinh(-KH)$
$c_2 = \cosh(-KH)$
$c_3 = \left[-\left(1 - \frac{\varepsilon\mu_0}{2\mu_2}\right)\chi^{(1)}(Z)\right]_{z=-H}$
$c_4 = \left[-\left(1 - \frac{\varepsilon\mu_0}{2\mu_2}\right)\chi^{(2)}(Z)\right]_{z=-H}$
$d_1 = \left[K\left(\mu - \frac{\varepsilon\mu_0}{2}\right)\cosh(Kz) - \mu_0bm(1 + bz)^{m-1}\sinh(Kz)\right]_{z=-H}$
$d_2 = \left[K\left(\mu - \frac{\varepsilon\mu_0}{2}\right)\sinh(Kz) - \mu_0bm(1 + bz)^{m-1}\cosh(Kz)\right]_{z=-H}$
$d_3 = \left[\mu_0bm(1 + bz)^{m-1}\chi^{(1)}(Z) - \mu\frac{\partial}{\partial z}\chi^{(1)}(Z)\right]_{z=-H}$
$d_4 = \left[\mu_0bm(1 + bz)^{m-1}\chi^{(2)}(Z) - \mu\frac{\partial}{\partial z}\chi^{(2)}(Z)\right]_{z=-H}$

Therefore the Eqs. (54) and (55) gives,

$$\begin{aligned}Pc_1 + Qc_2 + A_1c_3 + A_2c_4 &= 0\\ Pd_1 + Qd_2 + A_1d_3 + A_2d_4 &= 0\end{aligned} \tag{57}$$

The consistency of Eqs. (56) and (57) for a non-trivial solution of A_3 and A_4 implies that (Details calculation is available in **Appendix**)

$$\sum_{i=1}^{4} a_i c_i \sum_{j=1}^{4} b_j d_j = \sum_{k=1}^{4} a_k d_k \sum_{l=1}^{4} b_l c_l \tag{58}$$

This is the velocity equation for the Rayleigh wave propagation in the assumed medium.

3.1 Particular Cases

To check the validity of the result authors are considering two particular cases and used them in the final dispersion equation as follows.

3.1.1 Case-I

Authors are first cross verified the result in absence of substratum i.e. $H \to 0$, then equation (Eq. 58) reduces to

$$\Rightarrow \left(2 - \frac{c^2}{c_s^2}\right)^2 - \frac{\rho_1 g}{\mu_1 K}\left(\frac{c^2}{c_s^2}\right) = 4\sqrt{1 - \frac{c^2}{c_s^2}} \qquad \text{where, } c_s = \sqrt{\frac{\mu_1}{\rho_1}},$$

The above equation is the well known Rayleigh wave equation for an incompressible medium under gravity [5].

3.1.2 Case-II

Secondly authors are considered the result in absence of substratum and gravity field i.e. $H \to 0$ and $G \to 0$, which shows the equation (Eq. 58) reduces to

$$\Rightarrow \left(2 - \frac{c^2}{c_s^2}\right)^2 = 4\sqrt{1 - \frac{c^2}{c_s^2}} \qquad \text{where, } c_s = \sqrt{\frac{\mu_1}{\rho_1}},$$

In fact this equation is same as of the Rayleigh wave equation for an incompressible medium.

4 Results and Discussions

The velocity of Rayleigh waves being calculated from the final dispersion Eq. (58) for different values of m starting from 0 to 4 and observed the nature of the phase velocities for different values of $2\pi/KH$ in presence and absence of gravity. The parametric relations are shown in Table 1 for better understanding. The results are being presented at figure section for different values of the parameter m. As the authors are basically tried to understood the nature of the wave in respect of changing values for the unknown parameter, they are considered the values from 0 and increasing it by 0.5 unit to identify the nature in prominent way. In all the cases authors are considering 3 curves for different values of the Biot's gravity parameters $G = 0, 0.4, 0.6$ respectively but for the fixed values of rigidity and elasticity ratios of two layers i.e. $\frac{\rho_1}{\rho_0} = 1.2$ and $\frac{\mu_1}{\mu_0} = 1.6$ along with the value of $bH = 0.5$.

For the value of $m = 0$, the figure (Fig. 2) being clearly indicating that for the assumed layers the velocity of the Rayleigh waves in presence of gravity is more than that of the in absence of gravity.

Similarly for $m = 0.5$, the figure (Fig. 3) also shown the same result as for the value of $m = 0$ i.e. in the assumed conditions the velocity for Rayleigh waves in presence of gravity is more than in case of it's absence.

In fact when the authors are gone through the other values of m i.e. for 1.5, 2.5, 3, 3.5 and 4, the figures (Figs. 5, 7, 8, 9 and 10) are shown the same results, i.e. the velocity of Rayleigh wave has a variation in presence and in absence of the gravitational force.

On the other hand for the values of $m = 1$ and $m = 2$,i.e. in case of linearity and quadratic variation of m the observations shows opposite result (Figs. 4 and 6) as of the previous cases that is the velocity of Rayleigh wave in presence of gravity is smaller than the absence of gravity at the assumed layer and half space. These clearly indicates a new observations in respect of the work.

Another important observations authors are observed that for the assumed layer and in all the values of the parameter m; as $2\pi/K$ increases velocity of the wave increases.

5 Conclusion

From the above study authors are reveals the following fact regarding the propagation of Rayleigh waves in the heterogeneous substratum over homogeneous half-space under gravity as-

- The phase velocity of Rayleigh waves in an incompressible heterogeneous medium over an incompressible gravitating half space is increasing as gravity being increased for almost all variation of m except linearity and quadratic variation.
- The phase velocity of Rayleigh wave in absence of gravity is smaller than the presence of gravity in all cases except linear and quadratic variation in rigidity.
- For linearity and quadratic variation of m, the study reveals different fact that is, as the gravity being increased the phase velocity of the wave become decreased.
- Relation with phase velocity and the depth of the substratum being directly proportional to each other i.e. as depth increases velocity of the Rayleigh wave also increases.

In future any researcher can work with the large values of the parameter m , to obtained various observations. Use of Newland's method for shallow depth in this article reveals various unknown fact which researcher may be used for other surface waves for its better understanding.

Table for Calculation

Table 1 Parametric values for the figures

Figure no.	Curve no.	G	m	$\frac{\rho_1}{\rho_0}$	$\frac{\mu_1}{\mu_0}$	bH
Figure 2	1	0				
	2	0.4	0	1.2	1.6	0.5
	3	0.6				
Figure 3	1	0				
	2	0.4	0.5	1.2	1.6	0.5
	3	0.6				
Figure 4	1	0				
	2	0.4	1	1.2	1.6	0.5
	3	0.6				
Figure 5	1	0				
	2	0.4	1.5	1.2	1.6	0.5
	3	0.6				
Figure 6	1	0				
	2	0.4	2	1.2	1.6	0.5
	3	0.6				
Figure 7	1	0				
	2	0.4	2.5	1.2	1.6	0.5
	3	0.6				
Figure 8	1	0				
	2	0.4	3	1.2	1.6	0.5
	3	0.6				
Figure 9	1	0				
	2	0.4	3.5	1.2	1.6	0.5
	3	0.6				
Figure 10	1	0				
	2	0.4	4	1.2	1.6	0.5
	3	0.6				

Graphical Representation of the Results

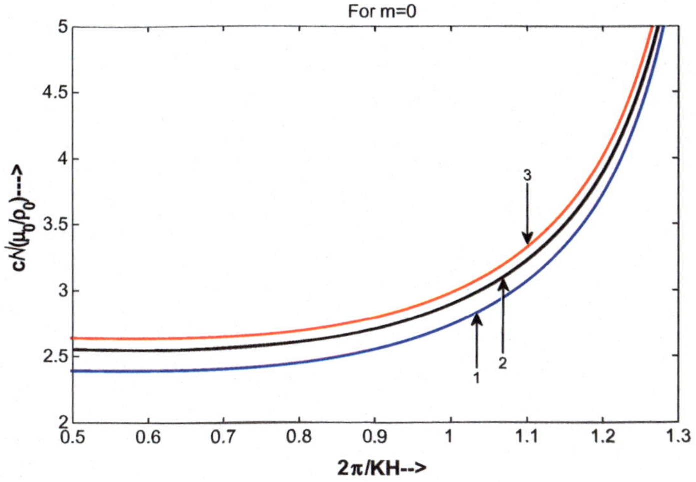

Fig. 2 Rayleigh wave dispersion for m = 0

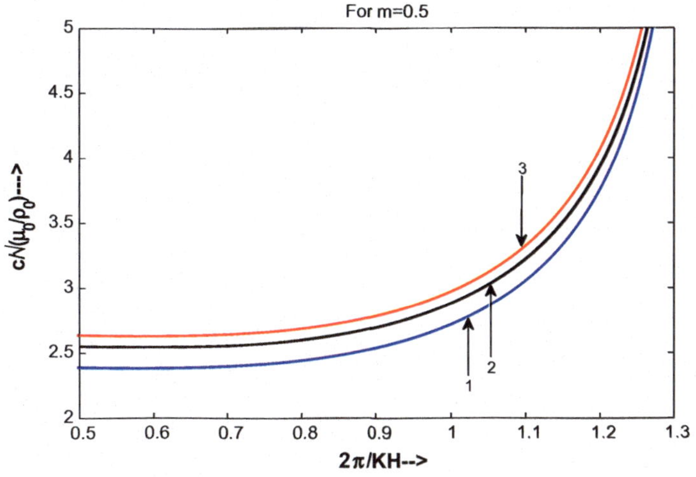

Fig. 3 Rayleigh wave dispersion for m = 0.5

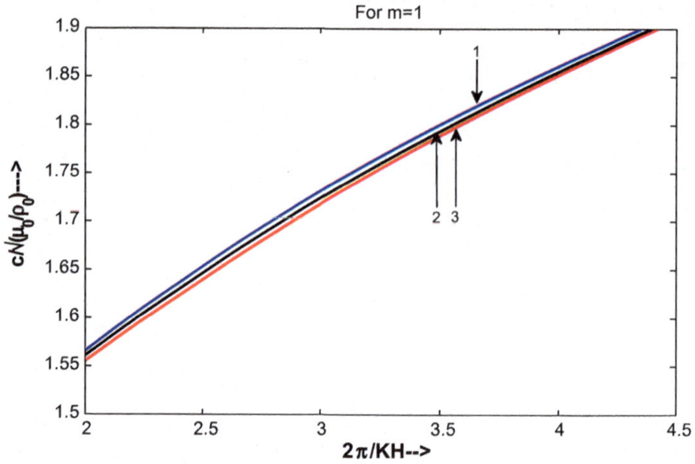

Fig. 4 Rayleigh wave dispersion for m = 1

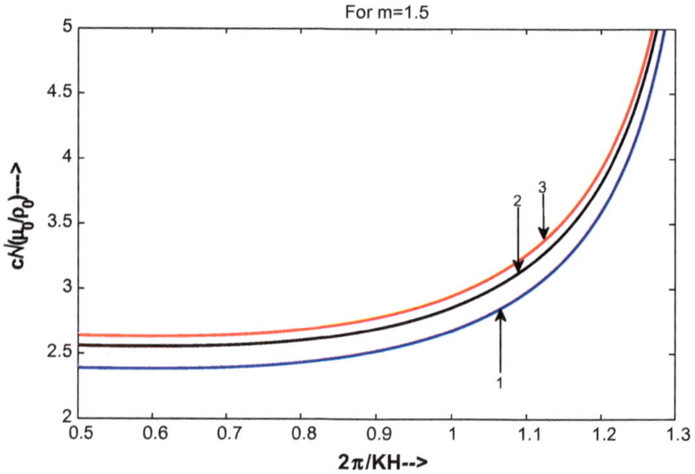

Fig. 5 Rayleigh wave dispersion for m = 1.5

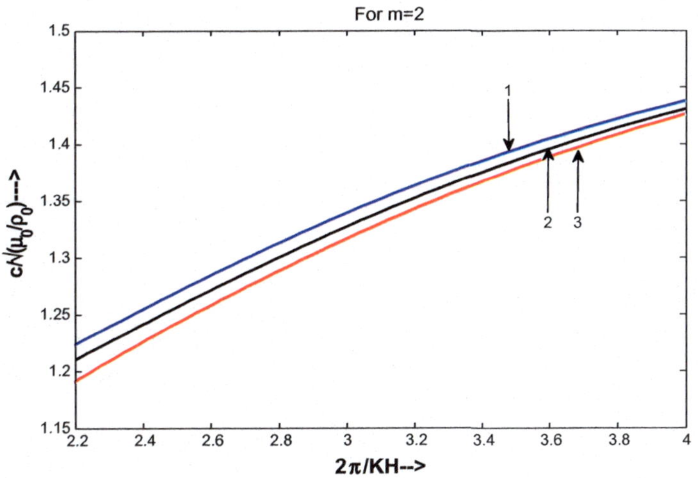

Fig. 6 Rayleigh wave dispersion for m = 2

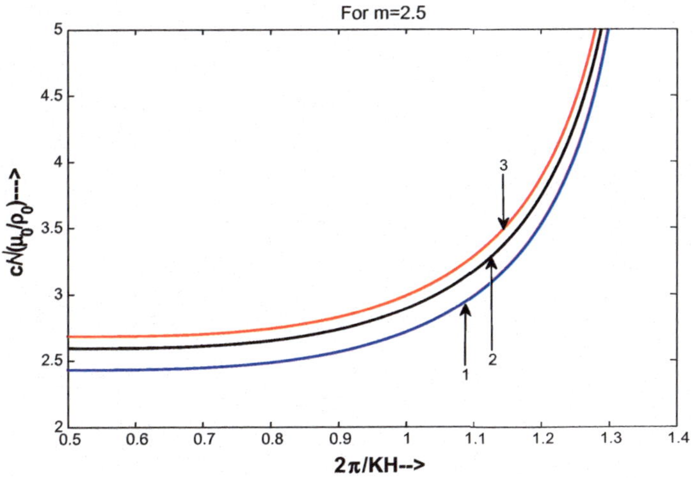

Fig. 7 Rayleigh wave dispersion for m = 2.5

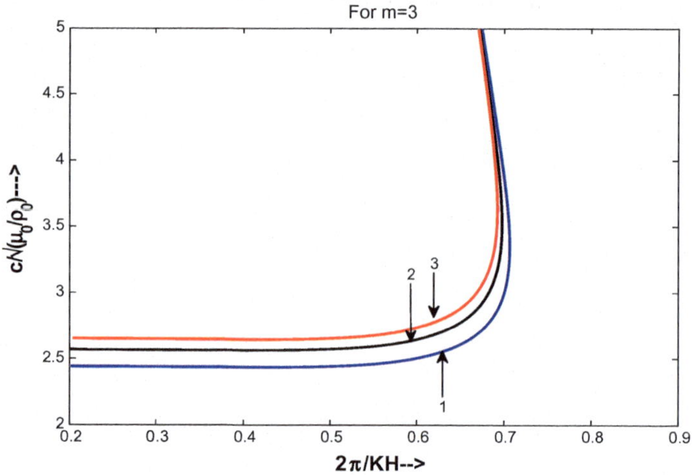

Fig. 8 Rayleigh wave dispersion for m = 3

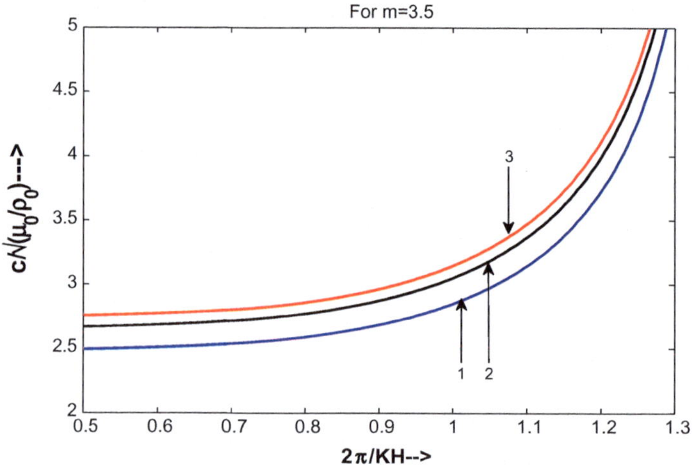

Fig. 9 Rayleigh wave dispersion for m = 3.5

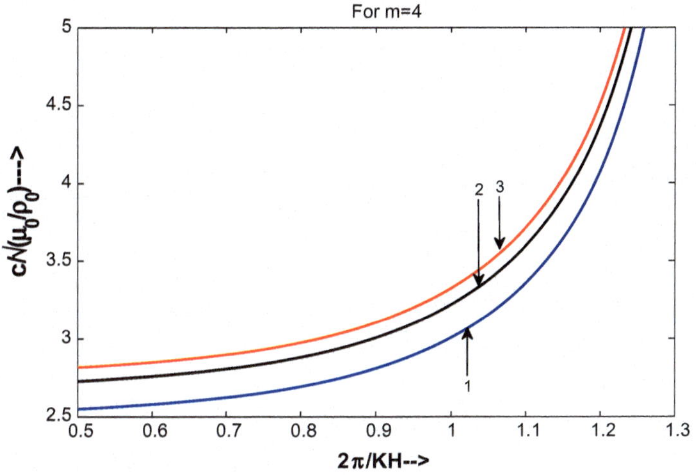

Fig. 10 Rayleigh wave dispersion for m = 4

Acknowledgements The authors are sincerely thankful to Brainware Group of Institutions-SDET, Kolkata; Asansol Engineering College, Asansol and Indian Institute of Technology (Indian School of Mines), Dhanbad for their all kind of support and encouragement. The authors are also sincerely thankful to the reviewers for their valuable comments to improve the quality of the article.

Appendix

From the Eqs. (56) and (57) the coefficient determinant will be-

$$\text{Let } \Delta = \begin{vmatrix} \frac{1}{2}\varepsilon & 0 & 0 & 0 & -a_1 & -b_1 \\ 0 & \frac{1}{2}\varepsilon & 0 & 0 & -a_2 & -b_2 \\ 0 & 0 & \frac{1}{2}\varepsilon & 0 & -a_3 & -b_3 \\ 0 & 0 & 0 & \frac{1}{2}\varepsilon & -a_4 & -b_4 \\ c_1 & c_2 & c_3 & c_4 & 0 & 0 \\ d_1 & d_2 & d_3 & d_4 & 0 & 0 \end{vmatrix} = 0$$

Expanding the determinant in terms of 1st row, we get,

$$\frac{1}{2}\varepsilon \begin{vmatrix} \frac{1}{2}\varepsilon & 0 & 0 & -a_2 & -b_2 \\ 0 & \frac{1}{2}\varepsilon & 0 & -a_3 & -b_3 \\ 0 & 0 & \frac{1}{2}\varepsilon & -a_4 & -b_4 \\ c_2 & c_3 & c_4 & 0 & 0 \\ d_2 & d_3 & d_4 & 0 & 0 \end{vmatrix} - a_1 \begin{vmatrix} 0 & \frac{1}{2}\varepsilon & 0 & 0 & -b_2 \\ 0 & 0 & \frac{1}{2}\varepsilon & 0 & -b_3 \\ 0 & 0 & 0 & \frac{1}{2}\varepsilon & -b_4 \\ c_1 & c_2 & c_3 & c_4 & 0 \\ d_1 & d_2 & d_3 & d_4 & 0 \end{vmatrix} + b_1 \begin{vmatrix} 0 & \frac{1}{2}\varepsilon & 0 & 0 & -a_2 \\ 0 & 0 & \frac{1}{2}\varepsilon & 0 & -a_3 \\ 0 & 0 & 0 & \frac{1}{2}\varepsilon & -a_4 \\ c_1 & c_2 & c_3 & c_4 & 0 \\ d_1 & d_2 & d_3 & d_4 & 0 \end{vmatrix} = 0$$

Now consider,

$$\Delta_1 = \begin{vmatrix} \frac{1}{2}\varepsilon & 0 & 0 & -a_2 & -b_2 \\ 0 & \frac{1}{2}\varepsilon & 0 & -a_3 & -b_3 \\ 0 & 0 & \frac{1}{2}\varepsilon & -a_4 & -b_4 \\ c_2 & c_3 & c_4 & 0 & 0 \\ d_2 & d_3 & d_4 & 0 & 0 \end{vmatrix},$$

$$\Delta_2 = \begin{vmatrix} 0 & \frac{1}{2}\varepsilon & 0 & 0 & -b_2 \\ 0 & 0 & \frac{1}{2}\varepsilon & 0 & -b_3 \\ 0 & 0 & 0 & \frac{1}{2}\varepsilon & -b_4 \\ c_1 & c_2 & c_3 & c_4 & 0 \\ d_1 & d_2 & d_3 & d_4 & 0 \end{vmatrix},$$

$$\Delta_3 = \begin{vmatrix} 0 & \frac{1}{2}\varepsilon & 0 & 0 & -a_2 \\ 0 & 0 & \frac{1}{2}\varepsilon & 0 & -a_3 \\ 0 & 0 & 0 & \frac{1}{2}\varepsilon & -a_4 \\ c_1 & c_2 & c_3 & c_4 & 0 \\ d_1 & d_2 & d_3 & d_4 & 0 \end{vmatrix}$$

Now from Δ_1 expanding in terms of 1st row, we get-

$$\Delta_1 = \frac{1}{2}\varepsilon \begin{vmatrix} \frac{1}{2}\varepsilon & 0 & -a_3 & -b_3 \\ 0 & \frac{1}{2}\varepsilon & -a_4 & -b_4 \\ c_3 & c_4 & 0 & 0 \\ d_3 & d_4 & 0 & 0 \end{vmatrix} + a_2 \begin{vmatrix} 0 & \frac{1}{2}\varepsilon & 0 & -b_3 \\ 0 & 0 & \frac{1}{2}\varepsilon & -b_4 \\ c_2 & c_3 & c_4 & 0 \\ d_2 & d_3 & d_4 & 0 \end{vmatrix} - b_2 \begin{vmatrix} 0 & \frac{1}{2}\varepsilon & 0 & -a_3 \\ 0 & 0 & \frac{1}{2}\varepsilon & -a_4 \\ c_2 & c_3 & c_4 & 0 \\ d_2 & d_3 & d_4 & 0 \end{vmatrix}$$

Again expanding in term of 4th column for each components of the determinant we get-

$$\Delta_1 = \frac{1}{2}\varepsilon b_3 \begin{vmatrix} 0 & \frac{1}{2}\varepsilon & -a_4 \\ c_3 & c_4 & 0 \\ d_3 & d_4 & 0 \end{vmatrix} - \frac{1}{2}\varepsilon b_4 \begin{vmatrix} \frac{1}{2}\varepsilon & 0 & -a_3 \\ c_3 & c_4 & 0 \\ d_3 & d_4 & 0 \end{vmatrix} + a_2 b_3 \begin{vmatrix} 0 & 0 & \frac{1}{2}\varepsilon \\ c_2 & c_3 & c_4 \\ d_2 & d_3 & d_4 \end{vmatrix}$$
$$- a_2 b_4 \begin{vmatrix} 0 & \frac{1}{2}\varepsilon & 0 \\ c_2 & c_3 & c_4 \\ d_2 & d_3 & d_4 \end{vmatrix} - b_2 a_3 \begin{vmatrix} 0 & 0 & \frac{1}{2}\varepsilon \\ c_2 & c_3 & c_4 \\ d_2 & d_3 & d_4 \end{vmatrix} + b_2 a_4 \begin{vmatrix} 0 & \frac{1}{2}\varepsilon & 0 \\ c_2 & c_3 & c_4 \\ d_2 & d_3 & d_4 \end{vmatrix}$$

Simplify the above expression we will get-

$$\Delta_1 = \tfrac{1}{2}\varepsilon[(c_3d_4 - c_4d_3)(a_3b_4 - a_4b_3) + (c_2d_3 - d_2c_3)(a_2b_3 - a_3b_2) + (c_2d_4 - d_2c_4)(a_2b_4 - a_4b_2)]$$

Similarly if we simplified the Δ_2 and Δ_3 we get the following expressions:

$\Delta_2 = \frac{\varepsilon^2}{4}[-b_4(c_1d_4 - d_1c_4) - b_3(c_1d_3 - d_1c_3) - b_2(c_1d_2 - d_1c_2)]$ and
$\Delta_3 = \frac{\varepsilon^2}{4}[-a_3(c_1d_3 - d_1c_3) - a_4(c_1d_4 - d_1c_4) - a_2(c_1d_2 - d_1c_2)]$

Now using these values of Δ_1, $\Delta_2 and \Delta_3$ in the value of Δ and simplifying for few steps we get the (Eq. 58) i.e. $\sum_{i=1}^{4} a_ic_i \sum_{j=1}^{4} b_jd_j = \sum_{k=1}^{4} a_kd_k \sum_{l=1}^{4} b_lc_l$.

References

1. Achenbach, J.D.: Wave propagation in Elastic Solids. North Holland Publi. Comp, Amsterdam, New York (1973)
2. Akbarov, S.D.: Dynamics of Pre-Strained bi-Material Elastic Systems: Linearized Three-Dimensional Approach. Springer, New York (2015)
3. Akbarov, S.D., Ozisik, M.: The influence of the third order elastic constants to the generalized Rayleigh wave dispersion in a pre-stressed stratified half-plane. Int. J. Eng. Sci. **41**, 2047–2061 (2003)
4. Abd-Alla, A.M., Abo-Dahab, S.M., Al-Thamali, T.A.: Propagation of Rayleigh waves in a rotating orthotropic material elastic half-space under initial stress and gravity. J. Mech. Sci. Technol. **26**(9), 2815–2823 (2012)
5. Biot, M.A.: Mechanics of Incremental Deformation. Wiley, New York (1965)
6. Bromwich, T.J.I.A.: On the influence of gravity on elastic waves and in particular on the vibration of elastic globe. Proc. Lond. Math. Soc. **30**, 98 (1968)
7. Chen, H., Ni, Y., Li, Y., Liu, F., Ou, S., Su, M., Peng, Y., Hu, Z., Uijttewaal, W., Suzuki, T.: Deriving vegetation drag coefficients in combined wave-current flows by calibration and direct measurement methods. Adv. Water Resour. **122**, 217–227 (2018)
8. Chandrasekharaiah, D.S.: Effects of surface stresses and voids on Rayleigh waves in elastic solid. Int. J. Eng. Sci. **25**, 205–211 (1987)
9. Burlak, G.N., Koshevaya, S.V., Hayakawa, M., Mondrag6n, J.S., Grimalsky, V.V.: Propagation of coupled Rayleigh-gravity waves on the ocean floor. Geofisica Internacional **38**(4), 261-268 (1999)
10. Cowin, S.C., Nunziato, J.W.: Linear elastic materials with voids. J. Elasticity **13**, 125 (1983)
11. Dey, S., Gupta, S.: Longitudinal elastic materials with voids. Pro. Indian Nat. Sci. Acad. **A53**, 554–563 (1987)
12. Dey, S. , Gupta, A.K., Gupta, S.: Propagation of Rayleigh waves in heterogeneous incompressible substratum over a homogeneous incompressible half space. Int. J. Num. Ana. Meth. Geomec. **20**, 365–375(1996)
13. Dey, S., Mukherjee, S.P.: Rayleigh waves in an initially stressed layer over a half-space under gravity. Acta Geophysica Polonica, XXXII **1**, 81–90 (1984)
14. Ewing, W.M., Jardetzky, W.S., Press, F.: Elastic Waves in Layered Media. McGraw-Hill, New York (1957)
15. Hu, Z., Wang, Z.B., Zitman, T.J., Stive, M.J.F., Bouma, T.J.: Predicting long-term and short-term tidal flat morphodynamics using a dynamic equilibrium theory. J. Geophys. Res. Earth Surface **120**, 1803–1823(2015)
16. Hu, Z., Yao, P., Wal1, D.V.D., Bouma, T.J.: Patterns and drivers of daily bed level dynamics on two tidal flats with contrasting waves exposure. Sci. Rep. (2017). https://doi.org/10.1038/s41598-017-075150y
17. John, J.P.: Wave propagation in two layered medium. J. Appl. Mech. **31**; J. Appl. Mech.
18. Kakar, R.: Analysis of the effect of gravity and non-homogeneity on Rayleigh waves on higher order elastic-viscoelastic hals-space. J. Mech. Behav. Mater. **23**, 3–4 (2014)
19. Lamb, H.: On the propagation of tremor over the surface of the elastic solid. Phil. Trans. R. Soc. **A**, 204 (1904)
20. Miklowitz, J.: The Theory of Elastic Waves and Waves Guides. North Holland Publi. Comp, Amsterdam, New York (1978)
21. Newlands, M.: Rayleigh waves in a two layer heterogeneous medium. Mon. Notices R. Astron. Soc. Geophys. (Suppl.) **6**, 109–125 (1950)
22. Novikova, T., Wen, K.L., Haung, B.S.: Amplification of gravity and Rayleigh waves in a layered water-soil model. Earth, Planets Space **52**(9), 579 (2000)
23. Rayleigh, J.W.S.: On Waves propagated along the plane surface of the elastic solids. Proc. Lond. Math. Soc. **17** (1887)

24. Suzuki, T., Hu, Z., Kumada, K., Phan, L.K., Zijlema, M.: Non-hydrostatic modeling of drag, inertia and porous effects in wave propagation over dense vegetation fields. Coastal Eng. **149**, 49–64 (2019)
25. Smith, L.M., Dahlen, F.A.: The azimuthal dependence of Love and Rayleigh wave propagation in a slightly anisotropic medium. J. Geophys. Res. **78**(17), 3321–3333 (1973)
26. Scalia, A.: Shock waves in viscoelastic materials with voids. Wave Motion **19**, 125–133 (1994)
27. Whittaker, E.T., Watson, G.N.: Modern Analysis, 2nd edn., vol. 33. Cambridge (1915)

On Defining Trigonometric Box Spline-Like Surface on Type-I Triangulation

Hrushikesh Jena and Mahendra Kumar Jena

Abstract Usually, a polynomial box spline surface is defined with the help of distributions, convolutions, Fourier transforms and recurrence relations. A better alternative to define the box spline surface is by subdivision method. In this paper, a trigonometric box spline surface on type-I triangulation is defined by introducing a new non-stationary subdivision scheme. This new subdivision scheme takes help of the previously defined non-stationary subdivision scheme in (Jena et al., A non-stationary subdivision scheme for generalizing trigonometric spline surfaces to arbitrary meshes, Computer Aided Geom. Design, 20, (2003), 61–77). The limit surface obtained by the repeated application of this new scheme to an initial regular triangular mesh, is a trigonometric box spline like surface. This can be considered as an initial attempt to define the trigonometric box spline surfaces by subdivision process. Besides, having a nice algorithm, the limit surface is compactly supported, satisfies the convex hull property and is uniformly continuous. We illustrate the performance of this scheme with some examples.

Keywords Type-I triangulation · Trigonometric box spline · Non-stationary subdivision scheme · Arbitrary topology · Subdivision matrix · Convergence · Continuity · Convex hull · Trigonometric spline

2010 MSC 65D07 · 65D17

1 Introduction

Subdivision is a very popular geometric modeling tool. Subdivision algorithms are widely used in Computer Graphics and Computer Aided Geometric Design (CAGD), due to their efficiency, flexibility and simplicity [1]. The *subdivision schemes* [2–15]

H. Jena (✉) · M. Kumar Jena
Department of Mathematics, Veer Surendra Sai University of Technology, Burla, Sambalpur 768018, Odisha, India
e-mail: mkjena_math@vssut.ac.in

J. Singh et al. (eds.), *Methods of Mathematical Modelling and Computation for Complex Systems*, Studies in Systems, Decision and Control 373,
https://doi.org/10.1007/978-3-030-77169-0_10

use some iterative algorithms to produce smooth curves and surfaces from an initial set of control points which are called as refinement rules. Initially, a control mesh is given. From this a new refined control mesh is generated by applying a subdivision scheme once. So, by repeated application of that subdivision scheme, a sequence of control meshes is obtained which converges to a smooth limit surface. In general, subdivision schemes can be categorized into two types: *interpolating schemes* and *approximating schemes*. In interpolating schemes, the initial control vertices are preserved in each levels of iteration whereas in approximating schemes, one obtains different vertices in different iteartion levels. Approximating schemes yield smoother curves with higher order continuity, comparable to interpolating schemes [16]. The scheme proposed in this article is an approximating scheme. A subdivision scheme is called a stationary subdivision scheme, if the set of subdivision masks remains the same in all levels of iteration. If it varies from one level to another level, then the subdivision scheme is called a non-stationary subdivision scheme. This paper is based on a binary non-stationary scheme. Some popular subdivision schemes for curve modeling are [2, 6, 11, 13–15] and the references therein. Similarly, some useful tensor product subdivision schemes for surface modeling can be found in [7, 10, 12]. In [7], a non-stationary variant of Doo-Sabin scheme [3] has been proposed. The subdivision scheme in [10] has been constructed by the authors by taking the tensor product of a three point curve scheme. In [12], the authors have proposed a nine-tic B-spline subdivision scheme which reduces the execution time needed for the computation of the subdivision process. There are so many research works can also be found on the applications and analysis of subdivision schemes [17–20]. Recently in [18], the authors have analyzed some geometric properties like curvature of limit curve, monotonicity preservation, convexity preservation etc. by considering a particular interpolating subdivision scheme.

One of the important categories of splines are trigonometric splines and they play an important role in shape designing and geometric modeling [21, 22]. They were first introduced by Schoenberg [23]. One way to express them is by linear combinations of trigonometric B-splines. Another way to study them is by non-stationary subdivision algorithm [6]. Normally, standard CAD geometries are represented in terms of tensor product B-splines and their rational version NURBS [24]. Tensor product representation is very efficient for computations but fails to deal with complicated shapes modeling and also in doing local mesh refinements. On the other hand, splines over the triangulations overcome such disadvantages. Box splines are attractive alternatives which combine several disadvantages from tensor product B-splines and splines over triangulations [25]. One of the important advantages of box splines is that they can handle more complex domains better than the corresponding tensor product counterparts. Box splines can be constructed by the help of convolutions, Fourier transforms, recurrence relations and subdivision schemes [26–28]. However, so far, the construction trigonometric box spline functions has not gained much attention. So we have taken an attempt to construct trigonometric box spline with the help of a non-stationary subdivision scheme.

Comparing to other meshes, triangular meshes are much more flexible to be adapted to regular topology or arbitrary topology. The main motivation of our work comes from the paper [7], where tensor product bi-quadratic spline surfaces are constructed over arbitrary topology. It has proposed a non-stationary subdivision scheme which is a tensor product scheme. But, in this article also we propose a non-stationary subdivision scheme by considering a special case of the subdivision algorithm for arbitrary topology defined in [7]. If the subdivision algorithm and connectivity rule of [7] are applied to an initial type-I triangular control mesh, a number of extraordinary vertices are produced. But in this work, after applying our subdivision algorithm to that initial mesh we again apply an intermediate averaging rule followed by a different connectivity rule. As a result, after each iteration of this new scheme, we always get a regular type-I triangular mesh. Thus, we get a limit surface which is free of extraordinary points. Since, box splines defined over triangulations are free of extraordinary points, this supports our attempt. The limit surfaces which are obtained by the application of our scheme are trigonometric box spline-like surfaces. Although, we can't directly call these as box splines, but these are bell shaped and satisfy several characteristics like compact support property, partition of unity and convex hull property. Also, we have shown that our new subdivision scheme converges and the limit surfaces produced by this scheme are continuous. The application of this scheme is illustrated with some example.

The paper is structured as follows. In Sect. 2, we briefly review the non-stationary scheme defined in [7]. Next, in Sect. 3, we describe the construction of our new non-stationary subdivision scheme which generates surfaces from an initial mesh of type-I triangulation. In Sect. 4, we give the convergence analysis of our proposed subdivision scheme. Here, it is emphasized that the convergence of our subdivision scheme is of the same order as that of [7]. In Sect. 5, we provide some examples which show the performance of our scheme over regular triangular meshes. Finally, we give a conclusion in Sect. 6.

2 Trigonometric Spline Surface on Arbitrary Topology

In this section, we briefly describe the subdivision scheme presented in [7] which produces trigonometric spline surfaces over arbitrary topological meshes. Let $\mathcal{X}^{(0)}$ be an initial control mesh with finite number of vertices and faces. Let $\mathcal{F}_0$ be a face of $\mathcal{X}^{(0)}$ made up of n number of ordered vertices $V_1, V_2, \ldots, V_n$. Then by implementing the subdivision scheme of [7], a set of new vertices associated to the face $\mathcal{F}^0$ are obtained. Indeed, these new vertices are the linear combinations of the old vertices V_is as given in the following subdivision rule.

$$a_i = \frac{1}{4}\,(\mathbb{P} + \mathbb{E}_{i-1} + \mathbb{E}_i + \mathbb{F}),\ i = 1, 2, \ldots n, \tag{1}$$

where

$$\mathbb{E}_i = \frac{1}{2}\ \sec h\ \sec^2(h/2)\ (V_i + V_{i+1}),\ \mathbb{P} = \sec^2(h/2)\ V_i,$$

and

$$\mathbb{F} = \frac{1}{n} \sec^2 h \sec^2(h/2) \sum_{i=1}^{n} V_i.$$

Here all the indices of the vertices are taken modulo n and h $(0 < h \leq \frac{\pi}{3})$ is the mesh size which also controls the shape of the surface. By simplifying (1), it is obtained that

$$a_i = \alpha(n, 1)\, V_i + \beta(n, 1)\, (V_{i-1} + V_{i+1}) + \gamma(n, 1)\, (V_{i+2} + \cdots + V_{i+n-2}), \quad (2)$$

where for $k \geq 1$,

$$\gamma(n, k) = \frac{1}{4n} \sec^2(h/2^k) \sec^2(h/2^{k-1}), \quad \beta(n, k) = \gamma(n, k) + \frac{1}{8} \sec^2(h/2^k) \sec(h/2^{k-1}),$$

$$\text{and} \quad \alpha(n, k) = \gamma(n, k) + \frac{1}{4} \sec^2(h/2^k)\, (1 + \sec(h/2^{k-1})). \quad (3)$$

After getting these new vertices, the Doo-Sabin's connectivity rule [3] is applied and by this a new control mesh $\mathcal{X}^{(1)}$ is obtained which is called as first iteration control mesh. By denoting $\mathcal{S}_1$ as the subdivision operator, this whole subdivision process is represented by $\mathcal{X}^{(1)} = \mathcal{S}_1 \mathcal{X}^{(0)}$. Similarly, $\mathcal{X}^{(2)}$ is obtained from $\mathcal{X}^{(1)}$ by applying the subdivision operator $\mathcal{S}_2$, where $\alpha(n, 1)$, $\beta(n, 1)$ and $\gamma(n, 1)$ are replaced by $\alpha(n, 2)$, $\beta(n, 2)$ and $\gamma(n, 2)$, respectively. Proceeding further, in general, $\mathcal{X}^{(k+1)}$ is obtained from $\mathcal{X}^{(k)}$ by applying the subdivision operator $\mathcal{S}_{k+1}$ as follows:

$$\mathcal{X}^{(k+1)} = \mathcal{S}_{k+1} \mathcal{X}^{(k)} = \mathcal{S}_{k+1} \mathcal{S}_k \ldots \mathcal{S}_1 \mathcal{X}^{(0)} = \mathcal{S}^{(k+1)} \mathcal{X}^{(0)}, \quad (4)$$

where $\mathcal{S}_{k+1}$ is associated with $\alpha(n, k+1)$, $\beta(n, k+1)$ and $\gamma(n, k+1)$, respectively. The connectivity rule used in [7] produces a new face of n vertices from an old face of n vertices; a new face of n vertices from an old interior vertex of valence n and new quadrilateral from an old interior edge. After each iteration, we get more and more number of quadrilaterals but the number of non-quadrilaterals remains fixed. So it also results in the decrement of the face size. However, the interior non-quadrilateral faces become surrounded by layers of quadrilaterals. As a result, the layers of quadrilaterals converge to a tensor product bi-quadratic spline surface and the non-quadrilateral faces shrink to extraordinary points. We illustrate (4) by providing a figure (see Fig. 1), where we have taken the initial value of $h = 0.25$. The initial control mesh $\mathcal{X}^{(0)}$ and first iteration control mesh $\mathcal{S}^{(1)} \mathcal{X}^{(0)}$ are given in (**a**) and (**b**) in Fig. 1, respectively. Similarly, second and third iteration control meshes, $\mathcal{S}^{(2)} \mathcal{X}^{(0)}$ and $\mathcal{S}^{(3)} \mathcal{X}^{(0)}$ are given in (**c**) and (**d**) in Fig. 1, respectively.

In the next section, we derive a new non-stationary subdivision scheme for type-I triangulation meshes which are regular i.e. the meshes which are free from extraordinary points.

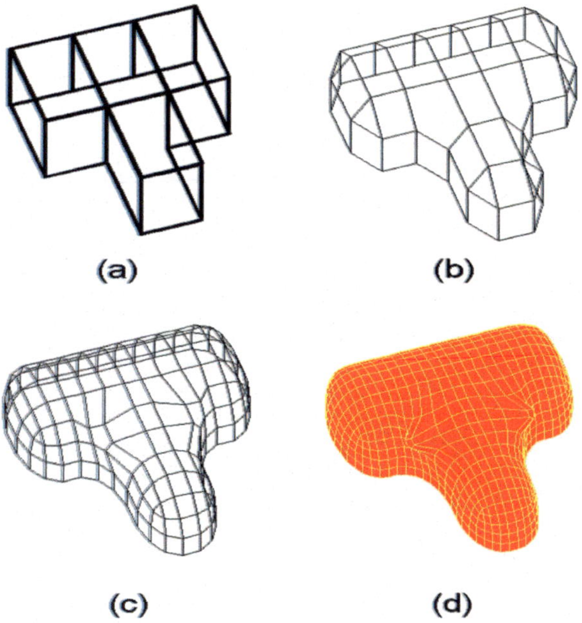

Fig. 1 **a** Initial control mesh $\mathcal{X}^{(0)}$. **b** First iteration control mesh $\mathcal{S}^{(1)}\mathcal{X}^{(0)}$. **c** Second iteration control mesh $\mathcal{S}^{(2)}\mathcal{X}^{(0)}$. **d** Third iteration control mesh $\mathcal{S}^{(3)}\mathcal{X}^{(0)}$

3 Construction of the New Non-stationary Subdivision Scheme

In this section, we construct a new non-stationary subdivision scheme which is different from the scheme of [7].

3.1 *Construction of the Scheme*

To construct our new scheme, we choose a type-I triangular mesh. Let $\mathcal{X}^{(0)}$ be such an initial control mesh which contains six triangular faces. These are joined as shown in Fig. 2. If we apply the subdivision scheme of [7] once to $\mathcal{X}^{(0)}$, we get a new control mesh which contains six triangular faces and one hexagonal face. Thus, repeated application of this subdivision scheme to $\mathcal{X}^{(0)}$ leads to the formation of a limit surface with 7 extraordinary points.

But in the case of our new subdivision scheme, we only apply the subdivision algorithm described in Eq. (2) but not the connectivity rule. By applying this subdivision algorithm with $n = 3$ (since all faces are triangular) to one of the triangular faces of $\mathcal{X}^{(0)}$ we get three new points (intermediate points). For example, the face $\{P_1, P_2, P_3\}$ gives new intermediate points q_1, q_7 and q_8 near the vertices P_1, P_2 and P_3, respectively (see Fig. 3). Similarly, other triangular faces also generate three new

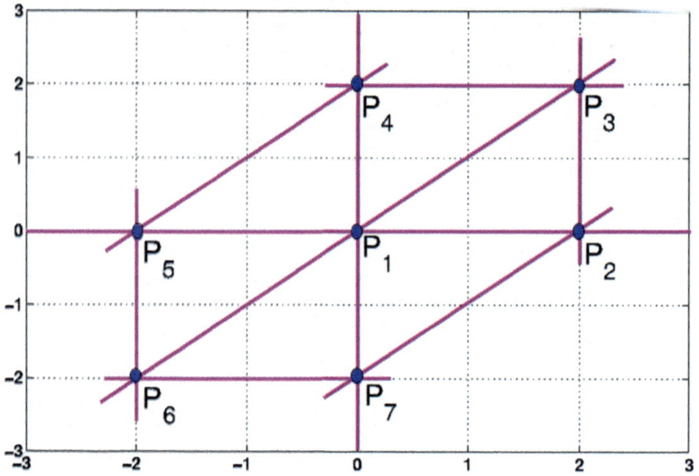

Fig. 2 Initial Control mesh $\mathcal{X}^{(0)}$

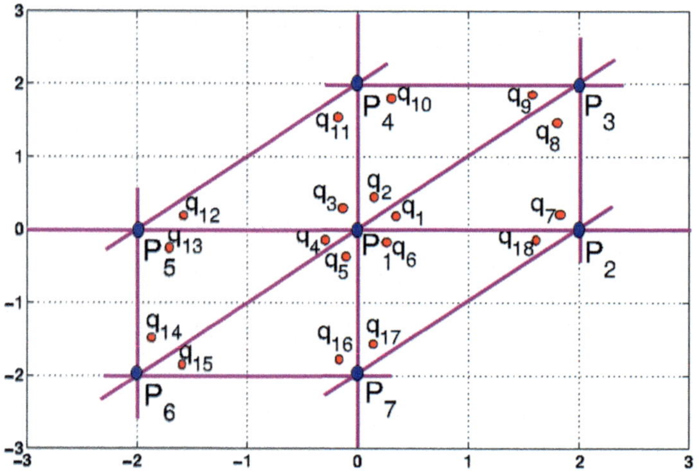

Fig. 3 $\mathcal{X}^{(0)}$ along with 18 intermediate points $(q_1 - q_{18})$

points each. In total, we get 18 new control points $(q_1 - q_{18})$ which are shown in the Fig. 3.

Out of the eighteen points, six intermediate points $(q_1 - q_6)$ are associated with the central vertex point P_1. These points are calculated as follows:

$$q_i = \alpha(3,1)P_1 + \beta(3,1)(P_{i+1} + P_{i+2}), \; i = 1, 2, \dots 6, \quad (P_8 = P_2). \tag{5}$$

Similarly, the remaining 12 points ($q_7 - q_{18}$) are associated to other vertex points and these are computed as follows:

$$
\begin{aligned}
q_7 &= \alpha(3,1)P_2 + \beta(3,1)(P_3 + P_1), \quad q_8 = \alpha(3,1)P_3 + \beta(3,1)(P_1 + P_2),\\
q_9 &= \alpha(3,1)P_3 + \beta(3,1)(P_4 + P_1), \quad q_{10} = \alpha(3,1)P_4 + \beta(3,1)(P_1 + P_3),\\
q_{11} &= \alpha(3,1)P_4 + \beta(3,1)(P_5 + P_1), \quad q_{12} = \alpha(3,1)P_5 + \beta(3,1)(P_1 + P_4),\\
q_{13} &= \alpha(3,1)P_5 + \beta(3,1)(P_6 + P_1), \quad q_{14} = \alpha(3,1)P_6 + \beta(3,1)(P_1 + P_5),\\
q_{15} &= \alpha(3,1)P_6 + \beta(3,1)(P_7 + P_1), \quad q_{16} = \alpha(3,1)P_7 + \beta(3,1)(P_1 + P_6),\\
q_{17} &= \alpha(3,1)P_7 + \beta(3,1)(P_2 + P_1), \text{ and } q_{18} = \alpha(3,1)P_2 + \beta(3,1)(P_1 + P_7) \qquad (6)
\end{aligned}
$$

where,

$$
\gamma(3,1) = \frac{1}{12}\sec^2(h/2)\sec^2 h,\ \beta(3,1) = \gamma(3,1) + \frac{1}{8}\sec^2(h/2)\sec h = \frac{2+3\cos h}{24\cos^2(h/2)\cos^2 h},
$$
$$
\text{and } \alpha(3,1) = \gamma(3,1) + \frac{1}{4}\sec^2(h/2)(1+\sec h) = \frac{1+3\cos h+3\cos^2 h}{12\cos^2(h/2)\cos^2 h}. \qquad (7)
$$

These are obtained by simply replacing $n = 3$ and $k = 1$ in (3). We take the help of the points q_1 to q_{18} to develop our scheme. Taking P_1 to P_7 as initial control points, the first iteration control points $\widetilde{Q_1}$ to $\widetilde{Q_7}$ are obtained by the following averaging rule. In the first iteration control mesh $\widetilde{Q_1}$ is the corresponding central vertex point, which is computed by,

$$
\widetilde{Q_1} = \frac{1}{6}\sum_{i=1}^{6} q_i = \widetilde{a_1}P_1 + \widetilde{a_2}(P_2 + \cdots + P_7). \qquad (8)
$$

Similarly, other first level control points are calculated as follows:

$$
\widetilde{Q_2} = \frac{\frac{1}{2}(q_1 + q_6) + \frac{1}{2}(q_7 + q_{18})}{2} = \widetilde{b_1}(P_1 + P_2) + \widetilde{b_2}(P_7 + P_3),
$$

and in similar fashion,

$$
\begin{aligned}
&\widetilde{Q_3} = \widetilde{b_1}(P_1 + P_3) + \widetilde{b_2}(P_2 + P_4),\ \widetilde{Q_4} = \widetilde{b_1}(P_1 + P_4) + \widetilde{b_2}(P_3 + P_5),\\
&\widetilde{Q_5} = \widetilde{b_1}(P_1 + P_5) + \widetilde{b_2}(P_4 + P_6),\ \widetilde{Q_6} = \widetilde{b_1}(P_1 + P_6) + \widetilde{b_2}(P_5 + P_7),\\
\text{and } &\widetilde{Q_7} = \widetilde{b_1}(P_1 + P_7) + \widetilde{b_2}(P_6 + P_2), \qquad (9)
\end{aligned}
$$

where, the weights are $\widetilde{a_1} = \frac{1+3\cos h+3\cos^2 h}{12\cos^2(h/2)\cos^2 h}$, $\widetilde{a_2} = \frac{2+3\cos h}{72\cos^2(h/2)\cos^2 h}$, $\widetilde{b_1} = \frac{4+9\cos h+6\cos^2 h}{48\cos^2(h/2)\cos^2 h}$ and $\widetilde{b_2} = \frac{2+3\cos h}{48\cos^2(h/2)\cos^2 h}$.

Note that here
$\widetilde{a_1} + 6\widetilde{a_2} = \frac{\cos^2(h/2)}{\cos^2 h}$,
and $2\widetilde{b_1} + 2\widetilde{b_2} = \frac{\cos^2(h/2)}{\cos^2 h}$.

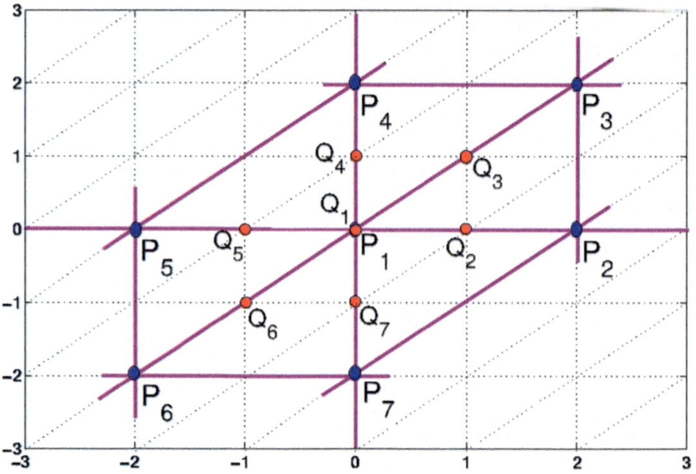

Fig. 4 Old control points P_1 to P_7 and new control points Q_1 to Q_7

Since, this sum is not unity, we **normalize** it by dividing the factor $\frac{\cos^2(h/2)}{\cos^2 h}$ to each weights. In this process, we get a new set of modified weights as follows:

$a_1 = \widetilde{a_1}/(\frac{\cos^2(h/2)}{\cos^2 h})$, $a_2 = \widetilde{a_2}/(\frac{\cos^2(h/2)}{\cos^2 h})$,
$b_1 = \widetilde{b_1}/(\frac{\cos^2(h/2)}{\cos^2 h})$ and $b_2 = \widetilde{b_2}/(\frac{\cos^2(h/2)}{\cos^2 h})$.
So, $a_1 + 6a_2 = 1$, and $2b_1 + 2b_2 = 1$. Now, we get our normalized non-stationary subdivision scheme. In this scheme, for given $\mathcal{X}^{(0)}$ as the initial control mesh and the first iteration control points Q_1 to Q_7 as follows:

$$
\begin{aligned}
Q_1 &= \frac{1}{6}\sum_{i=1}^{6} q_i = a_1 P_1 + a_2(P_2 + \cdots + P_7),\\
Q_2 &= b_1(P_1 + P_2) + b_2(P_7 + P_3),\ Q_3 = b_1(P_1 + P_3) + b_2(P_2 + P_4),\\
Q_4 &= b_1(P_1 + P_4) + b_2(P_3 + P_5),\ Q_5 = b_1(P_1 + P_5) + b_2(P_4 + P_6),\\
Q_6 &= b_1(P_1 + P_6) + b_2(P_5 + P_7),\ Q_7 = b_1(P_1 + P_7) + b_2(P_6 + P_2). \qquad (10)
\end{aligned}
$$

We categorize the newly obtained control points as *vertex points* (for e.g. Q_1) and *edge points* (for e.g. Q_2–Q_7). Each triangular face contains a vertex point and two edge points (see Fig. 4). For example, the face $\{P_1, P_2, P_3\}$ contains the new vertex point Q_1 and edge points Q_2 and Q_3.

Connectivity Rule: We follow the following connectivity rule. In each triangular face we join:
(i) vertex points to edge points.
and (ii) edge points to edge points.

For example, in the face $\{P_1, P_2, P_3\}$ join Q_1 to Q_2, Q_2 to Q_3 and Q_3 to Q_1. After this connectivity rule, we get a new control mesh $\mathcal{X}^{(1)}$.

3.2 Regular Subdivision Rules

Let $\mathcal{X}^{(k)}$ be the kth-level control mesh containing the set of vertices $\mathbf{p}^{(\mathbf{k})} := \{p_{\mathbf{i}}^{k},\ \mathbf{i} \in \mathbb{Z}^2\}$ and $\mathcal{X}^{(k+1)}$ be the $(k+1)$th-level control mesh given by the set of vertices $\mathbf{p}^{(\mathbf{k+1})} := \{p_{\mathbf{i}}^{k+1},\ \mathbf{i} \in \mathbb{Z}^2\}$. $\mathcal{X}^{(k+1)}$ is obtained from $\mathcal{X}^{(k)}$ by the following refinement rules:

$$p_{\mathbf{i}}^{k+1} = \sum_{\mathbf{j} \in \mathbb{Z}^2} a_{\mathbf{i}-2\mathbf{j}}^{(k)}\, p_{\mathbf{j}}^{k}, \quad \mathbf{i} \in \mathbb{Z}^2. \tag{11}$$

From (11), the finite set of coefficients are collected in kth-level subdivision mask

$$\mathbf{a}^{(\mathbf{k})} := \{a_{i,j}^{(k)},\ (i, j) \in \mathbb{Z}^2\}, \tag{12}$$

and the kth-level subdivision symbol is given by

$$\mathbf{a}^{(\mathbf{k})}(z_1, z_2) = \sum_{(i,j) \in \mathbb{Z}^2} a_{i,j}^{(k)} z_1^i z_2^j, \quad (z_1, z_2) \in (\mathbb{C} \setminus \{0\})^2. \tag{13}$$

Specifically, in our scheme

$$\mathbf{a}^{(\mathbf{k})} = \{a_{i,j}^{(k)},\ -2 \le i, j \le 2\} = \begin{bmatrix} 0 & 0 & w_1^{(k)} & w_3^{(k)} & w_1^{(k)} \\ 0 & w_3^{(k)} & w_2^{(k)} & w_2^{(k)} & w_3^{(k)} \\ w_1^{(k)} & w_2^{(k)} & w_0^{(k)} & w_2^{(k)} & w_1^{(k)} \\ w_3^{(k)} & w_2^{(k)} & w_2^{(k)} & w_3^{(k)} & 0 \\ w_1^{(k)} & w_3^{(k)} & w_1^{(k)} & 0 & 0 \end{bmatrix}.$$

Here the non-stationary subdivision masks $w_0^{(k)}, w_1^{(k)}, w_2^{(k)}$ and $w_3^{(k)}$ are the weights a_1, a_2, b_1 and b_2, respectively, in which h is replaced by $\frac{h}{2^k}$.

Furthermore, the kth-level subdivision symbol associated with our scheme is given by

$$\begin{aligned} \mathbf{a}^{(\mathbf{k})}(z_1, z_2) = & \frac{1}{144\, z_1^2 z_2^2} \big(w_1^{(k)}(1 + z_1^2 + z_2^2 + z_1^4 z_2^2 + z_1^2 z_2^4 + z_1^4 z_2^4) \\ & + w_3^{(k)}(z_1 + z_2 + z_1^3 z_2 + z_1^4 z_2^3 + z_1 z_2^3 + z_1^3 z_2^4) \\ & + w_2^{(k)}(z_1 z_2 + z_1^2 z_2 + z_1 z_2^2 + z_1^3 z_2^2 + z_1^2 z_2^3 + z_1^3 z_2^3) + w_0^{(k)} z_1^2 z_2^2\big). \end{aligned} \tag{14}$$

However, this Laurent polynomial is not factorizable. But it satisfies the following necessary condition for convergence.

$$\mathbf{a}^{(\mathbf{k})}(-1,1) = \mathbf{a}^{(\mathbf{k})}(1,-1) = \mathbf{a}^{(\mathbf{k})}(-1,-1) = 0, \ \mathbf{a}^{(\mathbf{k})}(1,1) = 4. \quad (15)$$

Applying this subdivision algorithm and then the connectivity rule repeatedly, we get a limit surface.

In matrix form this subdivision scheme is written as

$$\mathcal{X}^{(k+1)} = M_{k+1}\mathcal{X}^{(k)}, \quad k = 0, 1, 2, \ldots, \quad (16)$$

where M_{k+1} is the $(k+1)$th-level subdivision matrix which is given by

$$M_{k+1} = \begin{bmatrix} w_0^{(k)} & w_1^{(k)} & w_1^{(k)} & w_1^{(k)} & w_1^{(k)} & w_1^{(k)} & w_1^{(k)} \\ w_2^{(k)} & w_2^{(k)} & w_3^{(k)} & 0 & 0 & 0 & w_3^{(k)} \\ w_2^{(k)} & w_3^{(k)} & w_2^{(k)} & w_3^{(k)} & 0 & 0 & 0 \\ w_2^{(k)} & 0 & w_3^{(k)} & w_2^{(k)} & w_3^{(k)} & 0 & 0 \\ w_2^{(k)} & 0 & 0 & w_3^{(k)} & w_2^{(k)} & w_3^{(k)} & 0 \\ w_2^{(k)} & 0 & 0 & 0 & w_3^{(k)} & w_2^{(k)} & w_3^{(k)} \\ w_2^{(k)} & w_3^{(k)} & 0 & 0 & 0 & w_3^{(k)} & w_2^{(k)} \end{bmatrix}. \quad (17)$$

The matrix M_{k+1} has a simpler form:

$$M_{k+1} := \begin{bmatrix} a & \mathbf{b} \\ \mathbf{b}' & \mathbf{C} \end{bmatrix},$$

where $a = w_0^{(k)}$,

$\mathbf{b} = \left[w_1^{(k)} \ w_1^{(k)} \ w_1^{(k)} \ w_1^{(k)} \ w_1^{(k)} \ w_1^{(k)} \right]$,
$\mathbf{b}' = \left[w_2^{(k)} \ w_2^{(k)} \ w_2^{(k)} \ w_2^{(k)} \ w_2^{(k)} \ w_2^{(k)} \right]^{\mathbf{T}}$,

and $\mathbf{C} = \begin{bmatrix} w_2^{(k)} & w_3^{(k)} & 0 & 0 & 0 & w_3^{(k)} \\ w_3^{(k)} & w_2^{(k)} & w_3^{(k)} & 0 & 0 & 0 \\ 0 & w_3^{(k)} & w_2^{(k)} & w_3^{(k)} & 0 & 0 \\ 0 & 0 & w_3^{(k)} & w_2^{(k)} & w_3^{(k)} & 0 \\ 0 & 0 & 0 & w_3^{(k)} & w_2^{(k)} & w_3^{(k)} \\ w_3^{(k)} & 0 & 0 & 0 & w_3^{(k)} & w_2^{(k)} \end{bmatrix}$

which is a circulant matrix of order 6×6.

3.3 Subdivision Surface

Let the initial control mesh Y_0 be given by the Fig. 5 in which $p^0_{0,0} = 1$ and rest control points have values 0. We apply our subdivision scheme (subdivision algorithm + connectivity rule) to Y_0 a number of times to get a limit surface. For example, applying the subdivision scheme once to the initial control mesh Y_0 we get the control mesh Y_1, which is shown in Fig. 6. Similarly, applying the subdivision scheme three times to Y_0 we get the control mesh Y_3 which is shown in Fig. 7. Here we have taken the initial value of $h = 1$. It satisfies the convex hull property which is shown in the Fig. 8. The limit surface is bell shaped and has finite support. Here, we notice that the subsequent surfaces that we are getting are in fact box spline-like surfaces although these are not actually box splines.

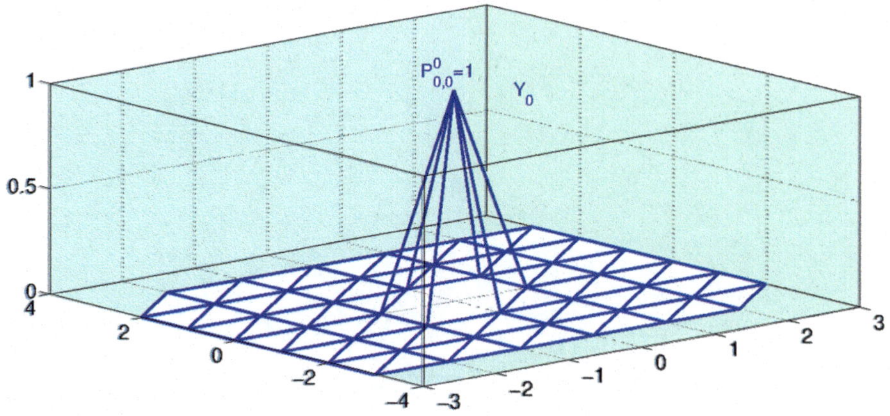

Fig. 5 Initial control mesh Y_0 where $p^0_{0,0} = 1$

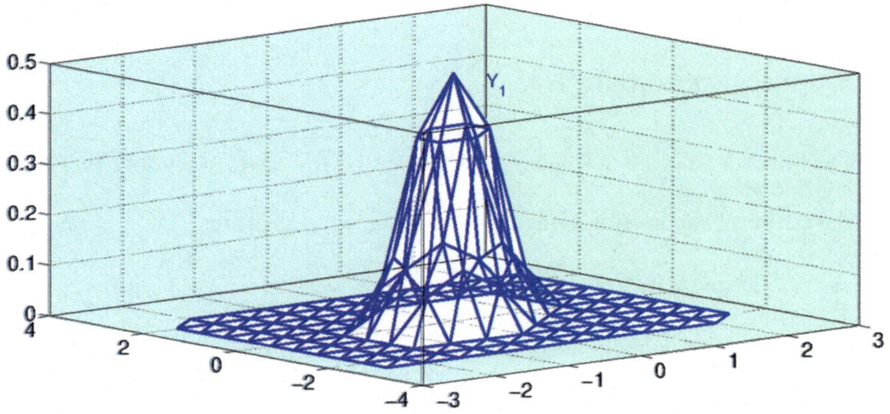

Fig. 6 The first iteration control mesh Y_1

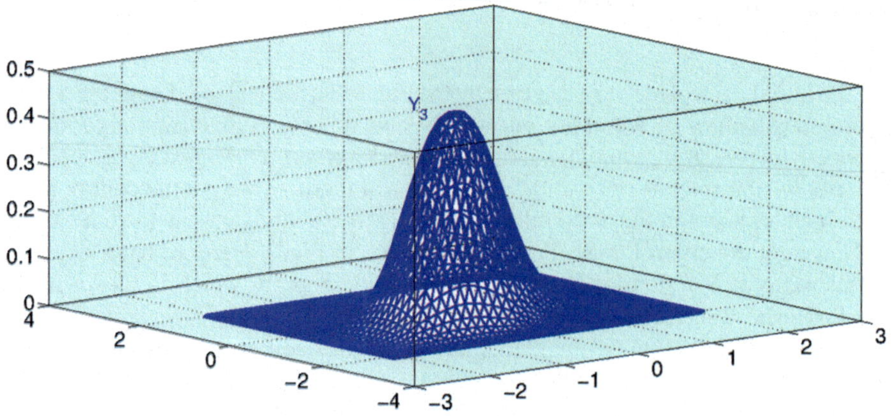

Fig. 7 The third iteration control mesh Y_3

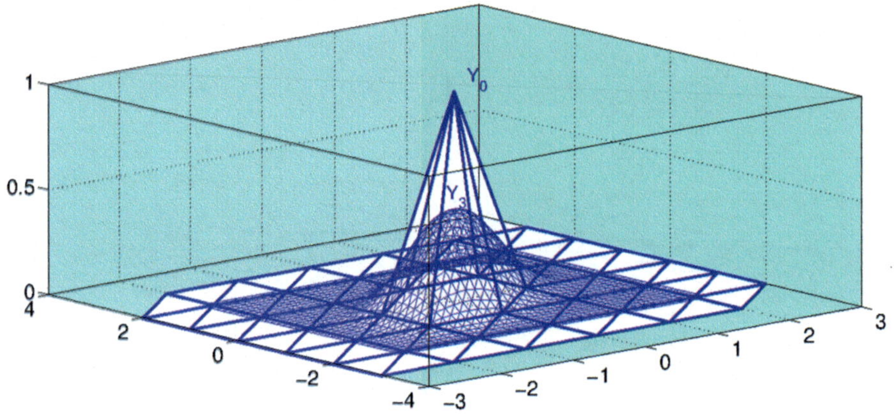

Fig. 8 The convex hull property of the limit surface (here Y_3)

4 Convergence Analysis

In this section, we present the convergence analysis [7, 20, 27] of our non-stationary subdivision scheme.

The kth-level subdivision symbol of our scheme (14) satisfies

$$\mathbf{a}^{(\mathbf{k})}(1,1) = 4,\ \mathbf{a}^{(\mathbf{k})}(1,-1) = \mathbf{a}^{(\mathbf{k})}(-1,1) = \mathbf{a}^{(\mathbf{k})}(-1,-1) = 0. \tag{18}$$

This is the necessary condition for the convergence of a bivariate scheme (see [27]). However, to describe the sufficient part of the convergence analysis, we follow a different theoretical approach. From (16),

$$\begin{aligned}\mathcal{X}^{(k+1)} &= M_{k+1}\mathcal{X}^{(k)}\\ &= M_{k+1}M_k\mathcal{X}^{(k-1)}\\ &= M_{k+1}M_k\ldots M_1\mathcal{X}^{(0)} \;=:\; M^{(k+1)}\mathcal{X}^{(0)}.\end{aligned} \tag{19}$$

We need the $(k+1)$th-level non-normalized subdivision matrix N_{k+1}, which is the subdivision matrix without normalization (where $w_0^{(k)} = \widetilde{a_1}$, $w_1^{(k)} = \widetilde{a_2}$, $w_2^{(k)} = \widetilde{b_1}$, and $w_3^{(k)} = \widetilde{b_2}$, and h is replaced by $h/2^k$). For the study of the convergence, we also need the stationary matrix M, which is obtained from M_k by putting $h = 0$. Let $S_{k+1} := M_{k+1} - M$. It is observed that entries of the matrix N_{k+1} are non-negative and non-increasing with k. The sum of all the entries in each row of N_{k+1} is $\frac{\cos^2(h/2^{k+1})}{\cos^2(h/2^k)}$, and each entry of N_{k+1} is greater than the corresponding entry of M. Overall, the non-normalized subdivision matrix N_{k+1} has a similar behavior like M_{k+1} in [7]. In our case, the sequence of subdivision matrices satisfy the following theorems which can be proved identically as described in [7].

For a 7×1 column vector x the norm of x is defined as

$$\|x\| = \max_{1\le k\le 7} |x_k|.$$

Theorem 1 *The stationary matrix M and the non-stationary matrix sequence $\{M^{(k)},\ k \in \mathbb{N}\}$ satisfy the following:*

(i) $\|Mx\| \le \|x\|$,
(ii) $\|M^k x\| \le \|x\|, \quad \forall\, k \in \mathbb{N}$.
(iii) $\|M^{(k)} x\| \le \|x\|, \quad \forall\, k \in \mathbb{N}$.

Theorem 2 *The non-normalized non-stationary matrices $\{N_k,\ k \in \mathbb{N}\}$ satisfy the following: $\|N_k x\| \le \frac{\cos^2(h/2^k)}{\cos^2(h/2^{k-1})}\|x\|$ and $\|N_k\| = \frac{\cos^2(h/2^k)}{\cos^2(h/2^{k-1})}$.*

Theorem 3 $\|S_k x\| \le \frac{C(h)}{4^k}\|x\|$, $\forall\, k$, *where* $C(h) = \frac{6h^2}{\cos^2 h}$.

Proof Note that

$$\|S_k x\| \;=\; \|M_k x - Mx\|.$$

Applying triangle inequality we get,

$$\|S_k x\| \;\le\; \|(M_k - N_k)x\| + \|N_k x - Mx\|.$$

Since M_k is the normalized version of N_k,

$$\begin{aligned}\|S_k x\| &\le \|\left(\frac{\cos^2(h/2^{k-1})}{\cos^2(h/2^k)}\right)N_k - N_k\|\|x\| + \|x\| \max_{1\le j\le N}\left(\frac{\cos^2(h/2^k)}{\cos^2(h/2^{k-1})} - 1\right)\\ &\le (\|N_k\||\frac{\cos^2(h/2^{k-1})}{\cos^2(h/2^k)} - 1| + |\frac{\cos^2(h/2^k)}{\cos^2(h/2^{k-1})} - 1|)\|x\|.\end{aligned}$$

By Theorem 2,

$$\begin{aligned}
\|S_k x\| &\le 2\,|\frac{\cos^2(h/2^k)}{\cos^2(h/2^{k-1})} - 1|\|x\| \\
&= |\frac{2(\cos(h/2^k) + \cos(h/2^{k-1}))(\cos(h/2^k) - \cos(h/2^{k-1}))}{\cos^2(h/2^{k-1})}|\|x\| \\
&\le \frac{2}{\cos^2 h}.2.2.\sin((\frac{h}{2^k} + \frac{h}{2^{k-1}})/2).|\sin((\frac{h}{2^k} + \frac{h}{2^{k-1}})/2)| \\
&\le \frac{8}{\cos^2 h}.\sin(\frac{3h}{2^{k+1}}).\sin(\frac{h}{2^{k+1}}) \\
&\le \frac{8}{\cos^2 h}.\frac{3h}{2^{k+1}}.\frac{h}{2^{k+1}} \\
&= \frac{6h^2}{\cos^2 h}.\frac{1}{4^k} \\
&= \frac{C(h)}{4^k}.
\end{aligned}$$

With the above results in our hand, we now ready to show the convergence of our subdivision scheme.

Theorem 4 *The sequence of control meshes* $\{\mathcal{X}^{(k)},\ k \in \mathbb{N} \cup \{0\}\}$ *converges to a limit surface and the limit surface is continuous.*

Proof We first take a 7×3 column vector $\mathbf{x}$ in order to analyze the convergence of the matrix sequence $\{M^{(k)}\mathbf{x}\}$. However, without loss of generality, we can study the convergence of $\{M^{(k)}x\}$, where x is a 7×1 column vector. Observe that

$$\begin{aligned}
\|(M^{(k+1)} - M^{k+1})x\| &= \|(M_{k+1}M^{(k)} - MM^k)x\| \\
&= \|((M + S_{k+1})M^{(k)} - MM^k)x\| \\
&= \|(MM^{(k)} - MM^k + S_{k+1}M^{(k)})x\|.
\end{aligned}$$

Now applying triangle inequality law, this produces

$$\|(M^{(k+1)} - M^{k+1})x\| \le \|M(M^{(k)} - M^k)x\| + \|S_{k+1}M^{(k)}x\|. \tag{20}$$

By Theorems 1 and 3, the above inequation implies

$$\|(M^{(k+1)} - M^{k+1})x\| \le \|(M^{(k)} - M^k)x\| + \frac{C(h)}{4^{k+1}}\|x\|. \tag{21}$$

Let $\epsilon_k = \|(M^{(k)} - M^k)x\|$. Then by (21)

$$\epsilon_{k+1} \le \epsilon_k + \frac{C(h)}{4^{k+1}}\|x\|,$$

which implies

$$|\epsilon_{k+1} - \epsilon_k| \leq \frac{C(h)}{4^{k+1}} \|x\|. \tag{22}$$

Now, for $m \geq k+1$

$$\begin{aligned}
\|\epsilon_m - \epsilon_k\| &\leq \|\epsilon_m - \epsilon_{m-1}\| + \|\epsilon_{m-1} - \epsilon_{m-2}\| + \cdots + \|\epsilon_{k+1} - \epsilon_k\| \\
&\leq C(h)\|x\| \left(\frac{1}{4^m} + \frac{1}{4^{m+1}} + \cdots + \frac{1}{4^{k+1}} \right) \\
&= C(h)\|x\| \left(\frac{1}{4^{m-k-1}} + \frac{1}{4^{m-k-2}} + \cdots + \frac{1}{4} + 1 \right) \frac{1}{4^{k+1}} \\
&= \frac{C(h)\|x\|}{4^{k+1}} \frac{1}{1 - \frac{1}{4}} \\
&= C(h)\|x\| \frac{1}{3.4^k}.
\end{aligned}$$

For given $\epsilon > 0$, we can find N such that $\frac{1}{4^k} < \epsilon$, for every $k > N$. Therefore,

$$|\epsilon_m - \epsilon_k| \leq \frac{C(h)}{3} \|x\| \epsilon, \quad \text{whenever } m,\ k > N. \tag{23}$$

This implies the sequence $\{\epsilon_k\}$ is Cauchy in $\mathbb{R}$ and hence converges.
Let $\epsilon_k \to e$. Then $\epsilon_k = e + o(1)$. This implies

$$\|(M^{(k)} - M^k)x\| = e + o(1).$$

Therefore,

$$(M^{(k)}\mathbf{x} - M^k\mathbf{x}) \to \mathbf{E} + \Theta(1),$$

where $\mathbf{E}$ is a fixed 7×3 matrix and $\Theta(1) \to 0$. Since $\{M^k\mathbf{x}\}$ converges, say to $\mathbf{Y}$, we have

$$X^{(k)} = M^{(k)}\mathbf{x} \to \mathbf{Y} + \mathbf{E} + \Theta(1).$$

This proves the theorem. □

In the above theorem $\mathbf{Y}$ is the limit surface of the associated stationary subdivision scheme. The surface $\mathbf{Y} + \mathbf{E}$ (see Fig. 7) is the limit surface of the non-stationary subdivision scheme which lies in the convex hull of initial mesh (see Fig. 8).

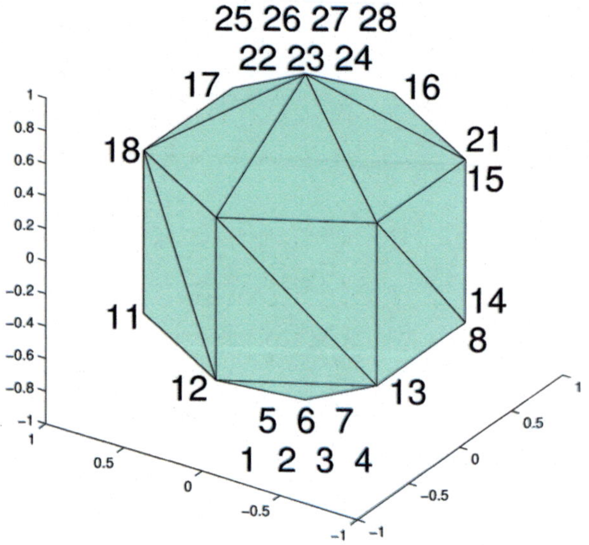

Fig. 9 Initial control mesh $\mathcal{X}^{(0)}$ consisting of twenty eight vertices

5 Application of the Proposed Scheme

In this section, we describe the application of our proposed non-stationary subdivision scheme. For this, we provide two examples of regular meshes below. we have used MATLAB software for the implementation of our proposed subdivision scheme.

Example 1 In first example, we take a regular (i.e. made of vertices of valence 6 only) initial control mesh $\mathcal{X}^{(0)}$ (see Fig. 9) which consists of twenty eight vertices of coordinates

$$[(\mathcal{X}^{(0)})_1]_{i,j} = \cos(p_{i-1})\cos(q_{j-1}),\ [(\mathcal{X}^{(0)})_2]_{i,j} = \cos(p_{i-1})\sin(q_{j-1}), [(\mathcal{X}^{(0)})_3]_{i,j} = \sin(p_{i-1}),$$

where $1 \leq j \leq 7, 1 \leq i \leq 4$, $p_k = -\frac{\pi}{2} + k\frac{\pi}{3},\ k = 0, 1, 2, 3$ and $q_l = l\frac{\pi}{3},\ l = 0, 1, 2, 3, 4, 5, 6$. p_k and q_l are equally spaced values with step size $\frac{\pi}{3}$.

However, this initial control mesh consists of only fourteen distinct points and the vertices are apparently of valence 5, while they have actual valence 6 due to the poles that are *multiple* vertices having coordinates $(0, 0, -1)$ and $(0, 0, 1)$. Since the subdivision methods are blind to the topological identification so they are virtually applied to this mesh with twenty eight vertices. By the repeated application of our non-stationary subdivision scheme this initial regular control mesh converges to a smooth limit surface. We have shown the surfaces obtained after first iteration, second iteration, third iteration and fourth iteration respectively, in Fig. 10. For this we have taken $h = \frac{\pi}{4}$.

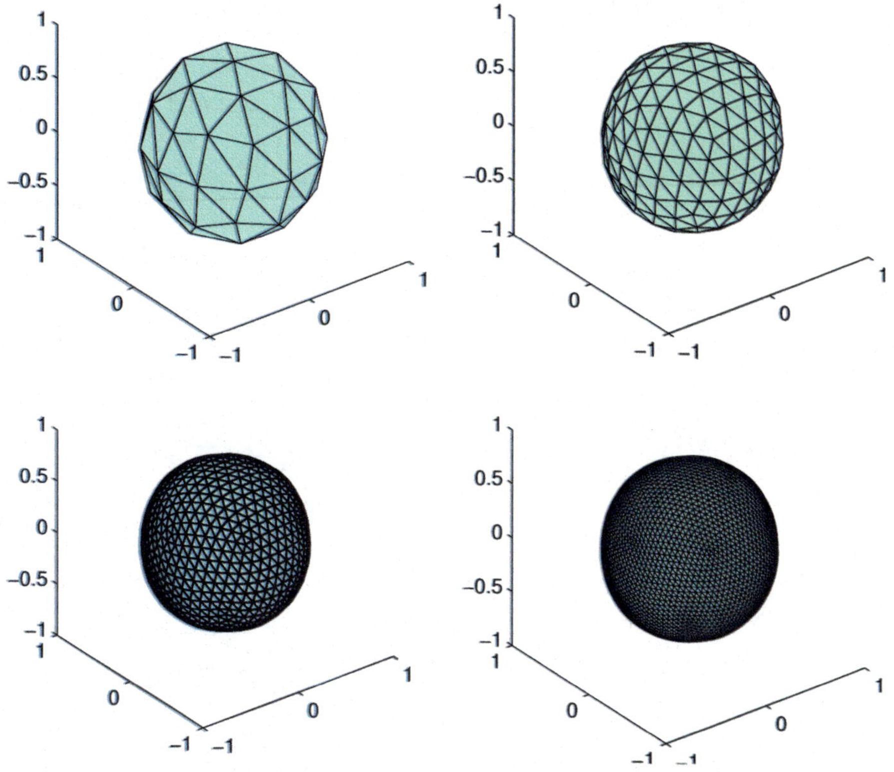

Fig. 10 Subsequent iteration surfaces $\mathcal{X}^{(1)}$, $\mathcal{X}^{(2)}$, $\mathcal{X}^{(3)}$ and $\mathcal{X}^{(4)}$ are shown starting from *left top* to *right bottom*, respectively

Example 2 In second example, we consider a NURBS surface. Generally many NURBS patches are joined to form a surface model. The NURBS is either a torus, a disk or a tube. Here we study the case of a *torus*. In its initial control mesh, we take, for e.g., 4 points in the minor circle direction and 8 points in the major circle direction. So it consists of thirty two number of vertices of coordinates

$$[(\mathcal{X}^{(0)})_1]_{i,j} = (R + r\cos(p_{i-1}))\cos(q_{j-1}),$$
$$[(\mathcal{X}^{(0)})_2]_{i,j} = (R + r\cos(p_{i-1}))\sin(q_{j-1}),$$
$$[(\mathcal{X}^{(0)})_3]_{i,j} = r\sin(p_{i-1}),$$

where $1 \le i \le 4, 1 \le j \le 8, p_k = \frac{\pi k}{2},\ k = 0, 1, 2, 3,$ and $q_l = \frac{\pi l}{2},\ l = 0, 1, 2, 3, 4, 5, 6, 7$. Suppose we choose $r = 1$ and $R = 1.8$, as the radii of minor and major circles, respectively. This initial control mesh (see Fig. 11) is regular i.e. all the vertices are of valence 6.

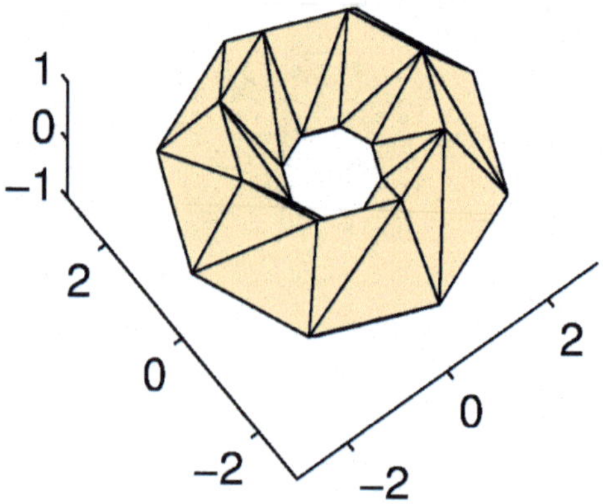

Fig. 11 Initial mesh of a torus, $\mathcal{X}^{(0)}$

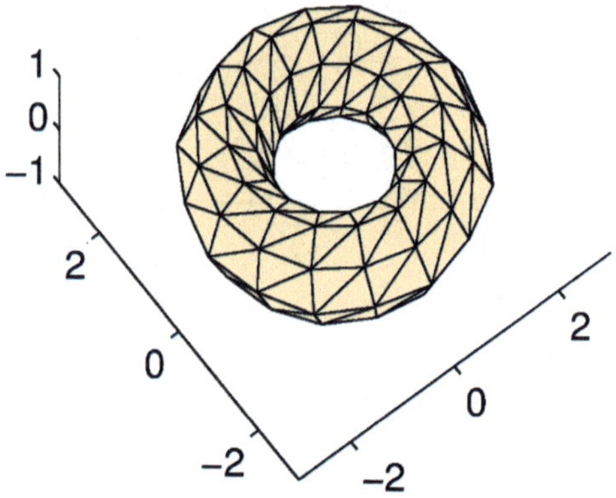

Fig. 12 First iteration mesh $\mathcal{X}^{(1)}$

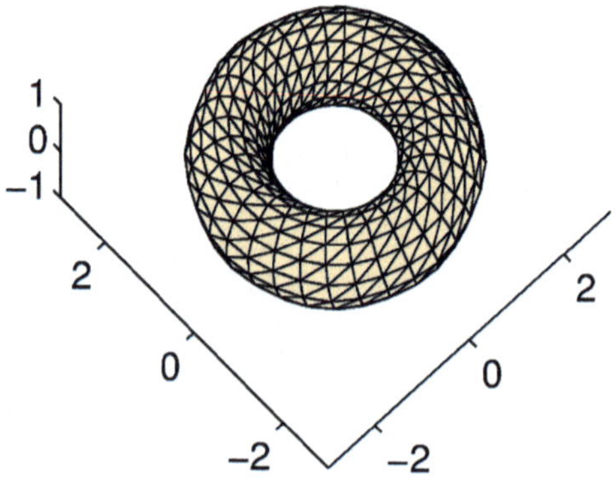

Fig. 13 Second iteration mesh $\mathcal{X}^{(2)}$

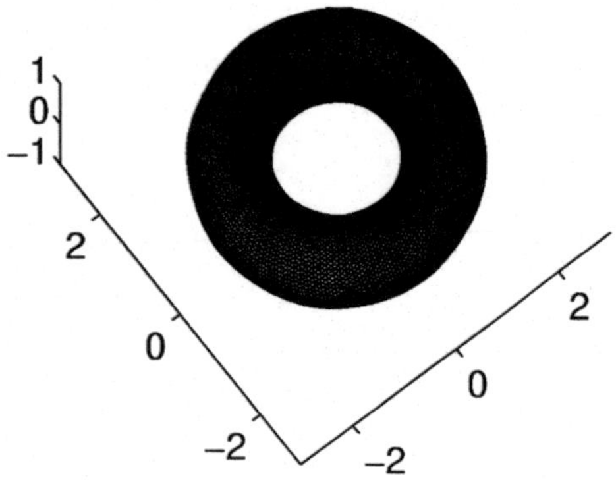

Fig. 14 Fourth iteration mesh $\mathcal{X}^{(4)}$

If we refine this initial mesh by our non-stationary scheme then we obtain comparatively smoother surface in each iteration level. The surfaces obtained after first, second and fourth iterations are displayed in Figs. 12, 13 and 14, respectively.

In the above two examples, one can notice that the subsequent surfaces obtained by successive refinements by our scheme are continuous at the joints. Sometimes after the deformation of NURBS surfaces, the cracks appear at the joints, which is not happening here. Our scheme also possesses some important properties such as compact support, affine invariance, smoothness and simplicity which are also properties of NURBS.

6 Conclusion

In this paper, a new non-stationary subdivision scheme has been derived. More specifically, this is a triangular subdivision scheme for regular meshes. We have described the application of this subdivision scheme over type-I triangulation. Moreover, the limit surfaces are trigonometric box spline-like surfaces. These are well shaped, compactly supported and satisfying convex hull property. Actually, polynomial box splines are well-defined through subdivision schemes whereas the trigonometric box splines are not. So we have taken an initial attempt to define trigonometric box splines with the help of a non-stationary subdivision scheme and we have become partially successful. For this we have used trigonometric polynomials as the subdivision masks. Although, the Laurent polynomial associated with this subdivision scheme is not factorizable but it is satisfying the necessary conditions for convergence. However, we have shown the convergence of this scheme by a different theoretical approach. Eventually, we have taken two examples to show the visual performance of our scheme. Since our scheme is only applicable to regular meshes, extending this to a mesh with arbitrary topology is an immediate requirement. Currently, this scheme is not capable to be implemented on any complex arbitrary geometries since there is a chance of presence of extraordinary vertices. After affine combinations of the masks with a little modifications, the associated symbol will be factorizable and the modified subdivision scheme can be applied to any complex arbitrary topological meshes. More works are needed to address these which are under observations currently.

Acknowledgements This research is supported by *Department of Science and Technology, Govt. of India*, bearing the Regd. No. *DST/INSPIRE Fellowship/2016/IF160366*. So the authors are grateful to Department of Science and Technology, India. The authors are also thankful to the anonymous referees and reviewers for their useful comments and suggestions for the improvement of this article.

References

1. Hoschek, J., Lasser, D., Schumaker, L.L.: Fundamentals of Computer Aided Geometric Design. AK Peters, Ltd. (1993)
2. Dyn, N., Levin, D., Gregory, J.A.: A 4-point interpolatory subdivision scheme for curve design. Comput. Aided Geom. Des. **4**(4), 257–268 (1987)
3. Doo, D., Sabin, M.: Behaviour of recursive division surfaces near extraordinary points. Comput. Aided Des. **10**(6), 356–360 (1978)
4. Catmull, E., Clark, J.: Recursively generated b-spline surfaces on arbitrary topological meshes. Comput. Aided Des. **10**(6), 350–355 (1978)
5. Loop, C.: Smooth subdivision surfaces based on triangles. *Master's thesis, University of Utah, Department of Mathematics* (1987)
6. Jena, M.K., Shunmugaraj, P., Das, P.C.: A subdivision algorithm for trigonometric spline curves. Comput. Aided Geom. Des. **19**(1), 71–88 (2002)
7. Jena, M.K., Shunmugaraj, P., Das, P.C.: A non-stationary subdivision scheme for generalizing trigonometric spline surfaces to arbitrary meshes. Comput. Aided Geom. Des. **20**(2), 61–77 (2003)
8. Kobbelt., L.: $\sqrt{3}$-subdivision. In: Proceedings of the 27th Annual Conference on Computer Graphics and Interactive Techniques, pp. 103–112. ACM Press/Addison-Wesley Publishing Co. (2000)
9. Badoual, A.: Novara, Paola, Romani, Lucia, Schmitter, Daniel, Unser, Michael: A non-stationary subdivision scheme for the construction of deformable models with sphere-like topology. Graphical Models **94**, 38–51 (2017)
10. Ghaffar, A., Mustafa, G., Qin, K.: Construction and Application of 3-Point Tensor Product Scheme (2013)
11. Ghaffar, A., Bari, M., Ullah, Z., Iqbal, M., Nisar, K.S., Baleanu, D.: A new class of 2q-point nonstationary subdivision schemes and their applications. Mathematics **7**(7), 639 (2019)
12. Ghaffar, A., Iqbal, M., Bari, M., Hussain, S.M., Manzoor, R., Baleanu, D.: Construction and application of nine-tic b-spline tensor product (2019)
13. Baleanu, D., Ghaffar, A., Ullah, Z., Bari, M., Nisar, K.S.: Family of odd point non-stationary subdivision schemes and their applications (2019)
14. Ghaffar, A., Ullah, Z., Bari, M., Al-Qurashi, M., Baleanu, D.: A new class of 2m-point binary non-stationary subdivision schemes (2019)
15. Mustafa, G., Khan, F., Ghaffar, A.: The m-point approximating subdivision scheme. Lobachevskii J. Math. **30**(2), 138–145 (2009)
16. Hassan, M.F., Ivrissimitzis, I.P., Dodgson, N.A., Sabin, M.A.: An interpolating 4-point c2 ternary stationary subdivision scheme. Comput. Aided Geom. Des. **19**(1), 1–18 (2002)
17. Zulkifli, N.A.B., Karim, S.A.A., Shafie, A.B., Sarfraz, M., Ghaffar, A.: Image interpolation using a rational bi-cubic ball (2019)

18. Ashraf, P., Nawaz, B., Baleanu, D., Nisar, K.S., Ghaffar, A., Ahmed Khan, M.A., Akram, S.: Analysis of geometric properties of ternary four-point rational interpolating subdivision scheme. Mathematics **8**(3), 338 (2020)
19. Ashraf, P., Sabir, M., Ghaffar, A., Nisar, K.S., Khan, I.: Shape-preservation of the four-point ternary interpolating non-stationary subdivision scheme. Front. Phys. **7**, 241 (2020). https://doi.org/10.3389/fphy
20. Conti, C., Donatelli, M., Novara, P., Romani, L.: A linear algebra approach to the analysis of non-stationary subdivision for 2-manifold meshes with arbitrary topology (2017). ArXiv:1707.01954
21. Schumaker, L.: Spline Functions: Basic Theory. Cambridge University Press, Cambridge (2007)
22. Lai, M.-J., Schumaker, L.L.: Spline Functions on Triangulations, vol. 110. Cambridge University Press (2007)
23. Schoenberg, I.J.: On trigonometric spline interpolation. J. Math. Mech. 795–825 (1964)
24. Rogers, D.F.: An Introduction to NURBS: With Historical Perspective. Elsevier (2000)
25. Morin, G., Warren, J., Weimer, H.: A subdivision scheme for surfaces of revolution. Comput. Aided Geom. Des. **18**(5), 483–502 (2001)
26. De Boor, C., Höllig, K., Riemenschneider, S.: Box Splines, vol. 98. Springer Science & Business Media (2013)
27. Cavaretta, A.S., Dahmen, W., Micchelli, C.A.: Stationary Subdivision, vol. 453. American Mathematical Soc. (1991)
28. Farin, G.: Curves and Surfaces for Computer-aided Geometric Design: A Practical Guide. Elsevier (2014)

Mathematical Modelling for Perishable Product Supply Chain Under Inflation and Variable Lead Time

Ritu Agarwal and Chandrakumar M. Badole

Abstract This study presents an inventory model for deteriorating items under a real time situation where the lead time varies with time. A mathematical model has been developed for finding the total cost and order quantity in a finite planning horizon containing m number of cycles. The effects of the inflation of currency, shortages and lead time along with the information technology on the lead time have been considered. Special cases for the complete backlogging and instantaneous deterioration have been obtained. Demand rate is a arbitrary log-concave function of time. The study has been illustrated for the error function as demand function. The effect of different parameters like deterioration rate and backlogging parameter is studied on the order quantity and the total supply chain cost for this function. Data and sensitivity analysis is carried out using MATLAB software.

Keywords Perishable product supply chain · Planning horizon · Lead time · Shortages · Inflation

1 Introduction

Supply of any commodity products with desired quality, cost and lead time is already very difficult due to globalization and high level of competition in the market. It becomes even more difficult for the perishable products due to the short shelf life and changing customer demand. Uncertain nature of demand gives rise to the conditions of either excess inventory or stock out scenario. Both these situations are barriers to the survival of supply chains [25]. The supply chains need to compete with growing variety of products, short delivery time, higher cycle service level, high quality, and

R. Agarwal (✉)
Department of Mathematics, Malaviya National Institute of Technology, Jaipur, India
e-mail: ragarwal.maths@mnit.ac.in

C. M. Badole
Department of Mechanical Engineering, Dr. Babasaheb Ambedkar Technological University, Lonere, India

J. Singh et al. (eds.), *Methods of Mathematical Modelling and Computation for Complex Systems*, Studies in Systems, Decision and Control 373,
https://doi.org/10.1007/978-3-030-77169-0_11

lower cost. Perishability imposes an intense pressure on managers to manage the supply of products in a timely manner. It adds an additional cost of disposal of outdated items, leading to out-of-stock situations and also loss of customer faith, if not managed effectively. The product becomes obsolete, if it is not getting consumed within its shelf life. In India about 32% of the fruits and vegetable products go waste due to obsolescence [28]. This waste is large and expensive for both supply chains as well as to the society. The management of optimum amount of perishable inventory is another challenge to satisfy customers. While higher amount of inventory leads to the obsolescence, stock outs can have serious impact on the reputation of a firm. Further, the customer, market fragmentation and specialization causes a rapid increase in the product demand variation [28]. However, the total market volume does not necessarily rise as fast as the number of products that are being offered. This leads to a decreased in volume per product type. Other sources of demand variations include seasonality of production, weather conditions, and biological nature of products. This results in input variation and unpredictability. Thus, it can create a big problem for the retailers in the terms of satisfying customer demand.

Modern industrial companies are operating in a fast changing world. We are under-stressed by global competition, control worldwide sourcing and dynamic markets oversee geographically distributed production facilities and aim to produce excellent products and high-quality customer service [16]. As a result, corporations have made enormous efforts to streamline their internal business processes. Fundamental changes are required in the role and form of supply chains and the way in which they innovate and conduct business and research. Efforts have also been made to recognize and improve key product value chain activities and massive investments have been made in new information and communication channels within the business such as data warehouse or enterprise resource planning (ERP) systems [31]. Properly designed supply chains would help industries to relieve their difficulties to a certain extent if the supply cabins are adequately implemented.

Supply chain management (SCM) is characterized as the integration of key end-user business processes by suppliers that supply goods, services and information and add value to customers and stakeholders [24]. The concept of SCM has evolved around the customer focused corporate vision, which drives changes throughout a firm's internal and external linkages and then captures the cluster of inter-functional and inter-organizational integration and coordination [36]. However, as a new way of conducting business, firms have also begun to realize the strategic importance of planning, controlling, and designing supply chains with an objective to enhance the operational efficiency, responsiveness, and profitability of the firms and its supply chain partners. The only way to achieve the objective of SCM is to develop and implement appropriate mathematical models in right directions [26].

2 Literature Review

The research on the *Perishable Product Supply Chain* (PPSC) dates back to 1950. But even today in the age of technological advancement, the challenges for PPSCs are multiplying day by day. The important early research contributors in this area were Ghare and Shrader [15], Covert and Philip [12] and Shah [32], who modelled economic order quantity for perishable inventory. Chew et al. [10] have developed an inventory model with deterministic yet linearly changing demand rate, constant deterioration rate and limited horizon of planning.

Whereas, to allow shortages, Sachan [30] expanded the model of Dave and Patel, Datta and Pal [13] have presented an *Economic Order Quantity* (EOQ) model considering variables such as deterioration and power demand pattern. Research work on models of deteriorating items, time-varying demand and shortages continues with a great dynamism and provides new dimensions for studies on supply chains (see, e.g. Benkherouf [3] and Hariga [17]). The common feature of all the above papers is that they shortages have been allowed while unsatisfied demand is completely backlogged [33].

The trend has altered in the early twentieth century, and scientists are inclined to model pricing strategies. To account for inventory loss due to obsolescence, the idea of dynamic pricing was implemented. In this approach, the scientists recommend that product prices are lowered as long as the shelf life, product quality is degraded every day (see, e.g. Dye et al. [14] and Chew et al. [10]). This approach was useful in boosting product demand and improving income generation that could have been lost in spoilage, as well as loss of business goodwill. Further, today's customers are very weary of the product quality. They prefer to buy a product which is "last in first out (LIFO)" over the one that has been in the shelf for long time [11]. In such situations, the inventory loss can only be reduced by managing and maintaining adequate quantity of perishable product. In order to do so, we need an information tracking system that can provide the information on remaining product life and movement of inventory in the supply chain.

A perishable product gets spoiled and its useful life reduces if not handled properly during transportation. If the reduced life information of items is not updated, an outdated item may get delivered to the customer. In such a case, there may be an additional cost of replacement of item and also loss of goodwill of company. Such spoilage could be reduced significantly if we can automate inventory management system by using RFID technology for product identification, while it moves through the supply chain [34]. RFID system can track the items in real time without product movement, scanning or human involvement. Using active RFID tags, it is possible to update information on it dynamically. RFID system gives a complete visibility of product movement in the supply chain [39]. This may help to make early decisions about inventory control in case there is any interruption in the supply. It partially or completely eliminates time and effort required for counting while loading, and unloading the items. This results into reduction of total lead-time for arrival of an order [22].

RFID technology can increase a company's efficiency and provide financial benefits to both companies and consumers. However, RFID, like any newly implemented technology, presents management with issues of new system threats and decisions about incorporating adequate controls over the new technology [19]. On the other hand, application fields and opportunities are vast. The key driver is that even in largely distributed, more stochastic than deterministic business environments, adaptive organizations and enterprises must react to demands quickly, else a competitor will take the business. Therefore, they must reduce waste and improve efficiency at all fronts. The most important aspect of this strategy is to know exactly what quantity of products they have in stock, exactly where these products are, and in what condition [29]. Furthermore, major distributors dealing with complex, global supply chains must be able to trace their shipments in detail, for security, safety, quality degradation or any other reasons. The improved information accuracy through RFID application will allow companies to substantially reduce out-of-stocks and back orders. They are also likely to find themselves with higher overall average inventory. This suggests remarkable improvement opportunity, namely that companies can potentially reduce reorder quantities and target inventory levels without hurting customer service levels.

Over the past decades, there has been a notable amount of contribution by various researchers in the area of perishable product supply chains. Their works specifically have focused towards developing optimal order quantity policies, dynamic pricing, and inventory management policies. However, prior researchers' works towards studying the impact and implementation of newer technologies like RFID in supply chain of perishable products are still at very preliminary stage. Because of the complexity, much of the existing literature available on short life products focus on the order quantity as a decision variable and do not incorporate lead time into their ordering policy decisions [27]. Lead time can be controlled by incorporating newer technology like RFID in the supply chain system. RFID can drastically reduce the lead time leading to reduced order quantity and reduced uncertainty but with an added extra cost of RFID which is termed here as crashing cost. Overall it is observed that RFID can make pronounced impact on reduction of lead time and minimize uncertainties in supply chain decisions [22, 29]. However, prior researchers have assumed a zero lead-time while modeling the inventory management for perishable products supply chain [33, 38]. Interestingly, impact of RFID on perishable product supply chain under realistic situation (like non-zero lead time) has not been adequately considered by earlier researchers.

Inflation has become an integral part of the world economy since the 1970s power crisis, and large inflation rates were not unusual in many countries. Chen [9]. For example, the annual inflation rate in China between 1993 and 1995 was roughly 17% on average. In most existing models, however, the effect of inflation and the time value of money are not explicitly considered as parameters. The attention towards the decreasing value of money credits to Buzacott [6], who initiated a pioneer research in this direction. He developed an economic order quantity (EOQ) model with inflation subject to various types of pricing policies. Chandra and Bahner [8] researched the effects of time discounting and inflation on the order-level system with shortages decision variables and the economic lot-size system with a finite replenishment rate. Bose et al. [5], Hariga [18] and Hou [20] have suggested two of the most general models combining the effects of inflation and the time value of money. These models assume the deterioration rate is a constant over time, demand rate is time-proportional, and shortages are allowed. Yang [39] suggested two warehouse models in which the impact of inflation on perishable products was studied considering partial backlogging. Hsich and Dye [21] has modelled the pricing and EOQ considering the effect of inflation with partial backlogging which prove that inflation affects price and order quantity adversely. However Jen-Ming Chen [9] in his model, assumed that the demand rate was time-proportionate and shortages were allowed while the effects of inflation and the time value of money were taken into account.

3 Research Gap

One of the major draw-back of the research cited above is that non consideration of realistic situations that pertains at every node of supply chain. Almost researchers have considered the condition of zero lead time while formulation of model. Another fact was observed that research in this area of perishable product supply chain lacks in consideration of importance of information technology in PPSCs. The research in this area is still in preliminary stage and needs to design and develop the sophisticated models on utilization of newer information technologies to overcome day to day supply chain problems [2]. Information at all stages of supply chain plays a very vital role in reducing bull whip effect and improving the decision capabilities in right direction [1]. Most of the problems in the supply chain results because of inadequate information regarding point of scale (PoS) data at different nodes of supply chains [35]. The lead time and its variability impose tremendous problems in supply chains and can be controlled using the information technology. The incorporation of adequate information technology can reduce lead time that is a requirement of every supply chains [23]. Nevertheless the effect of inflation of money have not been focussed to the depth as it was required.

In this paper efforts have been made to answer few of the research questions: does information technology helps resolving real time problems of supply chain; does inflation of money affects the overall performance of SCM; does consideration of real time situations in modelling SCM helps make it more effective and responsive?

The most realistic situations have been considered in this work such as variable lead time, deterioration rate, effect of inflation and varying demand that pertain at every supply chain nodes. And also the effect of inflation of money is considered while formulating the total cost. The remaining of the paper is organized in the following manner. Section 1 describes RFID framework how it helps in acquiring point of scale data at every node of supply chain. Section 2 describes the framework and model development. Section 3 presents results and discussion. The conclusions and future scope have been presented in the Sect. 4.

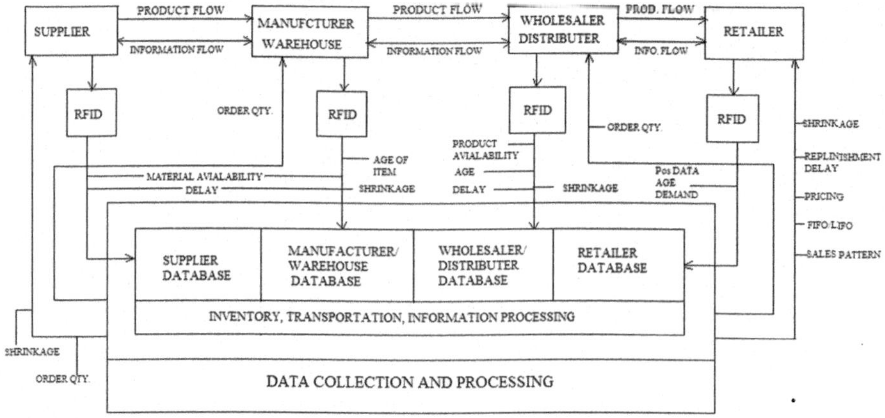

Fig. 1 RFID based architecture of supply chain of perishable product

1. Proposed RFID Based Supply Chain Architecture

For better understanding of the impact of RFID on supply chain, RFID based supply chain architecture is proposed as shown in Fig. 1 by the second author in his doctoral thesis [7]. It describes how material and information flow throughout the supply chain using RFID database. All the stages are interconnected to each other using RFID database; hence any stage can avail all pertinent information about demand, order quantity, age and quality of product, delay and order status at any stage. Thus, the point of sale (PoS) data can be used for planning and strategically decision making. It will eventually reduce the uncertainty in demand at every stage of the supply chain. The reduced uncertainty leads to better management of inventory, reduced order quantity, reduced (optimum) safety stock in supply chain, reduced shrinkage thereby leading to the reduction of overall cost. Using RFID sensors the quality of the product can be judged and the remaining available shelf life of product can be determined. This allows managers to formulate newer pricing strategies based on the lifecycle information, which will help minimize the loss due to obsolescence. Furthermore, it also provides a good data to implement an appropriate sales and marketing strategy (such as FIFO or LIFO) based on the nature and purchasing habits of the customers.

In order to quantify the impact of RFID on perishable product supply chain management, a mathematical model is developed and presented in the following section.

4 The Model Framework

In this section, we present the mathematical model of the problem with all the assumptions and notations.

A. *Assumption*

- A finite planning horizon is considered that consists of m number replenishment cycles. Formulation is carried out per cycle basis.
- Cycle starts with replenishment and ends with negative inventory due shortage. So at the start of cycle, inventory is maximum.
- The product has some initial freshness value and it starts deteriorating after certain time t_d after replenishment (Wu et al. [37]). Hence for the period of t_0 to t_d there is no deterioration and inventory varies as a function of demand. After t_d, inventory varies as cumulative effect of deterioration and demand.
- Replenishment is considered instantaneous.
- The finite lead time $T_l \neq 0$ is considered.
- The deterioration rate θ is considered as constant per unit time ($0 \leq \theta \leq 1$) as the product is maintained under the same set of conditions. The time t_d at which deterioration starts is function of θ.
- The demand is continuous log concave function of time $f(t)$ such that $f'(t) \neq 0$.
- The unsatisfied demands are backlogged partially with the rate exp (-αt) where t is the time between start of shortage till the next replenishment and $\alpha > 0$ is the cost dependent parameter.
- Information system do not have (except cold chains) impact on deterioration rate of perishable product and deterioration continues.

B. *Notations Used*

I	Inventory at any instant
θ	Rate of deterioration
$f(t)$	Demand rate, which is log concave function of time
α	Backlogging parameter
A	Fixed set-up replenishment cost
C_q	Additional replenishment cost paid per unit of order quantity
C_h	Inventory holding cost per unit
C_d	Deterioration cost per unit of deteriorated item
C_s	Backlogging cost per unit of backlogged inventory
C_o	Opportunity or lost sale cost per unit.
C_c	Crashing cost per unit.
L	Lead time = $T - t_r$, the time lapse between the reorder point and the replenishment point
I_{max}	Maximum inventory level at the start of cycle
I_s	Safety inventory level at reorder point
R	r-i = net discount rate of inflation is constant
H	Planning Horizon
Q	Lot size
T	Replenishment cycle
m	Number of replenishments during the planning horizon = H/T
k ($0 < k < 1$)	Shortage parameter, a fraction of the scheduling period determined for which there are no shortages

$T_j = jT$	The total time that is elapsed up to and including the jth replenishment cycle, j = 1, 2, …, m

where, $T_0 = 0$, $T_1 = T$ and $T_m = H$

t_d	log(1/ θ), Time of freshness of product, till this time there is no deterioration.
t_r	Time of placing the order, reorder point.
$t_j = (j+k)T$	Time at which the inventory level in the jth replenishment cycle drops to zero.
$T_{j+1}-t_j$	Period of shortage.
$p = t_r/t_1,\ 0 \leq p \leq 1$	as a fraction of t_1.

The period for which there is no shortage in each interval $[jT, (j+1)T]$ is a fraction of the scheduling period determined by the *shortage parameter* 'k' ($0 < k < 1$) and is $(j+k)T$ for the interval $[jT, (j+1)T]$. Shortages occur at time $t_j = (j+k)T$ and accumulated until time $t = (j+1)T$, $j = 0,1, 2, \ldots$, m before they are backordered. If $k \to 1$, the interval $[jT, (j+1)T]$ has no shortage at all (Fig. 2).

When θ is high, i.e. item is highly perishable, it starts deteriorating just after entering in the supply chain and hence t_d is low and vice versa. E.g. Mushroom being highly perishable t_d is small for it, while for a product like apple t_d is comparatively high. Thus, for highly perishable items, $t_d = 0$ and $\theta = 1$ and for non-perishing items, $\theta \to 0$ and $t_d \to \infty$. We define, $t_d = \log\left(\frac{1}{\theta}\right)$ as this function satisfies both

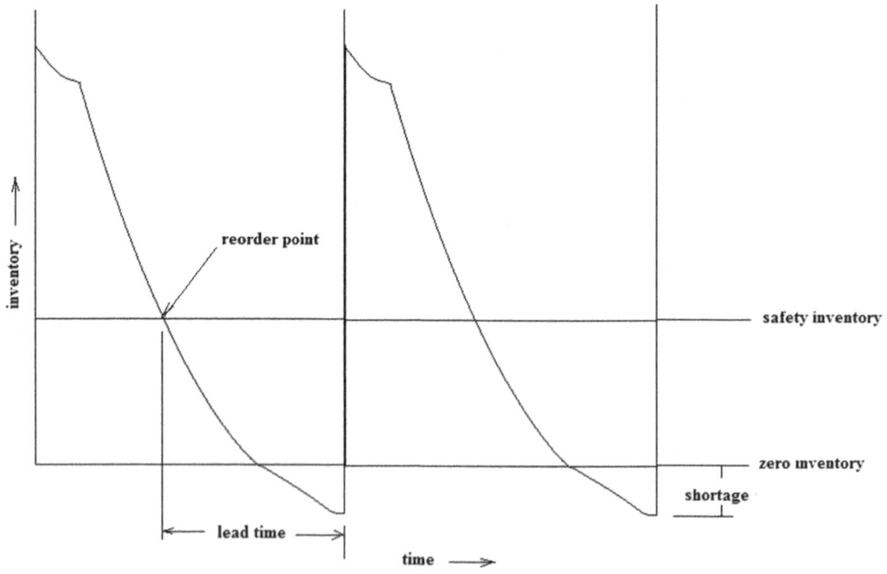

Fig. 2 Variation of inventory with respect to time

the boundary conditions. Proportionality constant can also be multiplied in case it is required. In the current work, it has been considered unity.

With this data base the inventory variation according to the demand and perishability effect, graph is plotted as in Fig. 1. In Fig. 1 the inventory variation is divided in three regions which are: t_0 to t_d, t_d to t_1 and t_1 to T. Inventory for all three regions is formulated first and then total cost per cycle is found out.

The mathematical formulation of the problem is as under.

C. *Mathematical formulation.*

Let us define inventory level I(t) at any time t as follows:

$$I(t) = \begin{cases} I_1(t), & T_0 \le t \le t_d \\ I_2(t), & t_d \le t \le t_1 \\ I_3(t), & t_1 \le t \le T \end{cases} \tag{1}$$

where, $I(T_0) = I_{max}$ and $I(t_1) = 0$.

Case (i) $T_0 \le t \le t_d$: The items start perishing after time t_d. Before that, we assume that the product is fresh. Therefore, the rate of change of inventory will depend on the demand.

$$\frac{dI_1(t)}{dt} = -f(t),\ I_1(T_0) = I_{max} \tag{2}$$

Thus, the inventory $I_1(t)$ is given by.

$$I_1(t) = I_{max} - \int_{T_0}^{t} f(u)du,\ T_0 \le t \le t_d \tag{3}$$

Observe that, the inventory level is decreasing with time.

Present value of the inventory carried under inflation during $T_0 \le t \le t_d$ is given by.

$$\begin{aligned} C_h \int_{T_0}^{t_d} e^{-Rt} I_1(t)dt &= C_h \int_{T_0}^{t_d} e^{-Rt} \left(I_{\max} - \int_{T_0}^{t} f(u)du \right) dt \\ &= C_h \left\{ \frac{I_{max}}{R}(e^{-RT_0} - e^{-Rt_d}) - \int_{T_0}^{t_d} \frac{e^{-Ru} - e^{-Rt_d}}{R} f(u)du \right\} \end{aligned} \tag{4}$$

Case (ii) $t_d \le t \le t_1$: During this period, perishability also comes into effect with demand. Therefore, the rate of change of inventory is given by.

$$\frac{dI_2(t)}{dt} + \theta\, I_2(t) = -f(t),\; I_2(t_1) = 0 \tag{5}$$

Thus, the inventory $I_2(t)$ is given by.

$$I_2(t) = \int_t^{t_1} e^{\theta(u-t)} f(u)du, \quad t_d \le t \le t_1 \tag{6}$$

Present value of inventory carried under inflation during $t_d \le t \le t_1$ is given by.

$$\int_{t_d}^{t_1} e^{-Rt} I_2(t)dt = C_h \int_{td}^{t_1} \left(e^{-(\theta+R)\,t} \int_t^{t_1} e^{\theta\, u} f(u)du \right) dt$$

$$= \frac{C_h}{\theta + R} \int_{t_d}^{t_1} \left(e^{(u-t_d)\theta - Rt_d} - e^{-Ru}\right) f(u)du \tag{7}$$

Also, the present value of deteriorated items is given by.

$$C_d \frac{\theta}{\theta + R} \int_{t_d}^{t_1} \left(e^{(u-t_d)\theta - Rt_d} - e^{-Ru}\right) f(u)du \tag{8}$$

Considering the continuity of I(t) at t_d, i.e. $I_1(t_d) = I_2(t_d)$, we get.

$$I_{\max} = \int_{T_0}^{t_d} f(u)du + \int_{t_d}^{t_1} \left(e^{(u-t_d)\theta}\right) f(u)du \tag{9}$$

Case (iii) $t_1 \le t \le T$: During this period, inventory goes negative and the unsatisfied demand is backlogged partially with the rate $\exp(-\alpha(T-t))$. Therefore, the rate of change of inventory is given by.

$$\frac{dI_3(t)}{dt} = -e^{-\alpha(T-t)} f(t), \quad I_3(t_1) = 0 \tag{10}$$

The amount of backlogged inventory I_3(t) is given by.

$$I_3(t) = \int_{t_1}^{t} e^{-\alpha(T-u)} f(u)du, \quad t_1 \le t \le T \tag{11}$$

Therefore the amount of shortage inventory is inventory backlogged.

The present value of backlogging cost during $t_1 \le t \le T$ is given by.

$$C_s \int_{t_1}^{T} e^{-Rt} I_3(t)dt = C_s \int_{t_1}^{T} e^{-Rt} \left(\int_{t_1}^{t} e^{-\alpha(T-u)} f(u)du \right) dt$$

$$= C_s \int_{t_1}^{T} e^{-\alpha(T-u)} \left(\frac{e^{-Ru} - e^{-RT}}{R} \right) f(u)du \tag{12}$$

The amount of lost sales is the difference of the demand arising and the demand backlogged during $t_1 \le t \le t_2$. Mathematically,

$$Lost\ Sales = \int_{t_1}^{T} f(u)du - \int_{t_1}^{T} e^{-\alpha(T-u)} f(u)du = \int_{t_1}^{T} \left(1 - e^{-\alpha(T-u)}\right) f(u)du \tag{13}$$

Present value of lost sales $= C_o \int_{t_1}^{T} e^{-Ru}\left(1 - e^{-\alpha(T-u)}\right) f(u)du.$

Lot Size: First replenishment lot size $Q_1 = I_{max} + \int_{t_1}^{T} e^{-\alpha(T-u)} f(u)du$. Second, third, …, mth lot size is.

$$Q_m = (I_{max} + S)_m = \int_{(m-1)T}^{t_{dm}} f(u)du + \int_{t_{dm}}^{(m+k-1)T} \left(e^{(u-t_{dm})\theta}\right) f(u)du$$

$$+ \int_{(m+k-1)T}^{mT} (T-u) e^{-\alpha(T-u)} f(u)du$$

$$= \left[\int_{T_0}^{t_d} f(u)du + \int_{t_d}^{kT} \left(e^{(u-t_d)\theta} - 1\right) f(u)du + \int_{kT}^{T} e^{-\alpha(T-u)} f(u)du \right] = Q_1 \tag{14}$$

Since $f(t)$ is periodic with period T and $t_{dm} = t_d + (m\text{-}1)T$.

Lead Time: The lead time $T_l = T - t_r$ where $t_d < t_r < t_1$. Total crashing cost depends on the inventory maximum and safety inventory I_s during $t_r < t < T$. Higher the safety inventory, lower the value of t_r and hence lower the crashing cost. As the point of zero inventory ($t = t_1$) moves towards replenishment point ($t = T$), the lead time decreases and crashing cost increases.

Since reorder point depends on the duration of shortages in the cycle, we here define reorder point t_r as a fraction of t_1:

$$t_r = p \times t_1 = \frac{p \times k \times H}{m}, \quad 0 \le p \le 1$$

Safety inventory I_s at time t_r is given by

$$I_s = \int_{t_r}^{t_1} e^{\theta(u-t_r)} f(u)du \tag{15}$$

Present value of crashing cost is.

$$C_c(I_{\max} - I_s) = C_c \left\{ \int_{T_0}^{t_d} f(u)du + \int_{t_d}^{t_1} \left(e^{(u-t_d)\theta}\right) f(u)du - \int_{t_r}^{t_1} e^{\theta(u-t)} f(u)du \right\} \tag{16}$$

Total cost function for one cycle

Now, we have all necessary quantities to formulate the total inventory cost function for one cycle:

TRC = Lot size×(Additional replenishment cost paid per unit of ordered item) +Holding cost +Deterioration cost +Backlogging cost +Opportunity cost + Crashing cost +Fixed setup replenishment cost

$$= \left(C_h \left\{ \frac{I_{max}}{R}\left(e^{-RT_0} - e^{-Rt_d}\right) - \int_{T_0}^{t_d} \frac{e^{-Ru} - e^{-Rt_d}}{R} f(u)du \right\} + \frac{C_h}{\theta + R} \int_{t_d}^{t_1} \left(e^{(u-t_d)\theta - Rt_d} - e^{-Ru}\right) f(u)du \right)$$
$$+ C_d \frac{\theta}{\theta + R} \int_{t_d}^{t_1} \left(e^{(u-t_d)\theta - Rt_d} - e^{-Ru}\right) f(u)du + C_s \int_{t_1}^{T} e^{-\alpha(T-u)} \left(\frac{e^{-Ru} - e^{-RT}}{R}\right) f(u)du$$
$$+ C_q e^{-RT} \left(I_{\max} + \int_{t_1}^{T} e^{-\alpha(T-u)} f(u)du \right) + C_o \int_{t_1}^{T} e^{-Ru}\left(1 - e^{-\alpha(T-u)}\right) f(u)du$$
$$+ C_c e^{-RT}(I_{\max} - I_s) + Ae^{-RT}$$

Rearranging the terms, we obtain the present worth of the total cost during cycle $[T_0, T]$ is given by:

$$TRC = \left(C_h \frac{(e^{-RT_0} - e^{-Rt_d})}{R} + c_q e^{-RT} \right) \left\{ \int_{T_0}^{t_d} f(u)du + \int_{t_d}^{t_1} \left(e^{(u-t_d)\theta} - 1\right) f(u)du \right\}$$
$$- C_h \int_{T_0}^{t_d} \frac{e^{-Ru} - e^{-Rt_d}}{R} f(u)du + \left(\frac{C_h + C_d\theta}{\theta + R} \right) \int_{t_d}^{t_1} \left(e^{\theta u - (\theta + R)t_d} - e^{-Ru}\right) f(u)du$$
$$+ C_c e^{-R(T-t_r)} \left(\int_{t_r}^{t_1} \left(e^{(u-t_r)\theta} - 1\right) f(u)du \right) + Ae^{-RT}$$

$$+\int_{t_1}^{T}\left\{C_o e^{-Ru}+e^{-\alpha(T-u)}\left(\frac{e^{-Ru}-e^{-RT}}{R}C_s-C_o e^{-Ru}+C_q e^{-RT}-C_c(T-u)e^{-R(T-t_r)}\right)\right\}f(u)du$$

Writing $T_0 = 0$, $T = H/m$ and $t_1 = kH/m = kT$, $(0 < k < 1)$,

$$TRC(m,k)=\left(C_h\frac{(1-e^{-Rt_d})}{R}+c_q e^{-RH/m}\right)\left\{\int_0^{t_d}f(u)du+\int_{t_d}^{kH/m}\left(e^{(u-t_d)\theta}-1\right)f(u)du\right\}$$

$$-C_h\int_0^{t_d}\frac{e^{-Ru}-e^{-Rt_d}}{R}f(u)du+\left(\frac{C_h+C_d\theta}{\theta+R}\right)\int_{t_d}^{kH/m}\left(e^{\theta u-(\theta+R)t_d}-e^{-Ru}\right)f(u)du$$

$$+C_c e^{-R\left(\frac{H}{m}-t_r\right)}\left(\int_{t_r}^{kH/m}\left(e^{(u-t_r)\theta}-1\right)f(u)du\right)+Ae^{-RH/m}$$

$$+\int_{kH/m}^{H/m}\left\{C_o^{-Ru}+e^{-\alpha\left(\frac{H}{m}-u\right)}\left(\frac{e^{-Ru}-e^{-RH/m}}{R}C_s-C_o e^{-Ru}+C_q e^{-RH/m}-C_c(T-u)e^{-R(\frac{H}{m}-t_r)}\right)\right\}f(u)du$$

The present value of the total cost of system over a planning horizon H is.

$$TC(m,k)=\sum_{j=0}^{m-1}TRC(m,k)\,e^{-jTR}=TRC(m,k)\left(\frac{1-e^{-RH}}{1-e^{-RH/m}}\right),\ \text{where}T=H/m.$$

The present value of total cost $TC(m, k)$ is a function of discrete variable m and continuous variable k, the shortage parameter.

D. *Special Cases.*

(i) **Model when technology is implemented to the full extent** ($T \to t_1$ or $k \to 1$).

When technology is implemented to the full extent, the time of replenishment tends to the time where physical inventory becomes zero. That indicates that the system is updated in such fashion that the replenishment lot is available as the inventory of earlier cycle reaches to zero and there is no shortage in the system. In this case, total cost for one cycle reduces to

$$TRC(m)=\left(C_h\frac{(1-e^{-Rt_d})}{R}+C_q e^{-RH/m}\right)\left\{\int_0^{t_d}f(u)du+\int_{t_d}^{H/m}\left(e^{(u-t_d)\theta}-1\right)f(u)du\right\}$$

$$-C_h\int_0^{t_d}\frac{e^{-Ru}-e^{Rt_d}}{R}f(u)du+\left(\frac{C_h+C_d\theta}{\theta+R}\right)\int_{t_d}^{H/m}\left(e^{\theta u-(\theta+R)t_d}-e^{-Ru}\right)f(u)du$$

$$+C_c e^{-R\left(\frac{H}{m}-t_r\right)}\left(\int_{t_r}^{H/m}\left(e^{(u-t_r)\theta}-1\right)f(u)du\right)+Ae^{-RH/m}$$

(ii) **Model with complete backlogging**
If we set the parameter $\alpha = 0$, we revert to the model with complete backlogging of the unsatisfied demand. For this special case, the total relevant cost for one cycle is given by

$$TRC(m,k) = \left(C_h\frac{(1-e^{-Rt_d})}{R} + c_q e^{-RH/m}\right)\left\{\int_0^{t_d} f(u)du + \int_{t_d}^{kH/m}\left(e^{(u-t_d)\theta} - 1\right)f(u)du\right\}$$
$$- C_h\int_0^{t_d}\frac{e^{-Ru} - e^{Rt_d}}{R}f(u)du + \left(\frac{C_h + C_d\theta}{\theta + R}\right)\int_{t_d}^{kH/m}\left(e^{\theta u-(\theta+R)t_d} - e^{-Ru}\right)f(u)du$$
$$+ C_c e^{-R(\frac{H}{m}-t_r)}\left(\int_{t_r}^{kH/m}\left(e^{(u-t_r)\theta} - 1\right)f(u)du\right) + Ae^{-RH/m}$$
$$+ \int_{kH/m}^{H/m}\left\{\left(\frac{e^{-Ru} - e^{-RH/m}}{R}C_s + C_q e^{-RH/m} - C_c(T-u)e^{-R(\frac{H}{m}-t_r)}\right)\right\}f(u)du.$$

In this model, the decision maker has some flexibility to choose a suitable value of α ranging from 0 to almost any desired rate of partial backlogging, which fits better in realistic situations.

(iii) **Model with instantaneous deterioration and complete backlogging**
If $t_d \to t_0$ (i.e. $\theta \to 0$) and $\alpha = 0$ in the proposed model, we can obtain the corresponding inventory model for the instantaneous deterioration. In this case, the total relevant cost for one cycle is given by

$$TRC(m,k) = \left(\frac{C_h}{R}\right)\int_0^{kH/m}\left(1 - e^{-Ru}\right)f(u)du + Ae^{-RH/m}$$
$$+ \int_{kH/m}^{H/m}\left\{\left(\frac{e^{-Ru} - e^{-RH/m}}{R}C_s + C_q e^{-RH/m} - C_c(T-u)e^{-R(\frac{H}{m}-t_r)}\right)\right\}f(u)du.$$

(iv) **Model with instantaneous deterioration and full RFID implementation**
If $t_d \to t_0$ (i.e. $\theta \to 0$) and $\alpha \to \infty$ in the proposed model, we can obtain the corresponding inventory model for the instantaneous deterioration without shortages. In this case, the total relevant cost for one cycle is given by

$$TRC(m,k) = \left(\frac{C_h}{R}\right)\int_{t_d}^{kH/m}(1 - e^{-Ru})f(u)du + Ae^{-RH/m}$$

$$+ C_o \int_{kH/m}^{H/m} e^{-Ru} f(u)du.$$

5 Result and Discussion

For analysis, the demand rate function is considered as $f(t) = \exp\left(-\frac{t^2}{2}\right)$, a log-concave function of time, so that demand at any time t is $F(t) = \sqrt{\frac{\pi}{2}} erf\left(\frac{t}{\sqrt{2}}\right)$. The graph of the above function is drawn for one cycle using MATLAB programming which is as shown in the Fig. 3.

It is to be understood that, the consideration of this demand rate function for this research work, suits the realistic situation that persists at the retail outlet for the perishable product. The demand rate decreasing concavely which is true for perishable product which is replenished every day and its shelf life decreases day by day. And as the shelf life of product ends i.e. the product approaches to obsolescence, the demand for the product also approaches to zero. The cumulative demand is the demand of the product till that time.

Hence the inventory level at any time t is given by

$$I(t) = \begin{cases} I_{\max} - \sqrt{\frac{\pi}{2}} erf\left(\frac{t}{\sqrt{2}}\right), & 0 \le t \le t_d \\ \sqrt{\frac{\pi}{2}} \exp\left(\frac{\theta^2}{2} - \theta\, t\right)\left\{erf\left(\frac{t_1 - \theta}{\sqrt{2}}\right) + erf\left(\frac{\theta - t}{\sqrt{2}}\right)\right\}, & t_d \le t \le t_1 \\ \sqrt{\frac{\pi}{2}} \exp\left(\frac{\alpha^2}{2} + 3\alpha\right)\left\{erf\left(\frac{\alpha + t_1}{\sqrt{2}}\right) - erf\left(\frac{\alpha + t}{\sqrt{2}}\right)\right\}, & t_1 \le t \le t_2 \end{cases} \tag{17}$$

Therefore, Maximum initial inventory,

$$I_{\max} = \int_{T_0}^{t_d} f(u)du + \int_{t_d}^{t_1} \left(e^{(u - t_d)\theta}\right) f(u)du \tag{18}$$

From Eq. (2.11), the maximum shortage inventory ($t = T$) can be obtained for as

$$S = \int_{kT}^{T} e^{-\alpha(T-u)} f(u)du = \sqrt{\frac{\pi}{2}} \exp\left(\frac{\alpha^2}{2} - \alpha T\right)\left\{erf\left(\frac{\alpha - T}{\sqrt{2}}\right) - erf\left(\frac{\alpha - kT}{\sqrt{2}}\right)\right\} \tag{19}$$

Using Eq. (2.15), we get safety inventory as

$$I_s = \int_{t_r}^{t_1} e^{\theta(u-t_r)} f(u)du = \sqrt{\frac{\pi}{2}} \exp\left(\frac{\theta^2}{2} - \theta\, t_r\right)\left\{erf\left(\frac{t_1 - \theta}{\sqrt{2}}\right) + erf\left(\frac{\theta - t_r}{\sqrt{2}}\right)\right\} \tag{20}$$

and order quantity is given by $Q = I_{max} + S$.

With this formulation the inventory variation function was programmed using Matlab and the graphs obtained are shown in the Fig. 4. It is seen the inventory variation graph (tri-colour) resembles the theoretical graph shown in the Fig. 2. This justifies the accuracy and correctness of our mathematical frame work and model formulation.

In order to validate the model, we considered various cost of supply chain of a perishable product as considered by early researchers [14] as $C_h = 0.5$; $C_d = 5$ and $C_o = 0.5$. Furthermore, to suit our framework, we assumed the values for other parameters as: $\theta = 0.5$; $\alpha = 0.8$; $t_d = 0.67$; $t_1 = 1.5$, $A = 10$; $Cc = 1.5$; $C_q = 0.2$ and $C_s = 3.5$ with the notations used in the framework. The variables θ and α are the decision variables and each of them influences the managerial decision. This clearly indicates that the model can be applied to all range of perishable products equally ranging from highly perishable to the least perishable product, in addition to the varying decision level of backlogging. The simulation was carried out using MATLAB platform and is presented as under (Table 1).

Effect of decrease in period of shortage (i.e. $k \to 1$) on the Total Cost

In the Fig. 4, the effect of varying 'k' on different costs of perishable product supply chain is shown. It is observed that the holding cost is increasing as the value of k increase from lowest value to the highest value i.e. from point of zero inventories to the point of replenishment. This happens because, as t_1 reaches to T, physical inventory lasts for longer time in the cycle leading to the raise in holding cost. For the same cost it is also inferred that the rate of increase in holding cost reduces as the

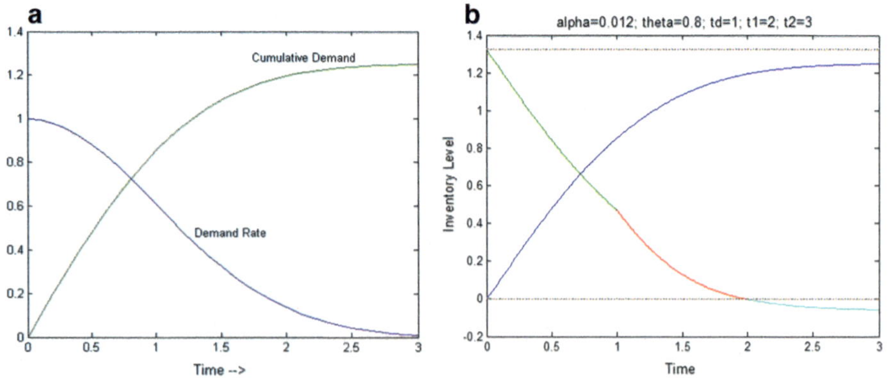

Fig. 3 Graphical presentation of **a** Demand function and demand rate **b** Inventory level and demand

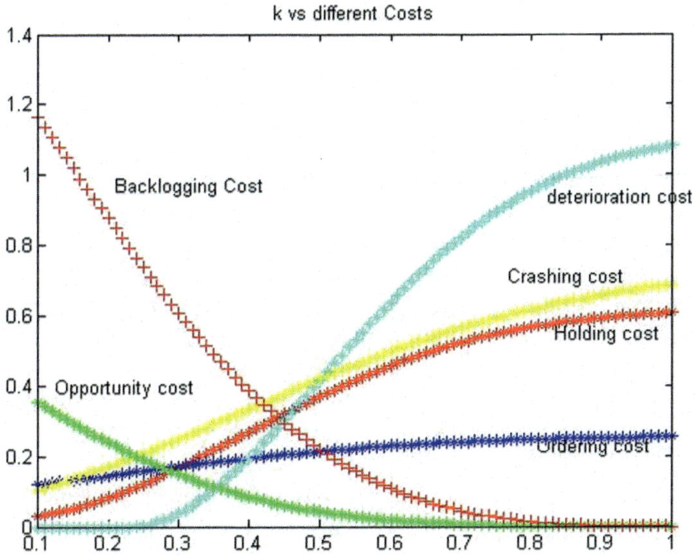

Fig. 4 A closer look in the change in different costs w.r.to k (Theta = 0.5)

Table 1 Effect of *k* on various costs and order quantity (Theta = 0.5, alpha = 0.8, m = 10, H = 30, p = 0.6)

k	Q	HC	QC	DC	CC	OC	BC	TC
0.10	30.61	0.375	0.945	0.16	0.104	0.355	1.164	17.3
0.15	24.9	0.348	0.769	0.06	0.136	0.295	1.019	14.643
0.20	20.29	0.334	0.626	0.008	0.171	0.24	0.876	12.577
0.25	16.66	0.332	0.514	0.003	0.208	0.191	0.738	11.084
0.30	13.9	0.342	0.429	0.038	0.249	0.149	0.609	10.121
0.35	11.87	0.36	0.366	0.105	0.291	0.113	0.491	9.624
0.40	**10.42**	**0.385**	**0.322**	**0.196**	**0.334**	**0.084**	**0.385**	**9.512**
0.45	9.44	0.414	0.291	0.303	0.376	0.06	0.294	9.701
0.50	8.8	0.445	0.272	0.417	0.418	0.042	0.218	10.106
0.55	8.42	0.475	0.26	0.53	0.458	0.029	0.157	10.649
0.60	8.21	0.505	0.253	0.638	0.496	0.019	0.108	11.262
0.65	8.11	0.531	0.25	0.736	0.531	0.012	0.071	11.891
0.70	**8.09**	0.555	0.25	0.822	0.563	0.007	0.045	12.497
0.75	8.1	0.574	0.25	0.894	0.592	0.004	0.026	13.051
0.80	8.13	0.591	0.251	0.953	0.617	0.002	0.014	13.538
0.85	8.16	0.603	0.252	1.000	0.639	0.001	0.006	13.953
0.90	8.2	0.613	0.253	1.036	0.658	0.000	0.002	14.295
0.95	8.23	0.621	0.254	1.063	0.674	0.000	0.000	14.571
1.00	8.25	0.626	0.255	1.083	0.687	0.000	0.000	14.788

value of k increases. This is true due to the reduced effect of inflation as time span reduced from t_1 to T.

Next, as k increases from lower value to the higher value, the I_{max} is seen increasing. This leads to increase in deteriorated items proportionately. As deteriorated items increases reorder quantity and hence the ordering cost increases. Backlogging cost is seen decreasing fast up to certain value of k and then remains constant. This is because in the beginning of cycle more and more orders will be backlogged as a result of exponentially decreasing backlogging than at the end of cycle and the rate of backlogging decreases as k approaches 1, i.e. t_1 tends to T. In other words backlogging is more effective when the shortage period is more.

Opportunity cost is seen simply decreasing as the effect of reduced lost sales. As inventory lasts for longer time, shortages are reduced and the demand arise is satisfied from the available inventory leading to the reduced opportunity cost.

Effect of perishability on Total Cost

In Fig. 4, the deterioration cost is observed to be increasing with the increase in value of 'k' (or t_1 is tending to t_2). This is because longer the time inventory present in cycle, more items will get deteriorated and the deterioration cost will be higher. Same thing happens here, as inventory lasts for longer time as t_1 reaches to t_2, more and more inventory gets deteriorated and cost increases. Even though it is true but finally deterioration cost will be affected by value of theta as explained mathematically in Eq. (8). In Fig. 5b, the cumulative effect of all the costs involved on total cost is seen that as k increases the total cost decreases to certain minimum value of k and then starts increasing as value of k increases. The deterioration rate has found pronounced effect on total cost. It can be seen from Fig. 5a, as theta increases the total cost also seen increasing at higher rate, however for lower value of theta (Fig. 5b) total cost is decreasing. For reduced value of theta, minima is seen shifting to right side mean towards higher value of k.

Crashing cost and reorder point

Figure 6 illustrates effect of k on crashing cost. When reorder point reaches too close to the cycle time (i.e. $k \to 1$), crashing cost increases to a maximum. Crashing cost account for the reduction of lead time due to incorporation of better technology and facilities, which reduces the amount of safety inventory. Better the technology incorporated, higher would be the crashing cost and lower would be the lead time. For higher values of p, reorder point is moving towards the cycle time for fixed value of k leading to rise in the crashing cost and consequently reduction in lead time.

Effect of number of replenishments on the Total Cost

Figure 7 explains the effect of number of replenishments m on the total cost. It is observed that as the value of m increases the total cost is increasing drastically. This is because of increased fixed setup replenishment cost. It is also observed that for higher value of m the point of minima of total cost is shifting to right that is towards higher value of k. This may due to the cumulative effect of reduction in backlogging and opportunity cost dominates the effect or raise in ordering cost.

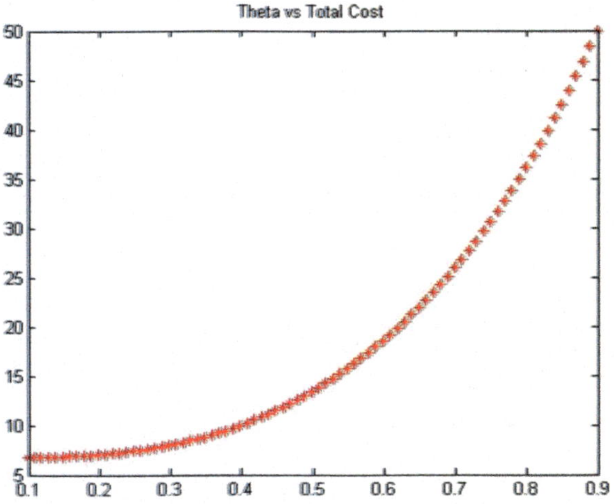

(a)

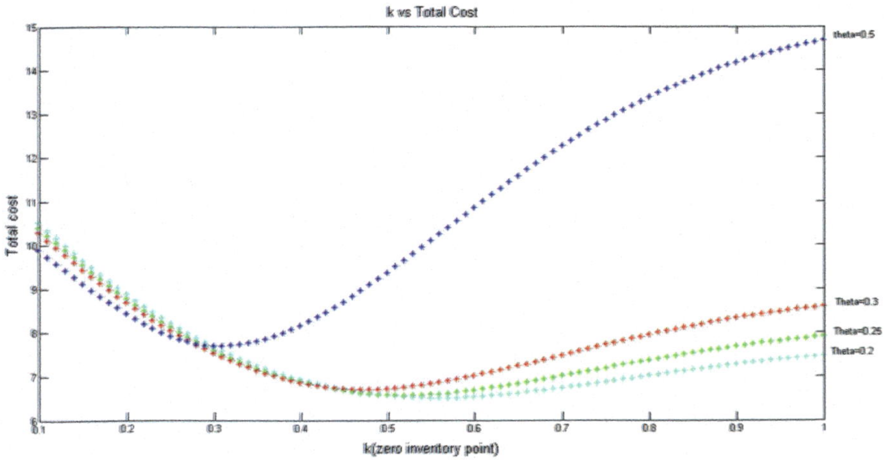

(b)

Fig. 5 Effect of **a** Perishability on total cost **b** k on various costs with different values of perishability parameter theta

Major paper contribution

The major contributions of this research work are summarized below:

- An exhaustive literature review of modelling-based and supporting research papers was conducted. And the surveyed literature is analysed chronologically according to the stages of development of supply chain of perishable product which reflects the research gap in scm of perishable products.
- Development of RFID-interfaced perishable product supply chain diagram to understand the potential benefits of RFID in a supply chain.

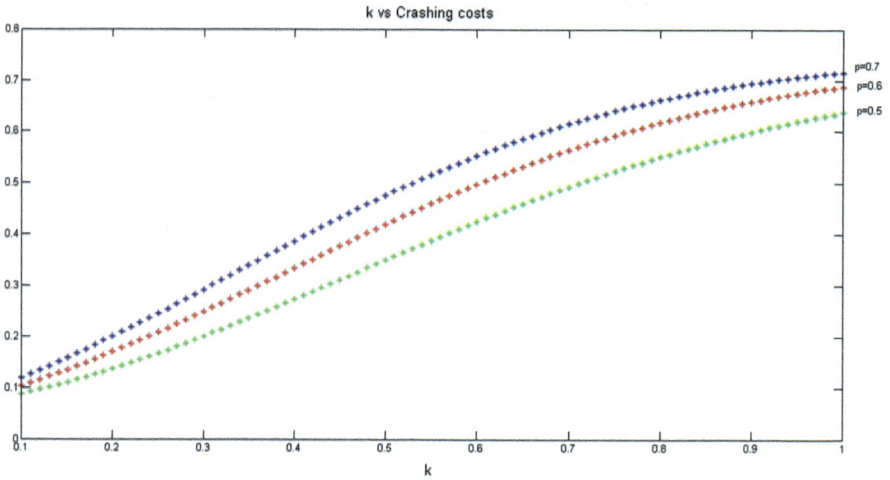

Fig. 6 Effect of reorder point on Total Cost

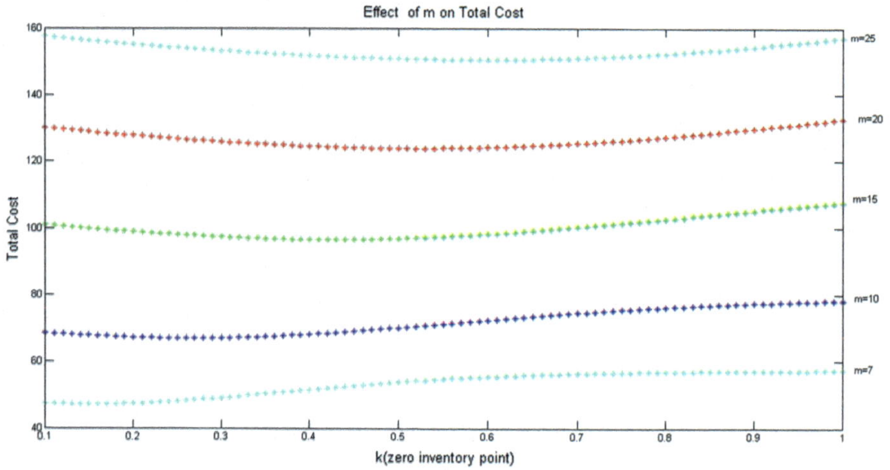

Fig. 7 Effect of the number of replenishments (m) and point of zero inventory (kT) on the Total Cost

- Development of a mathematical model for perishable product supply chain considering the most realistic conditions prevailing at any node of supply chain. The developed model was used for calculating the order quantity, shortage, maximum cycle inventory and the total cost for all members of supply chain. The peculiarity of this model is that it can be used to calculate the total cost for many sets of conditions as explained above.
- Cost effectiveness of RFID in perishable product supply chain indicates that the incorporation of newer technologies reduces the overall cost of supply chain. This

widens the scope of making better decisions and hence improving the efficiency and service level of supply chain.

- The capability of dealing with the effect of stock outs and backlogging in perishable product supply chain are explored in detail.
- The utilization of RFID as a newer and promising technology in supply chain is promoted to form an information hub that will facilitate better decisions in supply chain and will be helpful in reducing the bullwhip effect.
- Apart from other variables, the effect of inflation of money on total cost considered in this research paper adds its own importance in the field of supply chain management of perishable products.

5.1 Future Scope and Limitations of the Research Work

Although the current study presents an approach to understand the potential benefits of RFID for improving all major problems into a single framework, the enormity of the task itself renders the framework with many shortcomings that leaves a lot of scope for future research. The existing format as well as the capabilities of the framework can be significantly improved if the current research work is appropriately extended. The following future research directions have been proposed:

- This study considers a single product for modelling and calculations; however, the findings can be extended for multiple products consideration.
- The proposed framework can be extended to consider gradual replenishment and the results may be verified using the findings of the current study wherein instantaneous replenishment has been considered.
- The proposed work can also be extended to consider different demand functions like increasing or linear log concave and the results may be compared with those of the current study.

This work can be extended to consider the actual cost of RFID infrastructure and cost analysis may be conducted.

6 Conclusion

This model is proposed to see the cumulative effect of money inflation rate and technological advancements on perishable product supply chain. This work adds new value in this research area as results achieved are positive in terms of total cost of supply chain. The model may provide a road map and an opportunity for decision making to those planning to incorporate newer information technologies in their organization. Also the situations considered while developing the framework widens the scope of implementing real facts that pertains at different nodes of supply chain. The strength of this model lies in consideration of varying lead time, perishability effect, log-concave demand function, effect of information technology, shortages and also the rate of inflation all together.

For no inflation, we obtain the model studied by Badole [7] in his thesis. Further, if crashing cost is not considered, the model studied by Skouri et al. [33] follows, which further reduces to the model studied by Bhunia and Maiti [4] where backlogging rate is taken zero. Also these authors have considered the demand function *f(t)* to be linear only.

There is tremendous scope for research in this field of supply chain management as research on incorporation of technologies and its cost effectiveness is in the preliminary stage. This research can be continued considering probabilistic demand function which would be more realistic. Also, the results can be improved and optimized by using different evolutionary tools.

Acknowledgements The authors are thankful to the referee for his valuable comments and suggestions that were helpful in improving the quality of the paper.

References

1. Badole, C.M., Rathore, A.P.S., Jain, R., Nepal, B., Agarwal, R.: RFID and Perishable product supply chain with shortages. In: Proceedings of ICRTES'13, pp. 397–402. published by Elsevier Science (2013)
2. Badole, C.M., Jain, R., Rathore, A.P.S., Nepal, B.: Research and opportunities in supply chain modeling: a review. Int. J. Supply Chain Manag. **3**, 63–86 (2012)
3. Benkherouf, L.: On an inventory model with deteriorating items and decreasing time-varying demand and shortages. Euro. J. Oper. Res. **86**, 293–299 (1995)
4. Bhunia, A.K., Maiti, M.: An inventory model of deteriorating items with lot-size dependent replenishment cost and a linear trend in demand. Appl. Math. Model. **23**, 301–308 (1999)
5. Bose, S., Goswami, A., Chaudhuri, K.S.: An EOQ model for deteriorating items with linear time-dependent demand rate and shortages under inflation and time discounting. J. Oper. Res. Soc. **46**, 771–782 (1995)
6. Buzacott, J.A.: Economic order quantities with inflation. Oper. Res. Q. **26**, 553–558 (1975)
7. C. M. Badole, *Development of an Integrated Framework of RFID Based Supply Chain of Perishable Products*, Research Thesis, MNIT Jaipur, (2016), 1–156.
8. Chandra, M.J., Bahner, M.L.: The effects of inflation and the time value of money on some inventory systems. Int. J. Prod. Res. **23**, 723–730 (1984)
9. Chen, J.M.: An inventory model for deteriorating items with time-proportional demand and shortages under inflation and time discounting. Int. J. Prod. Econ. **55**, 21–30 (1998)
10. Chew, Ek P., Lee, C., Liu, R.: Joint inventory allocation and pricing decisions for perishable products. Int. J. Prod. Econ. **120**, 139–150 (2009)
11. Cohen, M.A., Prastacos, G.P.: Critical number ordering policy for LIFO perishable inventory systems. Comput. Oper. Res. **8**(3), 185–195 (1981)
12. Covert, R.P., Philip, G.C.: An EOQ model for items with Weibull distribution deterioration. Am. Inst. Ind. Eng. Trans. **5**, 323–326 (1973)
13. Datta, T.K., Pal, A.K.: Order level inventory system with power demand pattern for items with variables rate of deterioration. Indian J. Pure Appl. Math. **19**, 1043–1053 (1988)
14. Dye, C-Y., Hsieh, T.-P., Ouyang, L.-Y.: Determining optimal selling price and lot size with a varying rate of deterioration and exponential partial backlogging. Eur. J. Oper. Res. **181** (2007), 668–678

15. Ghare, P.M., Shrader, G.F.: A Model for exponentially decaying inventories. J. Indus. Eng. **14**, 238–243 (1963)
16. Gigler, J.K., Hendrix, E.M.T., Heesen, R.A., Van den Hazelkamp, V.G.W., Meerdink, G.: On optimization of agri chains by dynamic programming. Eur. J. Oper. Res. **139**, 613–625 (2002)
17. Hariga, M.: Optimal EOQ models for deteriorating items with time-varying demand. J. Oper. Res. Soc. **47**, 1228–1246 (1996)
18. Hariga, M.A.: Economic analysis of dynamic inventory models with non-stationary costs and demand. Int. J. Prod. Econ. **36**, 255–266 (1994)
19. Higgins, L.N., airney, T.: RFID opportunities and risks. J. Corporate Account. Finance 51–57 (2006)
20. Hou, K.L.: An inventory model for deteriorating items with stock dependent consumption rate and shortages under inflation and time discounting. Euro. J. Oper. Res. **168**, 463–474 (2006)
21. Hsieh, T.P., Dye, C.Y.: Pricing and lot-sizing policies for deteriorating items with partial backlogging under inflation. Expert Syst. Appl. **37**, 7234–7242 (2010)
22. Jones, P., Colin, C.-H., David, H., Daphne, C.: The benefits, challenges and impacts of radio frequency identification technology (RFID) for retailers in the UK. Mark. Intell. Plan. **23**(4), 395–402 (2005)
23. G. S. Joseph, T. Vera and Jiang Z., Analytical model of adoption of item level RFID in a two-echelon supply chain with shelf-space and price-dependent demand, *Decision Support Systems*, **51** (2011), 833–841.
24. Lambert, D.M., Cooper, M.C.: Issues in supply chain management. Ind. Mark. Manage. **28**, 65–83 (2000)
25. Lian, Z., Liu, L.: Continuous review perishable inventory systems: models and heuristics. IIE Trans. **33**, 809–822 (2001)
26. Min, H., Zhou, G.: Supply chain modeling: past, present and future. Comput. Ind. Eng. **43**, 239–249 (2002)
27. Pan, J.C.H., Yang, J.S.: A study of an integrated inventory with controllable lead time. Int. J. Prod. Res. **40**(5), 1263–1273 (2002)
28. Rajurkar, S.W., Jain, R.: Food supply chain management: review, classification and analysis of literature. Int. J. Integr. Supply Manag. **6**(1), 33–72 (2010)
29. Ranky, P.G.: An introduction to radio frequency identification (RFID) methods and solution. Assem. Autom. **26**(1), 28–33 (2006)
30. Sachan, R.S.: On (T, Si) policy inventory model for deteriorating items with time proportional demand. J. Oper. Res. Soc. **35**, 1013–1019 (1984)
31. Sergio, T., Sergio, C.: Simulation in the supply chain context: a survey. Comput. Ind. **53**, 3–16 (2004)
32. Shah, Y.K.: An order-level lot-size inventory for deteriorating items. Am. Inst. Eng. Trans. **9**, 108–112 (1977)
33. Skouri, K., Papachristos, S.: A continuous review inventory model, with deteriorating items, time-varying demand, linear replenishment cost, partially time-varying backlogging. Appl. Math. Model. **26**, 603–617 (2002)
34. Twist, D.C.: The impact of radio frequency identification on supply chain facilities. J. Facil. Manag. **3**(3), 226–239 (2005)
35. Ustundag, A., Kılınc, M.S., Cevikcan, E.: Fuzzy rule-based system for the economic analysis of RFID investments. Expert Syst. Appl. **37**, 5300–5306 (2010)
36. Wang, X., Liu, L.: Coordination in a retailer-led supply chain through option Contract. Int. J. Prod. Econ. **110**, 115–127 (2007)
37. Wu, K.S., Ouyang, L.Y., Yang, C.T.: An optimal replenishment policy for non-instantaneous deteriorating items with stock dependent demand and partial backlogging. Int. J. Prod. Econ. **101**, 369–384 (2006)
38. Yang, H.L.: Two-warehouse partial backlogging inventory models for deteriorating items under inflation. Int. J. Prod. Econ. **103**, 362–370 (2006)
39. Zhou, W.: RFID and item-level information visibility. Eur. J. Oper. Res. **198**, 252–258 (2009)

Mexican Hat Wavelet Transform and Its Applications

Abhishek Singh, Aparna Rawat, and Nikhila Raghuthaman

Abstract In this chapter, we discuss a unique method to time-frequency analysis which gives a centralized way to represent discrete and continuous time-frequency. This serves as a straightforward way to include all possible (countable) discrete and continuous time scales in one model. We consider the Mexican hat wavelet which is one of the basic wavelet functions formulated by the second derivative of Gaussian function to define the Mexican hat wavelet transform (MHWT). Further, the theory of MHWT is implemented to obtain the Mexican hat wavelet Stieltjes transform (MHWST) of a bounded variation function. Some convenient properties of MHWST are also presented. Further, a standard method is introduced for representing functions of class $B(m, n)$. Besides, an integral transform is constructed with the help of the Fourier summation kernel. This construction results in a flexible way to present some conditions that are necessary and sufficient for a function of class $B(m, n)$ to be Mexican hat wavelet and MHWST.

Keywords Weierstrass transform · Wavelet transform · Mexican hat wavelet transform (MHWT) · Numerical methods for wavelets

Mathematics Subject Classification (2010) 46F12 · 42C40 · 44A15 · 65T60

1 Introduction

Wavelets are a new advancement in different fields of pure mathematics and applied sciences. The theory of wavelet analysis developed from the Fourier analysis of fractional calculus. The relationship between Fourier and wavelet transform has been used to investigate many properties of the wavelet transform like Parseval relation and reconstructional formula. Also, the L^2-theory of Fourier transform has been used to obtain inversion formula for the wavelet transform. In [16], the wavelet transform is represented as a double Fourier transform to obtain PaleyWienerSchwartz type

A. Singh (✉) · A. Rawat · N. Raghuthaman
Department of Mathematics and Statistics, Banasthali Vidyapith, Banasthali, India

J. Singh et al. (eds.), *Methods of Mathematical Modelling and Computation for Complex Systems*, Studies in Systems, Decision and Control 373,
https://doi.org/10.1007/978-3-030-77169-0_12

theorem for the wavelet transform. Using this representation many properties of the wavelet transform can be derived from those of the Fourier transform in $\mathbb{R}^2$.

The theory of wavelet transform in L^p-spaces $(1 \leq p < \infty)$ is formulated by exploiting the theory of classical and distributional Fourier and Hilbert transforms. For the past two decades, the wavelet transform is rising as an important mathematical tool and has contributed significantly to signal analysis. The primary reason is the representation of time-dependent functions in a time-frequency plane and also identifying frequency in the temporary or spatial domain. Hence, the wavelet transform acts as a time and frequency localization operator. In particular, wavelets can adjust on long and short time intervals to achieve low and high-frequency components which would help to enhance the analysis of signals with localized impulses and oscillations. In a way, wavelets have a window that naturally modifies to provide a relevant resolution. Wavelet transform uses a function called the mother wavelet. This function has zero mean as it distinctly decays in an oscillatory fashion. The performance of various physical and biological processes can be explained by using fractional-order models. Recently, through a new technique, the EEG signal can be modeled accurately, by using wavelet support vector machines and using wavelet kernel functions such as Mexican hat wavelet and Morlet wavelet. The parameters retrieved from modeling are then utilized for analysis and classification of signals through vector machines (SVM).

In signal analysis mostly the information regarding the signal is transmitted by irregular structures and temporary events. In the medical field, the edge feature fusion is a significant part of medical image processing. Wavelet transform is used to develop different techniques of image fusion. Lifting wavelet transform domain, which is a multiresolution analysis, permits the identification of fuse image features at different scales and orientations. It produces large coefficients near edges, thus revealing salient information. Also, it is used for the formation of the second generation wavelets.

In the medical field, the broadcast of medical data poses many chances of threat that can rigorously alter its integrity, authenticity, and confidentiality. Hence a medical watermarking scheme was required to dodge prompting attention and prevent an unintended recipient to reach [21]. This needs utmost care when embedding additional data is done because the additional data must not modify the image quality. Confidentiality, authentication, integrity, and availability are important security requirements with Electronic Patient Record (EPR) data exchange through open channels. All of that can be achieved by applying suitable watermarks. The general watermarking method needs to keep factors like imperceptibility, robustness, capacity, and security reasonably very high. The image watermarking methods are based on two domain methods: Spatial domain method and Transform domain method. The latter is established on the wavelet transform. The primary benefits of the wavelet transform domain for watermarking applications are multi-resolution representation, space-frequency localization, adaptability, multi-scale analysis, and linear complexity. Further, the wavelet transform also allows localized watermarking of the image.

New representations of the wavelet transform and its relation with the Hilbert transform, fractional integral operators, and Watson transform was exploited to derive certain boundedness results, approximation results, and some general Parseval formulae for the wavelet transform [11]. Some of the results are even extended to distributions. These relations enhance the concept of the wavelet transform to give a better way for additional research and applications. For instance, the Hilbert transform has applications in signal processing, aerofoil problems, high energy physics, dispersion relations, and others. Therefore, using the aforesaid relation the wavelet transform can also be applied to tackle all such problems. In [17], Pathak and Singh expanded the theory of Dziubanski and Hernandez and investigated wavelets of infraexponential decay whose Fourier transforms have compact support. Further, they developed the theory of wavelet transform on the ultradistribution theory of Beurling and Bjorck which involved wavelets of infraexponential decay.

Wavelets are generated by one single function, called mother wavelet. The wavelet transform at a particular translation and dilation represents how well the original signal scaled and translated by the mother wavelet. So, a mother wavelet can be visualized as a windowing function [8]. There is an inverse relationship between the wavelet scales and frequency, such that a smaller scale corresponds to a compressed wavelet, which is high in frequency, while larger scales correspond to a stretched wavelet, denoting lower frequency.

By translation and dilation of the mother wavelet $\psi \in L^2(\mathbb{R})$, the wavelet $\psi_{b,a}(t)$ is given by

$$\psi_{b,a}(t) = (\sqrt{a})^{-1}\psi\left(\frac{t-b}{a}\right), \quad b, t \in \mathbb{R}, \quad a > 0. \tag{1}$$

The continuous wavelet transform of a square integrable function f, with respect to $\psi_{b,a}(t)$, is given by

$$W(b,a) = \int_{\mathbb{R}} f(t)\overline{\psi_{b,a}(t)}\,dt \tag{2}$$

and its inversion is as follows

$$\frac{2}{C_\psi}\int_0^\infty \left[\int_{-\infty}^{\infty} (\sqrt{a})^{-1} W(b,a)\psi\left(\frac{z-b}{a}\right) db\right]\frac{da}{a^2} = f(z), \quad z \in \mathbb{R}, \tag{3}$$

where

$$\frac{C_\psi}{2} = \int_0^\infty \frac{|\hat{\psi}(u)|^2}{|u|}du = \int_0^\infty \frac{|\hat{\psi}(-u)|^2}{|u|}du < \infty \qquad [2,\ \text{p. } 64].$$

Recently among many authors, the researches carried out by Pathak [11, 13–16] explored possible applications of distribution spaces in the study of wavelet and integral transforms. The study of the wavelet transform of distribution, ultra-distributions, and tempered distributions have expanded the applications of the wavelet transform. Pilipovi et al. founded some results of wavelet transform on Gelfand-Shilov type spaces and dual spaces and described the global and local behaviour of the wavelet transform of ultra-differentiable functions. These theories help to provide applications in several physical and engineering problems [5, 12, 23].

Mexican hat wavelet (Fig. 1), is considered as an even wavelet and is obtained by differentiating the Gaussian function twice:

$$\psi(t) = exp\left(\frac{-t^2}{2}\right)(1-t^2) = -\frac{d^2}{dt^2}exp\left(\frac{-t^2}{2}\right). \tag{4}$$

Therefore,

$$\psi_{b,a}(t) = -a^{3/2}D_t^2 exp\left(-\frac{(b-t)^2}{2a^2}\right), \quad \left(D_t = \frac{d}{dt}\right). \tag{5}$$

Thus (2) can be written as:

$$W(b,a) = -a^{3/2}\int_{\mathbb{R}} f(t)\; D_t^2\; exp\left(-\frac{(b-t)^2}{2a^2}\right)dt, \qquad a > 0. \tag{6}$$

Then, with some conditions on f, we get

$$W(b,a) = -a^{3/2}\int_{\mathbb{R}} f^{(2)}(t)\; exp\left(-\frac{(b-t)^2}{2a^2}\right)dt, \qquad a > 0. \tag{7}$$

From the above two equations it follows that the MHWT can also be considered as the Weierstrass transform of $\left(\frac{d}{dt}\right)^2 f(t)$. Hence, various properties of MHWT can be obtained from those of the Weierstrass transform [10]. We define the generalized MHWT by

$$W(b,a) = -a^{3/2}\left\langle f(t),\; D_t^2 exp\left(-\frac{(b-t)^2}{2a^2}\right)\right\rangle, \qquad \frac{\alpha}{\gamma} < \text{Re}\, b < \frac{\beta}{\gamma}. \tag{8}$$

$W(b,a)$ turns out to be an analytic function for $\frac{\alpha}{\gamma} < \text{Re}\, b < \frac{\beta}{\gamma}$.

A function $k(b,a)$ defined by [10]

$$k(b,a) = \frac{1}{\sqrt{2\pi a}}exp\left(\frac{-b^2}{2a}\right), \tag{9}$$

where $a \in (0,\infty)$ and $b \in \mathbb{R}$. Clearly,

$$D_t^2 k(b-t, a^2) = \frac{1}{\sqrt{2\pi}a}D_t^2\left(exp\left(-\frac{(b-t)^2}{2a^2}\right)\right). \tag{10}$$

Hence by (5),

$$\psi_{b,a}(t) = -\sqrt{2\pi}a^{\frac{5}{2}} D_t^2 k(b-t, a^2). \tag{11}$$

The MHWT of a function $f(t)$ is given by

$$W_f(b,a) = a^{3/2} \int_{\mathbb{R}} exp\left(-\frac{(t-b)^2}{2a^2}\right) f^{(2)}(t)dt. \tag{12}$$

In particular, the definition of MHWT can be extended to complex values of b whenever necessary. In [19], asymptotic properties of MHWT are discussed for an appropriate testing function space $\mathcal{W}^{\gamma}_{\alpha,\beta}$ on the real line $\mathbb{R}$ which is a generalization of the Zemanian space. Further, the asymptotic behaviour of the distributional MHWT was obtained and a real inversion formula was derived by constructing a structural formula. Also, the tauberian result related to the transform is discussed. Pathak and Singh in [15] studied convolution theory in $K'\{M_p\}$ space and obtained some boundedness result for the wavelet transform. They derived Calderóns formula in distribution sense as an application of the wavelet transform and used it to obtain an inversion formula for the wavelet transform of generalized functions. Further in [18], the authors studied the mutation phase by joining the BBO algorithm and the MHWT. The MHWT results in a reduction of the mean error and improves investigation and premature convergence. Secondly, they introduced the BMDDSF framework (BBO-Mexican hat wavelet-dragonfly dynamic scheduling framework) for the dynamic scheduling of tasks in the cloud computing environment. The benefit of working with the MHWT is the efficiency of the short term signal analysis. The introduction of this set of wavelet leads to productive analysis that may find use in a wide range of fields. Also, it provides a unified framework for representing time-dependent functions in a time-frequency plane including local frequency data.

Wavelet transform has been generalized to Wavelet Stieltjes transform of bounded variation function in [1]. This approach for time-frequency localization provides a unified framework to analyze both continuous and discrete-time signals in the same way as the distribution function allows a unified treatment of discrete and continuous cases. Let us begin with the definition of function of bounded variation.

Definition 1 Let

$$BV(1,2) = \left\{ F | F : \mathbb{R} \to \mathbb{R}, \ F(\cdot) = \int_{-\infty}^{\cdot} f(t)dt + \sum_{s \le \cdot} \rho(s), \right.$$
$$\left. f \in L^1(\mathbb{R}) \cap L^2(\mathbb{R}), \ \rho \in l_1 \cap l_2 \right\},$$

where summation convention is summed over countable discontinuities of F [1].

Note that any function F in $BV(1,2)$ is of bounded first variation, where the integral part is absolutely continuous component of F and the sum part is the jump component of F.

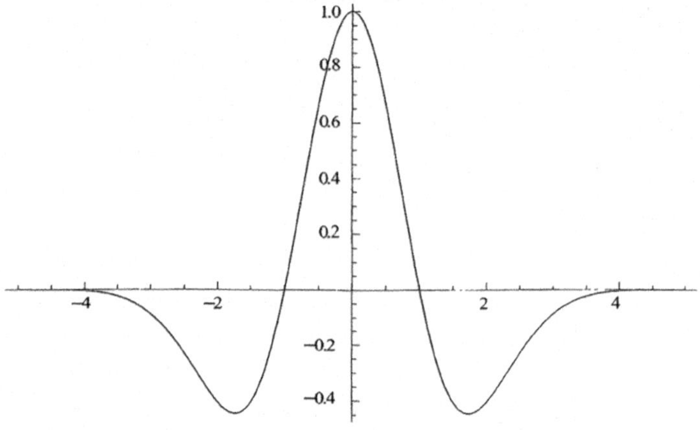

Fig. 1 Mexican hat wavelet

Definition 2 For $F \in BV(1, 2)$, define

$$\|F\|_1 := \int_{\mathbb{R}} |f(t)|dt + \sum_{-\infty<s<\infty} |\rho(s)| \tag{13}$$

and

$$\|F\|_2 := \left[\int_{\mathbb{R}} |f(t)|^2 dt + \sum_{-\infty<s<\infty} |\rho(s)|^2 \right]^{1/2}. \tag{14}$$

In fact, $\| \cdot \|_1$ and $\| \cdot \|_2$ are norms on $BV(1, 2)$.

The MHWST of F with kernel $\psi_{b,a}$ is given by

$$W_F S(b, a) = a^{3/2} \int_{\mathbb{R}} exp\left(-\frac{(t-b)^2}{2a^2}\right) dF^{(2)}(t), \tag{15}$$

such that for all $b = \sigma + i\omega$, the integrals (12) and (15) converges.

We shall now assume that $f(t)$ is Lebesgue integrable in every finite interval and that $F(t)$ is a function of bounded variation. Stieltjes transform has important applications in the theory of moments, queuing, orthogonal polynomials, and in mathematical physics as they are a representing class for finite measures.

The region of convergence for the MHWT is an interval where $b = \sigma + i\omega$, then we have a vertical strip for convergence [22]. Moreover, every MHWT which converges for $m < \sigma < n$ is also holomorphic in $m < Re(b) < n$. The representation theory for Laplace transform was obtained by Cooper [3]. In this chapter, we present representation theories for the MHWT and MHWST in connection with the integral transform [20]:

$$f_\lambda^{(2)}(t,\sigma) = \frac{1}{\sqrt{2\pi}a}\left(\frac{a^{-5/2}}{\sqrt{2\pi}}\right)\int_{\mathbb{R}} k_1(\omega,\lambda)exp\left(\frac{(b-t)^2}{2a^2}\right)W_f(b,a)d\omega$$

$$= \frac{1}{2\pi a^{7/2}}\int_{\mathbb{R}} k_1(\omega,\lambda)exp\left(\frac{(b-t)^2}{2a^2}\right)W_f(b,a)d\omega, \quad (16)$$

where $m < \sigma < n$ and $k_1(\omega,\lambda)$ is a Fourier summation kernel satisfying the following conditions:
The Fourier transform of $k_1(\omega,\lambda)$ denoted by $K(t,\lambda)$ exists and $k_1(\omega,\lambda)$ is regular, given by

$$k_1(\omega,\lambda) = \frac{1}{\sqrt{2\pi}}\int_{\mathbb{R}} exp\,(i\omega t)\,K(t,\lambda)dt \quad (17)$$

and

$$\int_{\mathbb{R}} |K(t,\lambda)|dt \le M', \quad (18)$$

where M' is a constant and independent of λ [3].

Lemma 1 *If (18) holds, then the set of transformations T_λ defined by*

$$(T_\lambda h)(\sigma) = \int_{\mathbb{R}} K\left(\frac{\sigma-\omega}{2},\lambda\right)h(\omega)d\omega,$$

forms a bounded set of transformations from $L^p(\mathbb{R})$ $(1 \le p < \infty)$ to itself.

Definition 3 A function $W_f(b,a)$ defined on (m,n) is said to be in class A if it is extended analytically into the complex plane satisfying

(i) $W_f(b,a)$ is holomorphic in the strip $m < x < n$,

(ii) $W_f(b,a) = \mathrm{O}\left(exp\left(\frac{\omega^2}{2a^2}\right)\right)$, $|\omega| \to \infty$, $(b = \sigma + i\omega)$, $m < \sigma < n$.

Definition 4 A function $W_f(b,a)$ defined on (m,n) is said to be in class B if it is extended analytically into the complex plane satisfying

(i) $W_f(b,a)$ is holomorphic in the strip $m < x < n$,

(ii) $W_f(b,a) = \mathrm{o}\left(exp\left(\frac{\omega^2}{2a^2}\right)\right)$, $|\omega| \to \infty$, $(b = \sigma + i\omega)$, $m < \sigma < n$.

2 Necessary Conditions for Mexican Hat Wavelet and MHWST

Theorem 1 *If* $W_f(b,a) = a^{3/2} \int_{\mathbb{R}} exp\left(-\frac{(t-b)^2}{2a^2}\right) f^{(2)}(t)\, dt$, *with*

$$f^{(2)}(t)exp\left(\frac{-(t-\sigma)^2}{2a^2}\right) \in L^p(\mathbb{R}),\ 1 < p \leq 2,\ m < \sigma < n \ \ and\ if\ for$$

each $\lambda > 0,\ k_1(\omega,\lambda) \in L^p(\mathbb{R})$ *then we have,* $W_f(b,a) \in B(m,n)$ *and*

$$\|exp\left(\frac{-(t-\sigma)^2}{2a^2}\right) f_\lambda^{(2)}(t,\sigma)\|_p \leq M,\ \ (1 < p \leq 2) \tag{19}$$

where M *is independent of* λ.

Proof As $W_f(b,a)$ exists and belongs to $B(m,n)$ for all $m < \sigma < n$, we show that (19) is satisfied. Consider

$$\begin{aligned}
f_\lambda^{(2)}(t,\sigma) &= \frac{1}{2\pi a^{7/2}} \int_{\mathbb{R}} k_1(\omega,\lambda) exp\left(\frac{(b-t)^2}{2a^2}\right) W_f(b,a) d\omega \\
&= \frac{1}{2\pi a^2} \int_{\mathbb{R}} k_1(\omega,\lambda) exp\left(\frac{(\sigma - t + i\omega)^2}{2a^2}\right) d\omega \\
&\quad \times \int_{\mathbb{R}} exp\left(\frac{-(u-\sigma-i\omega)^2}{2a^2}\right) f^{(2)}(u) du \\
&= \frac{1}{\sqrt{2\pi} a^2} exp\left(\frac{(t-\sigma)^2}{2a^2}\right) \frac{1}{\sqrt{2\pi}} \int_{\mathbb{R}} k_1(\omega,\lambda) exp\left(\frac{-(t-u)i\omega}{a^2}\right) d\omega \\
&\quad \times \int_{\mathbb{R}} exp\left(\frac{-(u-\sigma)^2}{2a^2}\right) f^{(2)}(u) du \\
&= \frac{1}{\sqrt{2\pi} a^2} exp\left(\frac{(t-\sigma)^2}{2a^2}\right) \int_{\mathbb{R}} K\left(\frac{t-u}{a^2},\lambda\right) exp\left(\frac{-(u-\sigma)^2}{2a^2}\right) f^{(2)}(u) du,
\end{aligned} \tag{20}$$

where the interchange of the order of integration is justified by Fubini's theorem. Therefore, by Lemma 1

$$\|exp\left(\frac{-(t-\sigma)^2}{2a^2}\right) f_\lambda^{(2)}(t,\sigma)\|_p \leq M.$$

Hence the proof is complete. □

Theorem 2 *If* $W_F S(b,a) = a^{3/2} \int_{\mathbb{R}} exp\left(-\frac{(t-b)^2}{2a^2}\right) dF^{(2)}(t)$ *with*

$$\int_{\mathbb{R}} exp\left(\frac{-(t-\sigma)^2}{2a^2}\right) |dF^{(2)}(t)| < \infty, \quad (m < \sigma < n) \tag{21}$$

and if $k_1(\omega,\lambda) \in L^1(\mathbb{R})$ *for each* $\lambda > 0$*, then* $W_F S(b,a) \in A(m,n)$ *and*

$$\|exp\left(\frac{-(t-\sigma)^2}{2a^2}\right) f_\lambda^{(2)}(t,\sigma)\|_1 \le M.$$

Proof Let $W_F S(b,a)$ exists for all $m < \sigma < n$ and $W_F S(b,a) \in A(m,n)$, then

$$f_\lambda^{(2)}(t,\sigma) = \frac{1}{2\pi a^{7/2}} \int_{\mathbb{R}} k_1(\omega,\lambda) exp\left(\frac{(b-t)^2}{2a^2}\right) W_F S(b,a) d\omega$$

$$= \frac{1}{2\pi a^2} \int_{\mathbb{R}} k_1(\omega,\lambda) exp\left(\frac{(\sigma-t+i\omega)^2}{2a^2}\right) d\omega \times \int_{\mathbb{R}} exp\left(\frac{-(u-\sigma-i\omega)^2}{2a^2}\right) dF^{(2)}(u)$$

$$= \frac{1}{\sqrt{2\pi} a^2} exp\left(\frac{(t-\sigma)^2}{2a^2}\right) \frac{1}{\sqrt{2\pi}} \int_{\mathbb{R}} k_1(\omega,\lambda) exp\left(\frac{-(t-u)i\omega}{a^2}\right) d\omega \times \int_{\mathbb{R}} exp\left(\frac{-(u-\sigma)^2}{2a^2}\right) dF^{(2)}(u)$$

$$= \frac{1}{\sqrt{2\pi} a^2} exp\left(\frac{(t-\sigma)^2}{2a^2}\right) \int_{\mathbb{R}} K\left(\frac{t-u}{a^2},\lambda\right) exp\left(\frac{-(u-\sigma)^2}{2a^2}\right) dF^{(2)}(u),$$

where the interchange of the order of integration is justified by Fubini's theorem. Therefore, by using Lemma 1, we have

$$\|exp\left(\frac{-(t-\sigma)^2}{2a^2}\right) f_\lambda^{(2)}(t,\sigma)\|_1 \le M.$$

This completes the proof of the theorem. □

3 Sufficient Conditions for Mexican Hat Wavelet and MHWST

Theorem 3 *If* $W_f(b,a) \in B(m,n)$, *with* $\|exp\left(\frac{-(t-\sigma)^2}{2a^2}\right) f_\lambda^{(2)}(t,\sigma)\|_p \leq M$

$$and\ \ k_1(\omega,\lambda) exp\left(\frac{-(\omega)^2}{2a^2}\right) W_f(b,a) \in L^1(\mathbb{R}),\ \ where k_1(\omega,\lambda) \to 1 as \lambda \to \infty$$

uniformly in ω *for every finite interval, then there exist a function* f *such that*

$$for\ \left(exp\left(\frac{-(\sigma-t)^2}{2a^2}\right) f^{(2)}(t)\right) \in L^p(\mathbb{R}),\ \ 1 < p \leq 2,\ \sigma \in (m,n),\ we\ have$$

$$W_f(b,a) = a^{\frac{3}{2}} \int\limits_{-\infty}^{\infty} exp\left(\frac{-(b-t)^2}{2a^2}\right) f^{(2)}(t)dt.$$

Proof Let $\sigma_0 \in (m,n)$, then by Theorem 1, the family of functions

$$\left\{exp\left(\frac{-(\sigma_0-t)^2}{2a^2}\right) f_\lambda^{(2)}(t,\sigma_0)\right\}$$

is bounded in $L^p(\mathbb{R})$. As defined in [22], there exists a subsequence $\{\lambda_k\}_{k=1}^{\infty}$ with $\lim_{k\to 0}\lambda_k = \infty$ and for a function $f(t,\sigma_0)$, we have

$$exp\left(\frac{-(\sigma_0-t)^2}{2a^2}\right) f^{(2)}(t,\sigma_0) \in L^p(\mathbb{R})$$

such that

$$\lim_{k\to\infty} \frac{1}{\sqrt{2\pi}a} \int\limits_{\mathbb{R}} exp\left(\frac{-(\sigma_0-t)^2}{2a^2}\right) f_{\lambda_k}^{(2)}(t,\sigma_0)\overline{\phi}(t)dt$$

$$= \frac{1}{\sqrt{2\pi}a} \int\limits_{\mathbb{R}} exp\left(\frac{-(\sigma_0-t)^2}{2a^2}\right) f^{(2)}(t,\sigma_0)\overline{\phi}(t)dt \qquad (22)$$

for all $\phi \in L^{p'} \cap L^1$, whose Fourier transforms φ are in L^p. Thus,

$$\frac{1}{\sqrt{2\pi}a}\int_{\mathbb{R}} exp\left(\frac{-(\sigma_0-t)^2}{2a^2}\right) f^{(2)}_{\lambda_k}(t,\sigma_0)\overline{\phi}(t)dt = \frac{1}{\sqrt{2\pi}a}\int_{\mathbb{R}} exp\left(\frac{-(\sigma_0-t)^2}{2a^2}\right)\overline{\phi}(t)dt$$

$$\times\left\{\frac{1}{2\pi a^{7/2}}\int_{\mathbb{R}} k_1(\omega,\lambda_k)exp\left(\frac{(\sigma_0-t+i\omega)^2}{2a^2}\right)W_f(\sigma_0+i\omega,a)d\omega\right\}$$

$$= \frac{1}{2\pi a^{9/2}}\int_{\mathbb{R}} exp\left(\frac{-\omega^2}{2a^2}\right) exp\left(\frac{i\sigma_0\omega}{a^2}\right) k_1(\omega,\lambda_k)W_f(\sigma_0+i\omega,a)d\omega$$

$$\times\left\{\frac{1}{\sqrt{2\pi}}\int_{\mathbb{R}}\overline{\phi}(t)exp\left(\frac{-it\omega}{a^2}\right)dt\right\}$$

$$= \frac{1}{2\pi a^{9/2}}\int_{\mathbb{R}} exp\left(\frac{-\omega^2}{2a^2}\right) exp\left(\frac{i\sigma_0\omega}{a^2}\right) k_1(\omega,\lambda_k)W_f(\sigma_0+i\omega,a)\overline{\varphi}\left(\frac{\omega}{a^2}\right)d\omega$$

$$= \frac{1}{2\pi a^{5/2}}\int_{\mathbb{R}} exp\left(\frac{-\omega^2a^2}{2}\right) exp\,(i\sigma_0\omega)\,k_1(\omega a^2,\lambda_k)W_f(\sigma_0+i\omega a^2,a)\overline{\varphi}(\omega)d\omega.$$

Now the functions

$$exp\left(\frac{-\omega^2a^2}{2}\right) exp\,(i\sigma_0\omega)\,k_1(\omega a^2,\lambda_k)W_f(\sigma_0+i\omega a^2,a) \tag{23}$$

are the Fourier transforms of $exp\left(\frac{-(\sigma_0-t)^2}{2a^2}\right) f^{(2)}_{\lambda_k}(t,\sigma_0)$.

This family of functions is bounded in L^p. Therefore by [7], the family (23) is bounded in $L^{p'}$. By the weak compactness of the $L^{p'}$-space, there exists a subsequence $\{\lambda_{k_j}\}_{j=1}^{\infty}$ with $\lim_{j\to\infty}\lambda_{k_j}=\infty$ such that for any $h\in L^p$, we have

$$\lim_{j\to\infty}\int_{\mathbb{R}} exp\left(\frac{-\omega^2a^2}{2}\right) exp\,(i\omega\sigma_0)\,k_1(\omega a^2,\lambda_{k_j})W_f(\sigma_0+i\omega a^2,a)h(\omega)d\omega$$

$$= \int_{\mathbb{R}} exp\left(\frac{-\omega^2a^2}{2}\right) exp\,(i\omega\sigma_0)\,W_f(\sigma_0+i\omega a^2,a)h(\omega)d\omega,$$

where $exp\left(\frac{-\omega^2a^2}{2}\right) exp\,(i\omega\sigma_0)\,W_f(\sigma_{0+i\omega a^2},a)$ is the limiting point of

$exp\left(\frac{-\omega^2a^2}{2}\right) exp\,(i\omega\sigma_0)\,k_1(\omega a^2,\lambda_{k_j})W_f(\sigma_{0+i\omega a^2},a)$. Also, the functions

$\phi \in L^{p'} \cap L^1$ are dense in $L^{p'}$, so that $h(\omega) = \dfrac{1}{2\pi a^{5/2}}\overline{\varphi}(\omega)$.

Therefore,

$$\lim_{j\to\infty} \frac{1}{\sqrt{2\pi}a} \int_{\mathbb{R}} \sqrt{2\pi}a^{5/2} exp\left(\frac{-(\sigma_0 - t)^2}{2a^2}\right) f^{(2)}_{\lambda_{k_j}}(t, \sigma_0)\overline{\phi}(t)dt$$

$$= \frac{1}{2\pi a^{5/2}} \int_{\mathbb{R}} \sqrt{2\pi}a^{5/2}\overline{\varphi}(\omega) exp\left(\frac{-\omega^2 a^2}{2}\right) exp\left(i\omega\sigma_0\right) W_f(\sigma_0 + i\omega a^2, a)d\omega.$$

Now using [4], and the fact that the functions (23) are the Fourier transforms

of $\left(exp\left(\frac{-(\sigma_0 - t)^2}{2a^2}\right) f^{(2)}_{\lambda_k}(t, \sigma_0)\right)$, we obtain

$$\frac{1}{\sqrt{2\pi}a^{3/2}} exp\left(\frac{-\omega^2 a^2}{2}\right) exp\left(i\sigma_0\omega\right) W_f(\sigma_0 + i\omega a^2, a)$$

$$= \frac{1}{\sqrt{2\pi}} \int_{\mathbb{R}} exp\left(i\omega t\right) exp\left(\frac{-(\sigma_0 - t)^2}{2a^2}\right) f^{(2)}(t, \sigma_0)dt.$$

Hence, we have

$$W_f(\sigma_0 + i\omega a^2, a) = a^{3/2} \int_{\mathbb{R}} exp\left(\frac{-(\sigma_0 - t + i\omega a^2)^2}{2a^2}\right) f^{(2)}(t, \sigma_0)dt$$

i.e.,

$$W_f(\sigma_0 + i\omega, a) = a^{3/2} \int_{\mathbb{R}} exp\left(\frac{-(\sigma_0 + i\omega - t)^2}{2a^2}\right) f^{(2)}(t, \sigma_0)dt.$$

Now we show that the function $f^{(2)}(t, \sigma_0)$ is independent of $\sigma_0 \in (m, n)$. Since $W_f(b, a) \in B(m, n)$, therefore, $W_f(b, a) exp\left(\frac{(b - t)^2}{2a^2}\right)$ is holomorphic for $m < Re(b) < n$. Hence, using Cauchy's integral theorem for $m < \sigma_0 \leq Re(b) \leq \sigma_1 < n$, we have

$$\int_C W_f(b, a) exp\left(\frac{(b - t)^2}{2a^2}\right) db = 0,$$

where the contour C is determined by the vertices $\sigma_0 + iT,\ \sigma_1 + iT$ with $m < \sigma_0 < \sigma_1 < n$. Therefore,

$$0=\left\{\int_{\sigma_0-iT}^{\sigma_0+iT}+\int_{\sigma_0+iT}^{\sigma_1+iT}+\int_{\sigma_1+iT}^{\sigma_1-iT}+\int_{\sigma_1-iT}^{\sigma_0-iT}\right\}W_f(b,a)exp\left(\frac{(b-t)^2}{2a^2}\right)db$$

$$=I_1+I_2+I_3+I_4$$

and as $T\to\infty$, the integral $|I_2|\to\infty$ and $|I_4|\to\infty$. It follows that

$$\int_{\mathbb{R}}\left[W_f(\sigma_0+i\omega,a)exp\left(\frac{(\sigma_0+i\omega-t)^2}{2a^2}\right)\right.$$
$$\left.-W(\sigma_1+i\omega,a)exp\left(\frac{(\sigma_1+i\omega-t)^2}{2a^2}\right)\right]d\omega=0. \quad (24)$$

Now by the regularity of $k_1(\omega,\lambda)$, hypotheses and (24)

$$\lim_{\lambda\to\infty}\int_{\mathbb{R}}k_1(\omega,\lambda)exp\left(\frac{(\sigma_0+i\omega-t)^2}{2a^2}\right)W_f(\sigma_0+i\omega,a)d\omega$$

$$=\lim_{\lambda\to\infty}\int_{\mathbb{R}}k_1(\omega,\lambda)exp\left(\frac{(\sigma_1+i\omega-t)^2}{2a^2}\right)W_f(\sigma_1+i\omega,a)d\omega.$$

From the uniqueness of the weak limit

$$f^{(2)}(t,\sigma_0)=f^{(2)}(t,\sigma_1)=f^{(2)}(t).$$

Hence by (24),

$$W_f(\sigma+i\omega,a)=a^{\frac{3}{2}}\int_{-\infty}^{\infty}exp\left(\frac{-(t-\sigma-i\omega)^2}{2a^2}\right)f^{(2)}(t)dt.$$

This completes the proof of the theorem. □

Theorem 4 *If* $W_FS(b,a)\in A(m,n)$, $\|exp\left(\frac{-(t-b)^2}{2a^2}\right)f_\lambda^{(2)}(t,b)\|_1\leq M$ *is satisfied and* $k_1(\omega,\lambda)exp\left(\frac{-\omega^2}{2a^2}\right)W_FS(b,a)\in L_1(\mathbb{R})$ *where* $k_1(\omega,\lambda)\to 1$ *as* $\lambda\to\infty$, *uniformly in* ω *for any finite interval, then there exists a function* F

$$with\int_{\mathbb{R}}exp\left(\frac{-(\sigma-t)^2}{2a^2}\right)\|dF^{(2)}(t)\|<\infty,\ \ \forall\sigma\in(m,n)$$

such that $W_F S(b,a) = a^{3/2} \int_{\mathbb{R}} exp\left(\frac{-(t-b)^2}{2a^2}\right) dF^{(2)}(t)$.

Proof By hypotheses, $f_\lambda^{(2)}(t,\sigma)$ is well defined $\forall \sigma \in (m,n)$. Let

$$F_\lambda^{(2)}(t,\sigma) \equiv \int_0^t f_\lambda^{(2)}(u,\sigma)du,$$

then for arbitrary finite interval $[l_1, l_2]$ and fixed $\sigma_0 \in (m,n)$

$$\int_{l_1}^{l_2} |dF_\lambda{}^{(2)}(t,\sigma_0)| \le \int_{l_1}^{l_2} |f_\lambda^{(2)}(t,\sigma_0)|dt$$

$$\le \max_{l_1<t<l_2} exp\left(\frac{(\sigma_0-t)^2}{2a^2}\right) \int_{l_1}^{l_2} exp\left(\frac{-(\sigma_0-u)^2}{2a^2}\right) |f_\lambda^{(2)}(u,\sigma_0)|du,$$

uniformly in λ. Thus $\{F_\lambda(t,\sigma_0)\}$ is of uniformly bounded variation in $[l_1, l_2]$ and

$$|F_\lambda^{(2)}(l_1,\sigma_0)| \le \int_0^{l_1} |f_\lambda^{(2)}(u,\sigma_0)|du$$

$$\le \max_{0<t<l_1} exp\left(\frac{(\sigma_0-t)^2}{2a^2}\right) \int_0^{l_1} exp\left(\frac{-(\sigma_0-u)^2}{2a^2}\right) |f_\lambda^{(2)}(u,\sigma_0)|du < \infty.$$

Hence by [22], there exists an increasing unbounded sequence λ_k and a function $F(t,\sigma_0)$ of bounded variation in $l_1 \le t \le l_2$ such that

$$\lim_{k\to\infty} F_{\lambda_k}^{(2)}(t,\sigma_0) = F^{(2)}(t,\sigma_0). \tag{25}$$

Moreover, by [22], for any continuous function $g(t)$ in $[l_1, l_2]$,

$$\lim_{k\to\infty} \int_{l_1}^{l_2} g(t)dF_{\lambda_k}^{(2)}(t,\sigma_0) = \int_{l_1}^{l2} g(t)dF^{(2)}(t,\sigma_0).$$

Hence, in particular for $g(t) = exp\left(\frac{-(\sigma_0-t)^2}{2a^2}\right)$, we have

$$\int_{l_1}^{l_2} exp\left(\frac{-(\sigma_0 - t)^2}{2a^2}\right) |dF^{(2)}(t, \sigma_0)|$$
$$\leq \underline{\lim}_{k\to\infty} \int_{l_1}^{l_2} exp\left(\frac{-(\sigma_0 - t)^2}{2a^2}\right) |dF^{(2)}_{\lambda_k}(t, \sigma_0)|$$
$$\leq \underline{\lim}_{k\to\infty} \int_{\mathbb{R}} exp\left(\frac{-(\sigma_0 - t)^2}{2a^2}\right) |f^{(2)}_{\lambda}(t, \sigma_0)|dt < \infty.$$

Now, by (25)

$$F^{(2)}(t, \sigma_0) = \lim_{k\to\infty} \int_0^t f^{(2)}_{\lambda_k}(u, \sigma_0)du \tag{26}$$

$$= \lim_{k\to\infty} \frac{1}{2\pi a^{7/2}} \int_0^t du \int_{\mathbb{R}} k_1(\omega, \lambda) exp\left(\frac{(\sigma_0 + i\omega - u)^2}{2a^2}\right) W_F S(b_0 + i\omega, a)d\omega$$
$$= \lim_{k\to\infty} \frac{1}{2\pi a^{7/2}} \int_{\mathbb{R}} k_1(\omega, \lambda_k) W_F S(\sigma_0 + i\omega, a)d\omega \int_0^t exp\left(\frac{(\sigma_0 + i\omega - u)^2}{2a^2}\right) du,$$

where the interchange of order of integration is justified by Fubini's theorem.

It remains to show that $F(t, \sigma_0)$ is independent of $\sigma_0 \in (m, n)$. Since $W_F S(b, a) \in A(m, n)$,

$$W_F S(b, a) \int_0^t exp\left(\frac{(b - u)^2}{2a^2}\right) du$$

is holomorphic in the strip $a < Re(b) < b$, then by [9, Theorem 1], we obtain

$$\int_{\mathbb{R}} \left\{ W_F S(\sigma_0 + i\omega, a) \left[\int_0^t exp\left(\frac{(\sigma_0 + i\omega)^2}{2a^2}\right) du \right] \right.$$
$$\left. - W_F S(\sigma_1 + i\omega, a) \left[\int_0^t exp\left(\frac{(\sigma_1 + i\omega - u)^2}{2a^2}\right) du \right] \right\} d\omega = 0,$$

where $m < \sigma_0 < \sigma_1 < n$. By the regularity of $k_1(\omega, \lambda)$ and the fact that $k_1(\omega, \lambda)$ $W_F S(\sigma + i\omega, a) exp\left(\frac{-\omega^2}{2a^2}\right) \in L_1(\mathbb{R})$, it follows from (26) that

$$F^{(2)}(t,\sigma_0) = F^{(2)}(t,\sigma_1) = F^{(2)}(t).$$

Now, let $H(t)$ be a continuous function in $[-T, T]$ and zero outside the interval. If h denotes the Fourier transform of H, then

$$\frac{1}{\sqrt{2\pi}a}\int_{\mathbb{R}} H(t)exp\left(\frac{-(\sigma-t)^2}{2a^2}\right) f^{(2)}_{\lambda_k}(t,b)dt$$

$$= \frac{1}{\sqrt{2\pi}a}\int_{\mathbb{R}} H(t)exp\left(\frac{-(\sigma-t)^2}{2a^2}\right) dF^{(2)}_{\lambda_k}(t,b)$$

$$= \frac{1}{2\pi a^{9/2}}\left(\frac{1}{\sqrt{2\pi}a}\right)\int_{\mathbb{R}} H(t)exp\left(\frac{-(\sigma-t)^2}{2a^2}\right) dt$$
$$\times \int_{\mathbb{R}} k_1(\omega,\lambda_k)exp\left(\frac{(b-t)^2}{2a^2}\right) W_F S(b,a)d\omega$$

$$= \frac{1}{2\pi a^{9/2}}\int_{\mathbb{R}} k_1(\omega,\lambda_k)exp\left(\frac{2i\omega\sigma-\omega^2}{2a^2}\right) W_F S(b,a)d\omega$$
$$\times \frac{1}{\sqrt{2\pi}}\int_{\mathbb{R}} H(t)exp\left(\frac{-i\omega t}{a^2}\right) dt$$

$$= \frac{1}{2\pi a^{9/2}}\int_{\mathbb{R}} k_1(\omega,\lambda_k)exp\left(\frac{i\omega\sigma}{a^2}\right) exp\left(\frac{-\omega^2}{2a^2}\right) W_F S(b,a)h\left(\frac{\omega}{a^2}\right) d\omega$$

$$= \frac{1}{2\pi a^{5/2}}\int_{\mathbb{R}} k_1(\omega a^2,\lambda_k)exp\,(i\omega\sigma)\, W_F S(\sigma+i\omega a^2,a)h(\omega)d\omega.$$

Therefore, $k_1(\omega a^2,\lambda_k)exp\left(i\omega a^2 b\right) exp\left(\frac{-\omega^2 a^2}{2}\right) W_F S(b+i\omega a^2,a)$ is the Fourier transform of $\left(\frac{-(\sigma-t)^2}{2a^2}\right) f^{(2)}_{\lambda_k}(t,b)$. Thus,

$$\frac{a^{-3/2}}{\sqrt{2\pi}} k_1(\omega^2,\lambda_k)exp\left(i\omega a^2 b\right) exp\left(\frac{-\omega^2 a^2}{2}\right) W_F S(b+i\omega a^2,a)$$

$$= \frac{1}{\sqrt{2\pi}}\int_{\mathbb{R}} exp\,(i\omega t)\, exp\left(\frac{-(b-t)^2}{2a^2}\right) f^{(2)}_{\lambda_k}(t,b)dt.$$

Therefore, we have

$$W_F S(\sigma + i\omega a^2, a) = a^{3/2} \int_{\mathbb{R}} exp\left(\frac{-1}{2a^2}\left\{(t-b)^2 + 2ib\omega - \omega^2 - 2it\omega\right\}\right) dF_k^{(2)}(t, b).$$

Let $k \to \infty$, then a change of variable yields

$$W_F S(b, a) = a^{3/2} \int_{\mathbb{R}} exp\left(\frac{-(t-\sigma-i\omega)^2}{2a^2}\right) dF^{(2)}(t)$$

which is the required result. □

Conclusion

The principles of wavelet analysis are similar to those of Fourier analysis and any application using the Fourier transform can be formulated using wavelets to provide more accurate data. The wavelet transform is considered to have one of the most appropriate wavelet basis functions which have many applications in time-frequency signal analysis. In the same way, the MHWT also has an obvious local feature in the frequency domain. As the Mexican hat wavelet is formulated by Gaussian function, it satisfies the Gaussian decays in both space and frequency, which indicates that it can also extract information in a space-frequency window. Moreover, the Mexican hat wavelet acts as a smooth function and is more beneficial in signal extraction. Therefore, it is used in extracting the feature parameters of ECG. In contrast to any other wavelet, the Mexican hat wavelets emanated from the continuous wavelet transform, are symmetrical and give an exact time-frequency analysis of input functions. Therefore Mexican hat wavelet has been thought to be the most suitable mother wavelet function. In [6], the Mexican hat wavelet and its transforms are analyzed from a Fourier perspective, to obtain an explicit expression in the Fourier domain. In the Fourier domain, the Mexican hat wavelet acts as the product of the Laplace-Beltrami operator and the heat kernel. It is, therefore, a scaled differential operator continuously dilated through heat diffusion. Zhou *et al.* has implemented the MHWT in modeling earthquake accelerogram records in the form of a scalogram where the coefficients of the continuous MHWT is used to describe the signal energy in the time-scale domain [23]. Here we used the theory of MHWT to define the MHWST of bounded variation function which can analyze both continuous and discrete-time signals. This relation helps to broaden the application scope of the transform. Further, we developed some necessary and sufficient conditions for representing functions as MHWT and MHWST. The MHWT is derived from the heat kernel by taking the negative first-order derivative with respect to time and has localization in space and frequency. Therefore, it provides numerous applications in space-frequency analysis, solutions for differential and integro-differential equations, and other digital modulation. Moreover, these results can be extended to obtain an analytic solution of the hyperbolic heat conduction problem, mixed boundary value problems, approximation theory, mathematical modeling, and computation.

Acknowledgements This work is supported by SERB-DST, Govt. of India, through Major Research Project sanction No. ECR/2017/000394.

References

1. Bielecki, T.R., Chen, J., Lin, E.B., Yau, S.S.T.: Some remarks on wavelet transforms. In Proceedings of first IEEE Regional Conference on Aerospace Control Systems, pp. 148–150 (1993)
2. Chui, C.K.: An Introduction to Wavelets. Academic press, San Diego (1992)
3. Cooper JLB (1965) The representation of functions as Laplace transforms. Mathematische Annalen, 159(4), 223–233
4. Cooper, J.L.B.: Umkehrformelen für Fourier-Transformationen, Approximations-und Interpolationstheorie. On Approximation Theory/Über Approximationstheorie **5**, 60–71 (1964)
5. Daubechies, I.: Ten Lectures on Wavelets, vol. 61. Siam (1992)
6. Hou T and Qin H (2012) Continuous and discrete Mexican hat wavelet transforms on manifolds. Graphical Models, 74(4), 221–232
7. Heinig, H.P.: Representation of functions as Weierstrass-transforms. Canadian Mathematical Bulletin **10**(5), 711–722 (1967)
8. Kim, C.H., Aggarwal, R.: Wavelet transforms in power systems. Part 1: General introduction to the wavelet transforms. Power Eng. J. **14**(2), 81–87 (2000)
9. Nessel, R.J.: Über die Darstellung holomorpher Funktionen durch Weierstraβ-und Weierstraβ-Stieltjes-Intgrale. Journal für die reine und angewandte Mathematik **218**, 31–50 (1965)
10. Pathak, R.S., Singh, A.: Mexican hat wavelet transform of distributions. Integral Transforms and Special Functions **27**(6), 468–483 (2016)
11. Pathak, R.S.: The Wavelet Transform. Atlantis Press, World Scientific, Amsterdam, Paris (2009)
12. Pathak, R.S.: Integral transforms of generalized functions and their applications. Gordon and Breach Science Publishers, Amsterdam (1997)
13. Pathak RS, Singh A (2017) Wavelet transform of Beurling-Bjorck type ultradistributions. Rendiconti del Seminario Matematico della Universita di Padova, 137(1), 211–222
14. Pathak, R.S., Singh, A.: Distributional Wavelet Transform. Proc. Nat. Acad. Sci., Sec. A **86**(2), 273–277. Springer, Berlin (2016)
15. Pathak, R.S., Singh, A.: Wavelet transform of generalized functions in $K'\{M_p\}$ spaces. Proc. Math. Sci. **126**(2), 213–226 (2016)
16. Pathak, R.S., Singh, A.: Paley-Wiener-Schwartz-type theorem for the wavelet transform. Applicable Analysis **98**(7), 1324–1332 (2019) transform. Applicable Analysis **98**(7), 1324–1332 (2019)
17. Pathak RS, Singh SK (2008) Infraexponential decay of wavelets. Proc. Nat. Acad. Sci. India Sect. A, 78:155–162
18. Shirani, M.R., SafiEsfahani, F.: Dynamic scheduling of tasks in cloud computing applying dragonfly algorithm, biogeographybased optimization algorithm and Mexican hat wavelet. J. Supercomput. **7–020**,(2020). https://doi.org/10.1007/s1122
19. Singh, A.: Distributional Mexican hat wavelet transform. The Journal of Analysis. **28**(2), 533–544 (2020)

20. Singh, A., Nikhila, R., Rawat, A., Singh, J.: Representation theorems for the Mexican hat wavelet transform. Mathematical Methods in the Applied Sciences **43**(7), 3914–3924 (2020)
21. Singh, A.K., Kumar, B., Dave, M., Mohan, A.: Multiple watermarking on medical images using selective discrete wavelet transform coefficients. Journal of Medical Imaging and Health Informatics **5**(3), 607–614 (2015)
22. Widder, D.V.: The Laplace transformation. Princeton University Press, Princeton (1946)
23. Zhou, Z., Adeli, H.: Time-frequency signal analysis of earthquake records using Mexican hat wavelets. Computer-Aided Civil and Infrastructure Engineering **18**(5), 379–389 (2003)

Fractal Fractional Derivative Operator Method on MCF-7 Cell Line Dynamics

Kolade M. Owolabi, Albert Shikongo, and Abdon Atangana

Abstract In this chapter, the dynamics modeling breast cancer known as MCF-7 cell line by means of a system of ordinary differential equations are considered. The dynamics are extended to a system of fractal fractional partial differential equations. The well-posed, physiological level, and stability conditions for the system of fractal fractional partial differential equations dynamics are established as well. Since the extended dynamics are not solvable analytically, a novel fractal fractional numerical method is derived, implemented and the results are presented with respect to the derived stability conditions.

Keywords Breast cancer · MCF-7 cell line · Fractal fractional operator · Well-posed · Stability analysis · Numerical method

1 Introduction

When normal cells divide in an orderly way, normal cells die when they are worn out or damaged and then new cells take over. Cancer occurs when normal cells start to grow out of control, which implies that cancer cells crowd out normal cells, resulting in problems in a part of a body where the crowding out has occurred. If the part of a body where cancer has started is a breast, then the cancer is referred to as breast cancer. Breast cancer cells usually form as a tumor that can often be seen through

K. M. Owolabi (✉)
Department of Mathematical Sciences, Federal University of Technology, Akure 704, Ondo, Nigeria
e-mail: kmowolabi@futa.edu.ng

K. M. Owolabi · A. Atangana
Institute for Groundwater Studies, Faculty of Natural and Agricultural Sciences University of the Free State, Bloemfontein 9300, South Africa

A. Shikongo
Department of Mathematics, University of Namibia, Private Bag, Windhoek Bag 13301, Namibia
e-mail: ashikongo@unam.na

J. Singh et al. (eds.), *Methods of Mathematical Modelling and Computation for Complex Systems*, Studies in Systems, Decision and Control 373,
https://doi.org/10.1007/978-3-030-77169-0_13

an x-ray or felt as a lump. However, breast cancer include ductal carcinoma in situ, invasive ductal carcinoma, inflammatory breast cancer, and metastatic and/or stage 4 breast cancer. In this study, the focus is on metastatic and/or stage 4 breast cancer that occurs almost entirely in women even though men can get breast cancer too. The metastatic and/or stage 4 breast cancer is also known as MCF-7, commonly diagnosed in women worldwide. The MCF-7 is a breast cancer cell line isolated in 1970 from a 69 year old Caucasian woman. Hence, MCF-7 is an acronym of Michigan Cancer Foundation-7, referring to the institute in Detroit where the cell line was established in 1973 by Herbert Soule and co-workers. Thus, this chapter, follows the wonderful study carried out in [46], that MCF-7 is a serious threat to human breast. Therefore, the recent studies carried out in order to treat this deadly infection are as follow.

Parush in [41] mentioned that in oncology, hyperthermia is understood as a planned, controlled technique of heating cancerous changes in order to destroy or to stop their cells growth, whereas in clinical practice, hyperthermia is used in combination with radiotherapy, chemotherapy, or immunological therapy. Furthermore, they stated that during the hyperthermia, the affected tissue is typically exposed to a temperature in the range of (40–45) °C, with the exception of thermoablation in which the temperatures reach much higher values. They further explained that thermoablation is a process characterized by the use of high temperatures up to 90 °C, in which the electrode using the radio frequency is inserted into the central area of the tumor whereas, interstitial thermoablation is used to treat, among others, breast and brain cancer. Thus, they noted that the therapy is consisting of inducing coagulation necrosis in an area that is heated to very high temperatures. Therefore, based on their above understanding they designed the dynamics coupled with thermo-electric model, in which the electric field is described by means of the Laplace equation and Pennes equation. Henceforth, coupling occurs at the level of the additional source function in the Pennes equation. They were able to present that temperature field obtained through their dynamics enable them to determine the Arrhenius integral as a determinant of the destruction of biological tissue. Similar studies can be traced in [24].

The study that deals with the mathematical model of breast cancer at the initial growth stage known as ductal carcinoma in situ is considered in [15], in which a computational approach is developed based on an iterative procedure, space marching and mollification methods. Their stability and convergence results support the efficiency and ability of the proposed numerical approach.

In [47] the stability analysis of a mixed immunotherapy and chemotherapy of tumors' model, presented an indicator of the host's ability to fight a cancer of the tumor-free equilibrium are obtained. Their studies present that the immune system is able to control a small tumor, and the host's ability to fight a cancer depends on individual variation. Their numerical method based on the continuation technique for one-parameter bifurcation analysis with periodically pulsed therapies provided a good approximation of the maximum tumor burden as a function of the dosage. Wei in [47] was able to deduce the treatment failure through the numerical simulation of chemotherapy-induced lymphocyte. Hence, they were able to determined an efficient and safe combination of dosages for combined chemotherapy and immunotherapy treatment.

Wei in [46] derived the dynamics for metastatic and/or stage 4 breast cancer, in which the parameter values and functional forms are justified by experimental and clinical data. Since, MCF-7 cell line is the most common in vitro dynamic used for experiments on human breast cancer, then the dynamics in [46] govern the MCF-7 cell growth with interaction among tumor cells, estradiol, natural killer cells, cytotoxic T lymphocytes (CTLs) or CD8+ T cells, and white blood cells (WBCs), which are denoted by $\mathscr{T}, E, N, L, \mathscr{C}$ in that order. Thus, Wei in [46] was able to deduced the dynamics in the form of a system of ordinary differential equations (ODEs) as

$$\frac{d\mathbf{X}}{dt} = \mathbf{F} \text{ with } \mathbf{X_0} = \mathbf{X_c}, \tag{1}$$

where,

$$\mathbf{X} = [\mathscr{T}, N, L, \mathscr{C}]' \text{ and } \mathbf{F} = [F_1, F_2, F_3, F_4]',$$

in which,

$$\left.\begin{array}{l} F_1 = (a + I_{TE}(\mathscr{T}, E)\left(1 - \frac{\mathscr{T}}{K}\right) - I_{\mathscr{T}N}(\mathscr{T}, N) - I_{\mathscr{T}L}(\mathscr{T}, L), \\ F_2 = e\mathscr{C} - fN - p_2 N\mathscr{T} + I_{N\mathscr{T}}(\mathscr{T}, N), \\ F_3 = \left(p_4 L_N + \frac{p_5 I}{a_4 + I} L\right)\left(1 - \frac{L}{K_L}\right)\frac{\mathscr{T}}{a_5 + \mathscr{T}} - dL, \\ F_4 = \alpha - \beta\mathscr{C}, \end{array}\right\} \tag{2}$$

with the initial conditions,

$$\mathbf{X_c} = [T(0) = \mathscr{T}_c,\ N(0) = N_c,\ L(0) = L_c,\ C(0) = \mathscr{C}_c]',\ \mathscr{T}_c \geq 0, N_c \geq 0, L_c \geq 0, \mathscr{C}_c \geq 0.$$

In the first equation in (2), a, K, $I_{\mathscr{T}E}(\mathscr{T}, E)$, $I_{\mathscr{T}N}(\mathscr{T}, N)$, $I_{\mathscr{T}L}(\mathscr{T}, L)$, $I_{N\mathscr{T}}(\mathscr{T}, N)$ denote the intrinsic growth rate, carrying capacity of the tumor cells, stimulation of the proliferation of MCF-7 cells by estradiol, tumor lysis by natural killer cells, tumor lysis by cytotoxic T lymphocytes, the process of recruitment of natural killer cells in the presence of tumor cells. Consequently, Wei in [46] experimentally deduced the functional response terms

$$I_{\mathscr{T}E}(\mathscr{T}, E), I_{\mathscr{T}N}(\mathscr{T}, N), I_{\mathscr{T}L}(\mathscr{T}, L), I_{N\mathscr{T}}(\mathscr{T}, N),$$

as

$$I_{\mathscr{T}E}(\mathscr{T}, E) = \frac{cE\mathscr{T}^2}{1 + \alpha_1 E + \beta_1 \mathscr{T}^2}, \quad I_{\mathscr{T}N}(\mathscr{T}, N) = \frac{p_1 \mathscr{T} N^2}{1 + \alpha_2 \mathscr{T} + \beta_2 N^2},$$

$$I_{\mathscr{T}L}(\mathscr{T}, L) = \frac{p_6 \mathscr{T}^2 L}{1 + \alpha_6 \mathscr{T}^2 + \beta_6 L}, \quad I_{N\mathscr{T}}(\mathscr{T}, N) = \frac{p_3 N\mathscr{T}}{1 + \alpha_2 \mathscr{T} + \beta_2 N^2},$$

where, $E = \tilde{E}(t - n\tau)$ denotes the blood content of estradiol [48] for $t \in [n, (n+1)\tau]$, $\tau = 29, n \in \mathbb{N}$, in which c, p_j, $(j = 1, 3, 6)$, denote growth rate of the interactions of estradiol with WBCs cells, growth rate of the interactions of WBCs with natural killers, growth rate of the interactions of WBCs with lymphocytes, $\alpha_j, (j = 1, 2, 6)$ denote the birth rates of estradiol, WBCs and $\beta_j, (j = 1, 2, 6)$ denote the death rates of WBCs, natural killers and lymphocytes, respectively. In [46] it is mentioned that the growth rate of the natural killer cells depends on the concentration of WBCs, thus, in the second equation in Eq. (2) the parameter e denotes the growth rate of lymphocytes due to the presence of the functional response $I_{N\mathscr{T}}(\mathscr{T}, N)$, the parameter f denotes the rate of exhaustion of the natural killers due to significant reduction in perforin and granzymes B content and p_2 denotes the rate of inactivation of natural killers cells after some encounters with tumour cells. The third equation present the fact that mature lymphocytes cells increases progressively during young life and then remains relatively constant during their adulthood and the term $\frac{\mathscr{T}}{a_5+\mathscr{T}}$ presents the phenomenon that lymphocytes are activated in the presence of tumor cells and undergo apoptosis at the end of immune response, d denotes the rate of the lymphocytes that did not make it to the adulthood. The last equation in (2) denotes the constant growth rate α and their linear death rate β.

The first aim in this chapter is to extend the dynamics in Eq. (1) to a system of fractal fractional parabolic differential equations (FFPDEs) in the Caputo sense [1, 2] as

$$ {}^{CFF}D^{\mu,\kappa}_{a,t}(\mathbf{X}(t)) - \mathcal{D}\frac{1}{2}\left[{}^{C}D^{\mu}_{a,x}\mathbf{X} + {}^{C}D^{\mu}_{b,x}\mathbf{X}\right] = \mathbf{F} \text{ with } \mathbf{X_0} = \mathbf{X_c}, \mu, \kappa \in (0, 1), \quad (3) $$

on $(x, t) \in \Omega \times (0, \infty)$, with no-flux boundary conditions, where, $0 \leq a < b \in \mathbb{Z}^+$, ${}^{CFF}D^{\mu,\kappa}_{0,t}$ denotes fractal fractional derivative with respect to time (t) in the Caputo sense, ${}^{C}D^{\mu}_{a,x}$, ${}^{C}D^{\mu}_{b,x}$ denote a left-side and right-side Caputo fractional derivatives with respect to space x [11, 44], the initial conditions are as in Eq. (1). The system of FFPDEs in Eq. (3) has the spatial effects of the involved cells denoted by $\mathcal{D} = [\mathcal{D}_{\mathscr{T}}, \mathcal{D}_N, \mathcal{D}_L, \mathcal{D}_{\mathscr{C}}]'$. Fractal fractional derivative is a natural extension of an fractional derivative. Consequently fractional derivative is an extension of ordinary derivative, where integrals and derivatives are defined for arbitrary orders. Thus, excellent literature on the application of fractional derivative to model a range of real-life phenomena can be traced in [3, 5–9, 17, 28, 30–39], and references therein. However, very little has been done in the direction of fractal fractional derivative on real-life phenomena. These few application of fractal fractional on real life phenomena are highlighted below.

A novel operator known as fractal-fractional in the sense of Caputo derivative for banking data is considered in [19]. Their comparative analysis of rural and commercial banks data of Indonesia for the years 2004 to 2014 through some reasonable set of parameters for both the data present the best fitting to the data, for their well-posed dynamics.

Chen et al., in [12] developed a fractal derivative model of anomalous diffusion and the fundamental solution of the fractal derivative equation for their model. They mentioned that their new dynamics characterizes a clear power law as compared to the corresponding fractional derivative dynamics.

The second aim in this chapter is to derive a novel well-posed numerical method for the dynamics in Eq. (3). In most cases the dynamics such as the ones in Eq. (3) are solved by means of quadrature numerical methods [10], which in turn require extensive process to reach to the final presentation of the solution. However, the novel numerical method is easy and at the same time enables us to combine integer solution with the non-integer solution. Thus, it suffices to refer readers to [1, 12, 19]. The rest of the chapter is organized as follows. Section 2 deals with the mathematical analysis of the system of FFPDEs, whereas in Sect. 3 deals with the derivation and implemention of the well-posed fractal fractional operator numerical method and present the results in Sect. 4. Section 5 concludes the chapter.

2 Mathematical Analysis of the Models

In this section, the dynamics in Eq. (3) are analyzed, first by establishing the continuously dependency of unique solution on the data, physiological levels, equilibrium point(s) and asymptotic stability condition(s) of equilibrium point of the of the dynamics.

2.1 Preliminaries

In view of the equation in (3), it can be seen that the left-side fractal fractional derivative in the Caputo sense of order μ with respect to time t is

$$
{}^{CFF}D_{a,t}^{\mu,\kappa}(\mathbf{X}(t)) = \begin{cases} \frac{1}{\Gamma(1-\alpha)} \int\limits_{a}^{t} (t-\tau)^{n-\mu-1} \frac{d}{d\tau^{\kappa}}(\mathbf{X}(\tau))d\tau, \\ = \frac{t^{1-\kappa}}{\kappa\Gamma(1-\mu)} \int\limits_{0}^{t} (t-\tau)^{-\mu}\dot{\mathbf{X}}(\tau)d\tau, \text{ if } \mu, \kappa \in (0,1), \\ \dot{\mathbf{X}}, \text{ if } \mu = \kappa = 1, \end{cases} \tag{4}
$$

where,

$$
\frac{d}{d\tau^{\kappa}}(\mathbf{X}(\tau)) = \lim_{t \to \tau} \frac{\mathbf{X}(t) - \mathbf{X}(\tau)}{t^{\kappa} - \tau^{\kappa}}.
$$

Similarly, for $n-1 < \alpha \leq n, n \in \mathbb{N} \leq 2$ (in this case) on $[a, b]$, the left-side Caputo fractional operator is

$$
{}^{C}D_{a,x}^{\mu}\mathbf{X}(x,t) = \begin{cases} \frac{1}{\Gamma(n-\mu)}\int\limits_{a}^{x}(x-\varphi)^{n-\mu-1}\frac{\partial^n}{\partial\varphi^n}\mathbf{X}(\varphi,t)d\varphi, \text{ if } n-1<\mu<n, \\ \frac{\partial^n}{\partial\varphi^n}\mathbf{X}(\varphi,t), \text{ if } \mu = n, \end{cases} \tag{5}
$$

and the right-side Caputo fractional operator is

$$
{}^{C}D_{x,b}^{\mu}\mathbf{X}(x,t) = \begin{cases} \frac{(-1)^n}{\Gamma(n-\mu)}\int\limits_{x}^{b}(\varphi-x)^{n-\mu-1}\frac{\partial^n}{\partial\varphi^n}\mathbf{X}(\varphi,t)d\varphi, \text{ if } n-1<\mu<n, \\ (-1)^n\frac{\partial^n}{\partial\varphi^n}\mathbf{X}(\varphi,t)d\varphi, \text{ if } \mu = n. \end{cases} \tag{6}
$$

Let

$$
R(\Omega) = \left\{\mathbf{X}(x,t)\middle|\frac{d^2}{dx^2}\mathbf{X} \in C_1^2(\bar{\Omega}) \text{ and } \frac{d}{dx}\mathbf{X} \in C(\bar{\Omega})\right\}, \tag{7}
$$

where, $\bar{\Omega} = [0, T] \times [a, b]$ for some $T \in \mathbb{N} < \infty$. In view of [21], the hypothesis below follow.

Lemma 1 *If a function $f \in C^1[0, T]$ attains its maximum over the interval $[0, T]$ at a point $\tau = t_0 \in (0, T]$, then*

$$
0 \leq {}^{FF}D_{a,t_0}^{\mu,\kappa} f(t_0), \forall \mu, \kappa \in (0, 1]. \tag{8}
$$

Lemma 2 *Let $f \in C^2([a, b])$, such that it attains its maximum over the interval $[a, b]$ at a point $x_0 \in [a, b]$ and $f'(x_0) \geq 0$. Then*

$$
{}^{C}D_{a,x_0}^{\beta} f(x_0) := \left[{}^{C}D_{a,x_0}^{\mu} f(x_0) + {}^{C}D_{b,x_0}^{\mu} f(x_0)\right] \leq 0, \forall \beta \in (0, 2). \tag{9}
$$

Lemma 3 *Let $f \in C^2([a, b])$, such that it attains its maximum over the interval $[a, b]$ at a point $x_0 \in [a, b]$ and $f'(b) \leq 0$. Then*

$$
{}^{C}D_{a,x_0}^{\beta} f(x_0) \leq 0, \forall \beta \in (0, 2]. \tag{10}
$$

Proof The prove is guaranteed by the definition. □

Theorem 1 *Let $\mathbf{X} \in R(\Omega)$ denotes the solution for the system of FFPDEs in Eq. (3) in Ω, in which $\frac{d}{dx}\mathbf{X}|_{x=a} \geq 0$ and $\frac{d}{dx}\mathbf{X}|_{x=b} \leq 0$. Then either $\mathbf{X}(x, t) \leq 0, \forall (x, t) \in \bar{\Omega}$, or $\mathbf{X}$ attains its positive maximum on the bottom or back-side parts of the sides $S = \{[a, b] \times \{a\} \cup \{a\} \times [a, T] \cup \{b\} \times [a, T]\}$ of the boundary of the domain Ω.*

Proof Let a point (x_0, t_0) exists for $x_0 \in (a, b)$ and $t_0 \in (0, T]$ such that

$$
\mathbf{X}(x_0, t_0) \geq \left\{\mathbf{0}, \max_{a \leq x \leq b} \mathbf{X}(0), \mathbf{0}, \mathbf{0}\right\} = \mathbf{M} \geq \mathbf{0}.
$$

Let $\epsilon = \mathbf{X}(x_0, t_0) - \mathbf{M} > 0$, such that

$$\mathbf{z}(x,t) = \mathbf{X}(x,t) + \frac{\epsilon(T-t)}{2T}, \forall (x,t) \in \bar{\Omega},$$

which is equivalent to

$$\mathbf{z}(x,t) \leq \mathbf{X}(x,t) + \frac{\epsilon}{2}, \forall (x,t) \in \bar{\Omega}.$$

Thus,

$$\mathbf{z}(x_0,t_0) \geq \mathbf{X}(x_0,t_0) = \epsilon + \mathbf{M} \geq \frac{1}{2}\epsilon + \mathbf{z}(x,t), \forall (x,t) \in S,$$

implies that $\mathbf{z}$ cannot attains its maximum on S. Hence, let (x_1, t_1) denotes the maximum point of $\mathbf{z}$ over $\bar{\Omega}$, such that $x_1 \in (a,b)$, $t_1 \in (a,T]$ and

$$\mathbf{z}(x_1,t_1) \geq \mathbf{z}(x_0,t_0) \geq \epsilon + \mathbf{M} > \epsilon.$$

In view of Lemma 1 and Lemma 3 one obtains

$$^{CFF}D_{a,t}^{\mu,\kappa}\mathbf{w}|_{(x_1,t_1)} > 0, \left[{}^{C}D_{a,x}^{\mu}\mathbf{w}|_{(x_1,t_1)} + {}^{C}D_{b,x}^{\mu}\mathbf{w}|_{(x_1,t_1)}\right] \leq 0, \;\; \forall \mu, \kappa \in (0,1). \quad (11)$$

Applying the properties of the fractal fractional derivative operator in the Caputo sense to z, one obtains

$$P(D_t)\mathbf{X} = P(D_t)\mathbf{z} + \frac{\epsilon}{2T}\left(\frac{t^{(1-\mu)(1-\kappa)}}{\kappa\Gamma(2-\mu)} + \sum_{i=1}^{m}\varpi_i \frac{t^{(1-\mu)((1-\kappa))}}{\kappa\Gamma(2-\mu)}\right),$$

where, $\varpi_i \geq 0$ and $i = 1, 2, \ldots, m \in \mathbb{N}$. Therefore, in view of the system in Eq. (3) one finds that

$$\begin{aligned}\left(P(D_t)\mathbf{X} - \mathcal{D}\frac{1}{2}\left[{}^{C}D_{a,x}^{\mu}\mathbf{X} + {}^{C}D_{b,x}^{\mu}\mathbf{X}\right] - \mathbf{F}\right)|_{(x_1,t_1)} &= P(D_t)\mathbf{z} \\ &+ \frac{1}{2T}\epsilon\left(\frac{t_1^{(1-\mu)(1-\kappa)}}{\kappa\Gamma(2-\alpha)} + \sum_{i=1}^{m}\varpi_i\frac{t_1^{(1-\mu)(1-\kappa)}}{\kappa\Gamma(2-\alpha)}\right) \\ &- \mathbf{F}\left(z(x_1,t_1) - \epsilon\frac{(T-t_1)}{2T}\right) \\ &\geq \epsilon\frac{1}{2T}\left(\frac{t_1^{(1-\mu)(1-\kappa)}}{\kappa\Gamma(2-\alpha)} + \sum_{i=1}^{m}\varpi_i\frac{t_1^{(1-\mu)(1-\kappa)}}{\kappa\Gamma(2-\alpha)}\right) \\ &- \epsilon\mathbf{F}\left(1 - \frac{T-t_1}{2T}\right) > 0,\end{aligned}$$

which is a contradiction. □

Theorem 2 *Let $\mathbf{X} \in R(\Omega)$ denotes the solution for one component of the system of FFPDEs in Eq. (3) in Ω, in which $\frac{d}{dx}\mathbf{X}|_{x=a} \geq 0$ and $\frac{d}{dx}\mathbf{X}|_{x=b} \leq 0$. Then* *either* $\mathbf{X}(x,t) \le 0, \forall (x,t) \in \bar{\Omega}$, *or* $\mathbf{X}$ *attains its positive minimum on the bottom or back-side parts of the sides* $S = \{[a,b] \times \{a\} \cup \{a\} \times [a,T] \cup \{b\} \times [a,T]\}$ S *of the boundary of the domain* Ω.

Proof The prove to this theorem is similar to the prove of Theorem 1. □

2.2 Uniqueness and Continuous Dependence of the Solution

By means of the maximum principle and minimum principle, the following theorems hold.

Theorem 3 *Let* $\mathbf{X}$ *denotes a classical solution to the system in Eq. (3),* $F \in C(\bar{\Omega})$, $\frac{d}{dx}\mathbf{X}|_{x=a} = \mathbf{0}$ *and* $\frac{d}{dx}\mathbf{X}|_{x=b} = \mathbf{0}$. *Then,*

$$\|\mathbf{X}\|_{C(\bar{\Omega})} \le \max\{\mathbf{M}_0, \mathbf{M}_1, \mathbf{M}_2\} + \frac{2T^{\alpha}}{\Gamma(1+\alpha)}\mathbf{M}, \tag{12}$$

where, $\mathbf{M}_0 = \|\mathbf{X}(0)\|_{C(\bar{\Omega})}$, $\mathbf{M}_1 = \|\dot{\mathbf{X}}(a,t)\|$, $\mathbf{M}_2 = \|\dot{\mathbf{X}}_1(b,t)\|$, *and* $\mathbf{M} = \|\mathbf{F}\|_{C(\bar{\Omega})}$.

Proof The prove follows directly from the Neumann boundary conditions and the fact that the source term for the system in Eq. (3) is identically zero. □

Theorem 4 *Let* $\frac{d}{dx}\mathbf{X}|_{x=a} = \mathbf{0}$ *and* $\frac{d}{dx}\mathbf{X}|_{x=b} = \mathbf{0}$. *Then the system in Eq. (3) possesses at most one classical solution. This solution, if it exists, it continuously depends on the data associated the system in Eq. (3) in the sense that if*

$$\|\mathbf{F} - \tilde{\mathbf{F}}\| \le \epsilon,\ \|\mathbf{X}(0) - \tilde{\mathbf{X}}(0)\|_{C(\bar{\Omega})} \le \epsilon_0,\ \|\dot{\mathbf{X}}(a,t) - \dot{\tilde{\mathbf{X}}}(a,t)\| \le \epsilon_1,\ \|\dot{\mathbf{X}}(b,t) - \dot{\tilde{\mathbf{X}}}(b,t)\| \le \epsilon_2,$$

then

$$\|\mathbf{X} - \tilde{\mathbf{X}}\|_{C(\bar{\Omega})} \le \max\{\epsilon, \epsilon_0, \epsilon_1, \epsilon_2\} + \frac{2T^{\alpha}}{\Gamma(1+\alpha)}\epsilon,$$

holds.

Before the establishment of the equilibrium point(s) for the dynamics in Eq. (3), the physiological level of the extended dynamics in Eq. (3) is established and it coincides with the physiological level reported in [46].

2.3 Physiological Level for the Dynamics

When the dynamics in (3) is cancer free, then the following theorem holds.

Theorem 5 *The physiological level of the system of coupled FFPDEs in Eq. (3) is*

$$(N, L, \mathscr{C}) = \left(\frac{N_c^2}{f} + \frac{(\mathscr{C}_c + \alpha)}{\beta}, \frac{L_c}{d}, \frac{(\mathscr{C}_c + \alpha)}{\beta}\right).$$

Proof Since at the physiological level, the system of coupled FFPDEs in Eq. (3) reduces to,

$$\left.\begin{array}{l}\frac{\partial N}{\partial t} - \mathcal{D}_N \frac{\partial^2 N}{\partial x^2} = eC - fN, \\ \frac{\partial L}{\partial t} - \mathcal{D}_L \frac{\partial^2 L}{\partial x^2} = -dL, \\ \frac{\partial \mathscr{C}}{\partial t} - \mathcal{D}_{\mathscr{C}} \frac{\partial^2 \mathscr{C}}{\partial x^2} = \alpha - \beta\mathscr{C},\end{array}\right\} \tag{13}$$

then, let $\tilde{\mathscr{C}}(x, s) = \mathfrak{L}\{\mathscr{C}(x, s)\}(s)$. Applying the Laplace transform [18] to the last equation in equation in (13) one obtains

$$\begin{aligned} s\tilde{\mathscr{C}}(x, s) - \mathscr{C}_c &= \mathcal{D}_{\mathscr{C}} \frac{d^2\tilde{\mathscr{C}}(x, s)}{dx^2} + \alpha - \beta\tilde{\mathscr{C}}(x, s), \\ \Rightarrow \mathcal{D}_{\mathscr{C}} \frac{d^2\tilde{\mathscr{C}}(x, s)}{dx^2} &- (\beta + s)\tilde{\mathscr{C}}(x, s) = -\mathscr{C}_c - \alpha. \end{aligned} \tag{14}$$

Equation in Eq. (14) is a nonhomogeneous, linear second order ordinary differential equation (ODE) [18]. Applying the method of variation of parameters [18] to the ODE in Eq. (14), one finds

$$P_1(x, s) = \int^x \frac{\exp(-(\beta + s)z)(\mathscr{C}_c + \alpha)}{W(u_1, u_2)} dz, \quad P_2(x, s) = -\int^x \frac{\exp((\beta + s)z)(\mathscr{C}_c + \alpha)}{W(u_1, u_2)} dz, \tag{15}$$

where, P_1, P_2 denote some arbitrary parameters, $u_1 = \exp((\beta + s)x), u_2 = \exp(-(\beta + s)x)$ and $W(u_1, u_2)$ denote the Wronskian of u_1, u_2 [18]. Thus, from Eq. (15), one easily obtains

$$P_1(x, s) = \frac{(\mathscr{C}_c + \alpha)}{2(\beta + s)^2} \exp(-(\beta + s)x), \quad P_2(x, s) = \frac{(\mathscr{C}_c + \alpha)}{2(\beta + s)^2} \exp((\beta + s)x). \tag{16}$$

Hence, the solution to the ODE in Eq. (14) is

$$\tilde{\mathscr{C}}(x, s) = k_1 \exp((\beta + s)x) + k_2 \exp(-(\beta + s)x) + \frac{(\mathscr{C}_c + \alpha)}{(\beta + s)^2}, \tag{17}$$

where k_1, k_2, are constants to be determined. Imposing the no flux boundary conditions to the solution in Eq. (17), one obtains

$$0 = k_1(\beta + s) - k_2(\beta + s), \Rightarrow k_1 = k_2 = 0, \tag{18}$$

Hence, the solution in Eq. (17) becomes

$$\tilde{\mathscr{C}}(x, s) = \frac{(\mathscr{C}_c + \alpha)}{(\beta + s)^2}, \tag{19}$$

Applying the inverse Laplace transform to equation in Eq. (19) one finds that

$$\mathscr{C}(x, t) = \mathfrak{L}^{-1}\left\{\frac{(\mathscr{C}_c + \alpha)}{(\beta + s)^2}\right\}, \tag{20}$$

in which,

$$\frac{(C_c + \alpha)}{(\beta + s)^2} = \frac{A}{\beta + s} + \frac{Bs}{\beta + s}, \tag{21}$$

where A, B, are constants to be determine as follows.

$$\mathscr{C}_c + \alpha = \beta A + sA + \beta Bs + Bs^2, \Rightarrow A = \frac{(\mathscr{C}_c + \alpha)}{\beta} \text{ and } B = 0. \tag{22}$$

Thus, equation in (20) becomes

$$\mathscr{C}(x, t) = \frac{(\mathscr{C}_c + \alpha)}{\beta}\mathfrak{L}^{-1}\left\{\frac{1}{\beta + s}\right\} = \frac{(\mathscr{C}_c + \alpha)}{\beta} \exp(-\beta t). \tag{23}$$

Similarly, for the second equation in Eq. (15), one can easily show that

$$L(x, t) = \frac{L_c}{d} \exp(-dt), \tag{24}$$

and using the same techniques, one deduced from equation in Eq. (19) that, the first equation in Eq. (13) is

$$\tilde{N}(x, s) = k_1 \exp((f + s)x) + k_2 \exp(-(f + s)x) + \frac{N_c}{(f + s)^2} + e\frac{(\mathscr{C}_c + \alpha)}{(\beta + s)^2}, \tag{25}$$

which, yields

$$N(x, s) = \mathfrak{L}^{-1}\left\{\frac{N_c}{(f + s)^2} + e\frac{(\mathscr{C}_c + \alpha)}{(\beta + s)^2}\right\} = \frac{N_c^2}{f} \exp(-ft) + \frac{(\mathscr{C}_c + \alpha)}{\beta} \exp(-\beta t). \tag{26}$$

Therefore, the physiological level for of coupled FFPDEs in Eq. (3) is at

$$(N, L, \mathscr{C}) = \left(\frac{N_c^2}{f}\exp(-ft) + \frac{(\mathscr{C}_c+\alpha)}{\beta}\exp(-\beta t), \frac{L_c}{d}\exp(-dt), \frac{(\mathscr{C}_c+\alpha)}{\beta}\exp(-\beta t)\right), \tag{27}$$

which is equivalent to

$$(N, L, \mathscr{C}) = \left(\frac{N_c^2}{f} + \frac{(\mathscr{C}_c+\alpha)}{\beta}, \frac{L_c}{d}, \frac{(\mathscr{C}_c+\alpha)}{\beta}\right), \tag{28}$$

as $t \to \infty$, which concludes the prove. □

2.4 Asymptotic Stability Conditions for the Dynamics

At the steady state the dynamics in Eq. (3) becomes,

$$-\mathcal{D}\frac{1}{2}\left[{}_a^C D_x^\mu \mathbf{X} + {}_x^C D_b^\mu \mathbf{X}\right] = \mathbf{F}, \tag{29}$$

which is equivalent to

$$\left.\begin{array}{l}
-\mathcal{D}_{\mathscr{T}}\frac{\left[{}_a^C D_x^\mu \mathscr{T} + {}_x^C D_b^\mu \mathscr{T}\right]}{2} = \left(\mathscr{T}a + \frac{cE\mathscr{T}^2}{1+\alpha_1 E+\beta_1\mathscr{T}^2}\right)\left(1-\frac{\mathscr{T}}{K}\right) - \frac{p_1\mathscr{T}N^2}{1+\alpha_2\mathscr{T}+\beta_2 N^2} - \frac{p_6\mathscr{T}^2 L}{1+\alpha_6\mathscr{T}^2+\beta_6 L},\\
-\mathcal{D}_N\frac{\left[{}_a^C D_x^\mu N + {}_x^C D_b^\mu N\right]}{2} = e\mathscr{C} - fN - p_2 N\mathscr{T} + \frac{p_3 N\mathscr{T}}{1+\alpha_3\mathscr{T}+\beta_3 N^2},\\
-\mathcal{D}_L\frac{\left[{}_a^C D_x^\mu L + {}_x^C D_b^\mu L\right]}{2} = \left(p_4 L_N + \frac{p_5 I}{a_4+I}L\right)\left(1-\frac{L}{K_L}\right)\frac{\mathscr{T}}{a_5+\mathscr{T}} - dL,\\
-\mathcal{D}_{\mathscr{C}}\frac{\left[{}_a^C D_x^\mu \mathscr{C} + {}_x^C D_b^\mu \mathscr{C}\right]}{2} = \alpha - \beta\mathscr{C}.
\end{array}\right\}$$

Theorem 6 *The system in Eq. (29) is*

(i) asymptotic stable at a tumor free equilibrium point if

$$\left|\mathcal{D}_L\left(\frac{a}{K} - \frac{p_1\bar{N}^2}{(1+\beta_2\bar{N}^2)} - \mathcal{D}_{\mathscr{T}}\right)(f+\mathcal{D}_N)\right| < 1. \tag{30}$$

(ii) asymptotic stable at a tumour equilibrium if

$$\begin{gathered}
cE + \beta_2 cE\bar{N}^2 + \alpha_2 Ka + \alpha_1\alpha_2 EKa + \beta_1 p_1\bar{N}^2K^2 > \alpha_2 cKE,\\
Ka + \alpha_1 EKa + \beta_2\bar{N}^2Ka + \alpha_1\beta_2 E\bar{N}^2Ka > cKE + \beta_2 cKE\bar{N}^2,\\
p_1\bar{N}^2K^2 + \alpha_1 p_1\bar{N}^2K^2E > Ka + \alpha_1 EKa + \beta_2\bar{N}^2Ka + \alpha_1\beta_2 E\bar{N}^2Ka + a.
\end{gathered}$$

Proof *(i):* Solving the system in Eq. (29) one obtains the tumor free equilibrium point

$$(\bar{\mathscr{T}}, \bar{N}, \bar{L}, \bar{\mathscr{C}})_1 = (0, \frac{e\alpha}{f\beta}, 0, \frac{\alpha}{\beta}). \tag{31}$$

Linearizing the system in Eq. (29) at the equilibrium point $(\bar{T}, \bar{N}, \bar{L}, \bar{\mathscr{C}})$, one finds the non-zero entries of the Jacobian matrix as

$$\left.\begin{aligned}
J_\mu(1,1) &= \tfrac{a}{K} + \tfrac{(2acE\bar{\mathscr{T}})(1+\alpha_1 E+\beta_1\bar{\mathscr{T}}^2)-(acE\bar{\mathscr{T}}^2)(2\beta_1\bar{\mathscr{T}})}{(1+\alpha_1 E+\beta_1\bar{\mathscr{T}}^2)^2} \\
&\quad - \tfrac{(3cE\bar{\mathscr{T}}^2)(1+\alpha_1 E+\beta_1\bar{\mathscr{T}}^2)-(2\beta_1\bar{\mathscr{T}})(cE\bar{\mathscr{T}}^3)}{(1+\alpha_1 E+\beta_1\bar{\mathscr{T}}^2)^2} \\
&\quad - \tfrac{(p_1\bar{N}^2)(1+\alpha_1\bar{\mathscr{T}}+\beta_2\bar{N}^2)-(p_1\bar{\mathscr{T}}\bar{N}^2)(\alpha_2)}{(1+\alpha_1\bar{\mathscr{T}}+\beta_2\bar{N}^2)^2} \\
&\quad - \tfrac{(2p_6\bar{\mathscr{T}}\bar{L})(1+\alpha_6\bar{\mathscr{T}}^2+\beta_6\bar{L})-(p_6\bar{\mathscr{T}}^2 L)(2\alpha_6)}{(1+\alpha_6\bar{\mathscr{T}}^2+\beta_6\bar{L})^2} - \mathcal{D}_{\mathscr{T}}, \\
J_\mu(1,2) &= -\tfrac{(2p_1\bar{\mathscr{T}}\bar{N})(1+\alpha_1\bar{\mathscr{T}}+\beta_2\bar{N}^2)-(p_1\bar{\mathscr{T}}\bar{N}^2)(\alpha_2)}{(1+\alpha_1\bar{\mathscr{T}}+\beta_2\bar{N}^2)^2}, \\
J_\mu(1,3) &= -\tfrac{(p_6\bar{\mathscr{T}}^2)(1+\alpha_6\bar{\mathscr{T}}^2+\beta_6\bar{L})-(p_6\bar{\mathscr{T}}^2\bar{L})(\beta_6)}{(1+\alpha_6\bar{\mathscr{T}}^2+\beta_6\bar{L})^2}, \\
J_\mu(2,1) &= -p_2\bar{N} + \tfrac{(p_3\bar{N})(1+\alpha_3\bar{\mathscr{T}}+\beta_3\bar{N}^2)-(p_3\bar{N}\bar{\mathscr{T}})(\alpha_3)}{(1+\alpha_3\bar{\mathscr{T}}+\beta_3\bar{N}^2)^2}, \\
J_\mu(2,2) &= -f - p_2\bar{\mathscr{T}} + \tfrac{(p_3\bar{\mathscr{T}})(1+\alpha_3\bar{\mathscr{T}}+\beta_3\bar{N}^2)-(p_3\bar{N}\bar{\mathscr{T}})(2\beta_3\bar{N})}{(1+\alpha_3\bar{\mathscr{T}}+\beta_3\bar{N}^2)^2} - \mathcal{D}_N,\ J_\mu(2,3) = e, \\
J_\mu(3,1) &= \left(p_4 L_N + \tfrac{p_5 I\bar{L}}{a_4+I}\right)\left(1-\tfrac{\bar{L}}{K_L}\right)\tfrac{(\alpha_5+\bar{\mathscr{T}})-\bar{\mathscr{T}}}{(\alpha_5+\bar{\mathscr{T}})^2}, \\
J_\mu(3,3) &= \left(\tfrac{p_5 I}{(\alpha_4+I)} - \tfrac{p_4 L_N}{K_L} - \tfrac{2p_5 I\bar{L}}{K_L(\alpha_4+I)}\right)\left(\tfrac{\bar{\mathscr{T}}}{\alpha_5+\bar{\mathscr{T}}}\right) - \mathcal{D}_L,\ J_\mu(4,4) = -\beta - \mathcal{D}_{\mathscr{C}}.
\end{aligned}\right\} \tag{32}$$

Evaluating the Jacobian matrix in (32) at the equilibrium point in Eq. (31), then the non-zero entries of the Jacobian matrix are

$$\left.\begin{aligned}
&J_\mu(1,1) = \tfrac{a}{K} - \tfrac{p_1\bar{N}^2}{(1+\beta_2\bar{N}^2)} - \mathcal{D}_{\mathscr{T}},\ J_\mu(2,1) = -p_2\bar{N} + \tfrac{p_3\bar{N}}{(1+\beta_3\bar{N}^2)},\ J_\mu(2,2) = -f - \mathcal{D}_N, \\
&J_\mu(2,3) = e,\ J_\mu(3,1) = p_4 L_N,\ J_\mu(3,3) = -\mathcal{D}_L,\ J_\mu(4,4) = -\beta - \mathcal{D}_{\mathscr{C}}.
\end{aligned}\right\} \tag{33}$$

Hence, the results follows easily from

$$|J_\mu(1,1)J_\mu(2,2)J_\mu(3,3)| < 1.$$

□

Proof *(ii):* At tumour equilibrium point one finds that

$$\begin{aligned}
&a + cKE\bar{\mathscr{T}} + \alpha_2\bar{\mathscr{T}}^2 cKE + \beta_2 cKE\bar{\mathscr{T}}\bar{N}^2 - cE\bar{\mathscr{T}}^2 - \alpha_2 cE\bar{\mathscr{T}}^3 - \beta_2 cE\bar{\mathscr{T}}^2\bar{N}^2 - \bar{\mathscr{T}}Ka \\
&-\alpha_1 E\bar{\mathscr{T}}Ka - \beta_1\bar{\mathscr{T}}^3 Ka - \alpha_2\bar{\mathscr{T}}^2 Ka - \alpha_1\alpha_2 E\bar{\mathscr{T}}^2 Ka - \alpha_2\beta_1\bar{\mathscr{T}}^4 Ka - \beta_2\bar{N}^2\bar{\mathscr{T}}Ka \\
&-\alpha_1\beta_2 E\bar{N}^2\bar{\mathscr{T}}Ka - \beta_1\beta_2\bar{\mathscr{T}}^3\bar{N}^2 Ka - p_1\bar{N}^2K^2 - \alpha_1 p_1\bar{N}^2K^2E - \beta_1 p_1\bar{N}^2K^2\bar{\mathscr{T}}^2 = 0.
\end{aligned} \tag{34}$$

Rearranging equation in (34), one obtains a quartic polynomial

$$\gamma\bar{\mathscr{T}}^4 + \zeta\bar{\mathscr{T}}^3 + \varpi\bar{\mathscr{T}}^2 + \varrho\bar{\mathscr{T}} + \psi = 0,$$

which is equivalent to

$$\bar{\mathscr{T}}^4 + \frac{\zeta}{\gamma}\bar{\mathscr{T}}^3 + \frac{\varpi}{\gamma}\bar{\mathscr{T}}^2 + \frac{\varrho}{\gamma}\bar{\mathscr{T}} + \frac{\psi}{\gamma} = 0, \tag{35}$$

where,

$$\begin{aligned}
\gamma &= -\alpha_2\beta_1 K a,\ \zeta = -(\beta_1\beta_2\bar{N}^2 K a + \beta_1 K a + \alpha_2 c E),\\
\varpi &= (\alpha_2 c K E - c E - \beta_2 c E\bar{N}^2 - \alpha_2 K a - \alpha_1\alpha_2 E K a - \beta_1 p_1 \bar{N}^2 K^2),\\
\varrho &= (c K E + \beta_2 c K E\bar{N}^2 - K a - \alpha_1 E K a - \beta_2\bar{N}^2 K a - \alpha_1\beta_2 E\bar{N}^2 K a),\\
\psi &= -p_1\bar{N}^2 K^2 - \alpha_1 p_1 \bar{N}^2 K^2 E + a.
\end{aligned}$$

The polynomial in Eq. (35) is stable [45] if

$$\frac{\zeta}{\gamma} > 0,\ \frac{\varpi}{\gamma} > 0,\ 0 < \frac{\varrho}{\gamma} < \frac{\psi}{\gamma}, \tag{36}$$

which concludes the prove. □

The stability conditions in Theorem 6 imply that the system of coupled FFPDEs in Eq. (3) converges to the physiological level in Theorem 5.

3 Construction of the Fractal Fractional Operator Numerical Method

In this section, the novel well-posed numerical method for discretising the fractal fractional and/or integer derivatives operators, such as the system of FFPDEs in Eq. (3) is derived. This method is based on the method known as the matrix strip approach [43]. Thus, discretizing the interval $[a, T]$ through the points

$$0 = a = t_0 < t_1 < t_2 < \cdots < T = t_f,$$

where, the step-size $\Delta t = t_{j+1} - t_j = T/S_t$, for $j = 0, 1, \ldots, S_t$ and $S_t \in \mathbb{Z}^+$. On the left hand side of Eq. (3), the fractal fractional derivative ${}^{CFF}D^{\alpha,\kappa}_{a,t}(\mathbf{X})$ in the Caputo sense is approximated by the backward finite difference operator as

$$\begin{aligned}
{}^{CFF}D^{\mu,\kappa}_{a,t_f}(\mathbf{X}) &:= \frac{1}{\Gamma(1-\mu)}\int_a^{t_f}(t_f-\tau)^{-\mu}\frac{d}{dt^{\kappa}}\mathbf{X}d\tau = \frac{t_f^{1-\kappa}}{\kappa\Gamma(1-\mu)}\int_a^{t_f}(t_f-\tau)^{-\mu}\mathbf{X}'(\tau)d\tau,\\
&= \frac{t_f^{1-\kappa}}{\kappa\Gamma(1-\mu)}\sum_{j=1}^{S_t}\int_a^{t_f}(t_f-\tau)^{-\mu}\left(\frac{\mathbf{X}_j-\mathbf{X}_{j-1}}{(\Delta t)} + \mathcal{O}^2(\Delta t)\right)d\tau.
\end{aligned} \tag{37}$$

Let $S_t = 1, 2, 3$. Then, respectively, one obtains

$$
\begin{aligned}
{}^{CFF}D^{\mu,\kappa}_{a,t_1}(\mathbf{X})\Big|_{t=t_1} &= \frac{t_1^{1-\kappa}}{\kappa\Gamma(1-\mu)}\left(\frac{\mathbf{X}_1-\mathbf{X}_0}{(\Delta t)}\right)\int_a^{t_1}(t_1-\tau)^{-\mu}d\tau,\\
&= \frac{t_1^{1-\kappa}}{\kappa\Gamma(1-\mu)}\left(\frac{\mathbf{X}_1-\mathbf{X}_0}{(\Delta t)}\right)\frac{(t_1-\tau)^{1-\mu}}{1-\mu}\Big|_a^{t_1},\\
&= \frac{t_1^{1-\kappa}}{\kappa(\Delta t)^{\mu}}\left(\frac{\mathbf{X}_1-\mathbf{X}_0}{(1-\mu)\Gamma(1-\mu)}\right) = \frac{t_1^{1-\kappa}}{\kappa(\Delta t)^{\mu}}(-1)^1\binom{\mu}{1}\mathbf{X}_{(1-0)},\\
{}^{CFF}D^{\mu,\kappa}_{a,t_2}(\mathbf{X})\Big|_{t=t_2} &= \frac{t_1^{1-\kappa}}{\kappa(\Delta t)^{\mu}}(-1)^1\binom{\mu}{1}\mathbf{X}_{(1-0)} + \frac{t_2^{1-\kappa}}{\kappa(\Delta t)^{\mu}}(-1)^2\binom{\mu}{2}\mathbf{X}_{(2-1)},\\
{}^{CFF}D^{\mu,\kappa}_{a,t_2}(\mathbf{X})\Big|_{t=t_3} &= \frac{t_1^{1-\kappa}}{\kappa(\Delta t)^{\mu}}(-1)^1\binom{\mu}{1}\mathbf{X}_{(1-0)} + \frac{t_2^{1-\kappa}}{\kappa(\Delta t)^{\mu}}(-1)^2\binom{\mu}{2}\mathbf{X}_{(2-1)}\\
&\quad + \frac{t_3^{1-\kappa}}{\kappa(\Delta t)^{\mu}}(-1)^3\binom{\mu}{3}\mathbf{X}_{(3-2)}.
\end{aligned}
\tag{38}
$$

Neglecting the error terms in Eq. (37), it follows from equation in (38) that equation in (37) is equivalent to

$$
{}^{CFF}D^{\mu,\kappa}_{a,t_f}(\mathbf{X}) = \frac{t_f^{1-\kappa}}{\kappa(\Delta t)^{\mu}}\sum_{j=0}^{k}(-1)^j\binom{\mu}{j}\mathbf{X}_{k-j}, \tag{39}
$$

where, $k = 1, 2, \cdots, S_t$, which is equivalent to

$$
\begin{bmatrix}
\frac{t_f^{1-\kappa}}{\kappa(\Delta t)^{\mu}}\nabla_-^{\mu}(\mathbf{X})(t_0)\\
\frac{t_f^{1-\kappa}}{\kappa(\Delta t)^{\mu}}\nabla_-^{\mu}(\mathbf{X})(t_1)\\
\frac{t_f^{1-\kappa}}{\kappa(\Delta t)^{\mu}}\nabla_-^{\mu}(\mathbf{X})(t_2)\\
\vdots\\
\frac{t_f^{1-\kappa}}{\kappa(\Delta t)^{\mu}}\nabla_-^{\mu}(\mathbf{X})(t_{S_{t-1}})\\
\frac{t_f^{1-\kappa}}{\kappa(\Delta t)^{\mu}}\nabla_-^{\mu}(\mathbf{X})(t_{S_t})
\end{bmatrix}
= \frac{t_f^{1-\kappa}}{\kappa(\Delta t)^{\mu}}\mathbf{B}
\begin{bmatrix}
(\mathbf{X})_0\\
(\mathbf{X})_1\\
(\mathbf{X})_2\\
\vdots\\
(\mathbf{X})_{S_{t-1}}\\
(\mathbf{X})_{S_t}
\end{bmatrix}, \tag{40}
$$

where,

$$\mathbf{B} = \begin{bmatrix} \omega_0^{(\mu)} & 0 & 0 & 0 & \cdots & 0 \\ \omega_1^{(\mu)} & \omega_0^{(\mu)} & 0 & 0 & \cdots & 0 \\ \ddots & \ddots & \ddots & \ddots & \cdots & \cdots \\ \omega_{S_t-1}^{(\mu)} & \ddots & \omega_2^{(\mu)} & \omega_1^{(\mu)} & \omega_0^{(\mu)} & 0 \\ \omega_{S_t}^{(\mu)} & \omega_{S_t-1}^{(\mu)} & \ddots & \omega_2^{(\mu)} & \omega_1^{(\mu)} & \omega_0^{(\mu)} \end{bmatrix} \text{ and } \omega_j^{(\mu)} = (-1)^j \binom{\mu}{j}, \; j = 0, 1, \cdots, S_t.$$

Applying similar approach to discretize the spatial interval $[a, b]$ through the points

$$0 = a = x_0 < x_1 < x_2 < \cdots < x_{S_x} = b = x_f,$$

where, the step-size $\Delta x = x_{i+1} - x_i = x_f / S_x$, $i = 0, 1, \ldots, S_x$, the spatial fractional derivative operator [26, 27, 43] is approximated by the sum $\frac{1}{2}\left[{}^C D_{a,x}^{\mu}\mathbf{X} + {}^C D_{b,x}^{\mu}\mathbf{X}\right]$ of the two fractional operators, namely backward and forward finite difference operators as

$$\begin{bmatrix} \frac{1}{h^\mu}(\nabla_+^\mu + \nabla_-^\mu)\mathbf{X}(t_0) \\ \frac{1}{h^\mu}(\nabla_+^\mu + \nabla_-^\mu)\mathbf{X}(t_1) \\ \frac{1}{h^\mu}(\nabla_+^\mu + \nabla_-^\mu)\mathbf{X}(t_2) \\ \vdots \\ \frac{1}{h^\mu}(\nabla_+^\mu + \nabla_-^\mu)\mathbf{X}(t_{S_t-1}) \\ \frac{1}{h^\mu}(\nabla_+^\mu + \nabla_-^\mu)\mathbf{X}(t_{S_t}) \end{bmatrix} = \frac{1}{h^\beta} \begin{bmatrix} \omega_0^{(\beta)} & \omega_1^{(\beta)} & \omega_2^{(\beta)} & \omega_3^{(\beta)} & \cdots & \omega_m^{(\beta)} \\ \omega_1^{(\beta)} & \omega_0^{(\beta)} & \omega_1^{(\beta)} & \omega_2^{(\beta)} & \cdots & \omega_{m-1}^{(\beta)} \\ \omega_2^{(\beta)} & \omega_1^{(\beta)} & \omega_0^{(\beta)} & \omega_1^{(\beta)} & \cdots & \omega_{m-2}^{(\beta)} \\ \ddots & \ddots & \ddots & \ddots & \cdots & \cdots \\ \omega_{m-1}^{(\beta)} & \ddots & \omega_2^{(\beta)} & \omega_1^{(\beta)} & \omega_0^{(\beta)} & \omega_1^{(\beta)} \\ \omega_m^{(\beta)} & \omega_{m-1}^{(\beta)} & \ddots & \omega_2^{(\beta)} & \omega_1^{(\beta)} & \omega_0^{(\beta)} \end{bmatrix} \begin{bmatrix} (\mathbf{X})_0 \\ (\mathbf{X})_1 \\ (\mathbf{X})_2 \\ \vdots \\ (\mathbf{X})_{S_t-2} \\ (\mathbf{X})_{S_t-1} \\ (\mathbf{X})_{S_t} \end{bmatrix}.$$

Therefore, following the derivations in [42] and the references therein, one obtains a well-posed discrete system of linear FFPDEs as

$$\left.\begin{aligned} \left[(\mathbf{B}_1)_{t_f\times t_f}^{(\mu)} \otimes \mathbf{I}_{x_f\times x_f} - \mathcal{D}_{\mathcal{T}}\mathbf{I}_{t_f\times t_f} \otimes (\mathbf{B}_1)_{x_f\times x_f}^{(\beta)}\right](\mathbf{X}_1)_{t_f,x_f} &= (\mathbf{F}_1)_{t_f,x_f}, \\ \left[(\mathbf{B}_2)_{t_f\times t_f}^{(\mu)} \otimes \mathbf{I}_{x_f\times x_f} - \mathcal{D}_{N}\mathbf{I}_{t_f\times t_f} \otimes (\mathbf{B}_2)_{x_f\times x_f}^{(\beta)}\right](\mathbf{X}_2)_{t_f,x_f} &= (\mathbf{F}_2)_{t_f,x_f}, \\ \left[(\mathbf{B}_3)_{t_f\times t_f}^{(\mu)} \otimes \mathbf{I}_{x_f\times x_f} - \mathcal{D}_{L}\mathbf{I}_{t_f\times t_f} \otimes (\mathbf{B}_3)_{x_f\times x_f}^{(\beta)}\right](\mathbf{X}_3)_{t_f,x_f} &= (\mathbf{F}_3)_{t_f,x_f}, \\ \left[(\mathbf{B}_4)_{t_f\times t_f}^{(\mu)} \otimes \mathbf{I}_{x_f\times x_f} - \mathcal{D}_{\mathcal{C}}\mathbf{I}_{t_f\times t_f} \otimes (\mathbf{B}_4)_{x_f\times x_f}^{(\beta)}\right](\mathbf{X}_4)_{t_f,x_f} &= (\mathbf{F}_4)_{t_f,x_f}, \end{aligned}\right\} \tag{41}$$

on which the auxiliary initial conditions [43] are imposed

$$\left.\begin{aligned} (\mathcal{E}_1)_{t_f,x_f}(x,t) &= (\mathbf{X}_1)_{t_f,x_f}(x,t) - (\mathbf{X}_1)_{t_f,x_f}(x,0), \\ (\mathcal{E}_2)_{t_f,x_f}(x,t) &= (\mathbf{X}_2)_{t_f,x_f}(x,t) - (\mathbf{X}_2)_{t_f,x_f}(x,0), \\ (\mathcal{E}_3)_{t_f,x_f}(x,t) &= (\mathbf{X}_2)_{t_f,x_f}(x,t) - (\mathbf{X}_3)_{t_f,x_f}(x,0), \\ (\mathcal{E}_4)_{t_f,x_f}(x,t) &= (\mathbf{X}_4)_{t_f,x_f}(x,t) - (\mathbf{X}_4)_{t_f,x_f}(x,0), \end{aligned}\right\} \tag{42}$$

due to the possibility of non-zero initial conditions

$$(\mathbf{X}_1)_{t_f,x_f}(x,0),\ (\mathbf{X}_2)_{t_f,x_f}(x,0),\ (\mathbf{X}_3)_{t_f,x_f}(x,0),\ (\mathbf{X}_4)_{t_f,x_f}(x,0).$$

However, the Caputo derivative for the auxiliary functions

$$(\mathcal{E}_1)_{t_f,x_f}(x,t),\ (\mathcal{E}_2)_{t_f,x_f}(x,t),\ (\mathcal{E}_3)_{t_f,x_f}(x,t),\ (\mathcal{E}_4)_{t_f,x_f}(x,t),$$

given in Eq. (42) are zero [22, 23], which in turn implies that initial and boundary conditions for the well-posed system of linear discrete FFPDEs in Eq. (41) are zero.

4 Numerical Results and Discussions

Based on the derived stability condition and in view of the numerical simulation and discussion in [46], in this section, it suffices to let an initial tumour burden to be $\mathscr{T}(0) = 10 \times 10^7$ and weaken initial immune surveillance to $N(0) = 10^6$, $L(0) = 10^6$, $C(0) = 0.0$ from which we see that the CTLs have not been activated yet. Our experiment is for ten days, the spatial distributions of the involved cells are $[\mathcal{D}_{\mathscr{T}}, \mathcal{D}_N, \mathcal{D}_L, \mathcal{D}_{\mathscr{C}}]' = [1 \times 10^{-4}, 1 \times 10^{-4}, 10^{-4}, 1 \times 10^{-4}]'$, and the adjusted parameter values are presented in Table 1 due to the requirements of stability conditions in Theorem 6. In Fig. 1, the results for $\mu = \kappa = 1$ and $\beta = 2$ are presented, whereas for other values of $\mu = 0.1$, $\kappa = 0.06$, $\beta = 2$ and $\mu = 0.07$, $\kappa = 0.5$, $\beta = 1.8$ the results are presented in Figs. 2 and 3, respectively.

In Fig. 1, all the cells grow linearly to their respective capacity. However the immune surveillance maintain control over the tumour cells throughout the experiment. In Fig. 2, all the cells grow steadily to their respective capacity, with the immune surveillance maintaining control over the tumour cells till eventually the immune surveillance eliminate the tumour cells. The profiles in Fig. 3 are similar to those in Fig. 2.

Table 1 Parameter values [46]

$a = 19 \times 10^7$	$\beta = 6.3 \times 10^{-3}$	$\alpha = 1.6 \times 10^3$	$c = 10.93 \times 10^6$	$f = 0.0693$
$L_N = 2.3 \times 10^8$	$e = 5 \times 10^5$	$d = 0.41$	$\beta_1 = 0.000041$	$\beta_2 = 5.4 \times 10^{-2}$
$\beta_6 = 4343$	$I = 2.3 \times 10^{-11}$	$\alpha_1 = 0.0507$	$\alpha_2 = 7 \times 10^6$	$\alpha_3 = 1.6 \times 10^{-5}$
$\alpha_5 = 1000$	$\alpha_6 = 10$	$p_1 = 30.9 \times 10^{-3}$	$p_2 = 3.42 \times 10^{-6}$	$p_3 = 2.87 \times 10^2$
$p_5 = 4.14 \times 10^{-3}$	$p_6 = 2.04 \times 10^{-2}$	$K = 2 \times 10^9$	$K_L = 8 \times 10^8$	$\beta_3 = 10^{-2}$
$\alpha_4 = 2.3 \times 10^{-11}$	$p_4 = 2$			

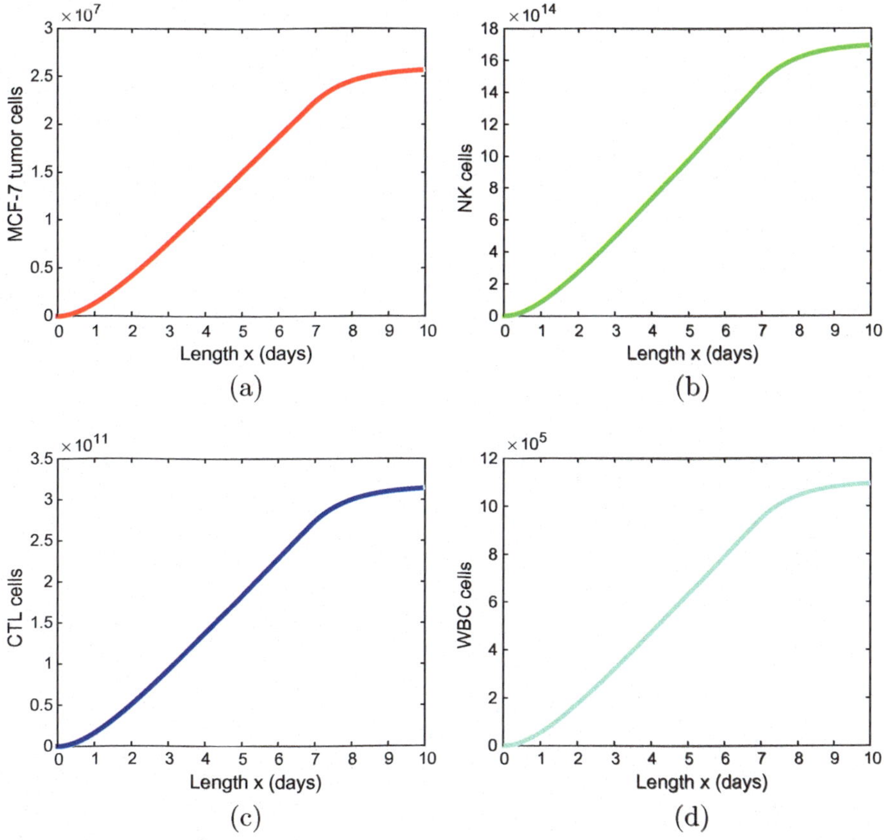

Fig. 1 Numerical solution for the dynamics in Eq. (41) presenting the spatial distributions of: **a** MCF-7 tumor cells **b** NK cells **c** CTL cells and **d** WBC cells

5 Conclusion

In this study, our main aim in this chapter, is to present the solution for the extended dynamics in Eq. (3) with respect to the derived asymptotic stability conditions in Theorem 6. Thus, in consideration of the numerical simulation and discussion in [46], we have weakened the initial effects of natural killer cells, cytotoxic T lymphocytes (CTLs) or CD8+ T cells, and white blood cells (WBCs) and strengthened the initial burden of tumour cells in a host in an effort to capture the full strength of tumour

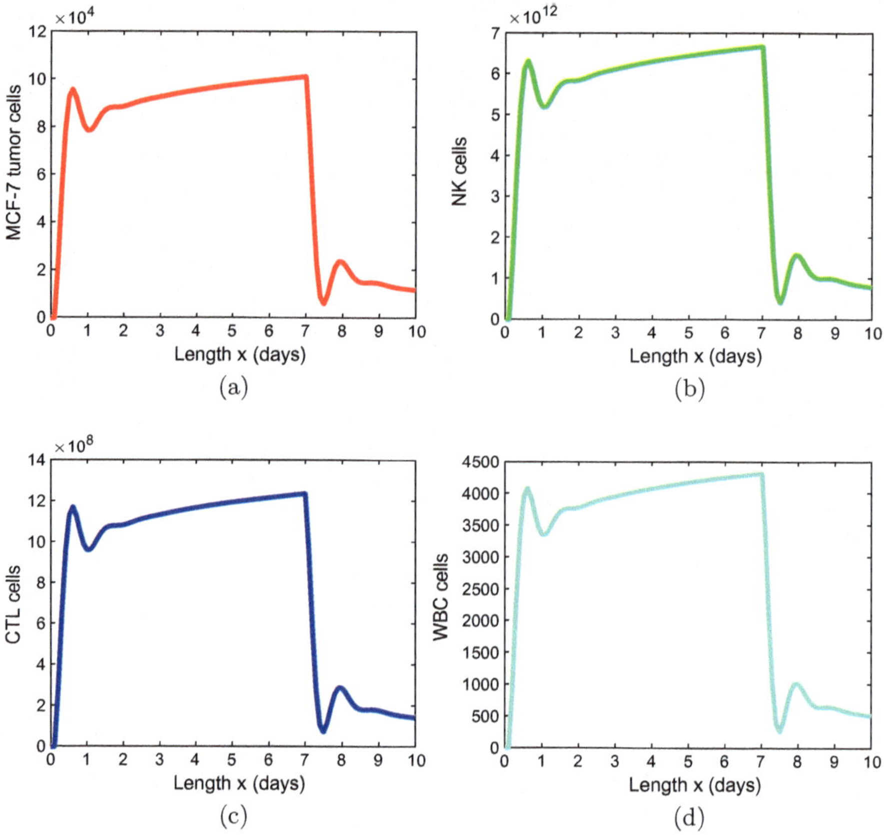

Fig. 2 Numerical solution for the dynamics in Eq. (41) presenting the spatial distributions of: **a** MCF-7 tumor cells **b** NK cells **c** CTL cells and **d** WBC cells

cells. These enable us to deal with the challenged posed by the tumour cells even though we have slightly reduced the value of the parameter related for the killing of cytotoxic T lymphocytes (CTLs) or CD8+ T cells. In view of the asymptotic stability conditions, one can easily see that rather then considering each cells with respect to an experiment as it is the case in [46], one rather consider the asymptotic stability conditions of a given dynamics. Thus, the stability conditions should then be tied to the clinical findings to enables one to obtain more informative and relevant results

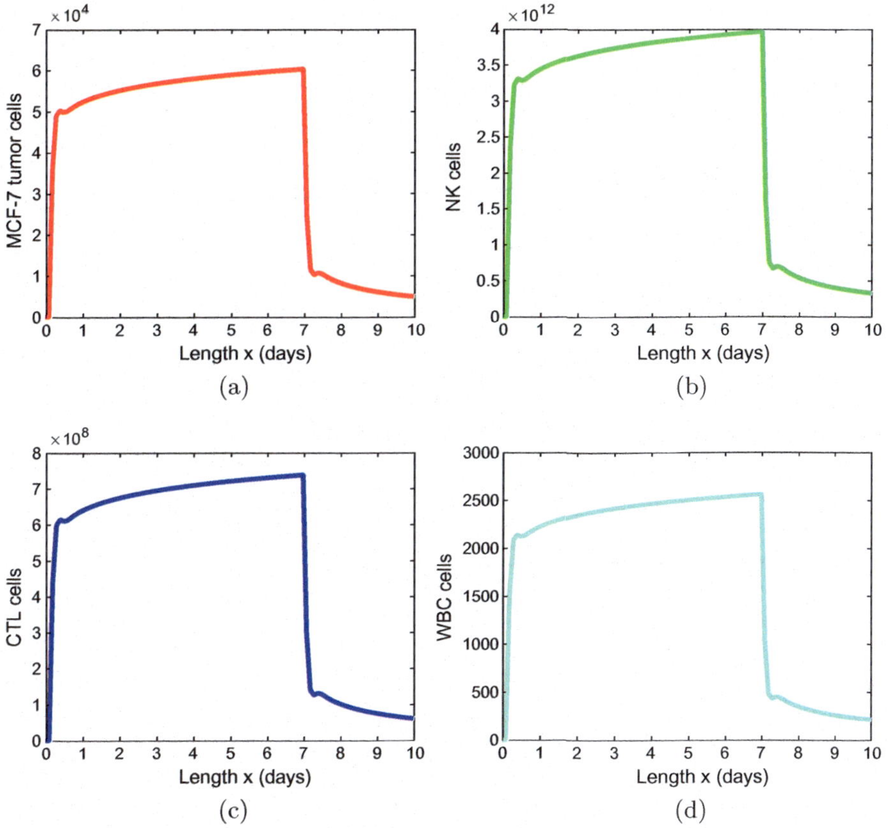

Fig. 3 Numerical solution for the dynamics in Eq. (41) presenting the spatial distributions of: **a** MCF-7 tumor cells **b** NK cells **c** CTL cells and **d** WBC cells

for all. Nevertheless, we also see that the numerical method's results based on the fractal fractional derivatives outperforms the results for the standard finite difference method (FDM). Hence, in this chapter, we believe that we have achieved our main objectives. Our future goal is to extend the study to higher dimensional spaces.

Acknowledgements This work was not funded.

References

1. Atangana, A.: Fractal-fractional differentiation and integration: connecting fractal calculus and fractional calculus to predict complex system. Chaos Solitons Fractals **102**, 396–406 (2017)
2. Atangana, A., Qureshi, S.: Modeling attractors of chaotic dynamical systems with fractal-fractional operators. Chaos Solitons Fractals **123**, 320–337 (2019)
3. Atangana, A., Baleanu, D.: New fractional derivatives with nonlocal and non-singular kernel: theory and application to heat transfer model. Thermal Sci. **20**, 763–769 (2016)
4. Atangana, A., et al.: Analysis of fractal fractional differential equations. Alexandria Eng. J. (2020). https://doi.org/10.1016/j.aej.2020.01.005
5. Atangana, A., Gómez-Aguilar, J.F.: Hyperchaotic behaviour obtained via a nonlocal operator with exponential decay and Mittag-Leffler laws. Chaos, Solitons Fractals **102**, 285–294 (2017)
6. Atangana, A., Gómez-Aguilar, J.F.: A new derivative with normal distribution kernel: theory, methods and applications. Physica A: Stat. Mech. Appl. **476**, 1–14 (2017)
7. Atangana, A., Gómez-Aguilar, J.F.: Fractional derivatives with no-index law property: Application to chaos and statistics. Chaos, Solitons Fractals **114**, 516–535 (2018)
8. Atangana, A.: Non validity of index law in fractional calculus: a fractional differential operator with Markovian and non-Markovian properties. Physica A **505**, 688–706 (2018)
9. Atangana, A., Owolabi, K.M.: New numerical approach for fractional differential equation. Math. Model. Nat. Phenomena **13**, 21 (2018). https://doi.org/10.1051/mmnp/2018010
10. Burden, R.L., Faires, J.D.: Numer. Anal. Brooks/Cole, USA (2011)
11. Caputo, M.: Elasticit$\grave{a}$ e dissipazione. Zanichelli, Bologna (1969)
12. Chen, W., Sun, H., Zhanga, X., Koro$\tilde{s}$ak, D.: Anomalous diffusion modeling by fractal and fractional derivatives. Comput. Math. Appl. **59**, 1754–1758 (2010)
13. Diethelm, K.: The Analysis of Fractional Differential Equations. Lecture Notes in Mathematics. Springer, Berlin (2004)
14. Faucette, W.M.: A geometric interpretation of the solution of the general quartic polynomial. Am. Math. Mon. **103**(1), 51–57 (1996)
15. Garshasbi, M.: Determination of unknown functions in a mathematical model of ductal carcinoma in situ. Numer. Methods Partial Differ. Equ. **35**(6), 2000–2016 (2019)
16. Jordan, D.W., Smith, P.: Nonlinear Ordinary Differential Equations. Clarendon Press, Oxford (1987)
17. Kilbas, A.A., Srivastava, H.M., Trujillo, J.J.: Theory and Applications of Fractional Differential Equations. Elsevier, Netherlands (2006)
18. King, A.C., Billingham, J., Otto, S.R.: Differential Equations. Cambridge University Press, New York (2003)
19. Li, Z., Liu, Z., Khan, M.A.: Fractional investigation of bank data with fractal-fractional Caputo derivative. Chaos, Solitons Fractals https://doi.org/10.1016/j.chaos.2019.109528
20. van Loan, C.F.: The ubiquitous Kronecker product. J. Comput. Appl. Math. **123**, 85–100 (2000)
21. Luchko, Y.: Maximum principle for the generalized time-fractional diffusion equation. J. Math. Anal. Appl. **351**, 218–223 (2009)
22. Meerschaert, M., Tadjeran, C.: Finite difference approximations for fractional advection-dispersion equations. J. Comput. Appl. Math. **172**(1), 65–77 (2004)
23. Metzler, R., Barkai, E., Klafter, J.: Deriving fractional Fokker-Planck equations from a generalized master equation. Europhys. Lett. **46**(4), 431–436 (1999)
24. Miaskowski, A., Subramanian, M.: Numerical model for magnetic fluid hyperthermia in a realistic breast phantom: calorimetric calibration and treatment planning. Int. J. Molecular Sci. **20**(18), 4644 (2019)
25. Oldham, K.B., Spanier, J.: The Fractional Calculus. Academic, New York (1974)
26. Ortigueira, M.D.: Riesz potential operators and inverses via fractional centred derivatives. Int. J. Math. Math. Sci. **2006**(48391), 1–12 (2006)
27. Ortigueira, M.D., Batista, A.G.: On the relation between the fractional Brownian motion and the fractional derivatives. Phys. Lett. A **372**, 958–968 (2008)

28. Owolabi, K.M.: Mathematical analysis and numerical simulation of patterns in fractional and classical reaction-diffusion systems. Chaos, Solitons Fractals **93**, 89–98 (2016)
29. Owolabi, K.M., Atangana, A.: Numerical approximation of nonlinear fractional parabolic differential equations with Caputo-Fabrizio derivative in Riemann-Liouville sense. Chaos, Solitons Fractals **99**, 171–179 (2017)
30. Owolabi, K.M.: Robust and adaptive techniques for numerical simulation of nonlinear partial differential equations of fractional order. Commun. Nonlinear Sci. Numer. Simul. **44**, 304–317 (2017)
31. Owolabi, K.M.: Mathematical modelling and analysis of two-component system with Caputo fractional derivative order. Chaos, Solitons Fractals **103**, 544–554 (2017)
32. Owolabi, K.M.: Numerical approach to fractional blow-up equations with Atangana-Baleanu derivative in Riemann-Liouville sense. Math. Model. Nat. Phenomena **13**, 7 (2018)
33. Owolabi, K.M., Atangana, A.: Modelling and formation of spatiotemporal patterns of fractional predation system in subdiffusion and superdiffusion scenarios. Euro. Phys. J. Plus **133**, 43 (2018)
34. Owolabi, K.M.: Modelling and simulation of a dynamical system with the Atangana-Baleanu fractional derivative. Euro. Phys. J. Plus **133**, 15 (2018)
35. Owolabi, K.M.: Efficient numerical simulation of non-integer-order space-fractional reaction-diffusion equation via the Riemann-Liouville operator. Euro. Phys. J. Plus **133**, 98 (2018)
36. Owolabi, K.M., Atangana, A.: Robustness of fractional difference schemes via the Caputo subdiffusion-reaction equations. Chaos, Solitons Fractals **111**, 119–127 (2018)
37. Owolabi, K.M., Atangana, A.: Chaotic behaviour in system of noninteger-order ordinary differential equations. Chaos, Solitons Fractals **115**, 362–370 (2018)
38. Owolabi, K.M.: Analysis and numerical simulation of multicomponent system with AtanganaBaleanu fractional derivative. Chaos, Solitons Fractals **115**, 127–134 (2018)
39. Owolabi, K.M.: Numerical patterns in system of integer and non-integer order derivatives. Chaos, Solitons Fractals **115**, 143–153 (2018)
40. Owolabi, K.M.: Numerical patterns in reaction-diffusion system with the Caputo and Atangana-Baleanu fractional derivatives. Chaos, Solitons Fractals **115**, 160–169 (2018)
41. Paruch, M.: Mathematical modeling of breast tumor destruction using fast heating during radiofrequency ablation. Materials **13**(1), 136 (2020)
42. Podlubny, I.: Fractional Differential Equations. Academic, San Diego (1999)
43. Podlubny, I., Chechkin, A.V., Skovranek, T., Chen, Y., Jara, B.M.V.: Matrix approach to discrete fractional calculus II: partial fractional differential equations. J. Comput. Phys. **228**(8), 3137–3153 (2009)
44. Saichev, A., Zaslavsky, G.: Fractional kinetic equations: solutions and applications. Chaos **7**(4), 753–764 (1997)
45. Séroul, R.: Stable Polynomials in Programming for Mathematicians, pp. 280–286. Springer, Berlin (2000)
46. Wei, H.C.: Mathematical modeling of tumor growth: the MCF-7 breast cancer cell line. Math. Biosci. Eng. **16**(6), 6512–6535 (2019)
47. Wei, H.C.: Mathematical and numerical analysis of a mathematical model of mixed immunotherapy and chemotherapy of cancer. Am. Inst. Math. Sci. **21**(4), 1279–1295 (2016)
48. Wu, C., Motosha, T., Abdel-Rahman, H.A.: Free and protein-bound plasma estradiol-17 during the menstrual cycle. J. Clin. Endocrinol. Metabol. **43**, 436–445 (1976)

The Exponentiated Half Logistic-Topp-Leone-G Power Series Class of Distributions: Model, Properties and Applications

Fastel Chipepa, Broderick Oluyede, Divine Wanduku, and Thatayaone Moakofi

Abstract We develop a new class of distributions, namely, the exponentiated half logistic-Topp-Leone-G power series (EHL-TL-GPS) class of distributions. We present some special classes in the proposed distribution. Structural properties were also derived including moments, entropy and maximum likelihood estimates. We conducted a simulation study to evaluate the consistency of the maximum likelihood estimates. We also present two real data examples to illustrate the applicability of the new class of distributions. The proposed model performs better than several non-nested models on selected data sets.

Keywords Topp-Leone-G distribution · Half Logistic-G distribution · Power series · Poisson distribution · Geometric distribution · Maximum likelihood estimation

F. Chipepa · B. Oluyede · T. Moakofi
Department of Mathematics and Statistical Sciences, Botswana International University of Science and Technology, Palapye, BW, Botswana
e-mail: chipepa.fastel@studentmail.biust.ac.bw

B. Oluyede
e-mail: oluyedeo@biust.ac.bw

F. Chipepa
Department of Applied Mathematics and Statistics, Midlands State University, Gweru, ZW, Zimbabwe

D. Wanduku (✉)
Department of Mathematical Sciences, Georgia Southern University, Statesboro, Georgia
e-mail: dwanduku@georgiasouthern.edu; wandukudivine@yahoo.com

J. Singh et al. (eds.), *Methods of Mathematical Modelling and Computation for Complex Systems*, Studies in Systems, Decision and Control 373,
https://doi.org/10.1007/978-3-030-77169-0_14

1 Introduction

Classical distributions have a limited range of behavior when it comes to modeling lifetime data. Modifications of the well-known distributions are highly recommended to gain flexibility. Numerous generalizations of classical distributions are available in the literature. For instance, the exponentiated Kumaraswamy-G family by Silva et al. [40], the modified odd Weibull family of distributions by Chesneau and El Achi [13], the Nadarajah-Haghighi Topp-Leone-G family by Reyad et al. [34], the Kumaraswamy Odd Lindley-G by Chipepa et al. [16], new Weibull-X family by Ahmad et al. [2], Kumaraswamy Marshall-Olkin family by Alizadeh et al. [4], the Topp-Leone-Marshall-Olkin-G family by Chipepa et al. [14], the exponentiated generalized (EG) family by Cordeiro et al. [20], Weibull-G by Bourguignon et al. [10], the odd exponentiated half-logistic-G family of distributions by Afify et al. [1], to mention a few.

Cordeiro et al. [18] developed the type I half-logistic-G (TIHL-G) family of distributions with cumulative distribution function (cdf) defined by

$$
\begin{aligned}
F(x;\lambda,\xi) &= \int_0^{-ln(1-G(x;\xi))} \frac{2\lambda e^{-\lambda t}}{(1+e^{-\lambda t})^2} dt \\
&= \frac{1-\left[1-G(x;\xi)\right]^{\lambda}}{1+\left[1-G(x;\xi)\right]^{\lambda}},
\end{aligned}
$$

for $\lambda > 0$. The half-logistic transformation was applied to several well-known distributions, for instance, Anwar and Zahoor [6] introduced the half-logistic-Lomax distribution for lifetime data modeling, Anwar and Bibi [5] developed the half-logistic generalized Weibull distribution, Chipepa et al. [15] introduced the odd generalized half logistic Weibull-G family of distributions, Muhammad and Yahaya [30] studied the half logistic-Poisson distribution, to mention a few.

Furthermore, Al-Shomrani et al. [7] proposed the Topp-Leone generated family of distributions with the cdf and probability density function (pdf) given by

$$
F_{TL-G}(x;b,\xi) = \left[1-\overline{G}^2(x;\xi)\right]^b,
$$

and

$$
f_{TL-G}(x;b,\xi) = 2b\left[1-\overline{G}^2(x;\xi)\right]^{b-1}\overline{G}(x;\xi)g(x;\xi),
$$

respectively, for $b > 0$ and parameter vector ξ. Some generalizations of the Topp-Leone-G family of distributions include the Topp-Leone-Marshall-Olkin-G family by Chipepa et al. [15], Type II power Topp-Leone generated family by Bantan et al. [8], Topp-Leone-Weibull by Rezaei et al. [33], Topp-Leone generalized exponential by Sangsanit and Bodhisuwan [36].

In this paper, we develop a new power series distribution, namely the exponentiated half logistic-Topp-Leone-G power series (EHL-TL-GPS) class of distributions. We are motivated by the desirable properties exhibited by the generalizations of both the half logistic-G and the Topp-Leone-G distributions. We are also motivated by the wide acceptability of power series distributions in the field of finance and actuarial science. We hope the new proposed model will receive attention from researchers in other areas of medicine, engineering, science, environmental science, and economics.

The cdf and pdf of the exponentiated half logistic-Topp-Leone-G (EHL-TL-G, Oluyede et al. [31]) distribution are given by

$$F(x;b,\beta,\xi)=\left[\frac{\left[1-\overline{G}^{2}(x;\xi)\right]^{b}}{1+\left(1-\left[1-\overline{G}^{2}(x;\xi)\right]^{b}\right)}\right]^{\beta} \tag{1}$$

and

$$f(x;b,\beta,\xi)=\frac{4\beta b g(x;\xi)[1-\overline{G}^{2}(x;\xi)]^{\beta b-1}\overline{G}(x;\xi)}{[1+(1-[1-\overline{G}^{2}(x;\xi)]^{b})]^{\beta+1}}, \tag{2}$$

respectively, for $b, \beta > 0$ and ξ is the parameter vector from the baseline distribution. Let N be a zero truncated discrete random variable having a power series distribution, whose probability mass function (pmf) is given by

$$P(N=n)=\frac{a_n\theta^n}{C(\theta)}, n=1,2,3,\ldots, \tag{3}$$

where $C(\theta)=\sum_{n=1}^{\infty}a_n\theta^n$ is finite, $\theta > 0$ and $\{a_n\}_{n\geq 1}$ a sequence of positive real numbers. The power series family of distributions includes binomial, Poisson, geometric and logarithmic distributions [26]. Several generalized distributions proposed in the literature involving the power series include a new generalized Lindley-Weibull class of distributions by Makubate et al. [27], the exponentiated power generalized Weibull power series family of distributions by Aldahlan [3], Weibull-power series distributions by Morais and Barreto-Souza [28], complementary exponential power series by Flores et al. [23], complementary extended Weibull-power series by Cordeiro and Silva [19], Burr XII power series by Silva and Cordeiro [39], extended Weibull-power series (EWPS) distribution by Silva et al. [38], the Burr-Weibull power series class of distributions by Oluyede et al. [29].

The rest of the paper is organized as follows: In Sect. 2, we present the new model and some of the statistical properties. We present some special cases of the proposed class of distributions in Sect. 3. A simulation study is presented in Sect. 4 and applications in Sect. 5 followed by concluding remarks.

2 The Model, Sub-Classes and Properties

In this section, we present the new model, some statistical properties which include expansion of the density function, hazard rate function, quantile function, sub-classes, moments, conditional moments and maximum likelihood estimation of model parameters.

2.1 The Model

Let $X_1, X_2, \ldots, X_N$ be N identically and independently distributed (iid) random variables following the EHL-TL-G distribution. Let $X_{(1)} = \min(X_1, X_2, \ldots, X_N)$, then the cdf of $X_{(1)}|N = n$ is given by

$$F_{X_{(1)}|N=n}(x; b, \theta, \xi) = 1 - \left(1 - \left[\frac{\left[1-\overline{G}^2(x;\xi)\right]^b}{1+\left(1-\left[1-\overline{G}^2(x;\xi)\right]^b\right)}\right]^\beta\right)^n, \quad (4)$$

for b, $\theta > 0$, $n \geq 1$ and parameter vector ξ. The exponentiated half logistic-Topp-Leone-G power series (EHL-TL-GPS) class of distributions denoted by EHL-TL-GPS(b, β, θ, ξ) is defined by the marginal distribution of $X_{(1)}$, that is,

$$F_{X_{(1)}}(x) = 1 - \frac{C\left(\theta\left(1-\left[\frac{\left[1-\overline{G}^2(x;\xi)\right]^b}{1+\left(1-\left[1-\overline{G}^2(x;\xi)\right]^b\right)}\right]^\beta\right)\right)}{C(\theta)}, \quad (5)$$

for b, β, θ, $x > 0$ and parameter vector ξ. The pdf is given by

$$f_{X_{(1)}}(x) = \frac{4\beta\theta b g(x;\xi)[1-\overline{G}^2(x;\xi)]^{\beta b-1}\overline{G}(x;\xi)}{[1+(1-[1-\overline{G}^2(x;\xi)]^b)]^{\beta+1}} \frac{C'\left(\theta\left(1-\left[\frac{\left[1-\overline{G}^2(x;\xi)\right]^b}{1+\left(1-\left[1-\overline{G}^2(x;\xi)\right]^b\right)}\right]^\beta\right)\right)}{C(\theta)}. \quad (6)$$

The hazard rate function (hrf) is given by

Table 1 Special families of the EHL-TL-GPS distribution

Distribution	$C(\theta)$	a_n	cdf
EHL-TL-G poisson	$e^{\theta}-1$	$(n!)^{-1}$	$1-\frac{\exp\left(\theta\left(1-\left[\frac{\left[1-\overline{G}^2(x;\xi)\right]^b}{1+\left(1-\left[1-\overline{G}^2(x;\xi)\right]^b\right)}\right]^{\beta}\right)\right)-1}{\exp(\theta)-1}$
EHL-TL-G geometric	$\theta(1-\theta)^{-1}$	1	$1-\frac{(1-\theta)\left(1-\left[\frac{\left[1-\overline{G}^2(x;\xi)\right]^b}{1+\left(1-\left[1-\overline{G}^2(x;\xi)\right]^b\right)}\right]^{\beta}\right)}{\left(1-\theta\left(1-\left[\frac{\left[1-\overline{G}^2(x;\xi)\right]^b}{1+\left(1-\left[1-\overline{G}^2(x;\xi)\right]^b\right)}\right]^{\beta}\right)\right)}$
EHL-TL-G logarithmic	$-\log(1-\theta)$	n^{-1}	$1-\frac{\log\left(1-\theta\left(1-\left[\frac{\left[1-\overline{G}^2(x;\xi)\right]^b}{1+\left(1-\left[1-\overline{G}^2(x;\xi)\right]^b\right)}\right]^{\beta}\right)\right)}{\log(1-\theta)}$
EHL-TL-G binomial	$(1+\theta)^m-1$	$\binom{m}{n}$	$1-\frac{\left(1+\theta\left(1-\left[\frac{\left[1-\overline{G}^2(x;\xi)\right]^b}{1+\left(1-\left[1-\overline{G}^2(x;\xi)\right]^b\right)}\right]^{\beta}\right)\right)^m-1}{(1+\theta)^m-1}$

$$h_F(x)=\frac{\frac{4\beta\theta b g(x;\xi)[1-\overline{G}^2(x;\xi)]^{\beta b-1}\overline{G}(x;\xi)}{[1+(1-[1-\overline{G}^2(x;\xi)]^b)]^{\beta+1}}C'\left(\theta\left(1-\left[\frac{\left[1-\overline{G}^2(x;\xi)\right]^b}{1+\left(1-\left[1-\overline{G}^2(x;\xi)\right]^b\right)}\right]^{\beta}\right)\right)}{C\left(\theta\left(1-\left[\frac{\left[1-\overline{G}^2(x;\xi)\right]^b}{1+\left(1-\left[1-\overline{G}^2(x;\xi)\right]^b\right)}\right]^{\beta}\right)\right)}. \quad (7)$$

Table 1 below present the special families of EHL-TL-GPS distribution when $C(\theta)$ is specified in Eq. (5).

2.2 *Quantile Function*

Let X be a random variable with cdf defined by Eq. (5). The quantile function $Q_{X_{(1)}}(u)$ is defined by $F_{X_{(1)}}(Q_{X_{(1)}}(u))=u, 0\leq u\leq 1$. Note that

$$1-\frac{C\left(\theta\left(1-\left[\frac{\left[1-\overline{G}^2(x;\xi)\right]^b}{1+\left(1-\left[1-\overline{G}^2(x;\xi)\right]^b\right)}\right]^{\beta}\right)\right)}{C(\theta)}=u,$$

so that

$$C\left(\theta\left(1-\left[\frac{\left[1-\overline{G}^{2}(x;\xi)\right]^{b}}{1+\left(1-\left[1-\overline{G}^{2}(x;\xi)\right]^{b}\right)}\right]^{\beta}\right)\right)=C(\theta)\,(1-u)\,.$$

This is equivalent to

$$\frac{\left[1-\overline{G}^{2}(x;\xi)\right]^{b}}{1+\left(1-\left[1-\overline{G}^{2}(x;\xi)\right]^{b}\right)}=\left(1-\frac{C^{-1}\left(C(\theta)\,(1-u)\right)}{\theta}\right)^{\frac{1}{\beta}},$$

that is,

$$\frac{2}{1-\overline{G}^{2}(x;\xi)}=1+\left(1-\frac{C^{-1}\left(C(\theta)\,(1-u)\right)}{\theta}\right)^{\frac{-1}{\beta}}.$$

The expression further simplifies to

$$G(x;\xi)=1-\left(1-2\left(1+\left(1-\frac{C^{-1}\left(C(\theta)\,(1-u)\right)}{\theta}\right)^{\frac{-1}{\beta}}\right)^{-1}\right)^{\frac{1}{2}}.$$

Therefore, we obtain the quantile values from the EH-TL-GPS class of distributions by solving the non-linear equation

$$Q_{X_{(1)}}(u)=G^{-1}\left[1-\left(1-2\left(1+\left(1-\frac{C^{-1}\left(C(\theta)\,(1-u)\right)}{\theta}\right)^{\frac{-1}{\beta}}\right)^{-1}\right)^{\frac{1}{2}}\right] \quad (8)$$

using Newton Raphson method with the aid of statistical software such as R, MATLAB and SAS.

2.3 Expansion of Density

Expansion of the density function of the EHL-TL-GPS class of distributions is presented in this sub-section. Equation (6) can be written as

$$f_{EHL-TL-GPS}(x) = \sum_{n=1}^{\infty} \frac{n a_n \theta^n}{C(\theta)} \left(1 - \left[\frac{\left[1-\overline{G}^2(x;\xi)\right]^b}{1+\left(1-\left[1-\overline{G}^2(x;\xi)\right]^b\right)}\right]^{\beta}\right)^{n-1}$$
$$\times \frac{4\beta b g(x;\xi)[1-\overline{G}^2(x;\xi)]^{\beta b-1}\overline{G}(x;\xi)}{[1+(1-[1-\overline{G}^2(x;\xi)]^b)]^{\beta+1}}.$$

Using the generalized binomial expansion

$$\left(1 - \left[\frac{\left[1-\overline{G}^2(x;\xi)\right]^b}{1+\left(1-\left[1-\overline{G}^2(x;\xi)\right]^b\right)}\right]^{\beta}\right)^{n-1} = \sum_{j=0}^{\infty} \binom{n-1}{j} (-1)^j$$
$$\times \frac{\left[1-\overline{G}^2(x;\xi)\right]^{\beta b j}}{\left(1+\left(1-\left[1-\overline{G}^2(x;\xi)\right]^b\right)\right)^{\beta j}},$$

we write the pdf of the EHL-TL-GPS class of distributions as

$$f_{EHL-TL-GPS}(x) = \sum_{j=0}^{\infty}\sum_{n=1}^{\infty} \binom{n-1}{j} (-1)^j \frac{n a_n \theta^n}{C(\theta)} 4b\beta \left[1-\overline{G}^2(x;\xi)\right]^{\beta b(j+1)-1}$$
$$\times \left(1+\left(1-\left[1-\overline{G}^2(x;\xi)\right]^b\right)\right)^{-(\beta(j+1)+1)} \overline{G}(x;\xi) g(x;\xi).$$

Furthermore, applying the generalized binomial expansion

$$\left(1+\left(1-\left[1-\overline{G}^2(x;\xi)\right]^b\right)\right)^{-(\beta(j+1)+1)} = \sum_{k=0}^{\infty} \frac{\Gamma(\beta(j+1)+1+k)}{\Gamma(\beta(j+1)+1)k!} (-1)^k$$
$$\times \left[1-\left[1-\overline{G}^2(x;\xi)\right]^b\right]^k$$
$$= \sum_{k,q=0}^{\infty} \frac{\Gamma(\beta(j+1)+1+k)}{\Gamma(\beta(j+1)+1)k!} \binom{k}{q}$$
$$\times (-1)^{k+q} \left[1-\overline{G}^2(x;\xi)\right]^{bq},$$

we further get

$$
\begin{aligned}
f_{EHL-TL-GPS}(x) &= \sum_{j,k,q=0}^{\infty}\sum_{n=1}^{\infty}\binom{n-1}{j}\frac{na_n\theta^n}{C(\theta)}4b\beta\left[1-\overline{G}^2(x;\xi)\right]^{\beta b(j+1)+bq-1} \\
&\times \frac{\Gamma(\beta(j+1)+1+k)}{\Gamma(\beta(j+1)+1)k!}\binom{k}{q}(-1)^{j+k+q}\overline{G}(x;\xi)g(x;\xi) \\
&= \sum_{j,k,q,p=0}^{\infty}\sum_{n=1}^{\infty}\binom{n-1}{j}(-1)^{j+k+q+p}\frac{na_n\theta^n}{C(\theta)}4b\beta\binom{k}{q} \\
&\times \binom{\beta b(j+1)+bq-1}{p}\frac{\Gamma(\beta(j+1)+1+k)}{\Gamma(\beta(j+1)+1)k!}\overline{G}^{2p+1}(x;\xi)g(x;\xi).
\end{aligned}
$$

Also, by applying the generalized binomial series expansion

$$
\overline{G}^{2p+1}(x;\xi) = (1-G(x;\xi))^{2p+1} = \sum_{m=0}^{\infty}\binom{2p+1}{m}(-1)^m\left(G(x;\xi)\right)^m,
$$

the pdf of the EHL-TL-GPS class of distributions simplifies to

$$
\begin{aligned}
f_{EHL-TL-GPS}(x) &= \sum_{j,k,q,p,m=0}^{\infty}\sum_{n=1}^{\infty}\binom{n-1}{j}(-1)^{j+k+q+p+m}\frac{na_n\theta^n}{C(\theta)}4b\beta \\
&\times \binom{\beta b(j+1)+bq-1}{p}\frac{\Gamma(\beta(j+1)+1+k)}{\Gamma(\beta(j+1)+1)k!}\binom{k}{q}\binom{2p+1}{m} \\
&\times \left(G(x;\xi)\right)^m g(x;\xi) \\
&= \sum_{j,k,q,p,m=0}^{\infty}\sum_{n=1}^{\infty}\binom{n-1}{j}(-1)^{j+k+q+p+m}\frac{na_n\theta^n}{C(\theta)}4b\beta \\
&\times \binom{\beta b(j+1)+bq-1}{p}\frac{\Gamma(\beta(j+1)+1+k)}{\Gamma(\beta(j+1)+1)k!}\binom{k}{q}\binom{2p+1}{m} \\
&\times \left(\frac{m+1}{m+1}\right)\left(G(x;\xi)\right)^m g(x;\xi) \\
&= \sum_{m=0}^{\infty}U_m g_m(x;\xi),
\end{aligned}
\tag{9}
$$

where $g_m(x;\xi) = (m+1)\left(G(x;\xi)\right)^m g(x;\xi)$ is the exponentiated-G (Exp-G) distribution with power parameter m and

$$
\begin{aligned}
U_m &= \sum_{j,k,q,p=0}^{\infty}\sum_{n=1}^{\infty}\binom{n-1}{j}(-1)^{j+k+q+p+m}\frac{na_n\theta^n}{C(\theta)}4b\beta\binom{\beta b(j+1)+bq-1}{p} \\
&\times \frac{\Gamma(\beta(j+1)+1+k)}{\Gamma(\beta(j+1)+1)k!}\binom{k}{q}\binom{2p+1}{m}\frac{1}{m+1}.
\end{aligned}
\tag{10}
$$

Thus, the pdf of EHL-TL-GPS class of distributions can be written as an infinite linear combination of Exponentiated-G (Exp-G) distribution.

2.4 Moments and Generating Function

If X follows the EHL-TL-GPS distribution and $Y \sim Exp - G(m)$. Then using equation (9) the rth raw moment, μ_r' of the EHL-TL-GPS class of distributions is obtained as

$$\mu_r' = E(X^r) = \int_{-\infty}^{\infty} x^r f(x) dx$$
$$= \sum_{m=0}^{\infty} U_m E(Y^r),$$

where U_m is given by Eq. (10). The moment generating function (MGF) $M(t) = E(e^{tX})$ is given by:

$$M_X(t) = \sum_{m=0}^{\infty} U_m M_Y(t),$$

where $M_Y(t)$ is the mgf of Y and U_m is given by Eq. (10).

2.5 Conditional Moments

The rth conditional moments of the EHL-TL-GPS class of distributions is given by

$$E(X^r | X \geq t) = \frac{1}{\overline{F}(a; b, \beta, \theta, \xi)} \int_{t}^{\infty} x^r f(x; b, \beta, \theta, \xi) dx$$
$$= \frac{1}{\overline{F}(a; b, \beta, \theta, \xi)} \sum_{m=0}^{\infty} U_m E\left(Y^r I_{\{Y^r \geq t\}}\right), \quad (11)$$

where

$$E\left(Y^r I_{\{Y^r \geq t\}}\right) = \int_t^\infty y^r g_m(y;\xi)dy$$

$$= m \int_{G(u;\xi)}^1 [Q_G(u;\xi)]^r u^m du, \tag{12}$$

for $b, \beta, \theta > 0$, and parameter vector ξ.

2.6 Mean Deviation, Lorenz and Bonferroni Curves

The mean deviation about the mean and mean deviation about the median as well as Lorenz and Bonferroni curves for the EHL-TL-GPS class of distributions are presented in this subsection.

2.6.1 Mean Deviations

The mean deviation about the mean D(μ) and the mean deviation about the median D(M), are defined as

$D(\mu) = \int_0^\infty |x - \mu| \, f_{EHL-TL-GPS}(x)dx$, $D(M) = \int_0^\infty |x - M| \, f_{EHL-TL-GPS}(x)dx$, respectively, where $\mu = E(X)$ and $M = Median(X) = F^{-1}(\frac{1}{2})$ is the median of $F_{EHL-TL-GPS}$. However, the following relationships can be used to evaluate the measures D(μ) and D(M):

$$D(\mu) = 2\mu F_{EHL-TL-GPS}(\mu) - 2\mu + 2\sum_{m=0}^{\infty} U_m E\left(Y I_{\{Y \geq \mu\}}\right),$$

and

$$D(M) = -\mu + 2\sum_{m=0}^{\infty} U_m E\left(Y_m I_{\{Y_m \geq M\}}\right)$$

where $E\left(Y I_{\{Y \geq M\}}\right)$ is given by (12) with $r = 1$ and M in place of t.

2.6.2 Lorenz and Bonferroni Curves

In this subsection, we present the Lorenz and Bonferroni curves for EHL-TL-GPS distribution. Lorenz and Bonferroni curves have several applications in different fields such as medicine, income and poverty, reliability, and insurance. Lorenz and

Bonferroni curves are given by

$$L((p)) = \frac{1}{\mu}\sum_{m=0}^{\infty} U_m \int_0^q x g_m(x;\xi)dx, \quad \text{and} \quad B(p) = \frac{1}{p\mu}\sum_{m=0}^{\infty} U_m \int_0^q x g_m(x;\xi)dx, \tag{13}$$

respectively, where $\int_0^q x g_m(x;\xi)dx$, is the first incomplete moment of Exp-G distribution and U_m is given in Eq. (10).

2.7 *Order Statistics and Rényi Entropy*

In this section, we present the distribution of the kth order statistic and Rényi entropy.

2.7.1 Distribution of Order Statistics

Le $X_1, X_2, \ldots, X_n$ be a random sample from EHL-TL-GPS class of distributions and suppose $X_{1:n} < X_{2:n}, \ldots < X_{n:n}$ denote the corresponding order statistics. The pdf of the kth order statistic is given by

$$f_{k:n}(x) = \frac{n!}{(k-1)!(n-k)!}\sum_{l=0}^{n-k}\binom{n-k}{l}(-1)^l f(x)\,[F(x)]^{k+l-1}. \tag{14}$$

Using Eqs. (5) and (6), we write

$$\begin{aligned} f(x)\,[F(x)]^{k+l-1} &= \sum_{n=1}^{\infty}\frac{n a_n \theta^n}{C(\theta)}\left(1-\left[\frac{\left[1-\overline{G}^2(x;\xi)\right]^b}{1+\left(1-\left[1-\overline{G}^2(x;\xi)\right]^b\right)}\right]^{\beta}\right)^{n-1} \\ &\times \frac{4\beta b g(x;\xi)[1-\overline{G}^2(x;\xi)]^{\beta b-1}\overline{G}(x;\xi)}{[1+(1-[1-\overline{G}^2(x;\xi)]^b)]^{\beta+1}} \\ &\times \left(1-\frac{C\left(\theta\left(1-\left[\frac{\left[1-\overline{G}^2(x;\xi)\right]^b}{1+\left(1-\left[1-\overline{G}^2(x;\xi)\right]^b\right)}\right]^{\beta}\right)\right)}{C(\theta)}\right)^{k+l-1}. \end{aligned}$$

Using the generalized binomial expansion

$$\left(1-\frac{C\left(\theta\left(1-\left[\frac{\left[1-\overline{G}^{2}(x;\xi)\right]^{b}}{1+\left(1-\left[1-\overline{G}^{2}(x;\xi)\right]^{b}\right)}\right]^{\beta}\right)\right)}{C(\theta)}\right)^{k+l-1}=\sum_{p=0}^{\infty}\binom{k+l-1}{p}(-1)^{p}\times\left(\frac{C\left(\theta\left(1-\left[\frac{\left[1-\overline{G}^{2}(x;\xi)\right]^{b}}{1+\left(1-\left[1-\overline{G}^{2}(x;\xi)\right]^{b}\right)}\right]^{\beta}\right)\right)}{C(\theta)}\right)^{p}$$

and the power series raised to a positive integer (see Gradshyten and Ryzhik [24] for details)

$$\left(\frac{C\left(\theta\left(1-\left[\frac{\left[1-\overline{G}^{2}(x;\xi)\right]^{b}}{1+\left(1-\left[1-\overline{G}^{2}(x;\xi)\right]^{b}\right)}\right]^{\beta}\right)\right)}{C(\theta)}\right)^{p}=\sum_{z=0}^{\infty}\frac{d_{z,p}\theta^{z}}{C^{z}(\theta)}\left(1-\left[\frac{\left[1-\overline{G}^{2}(x;\xi)\right]^{b}}{1+\left(1-\left[1-\overline{G}^{2}(x;\xi)\right]^{b}\right)}\right]^{\beta}\right)^{z},$$

where $d_{z,p}=(zb_0)^{-1}\sum_{h=1}^{z}[p(h+1)-z]b_h d_{z-h,p}$ and $d_{0,p}=b_0^p$, we get

$$f(x)\,[F(x)]^{k+l-1}=\sum_{p,z=0}^{\infty}\sum_{n=1}^{\infty}(-1)^{p}\frac{na_n\theta^{n}}{C^{z+1}(\theta)}\frac{4\beta bg(x;\xi)[1-\overline{G}^{2}(x;\xi)]^{\beta b-1}\overline{G}(x;\xi)}{[1+(1-[1-\overline{G}^{2}(x;\xi)]^{b})]^{\beta+1}}\times\binom{k+l-1}{p}\left(1-\left[\frac{\left[1-\overline{G}^{2}(x;\xi)\right]^{b}}{1+\left(1-\left[1-\overline{G}^{2}(x;\xi)\right]^{b}\right)}\right]^{\beta}\right)^{z+n-1}.$$

Considering the following generalized binomial series expansions

$$\left(1-\left[\frac{\left[1-\overline{G}^{2}(x;\xi)\right]^{b}}{1+\left(1-\left[1-\overline{G}^{2}(x;\xi)\right]^{b}\right)}\right]^{\beta}\right)^{z+n-1}=\sum_{q=0}^{\infty}(-1)^{q}\binom{z+n-1}{q}\times\frac{[1-\overline{G}^{2}(x;\xi)]^{b\beta q}}{(1+(1-[1-\overline{G}^{2}(x;\xi)]^{b}))^{\beta q}},$$

$$(1+(1-[1-\overline{G}^{2}(x;\xi)]^{b}))^{-(\beta(q+1)+1)}=\sum_{j,i=0}^{\infty}(-1)^{j+i}\frac{\Gamma(\beta(q+1)+1+j)}{\Gamma(\beta(q+1)+1)j!}\binom{j}{i}[1-\overline{G}^{2}(x;\xi)]^{bi},$$

$$[1-\overline{G}^2(x;\xi)]^{b(\beta q+\beta+i)-1} = \sum_{s=0}^{\infty}(-1)^s \binom{b(\beta q+\beta+i)-1}{s}\overline{G}^{2s}(x;\xi),$$

and

$$\overline{G}^{2s+1}(x;\xi) = \sum_{m=0}^{\infty}(-1)^m \binom{2s+1}{m} G^m(x;\xi),$$

yields

$$\begin{aligned} f(x)\,[F(x)]^{k+l-1} = & \sum_{p,z,q,j,i,s,m=0}^{\infty}\sum_{n=1}^{\infty} \frac{na_n\theta^n}{C^{z+1}(\theta)}(-1)^{p+q+j+i+s+m}4\beta b d_{z,p}\binom{k+l-1}{p} \\ & \times \frac{\Gamma(\beta(q+1)+1+j)}{\Gamma(\beta(q+1)+1)j!}\binom{j}{i}\binom{b(\beta q+\beta+i)-1}{s}\binom{2s+1}{m} \\ & \times \binom{z+n-1}{q} g(x;\xi)G^m(x;\xi). \end{aligned}$$

Therefore, the distribution of the kth order statistics from EHL-TL-GPS class of distributions is given by

$$\begin{aligned} f_{k:n}(x) = & \frac{n!4\beta b}{(k-1)!(n-k)!}\sum_{p,z,q,j,i,s,m=0}^{\infty}\sum_{n=1}^{\infty}\sum_{l=0}^{n-k}\binom{n-k}{l}\frac{na_n\theta^n}{C^{z+1}(\theta)}\binom{z+n-1}{q} \\ & \times (-1)^{p+q+j+i+s+m} d_{z,p}\binom{k+l-1}{p}\frac{\Gamma(\beta(q+1)+1+j)}{\Gamma(\beta(q+1)+1)j!}\binom{j}{i} \\ & \times \binom{b(\beta q+\beta+i)-1}{s}\binom{2s+1}{m}\left(\frac{m+1}{m+1}\right)g(x;\xi)G^m(x;\xi) \\ & \times \sum_{m=0}^{\infty} U_m^* g_m(x;\xi), \end{aligned} \tag{15}$$

where $g_m(x;\xi) = (m+1)g(x;\xi)G^m(x;\xi)$ is an Exp-G with power parameter m and the linear component

$$U_m^* = \frac{n!4\beta b}{(k-1)!(n-k)!} \sum_{p,z,q,j,i,s=0}^{\infty} \sum_{n=1}^{\infty} \sum_{l=0}^{n-k} \binom{n-k}{l} \frac{n a_n \theta^n}{C^{z+1}(\theta)} \binom{z+n-1}{q}$$
$$\times (-1)^{p+q+j+i+s+m} d_{z,p} \binom{k+l-1}{p} \frac{\Gamma(\beta(q+1)+1+j)}{\Gamma(\beta(q+1)+1)j!} \binom{j}{i}$$
$$\times \binom{b(\beta q+\beta+i)-1}{s} \binom{2s+1}{m} \left(\frac{1}{m+1}\right). \quad (16)$$

The tth moment of the distribution of the kth order statistic from EHL-TL-GPS class of distributions can be readily obtained from Eq. (15).

2.7.2 Rényi Entropy

In this subsection, Rényi entropy for EHL-TL-GPS class of distributions is derived. An entropy is a measure of uncertainty or variation of a random variable. Rényi entropy [35] is a generalization of Shannon entropy [37]. Rényi entropy is defined by

$$I_R(v) = \frac{1}{1-v} \log\left(\int_0^{\infty} [f_{EHL-TL-GPS}(x; b, \beta, \theta, \xi)]^v dx\right), v \neq 1, v > 0. \quad (17)$$

Note that

$$[f_{EHL-TL-GPS}(x; b, \beta, \theta, \xi)]^v = \frac{(4\beta\theta b)^v g^v(x;\xi)[1-\overline{G}^2(x;\xi)]^{(\beta b-1)v}\overline{G}(x;\xi)}{[1+(1-[1-\overline{G}^2(x;\xi)]^b)]^{(\beta+1)v}}$$
$$\times \left(\frac{C'\left(\theta\left(1-\left[\frac{[1-\overline{G}^2(x;\xi)]^b}{1+\left(1-[1-\overline{G}^2(x;\xi)]^b\right)}\right]^{\beta}\right)\right)}{C(\theta)} \right)^{v}.$$

Also, note that

$$\left(\frac{C'\left(\theta\left(1-\left[\frac{[1-\overline{G}^2(x;\xi)]^b}{1+\left(1-[1-\overline{G}^2(x;\xi)]^b\right)}\right]^{\beta}\right)\right)}{C(\theta)} \right)^{v} = \left(b_n \left(1-\left[\frac{[1-\overline{G}^2(x;\xi)]^b}{1+\left(1-[1-\overline{G}^2(x;\xi)]^b\right)}\right]^{\beta}\right)^{n-1} \right)^{v}$$

where $b_n = \sum_{n=1}^{\infty} \frac{a_n \theta^{n-1}}{C(\theta)}$. Applying the power series raised to a positive integer (see Gradshyten and Ryzhik [24] p. 17, for details), we get

$$[f_{EHL-TL-GPS}(x; b, \beta, \theta, \xi)]^{\nu} = \frac{(4\beta\theta b)^{\nu} g^{\nu}(x;\xi)[1-\overline{G}^2(x;\xi)]^{(\beta b-1)\nu}\overline{G}(x;\xi)}{[1+(1-[1-\overline{G}^2(x;\xi)]^b)]^{(\beta+1)\nu}}$$
$$\times\, d_{z,\nu} b_n^z \left(1-\left[\frac{\left[1-\overline{G}^2(x;\xi)\right]^b}{1+\left(1-\left[1-\overline{G}^2(x;\xi)\right]^b\right)}\right]^{\beta}\right)^{(n-1)z}$$

where $d_{z,\nu} = (zb_0)^{-1}\sum_{h=1}^{z}[\nu(h+1)-z]b_h d_{z-h,\nu}$ and $d_{0,\nu} = b_0^{\nu}$. Considering the following generalized binomial series expansions

$$\left(1-\left[\frac{\left[1-\overline{G}^2(x;\xi)\right]^b}{1+\left(1-\left[1-\overline{G}^2(x;\xi)\right]^b\right)}\right]^{\beta}\right)^{(n-1)z} = \sum_{p=0}^{\infty}(-1)^p \frac{[1-\overline{G}^2(x;\xi)]^{b\beta p}}{(1+(1-[1-\overline{G}^2(x;\xi)]^b))^{\beta p}} \times \binom{(n-1)z}{p},$$

$$(1+(1-[1-\overline{G}^2(x;\xi)]^b))^{-(\beta(p+\nu)+\nu)} = \sum_{j,i=0}^{\infty}(-1)^{j+i}\frac{\Gamma(\beta(p+\nu)+\nu+j)}{\Gamma(\beta(p+\nu)+\nu)j!}\binom{j}{i}[1-\overline{G}^2(x;\xi)]^{bi},$$

$$[1-\overline{G}^2(x;\xi)]^{b(\beta p+\beta\nu+i)-\nu} = \sum_{s=0}^{\infty}(-1)^s\binom{b(\beta p+\beta\nu+i)-\nu}{s}\overline{G}^{2s}(x;\xi),$$

and

$$\overline{G}^{2s+\nu}(x;\xi) = \sum_{m=0}^{\infty}(-1)^m\binom{2s+\nu}{m}G^m(x;\xi),$$

yields

$$[f_{EHL-TL-GPS}(x; b, \beta, \theta, \xi)]^{\nu} = \sum_{z,p,j,i,s,m=0}^{\infty}(4\beta\theta b)^{\nu} b_n^z d_{z,\nu}(-1)^{p+j+i+s+m}\binom{(n-1)z}{p}$$
$$\times \frac{\Gamma(\beta(p+\nu)+\nu+j)}{\Gamma(\beta(p+\nu)+\nu)j!}\binom{j}{i}\binom{b(\beta p+\beta\nu+i)-\nu}{s}$$
$$\times \binom{2s+\nu}{m} g^{\nu}(x;\xi)G^m(x;\xi).$$

Therefore, Rényi entropy for EHL-TL-GPS class of distributions is given by

$$I_R(v) = \frac{1}{1-v}\log\left(\sum_{m=0}^{\infty} u^* e^{(1-\nu)I_{REG}}\right), \tag{18}$$

where $I_{REG} = \int_0^\infty [(1+m/\nu)g(x;\xi)G^{m/\nu}]^\nu dx$ is Rényi entropy for an Exp-G distribution with power parameter m/ν and

$$u^* = \sum_{z,p,j,i,s=0}^{\infty} (4\beta\theta b)^\nu b_n^z d_{z,\nu}(-1)^{p+j+i+s+m}\binom{(n-1)z}{p}\frac{\Gamma(\beta(p+\nu)+\nu+j)}{\Gamma(\beta(p+\nu)+\nu)j!}\binom{j}{i} \times \binom{b(\beta p+\beta\nu+i)-\nu}{s}\binom{2s+\nu}{m}\frac{1}{(1+m/\nu)^\nu}. \tag{19}$$

Consequently, Rényi entropy for EHL-TL-GPS class of distributions can be obtained from Reényi entropy of the Exp-G distribution.

2.8 Maximum Likelihood Estimation

We obtain the maximum likelihood estimates of the parameters of the EHL-TL-GPS class of distributions in this section. Let $X_i \sim EHL-TL-GPS(b,\beta,\theta,\xi)$ and $\Delta = (b,\beta,\theta,\xi)^T$ be the parameter vector. The log-likelihood $\ell = \ell(\Delta)$ based on a random sample of size n is given by

$$\begin{aligned}\ell(\Delta) &= n\ln(4\beta\theta b) + (\beta b - 1)\sum_{i=1}^{n}\ln\left[1-\overline{G}^2(x_i;\xi)\right] + \sum_{i=1}^{n}\ln\left(\overline{G}(x_i;\xi)\right) + \sum_{i=1}^{n}\ln\left(g(x_i;\xi)\right)\\ &- n\ln(C(\theta)) + \sum_{i=1}^{n}\ln\left(C'\left(\theta\left(1-\left[\frac{\left[1-\overline{G}^2(x_i;\xi)\right]^b}{1+\left(1-\left[1-\overline{G}^2(x_i;\xi)\right]^b\right)}\right]^\beta\right)\right)\right)\\ &- (\beta+1)\sum_{i=1}^{n}\ln\left(1+(1-[1-\overline{G}^2(x_i;\xi)]^b)\right).\end{aligned}$$

The elements of the score vector are given by

$$\begin{aligned}\frac{\partial\ell}{\partial\beta} &= \frac{n}{\beta} + b\sum_{i=1}^{n}\ln\left[1-\overline{G}^2(x_i;\xi)\right] - \sum_{i=1}^{n}\frac{\theta\left(C''\left(\theta\left(1-\left[\frac{\left[1-\overline{G}^2(x_i;\xi)\right]^b}{1+\left(1-\left[1-\overline{G}^2(x_i;\xi)\right]^b\right)}\right]^\beta\right)\right)\right)}{\left(C'\left(\theta\left(1-\left[\frac{\left[1-\overline{G}^2(x_i;\xi)\right]^b}{1+\left(1-\left[1-\overline{G}^2(x_i;\xi)\right]^b\right)}\right]^\beta\right)\right)\right)}\\ &\times\left[\frac{\left[1-\overline{G}^2(x_i;\xi)\right]^b}{1+\left(1-\left[1-\overline{G}^2(x_i;\xi)\right]^b\right)}\right]^\beta \ln\left(\left[\frac{\left[1-\overline{G}^2(x_i;\xi)\right]^b}{1+\left(1-\left[1-\overline{G}^2(x_i;\xi)\right]^b\right)}\right]\right)\\ &- \sum_{i=1}^{n}\ln\left(1+(1-[1-\overline{G}^2(x_i;\xi)]^b)\right),\end{aligned}$$

$$\frac{\partial \ell}{\partial b} = \frac{n}{b} + \beta \sum_{i=1}^{n} \ln\left[1-\overline{G}^2(x_i;\xi)\right] + \left[\sum_{i=1}^{n} \frac{\beta\theta\left(C''\left(\theta\left(1-\left[\frac{\left[1-\overline{G}^2(x_i;\xi)\right]^b}{1+\left(1-\left[1-\overline{G}^2(x_i;\xi)\right]^b\right)}\right]^{\beta}\right)\right)\right)}{\left(C'\left(\theta\left(1-\left[\frac{\left[1-\overline{G}^2(x_i;\xi)\right]^b}{1+\left(1-\left[1-\overline{G}^2(x_i;\xi)\right]^b\right)}\right]^{\beta}\right)\right)\right)}\right.$$

$$\left. \times \frac{2\left[\frac{\left[1-\overline{G}^2(x_i;\xi)\right]^b}{1+\left(1-\left[1-\overline{G}^2(x_i;\xi)\right]^b\right)}\right]^{\beta-1}\left[1-\overline{G}^2(x_i;\xi)\right]^b \ln\left[1-\overline{G}^2(x_i;\xi)\right]}{\left[1+\left(1-\left[1-\overline{G}^2(x_i;\xi)\right]^b\right)\right]^2}\right]$$

$$+ (\beta+1)\sum_{i=1}^{n} \frac{\left[1-\overline{G}^2(x_i;\xi)\right]^b \ln\left[1-\overline{G}^2(x_i;\xi)\right]}{\left(1+(1-[1-\overline{G}^2(x_i;\xi)]^b)\right)},$$

$$\frac{\partial \ell}{\partial \theta} = \frac{n}{\theta} - \frac{nC'(\theta)}{C(\theta)} + \sum_{i=1}^{n} \frac{\left(C''\left(\theta\left(1-\left[\frac{\left[1-\overline{G}^2(x_i;\xi)\right]^b}{1+\left(1-\left[1-\overline{G}^2(x_i;\xi)\right]^b\right)}\right]^{\beta}\right)\right)\right)\left(1-\left[\frac{\left[1-\overline{G}^2(x_i;\xi)\right]^b}{1+\left(1-\left[1-\overline{G}^2(x_i;\xi)\right]^b\right)}\right]^{\beta}\right)}{\left(C'\left(\theta\left(1-\left[\frac{\left[1-\overline{G}^2(x_i;\xi)\right]^b}{1+\left(1-\left[1-\overline{G}^2(x_i;\xi)\right]^b\right)}\right]^{\beta}\right)\right)\right)},$$

and

$$\frac{\partial \ell}{\partial \xi_k} = -(\beta b - 1)\sum_{i=1}^{n} \frac{2\overline{G}(x_i;\xi)\frac{\partial \overline{G}(x_i;\xi)}{\partial \xi_k}}{\left[1-\overline{G}^2(x_i;\xi)\right]} + \sum_{i=1}^{n} \frac{\frac{\partial \overline{G}(x_i;\xi)}{\partial \xi_k}}{(\overline{G}(x_i;\xi))} + \sum_{i=1}^{n} \frac{\frac{\partial g(x_i;\xi)}{\partial \xi_k}}{g(x_i;\xi)}$$

$$+ \sum_{i=1}^{n} \frac{C''\left(\theta\left(1-\left[\frac{\left[1-\overline{G}^2(x_i;\xi)\right]^b}{1+\left(1-\left[1-\overline{G}^2(x_i;\xi)\right]^b\right)}\right]^{\beta}\right)\right)}{C'\left(\theta\left(1-\left[\frac{\left[1-\overline{G}^2(x_i;\xi)\right]^b}{1+\left(1-\left[1-\overline{G}^2(x_i;\xi)\right]^b\right)}\right]^{\beta}\right)\right)} \theta\beta \left[\frac{\left[1-\overline{G}^2(x_i;\xi)\right]^b}{1+\left(1-\left[1-\overline{G}^2(x_i;\xi)\right]^b\right)}\right]^{\beta-1}$$

$$\times \frac{4b\left[1-\overline{G}^2(x_i;\xi)\right]^{b-1}\overline{G}(x_i;\xi)\frac{\partial \overline{G}(x_i;\xi)}{\partial \xi_k}}{\left[1+\left(1-\left[1-\overline{G}^2(x_i;\xi)\right]^b\right)\right]^2} + (\beta+1)\sum_{i=1}^{n} \frac{2b\left[1-\overline{G}^2(x_i;\xi)\right]^{b-1}\overline{G}(x_i;\xi)\frac{\partial \overline{G}(x_i;\xi)}{\partial \xi_k}}{\left(1+(1-[1-\overline{G}^2(x_i;\xi)]^b)\right)}.$$

The maximum likelihood estimates of the parameters, denoted by $\hat{\Delta}$ is obtained by solving the nonlinear equation $(\frac{\partial \ell_n}{\partial \beta}, \frac{\partial \ell_n}{\partial b}, \frac{\partial \ell_n}{\partial \theta}, \frac{\partial \ell_n}{\partial \xi_k})^T = \mathbf{0}$, using a numerical method such as Newton-Raphson procedure. The multivariate normal distribution $N_{q+3}(\underline{0}, J(\hat{\Delta})^{-1})$, where the mean vector $\underline{\mathbf{0}} = (0, 0, 0, \underline{0})^T$ and $J(\hat{\Delta})^{-1}$ is the observed Fisher information matrix evaluated at $\hat{\Delta}$, can be used to construct confidence intervals and confidence regions for the individual model parameters and for the survival and hazard rate functions.

3 Some Special Classes of the EHL-TL-GPS Class of Distributions

In this section, special classes of EHL-TL-GPS class of distributions are presented by specifying the baseline cdf $G(x;\xi)$ and pdf $g(x;\xi)$ in Eqs. (5) and (6).

3.1 Exponentiated Half Logistic Topp-Leone-Log-Logistic Power Series (EHL-TL-LLoGPS) Class of Distributions

If the baseline cdf and pdf are given by $G(x;\lambda)=1-(1+x^{\lambda})^{-1}$ and $g(x;\lambda)=\lambda x^{\lambda-1}(1+x^{\lambda})^{-2}$, for $\lambda>0$, and $x>0$, then the cdf and pdf of the EHL-TL-LLoG power series class of distributions are given by

$$F_{EHL-TL-LLoGPS}(x)=1-\frac{C\left(\theta\left(1-\left[\frac{\left[1-(1+x^{\lambda})^{-2}\right]^{b}}{1+\left(1-\left[1-(1+x^{\lambda})^{-2}\right]^{b}\right)}\right]^{\beta}\right)\right)}{C(\theta)} \tag{20}$$

and

$$f_{EHL-TL-LLoGPS}(x)=\frac{4\beta\theta b\lambda x^{\lambda-1}\left(1+x^{\lambda}\right)^{-2}[1-\left(1+x^{\lambda}\right)^{-2}]^{\beta b-1}\left(1+x^{\lambda}\right)^{-1}}{[1+(1-[1-\left(1+x^{\lambda}\right)^{-2}]^{b})]^{\beta+1}} \times\frac{C'\left(\theta\left(1-\left[\frac{\left[1-(1+x^{\lambda})^{-2}\right]^{b}}{1+\left(1-\left[1-(1+x^{\lambda})^{-2}\right]^{b}\right)}\right]^{\beta}\right)\right)}{C(\theta)}, \tag{21}$$

respectively for β, b, θ, λ and $x>0$. The hrf is given by

$$h_{EHL-TL-LLoGPS}(x)=\frac{4\beta\theta b\lambda x^{\lambda-1}\left(1+x^{\lambda}\right)^{-2}[1-\left(1+x^{\lambda}\right)^{-2}]^{\beta b-1}\left(1+x^{\lambda}\right)^{-1}}{[1+(1-[1-\left(1+x^{\lambda}\right)^{-2}]^{b})]^{\beta+1}} \times\frac{C'\left(\theta\left(1-\left[\frac{\left[1-(1+x^{\lambda})^{-2}\right]^{b}}{1+\left(1-\left[1-(1+x^{\lambda})^{-2}\right]^{b}\right)}\right]^{\beta}\right)\right)}{C(\theta)} \times\left(\frac{C\left(\theta\left(1-\left[\frac{\left[1-(1+x^{\lambda})^{-2}\right]^{b}}{1+\left(1-\left[1-(1+x^{\lambda})^{-2}\right]^{b}\right)}\right]^{\beta}\right)\right)}{C(\theta)}\right)^{-1}. \tag{22}$$

3.1.1 Exponentiated Half Logistic-Topp-Leone-Log-Logistic Poisson (EHL-TL-LLoGP) Distribution

The cdf and pdf of EHL-TL-LLoGP distribution are given by

$$F_{EHL-TL-LLoGP}(x) = 1 - \frac{\exp\left(\theta\left(1-\left[\frac{\left[1-(1+x^{\lambda})^{-2}\right]^{b}}{1+\left(1-\left[1-(1+x^{\lambda})^{-2}\right]^{b}\right)}\right]^{\beta}\right)\right)-1}{\exp(\theta)-1}$$

and

$$f_{EHL-TL-LLoGP}(x) = \frac{4\beta\theta b\lambda x^{\lambda-1}(1+x^{\lambda})^{-3}[1-\left(1+x^{\lambda}\right)^{-2}]^{\beta b-1}}{[1+(1-[1-\left(1+x^{\lambda}\right)^{-2}]^{b})]^{\beta+1}} \times \frac{\exp\left(\theta\left(1-\left[\frac{\left[1-(1+x^{\lambda})^{-2}\right]^{b}}{1+\left(1-\left[1-(1+x^{\lambda})^{-2}\right]^{b}\right)}\right]^{\beta}\right)\right)}{\exp(\theta)-1},$$

respectively, for β, b, θ, λ and $x > 0$. The hrf is given by

$$h_{EHL-TL-LLoGP}(x) = \frac{4\beta\theta b\lambda x^{\lambda-1}(1+x^{\lambda})^{-3}[1-\left(1+x^{\lambda}\right)^{-2}]^{\beta b-1}}{[1+(1-[1-\left(1+x^{\lambda}\right)^{-2}]^{b})]^{\beta+1}} \times \frac{\exp\left(\theta\left(1-\left[\frac{\left[1-(1+x^{\lambda})^{-2}\right]^{b}}{1+\left(1-\left[1-(1+x^{\lambda})^{-2}\right]^{b}\right)}\right]^{\beta}\right)\right)}{\exp(\theta)-1} \times \left(\frac{\exp\left(\theta\left(1-\left[\frac{\left[1-(1+x^{\lambda})^{-2}\right]^{b}}{1+\left(1-\left[1-(1+x^{\lambda})^{-2}\right]^{b}\right)}\right]^{\beta}\right)\right)-1}{\exp(\theta)-1}\right)^{-1},$$

for β, b, θ, λ and $x > 0$.

Figure 1 shows the pdfs of the EHL-TL-LLoGP distribution. The pdf can take various shapes that include almost symmetric, reverse-J, left, or right-skewed. Furthermore, the hazard rate functions (hrfs) for the EHL-TL-LLoGP distribution exhibit increasing, decreasing, reverse-J, bathtub, and upside bathtub shapes.

3.1.2 Exponentiated Half Logistic-Topp-Leone-Log-Logistic Geometric (EHL-TL-LLoGG) Distribution

The cdf and pdf of EHL-TL-LLoGG distribution are given by

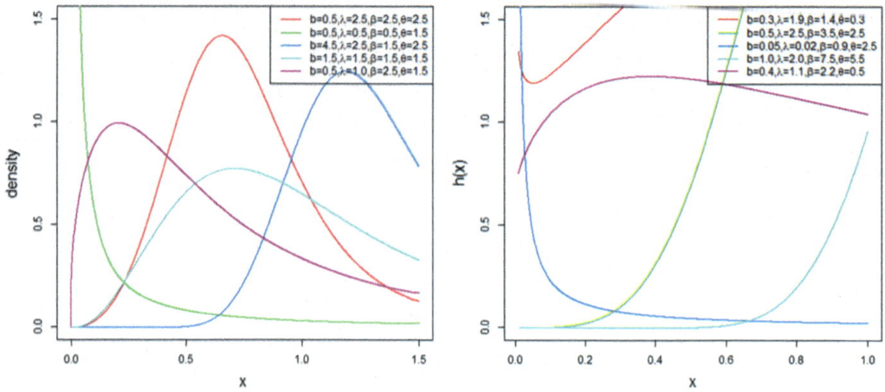

Fig. 1 Plots of the pdf and hrf for the EHL-TL-LLoGP distribution

$$F_{EHL-TL-LLoGG}(x) = 1 - \frac{(1-\theta)\left(\theta\left(1-\left[\frac{\left[1-(1+x^{\lambda})^{-2}\right]^{b}}{1+\left(1-\left[1-(1+x^{\lambda})^{-2}\right]^{b}\right)}\right]^{\beta}\right)\right)}{\left(1-\theta\left(1-\left[\frac{\left[1-(1+x^{\lambda})^{-2}\right]^{b}}{1+\left(1-\left[1-(1+x^{\lambda})^{-2}\right]^{b}\right)}\right]^{\beta}\right)\right)}$$

and

$$f_{EHL-TL-LLoGG}(x) = \frac{4\beta\theta b\lambda x^{\lambda-1}(1+x^{\lambda})^{-3}[1-(1+x^{\lambda})^{-2}]^{\beta b-1}}{[1+(1-[1-(1+x^{\lambda})^{-2}]^{b})]^{\beta+1}} \times \frac{\left(1-\theta\left(1-\left[\frac{\left[1-(1+x^{\lambda})^{-2}\right]^{b}}{1+\left(1-\left[1-(1+x^{\lambda})^{-2}\right]^{b}\right)}\right]^{\beta}\right)\right)^{-2}}{\theta(1-\theta)^{-1}},$$

respectively, for β, b, θ, λ and $x > 0$. The hrf is given by

$$h_{EHL-TL-LLoGG}(x) = \frac{4\beta\theta b\lambda x^{\lambda-1}(1+x^{\lambda})^{-3}[1-(1+x^{\lambda})^{-2}]^{\beta b-1}}{[1+(1-[1-(1+x^{\lambda})^{-2}]^{b})]^{\beta+1}} \times \frac{\left(1-\theta\left(1-\left[\frac{\left[1-(1+x^{\lambda})^{-2}\right]^{b}}{1+\left(1-\left[1-(1+x^{\lambda})^{-2}\right]^{b}\right)}\right]^{\beta}\right)\right)^{-2}}{\theta(1-\theta)^{-1}} \times \left(\frac{(1-\theta)\left(\theta\left(1-\left[\frac{\left[1-(1+x^{\lambda})^{-2}\right]^{b}}{1+\left(1-\left[1-(1+x^{\lambda})^{-2}\right]^{b}\right)}\right]^{\beta}\right)\right)}{\left(1-\theta\left(1-\left[\frac{\left[1-(1+x^{\lambda})^{-2}\right]^{b}}{1+\left(1-\left[1-(1+x^{\lambda})^{-2}\right]^{b}\right)}\right]^{\beta}\right)\right)}\right)^{-1}.$$

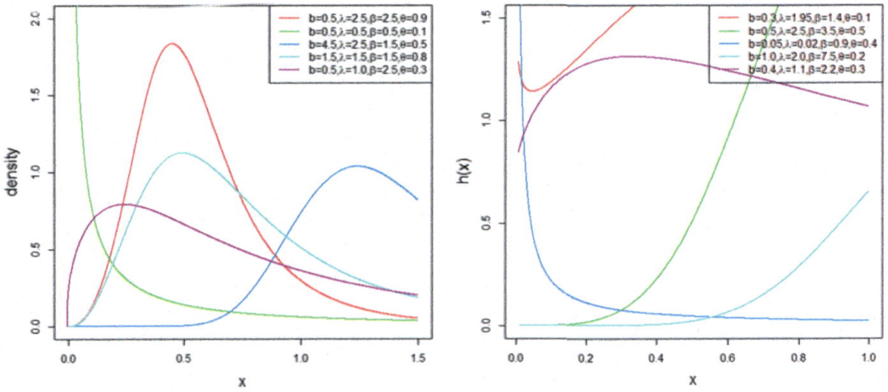

Fig. 2 Plots of the pdf and hrf for the EHL-TL-LLoGG distribution

Figure 2 shows the pdfs of the EHL-TL-LLoGG distribution. The pdf can take various shapes that include almost symmetric, reverse-J, left, or right-skewed. Furthermore, the hrfs for the EHL-TL-LLoGG distribution exhibit increasing, decreasing, reverse-J, bathtub, and upside bathtub shapes.

3.2 *Exponentiated Half Logistic-Topp-Leone-Weibull Power Series (EHL-TL-WPS) Class of Distributions*

Suppose the cdf and pdf of the Weibull distribution are given by $G(x; a) = 1 - \exp(-x^a)$, for $x \geq 0, a > 0$ and $g(x; a) = ax^{a-1}\exp(-x^a)$, for $a > 0$, and $x > 0$, then, the cdf and pdf of the EHL-TL-WPS class of distributions are given by

$$F_{EHL-TL-WPS}(x) = 1 - \frac{C\left(\theta\left(1-\left[\frac{[1-\exp(-2x^a)]^b}{1+\left(1-[1-\exp(-2x^a)]^b\right)}\right]^\beta\right)\right)}{C(\theta)} \tag{23}$$

and

$$f_{EHL-TL-WPS}(x) = \frac{4\beta\theta b a x^{a-1}\exp(-x^a)[1-\exp(-2x^a)]^{\beta b-1}\exp(-x^a)}{[1+(1-[1-\exp(-2x^a)]^b)]^{\beta+1}} \times \frac{C'\left(\theta\left(1-\left[\frac{[1-\exp(-2x^a)]^b}{1+\left(1-[1-\exp(-2x^a)]^b\right)}\right]^\beta\right)\right)}{C(\theta)},$$

respectively, for b, β, θ, $a > 0$ and $x > 0$. The hrf is given by

$$h_{EHL-TL-WPS}(x) = \frac{4\beta\theta b a x^{a-1}\exp(-x^a)[1-\exp(-2x^a)]^{\beta b-1}\exp(-x^a)}{[1+(1-[1-\exp(-2x^a)]^b)]^{\beta+1}}$$

$$\times \frac{C'\left(\theta\left(1-\left[\frac{[1-\exp(-2x^a)]^b}{1+\left(1-[1-\exp(-2x^a)]^b\right)}\right]^{\beta}\right)\right)}{C(\theta)}$$

$$\times \left(\frac{C\left(\theta\left(1-\left[\frac{[1-\exp(-2x^a)]^b}{1+\left(1-[1-\exp(-2x^a)]^b\right)}\right]^{\beta}\right)\right)}{C(\theta)}\right)^{-1}. \tag{24}$$

3.2.1 Exponentiated Half Logistic-Topp-Leone-Weibull Poisson (EHL-TL-WP) Distribution

The cdf and pdf of EHL-TL-WP distribution are given by

$$F_{EHL-TL-WP}(x) = 1 - \frac{\exp\left(\theta\left(1-\left[\frac{[1-\exp(-2x^a)]^b}{1+\left(1-[1-\exp(-2x^a)]^b\right)}\right]^{\beta}\right)\right)-1}{\exp(\theta)-1}$$

and

$$f_{EHL-TL-WP}(x) = \frac{4\beta\theta bax^{a-1}\exp(-x^a)[1-\exp(-2x^a)]^{\beta b-1}\exp(-x^a)}{[1+(1-[1-\exp(-2x^a)]^b)]^{\beta+1}}$$

$$\times \frac{\exp\left(\theta\left(1-\left[\frac{[1-\exp(-2x^a)]^b}{1+\left(1-[1-\exp(-2x^a)]^b\right)}\right]^{\beta}\right)\right)}{\exp(\theta)-1},$$

respectively, for b, β, θ, $a > 0$ and $x > 0$. The corresponding hrf is given by

$$h_{EHL-TL-WP}(x) = \frac{4\beta\theta bax^{a-1}\exp(-x^a)[1-\exp(-2x^a)]^{\beta b-1}\exp(-x^a)}{[1+(1-[1-\exp(-2x^a)]^b)]^{\beta+1}}$$

$$\times \frac{\exp\left(\theta\left(1-\left[\frac{[1-\exp(-2x^a)]^b}{1+\left(1-[1-\exp(-2x^a)]^b\right)}\right]^{\beta}\right)\right)}{\exp(\theta)-1}$$

$$\times \left(\frac{\exp\left(\theta\left(1-\left[\frac{[1-\exp(-2x^a)]^b}{1+\left(1-[1-\exp(-2x^a)]^b\right)}\right]^{\beta}\right)\right)-1}{\exp(\theta)-1}\right)^{-1}.$$

Figure 3 shows the pdfs of the EHL-TL-WP distribution. The pdf can take various shapes that include almost symmetric, reverse-J, left, or right-skewed. Furthermore, the hrfs for the EHL-TL-WP distribution exhibit increasing, decreasing, reverse-J, bathtub, and upside bathtub shapes.

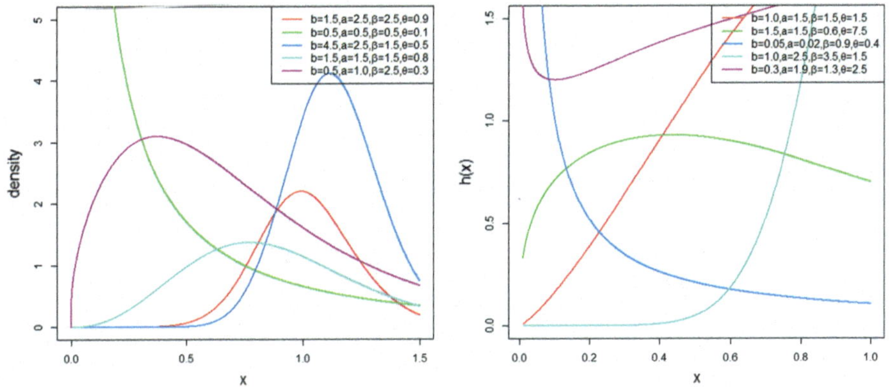

Fig. 3 Plots of the pdf and hrf for the EHL-TL-WP distribution

3.2.2 Exponentiated Half Logistic-Topp-Leone-Weibull Geometric (EHL-TL-WG) Distribution

The cdf, pdf and hazard rate function of EHL-TL-WG distribution are given by

$$F_{EHL-TL-WG}(x) = 1 - \frac{(1-\theta)\left(1-\left[\frac{[1-\exp(-2x^a)]^b}{1+\left(1-[1-\exp(-2x^a)]^b\right)}\right]^{\beta}\right)}{\left(1-\theta\left(1-\left[\frac{[1-\exp(-2x^a)]^b}{1+\left(1-[1-\exp(-2x^a)]^b\right)}\right]^{\beta}\right)\right)},$$

$$f_{EHL-TL-WG}(x) = \frac{4\beta\theta bax^{a-1}\exp(-x^a)\,[1-\exp(-2x^a)]^{\beta b-1}\exp(-x^a)}{[1+(1-[1-\exp(-2x^a)]^b)]^{\beta+1}} \times \frac{\left(1-\theta\left(1-\left[\frac{[1-\exp(-2x^a)]^b}{1+\left(1-[1-\exp(-2x^a)]^b\right)}\right]^{\beta}\right)\right)^{-2}}{\theta(1-\theta)^{-1}}$$

and

$$h_{EHL-TL-WG}(x) = \frac{4\beta\theta bax^{a-1}\exp(-x^a)\,[1-\exp(-2x^a)]^{\beta b-1}\exp(-x^a)}{[1+(1-[1-\exp(-2x^a)]^b)]^{\beta+1}} \times \frac{\left(1-\theta\left(1-\left[\frac{[1-\exp(-2x^a)]^b}{1+\left(1-[1-\exp(-2x^a)]^b\right)}\right]^{\beta}\right)\right)^{-2}}{\theta(1-\theta)^{-1}} \times \left(1-\frac{(1-\theta)\left(1-\left[\frac{[1-\exp(-2x^a)]^b}{1+\left(1-[1-\exp(-2x^a)]^b\right)}\right]^{\beta}\right)}{\left(1-\theta\left(1-\left[\frac{[1-\exp(-2x^a)]^b}{1+\left(1-[1-\exp(-2x^a)]^b\right)}\right]^{\beta}\right)\right)}\right)^{-1}.$$

for b, $\beta, a > 0$, $0 < \theta < 1$, and $x > 0$.

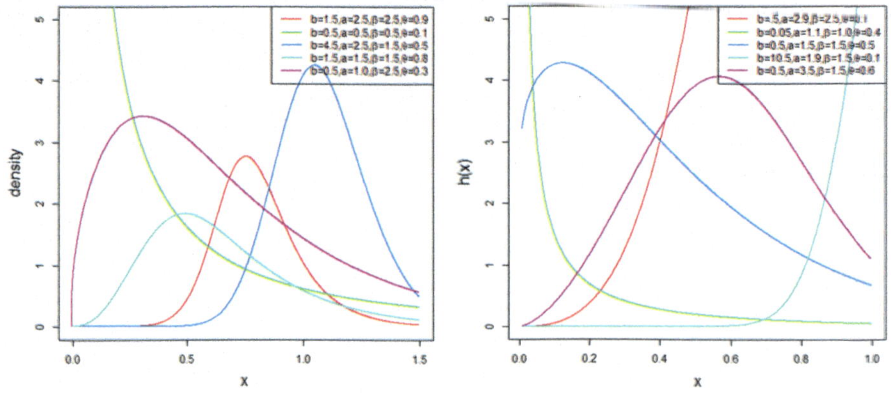

Fig. 4 Plots of the pdf and hrf for the EHL-TL-WG distribution

Figure 4 shows the pdfs of the EHL-TL-WG distribution. The pdf can take various shapes that include almost symmetric, reverse-J, left, or right-skewed. Furthermore, the hrfs for the EHL-TL-WG distribution exhibits increasing, decreasing, reverse-J and uni-modal shapes.

4 Simulation Study

In this section, a simulation study was conducted to assess consistency of the maximum likelihood estimators. We considered the EHL-TL-LLoGP distribution. We simulated for the sample sizes n= 60, 120, 240, 480, 960 and 1920, for N=1000 for each sample. We estimate the mean, root mean square error (RMSE), and average bias. We consider simulations for the following sets of initial parameters values (I: $b = 1.0, \theta = 0.05, \lambda = 1.0, \beta = 0.9$), (II: $b = 1.0, \theta = 0.05, \lambda = 0.8, \beta = 0.9$), (III: $b = 1.0, \theta = 0.05, \lambda = 1.5, \beta = 0.9$), (IV: $b = 1.0, \theta = 0.3, \lambda = 2.0, \beta = 1.0$), (V: $b = 1.2, \theta = 0.3, \lambda = 1.2, \beta = 0.9$) and (VI: $b = 1.2, \theta = 0.3, \lambda = 1.2, \beta = 1.1$). If the model performs better, we except the mean to approximate the true parameter values, the RMSE, and bias to decay toward zero for an increase in sample size. From the results in Table 2, the mean values approximate the true parameter values, RMSE and bias decay towards zero for all the parameter values.

5 Applications

In this section, we present real data examples to demonstrate the usefulness of the exponentiated half logistic-Topp-Leone-log-logistic Poisson (EHL-TL-LLoGP) distribution. We compared the EHL-TL-LLoGP distribution to several non-nested mod-

Table 2 Monte Carlo simulation results for EHL-TL-LLoGP distribution: Mean, RMSE and average bias

		I: $b = 1.0, \theta = 0.05, \lambda = 1.0, \beta = 0.9$			II: $b = 1.0, \theta = 0.05, \lambda = 0.8, \beta = 0.9$			III: $b = 1.0, \theta = 0.05, \lambda = 1.5, \beta = 0.9$		
Parameter	n	Mean	RMSE	Bias	Mean	RMSE	Bias	Mean	RMSE	Bias
b	60	2.707771	3.895996	1.707771	2.616078	3.778058	1.616078	2.702649	3.643779	1.702649
	120	1.866261	2.099757	0.866261	2.178139	2.817496	1.178139	1.993941	2.411196	0.993941
	240	1.541102	1.709914	0.541102	1.538611	1.503157	0.538611	1.603828	1.837097	0.603828
	480	1.332111	0.960462	0.332111	1.366298	0.994440	0.366298	1.345543	0.921921	0.345543
	960	1.152374	0.518358	0.152374	1.218686	0.555648	0.218686	1.161942	0.526746	0.161942
	1920	1.083691	0.404742	0.083691	1.144984	0.389172	0.144983	1.083449	0.386929	0.083449
θ	60	1.203214	1.838525	1.153214	1.105834	1.585565	1.055834	1.188181	1.849590	1.138181
	120	0.745443	1.148939	0.695443	0.784253	1.268127	0.734253	0.791602	1.298863	0.741601
	240	0.510365	0.720765	0.460365	0.511301	0.727407	0.461301	0.482082	0.618176	0.432082
	480	0.346360	0.409080	0.296360	0.354023	0.415578	0.304023	0.354770	0.414970	0.304770
	960	0.255502	0.276219	0.205502	0.275940	0.300502	0.225940	0.253721	0.277442	0.203721
	1920	0.188108	0.192184	0.138108	0.204440	0.209314	0.154440	0.186697	0.189730	0.136697
λ	60	0.924159	0.218247	−0.075841	0.744049	0.161258	−0.055951	1.388842	0.323155	−0.111158
	120	0.949921	0.132567	−0.050079	0.758556	0.113107	−0.041444	1.415592	0.215964	−0.084408
	240	0.953201	0.096541	−0.046799	0.760444	0.077621	−0.039556	1.433648	0.137620	−0.066352
	480	0.973110	0.065504	−0.026890	0.778482	0.051551	−0.021518	1.460386	0.098567	−0.039614
	960	0.983014	0.045723	−0.016986	0.787559	0.036504	−0.012441	1.473639	0.069001	−0.026361
	1920	0.984627	0.035406	−0.015373	0.790484	0.026794	−0.009516	1.477334	0.052305	−0.022666
β	60	1.540731	3.020547	0.640731	1.673115	4.547574	0.773115	1.448393	2.746813	0.548392
	120	1.264256	1.139948	0.364256	1.203038	1.119194	0.303038	1.285514	1.234584	0.385513
	240	1.178854	0.917142	0.278854	1.139770	0.858162	0.239770	1.158484	0.941201	0.258484
	480	1.073442	0.741656	0.173442	1.039735	0.670350	0.139735	1.040626	0.672228	0.140626
	960	1.000387	0.492108	0.100387	0.945099	0.413009	0.045099	0.992142	0.477380	0.092142
	1920	0.986689	0.379180	0.086689	0.909158	0.246220	0.009158	0.980289	0.376983	0.080289

(continued)

Table 2 (continued)

		I: $b = 1.0, \theta = 0.05, \lambda = 1.0, \beta = 0.9$			II: $b = 1.0, \theta = 0.05, \lambda = 0.8, \beta = 0.9$			III: $b = 1.0, \theta = 0.05, \lambda = 1.5, \beta = 0.9$		
Parameter	n	Mean	RMSE	Bias	Mean	RMSE	Bias	Mean	RMSE	Bias
b	60	2.864298	4.247031	1.864298	3.235427	4.406235	2.035427	3.607113	4.907022	2.407113
	120	2.253074	2.905382	1.253074	2.497023	3.150110	1.297023	2.971264	3.979075	1.771264
	240	1.629373	1.641907	0.629373	1.907341	2.138466	0.707341	2.022881	2.028169	0.822881
	480	1.406814	1.148629	0.406814	1.539093	1.103449	0.339093	1.774630	1.576765	0.574630
	960	1.226219	0.688901	0.226219	1.362305	0.708177	0.162305	1.501184	0.896564	0.301184
	1920	1.123525	0.503025	0.123525	1.304351	0.559881	0.104351	1.393844	0.628473	0.193844
θ	60	1.273390	1.721487	0.973390	1.287581	1.740959	0.987581	1.398593	1.804765	1.098593
	120	0.930847	1.353273	0.630847	0.981581	1.455336	0.681581	1.171474	1.672507	0.871474
	240	0.615520	0.651379	0.315520	0.625563	0.682115	0.325563	0.798746	1.091366	0.498746
	480	0.465039	0.371517	0.165039	0.494980	0.535500	0.194980	0.554417	0.693210	0.254417
	960	0.367484	0.250283	0.067484	0.369660	0.258058	0.069660	0.387658	0.282601	0.087658
	1920	0.303889	0.185294	0.003889	0.307401	0.185317	0.007401	0.331956	0.202224	0.031956
λ	60	1.869737	0.423274	−0.130263	1.113375	0.247294	−0.086625	1.105537	0.256097	−0.094463
	120	1.916266	0.312279	−0.083734	1.135389	0.197241	−0.064611	1.115182	0.215538	−0.084818
	240	1.935497	0.197860	−0.064503	1.154636	0.119589	−0.045364	1.139446	0.148439	−0.060554
	480	1.971491	0.133616	−0.028509	1.175399	0.090592	−0.024601	1.173814	0.098353	−0.026186
	960	1.991317	0.096030	−0.008683	1.189486	0.058465	−0.010514	1.192335	0.058866	−0.007665
	1920	1.997994	0.073577	−0.002006	1.196247	0.044473	−0.003753	1.196283	0.042858	−0.003717
β	60	1.893390	4.828248	0.893390	1.543535	3.024458	0.643535	2.620278	7.129344	1.520278
	120	1.291632	1.385679	0.291632	1.299866	1.533503	0.399866	1.615680	2.371200	0.515680
	240	1.229541	1.010470	0.229541	1.156181	0.925307	0.256181	1.406496	1.191610	0.306496
	480	1.166697	0.821974	0.166697	1.131661	0.833036	0.231661	1.275798	0.886425	0.175797
	960	1.104690	0.678891	0.104690	1.063246	0.699390	0.163246	1.251526	0.848147	0.151526
	1920	1.080899	0.546507	0.080899	0.998310	0.521495	0.098310	1.173797	0.608301	0.073797

els. We use R software to estimate model parameters and standard errors. We assessed model performance using -2loglikelihood (-2 log L), Akaike Information Criterion (AIC), Consistent Akaike Information Criterion (AICC), Bayesian Information Criterion (BIC), Cramer-von-Mises (W^*), and Andersen-Darling (A^*) (see Chen and Balakrishnan [12] for details), Kolmogorov-Smirnov (K-S) statistic (and its p-value).

Tables 3 and 4 shows model parameters estimates (standard errors in parentheses) and several goodness-of-fit statistics. We also provide fitted densities and probability plots (as described by Chambers et al. [11]) to demonstrate how well our model fits the selected data sets.

The non-nested models considered in this paper are the Kumaraswamy odd Lindley-Log logistic (KOL-LLoG) by Chipepa et al. [16], beta generalized Lindley (BGL) by Oluyede and Yang [32], beta-Weibull (BW) by Cordeiro et al. [21], Kumaraswamy-Weibull (KwW) by Cordeiro et al. [22] and Topp-Leone-Weibull-Lomax (TL-WLx) by Jamal et al. [25]. The pdfs of the non-nested models are as follows:

$$
\begin{aligned}
f_{KOL-LLoG}(x;a,b,\lambda,c) &= ab\left[\frac{\lambda^2}{(1+\lambda)}\frac{cx^{c-1}}{(1+x^c)^{-1}}\exp\left\{-\lambda\frac{1-(1+x^c)^{-1}}{(1+x^c)^{-1}}\right\}\right]\\
&\times\left[1-\frac{\lambda+(1+x^c)^{-1}}{(1+\lambda)(1+x^c)^{-1}}\exp\left\{-\lambda\frac{(1-(1+x^c)^{-1})}{(1+x^c)^{-1}}\right\}\right]^{a-1}\\
&\times\left(1-\left[1-\frac{\lambda+(1+x^c)^{-1}}{(1+\lambda)(1+x^c)^{-1}}\exp\left\{-\lambda\frac{(1-(1+x^c)^{-1})}{(1+x^c)^{-1}}\right\}\right]^{a}\right)^{b-1},
\end{aligned}
$$

for $a, b, \lambda, c > 0$,

$$
f_{BW}(x;a,b,\alpha,\beta) = \frac{\beta\alpha^{\beta}}{B(a,b)}x^{\beta-1}e^{-b(\alpha x)^{\beta}}(1-e^{-(\alpha x)^{\beta}})^{a-1},
$$

for $a, b, \alpha, \beta > 0$,

$$
f_{KwW}(x;a,b,\alpha,\beta) = ab\alpha^{\beta}x^{\beta-1}e^{-(\alpha x)^{\beta}}(1-e^{-(\alpha x)^{\beta}})^{a-1}(1-(1-e^{-(\alpha x)^{\beta}})^{a})^{b-1},
$$

for $a, b, \alpha, \beta > 0$,

$$
\begin{aligned}
f_{TL-WLx}(x;a,b,\alpha,\theta) &= 2\theta\alpha ab(1+bx)^{a\alpha-1}(1-(1+bx)^{-a})^{\alpha-1}\\
&\times\exp\left(-2\left(\frac{1-(1+bx)^{-a}}{(1+bx)^{-a}}\right)\right)\\
&\times\left[1-\exp\left(-2\left(\frac{1-(1+bx)^{-a}}{(1+bx)^{-a}}\right)\right)\right]^{\theta-1},
\end{aligned}
$$

for $a, b, \alpha, \theta > 0$, and

$$
g_{BGL}(x;\alpha,\lambda,a,b) = \frac{\alpha\lambda^2(1+x)\exp(-\lambda x)}{B(a,b)(1+\lambda)}\left[1-\frac{1+\lambda+\lambda x}{1+\lambda}\exp(-\lambda x)\right]^{a\alpha-1}
$$

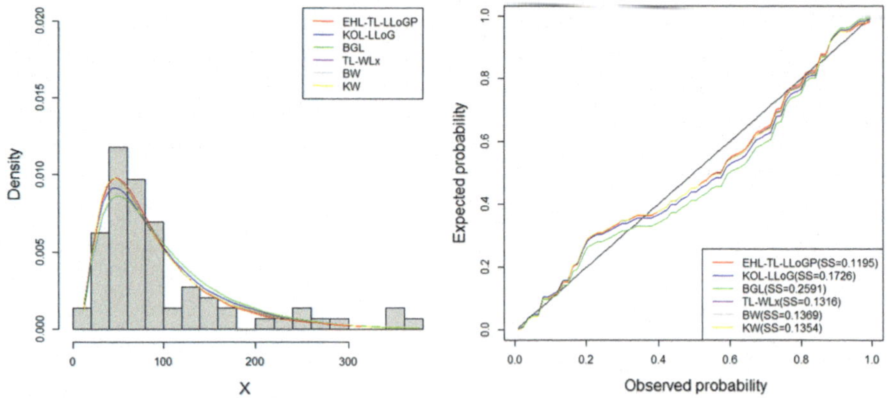

Fig. 5 Fitted pdfs and probability plots for guinea pigs survial times data set

$$\times \left[1 - \left[1 - \frac{1 + \lambda + \lambda x}{1 + \lambda}\right]^{\alpha}\right]^{b-1}, \tag{25}$$

for $\alpha, \lambda, a, b > 0$.

5.1 Guinea Pigs Survival Times Data

The data set was first analyzed by Bjerkedal [9] and represents the survival times of guinea pigs injected with different doses of tubercle bacilli. The observations are: 12, 15, 22, 24, 24, 32, 32, 33, 34, 38, 38, 43, 44, 48, 52, 53, 54, 54, 55, 56, 57, 58, 58, 59, 60, 60, 60, 60, 61, 62, 63, 65, 65, 67, 68, 70, 70, 72, 73, 75, 76, 76, 81, 83, 84, 85, 87, 91, 95, 96, 98, 99, 109, 110, 121, 127, 129, 131, 143, 146, 146, 175, 175, 211, 233, 258, 258, 263, 297, 341, 341, 376.

The estimated variance-covariance matrix is given by

$$\begin{bmatrix} 0.7863 & -3.9651 & 0.0636 & 0.2575 \\ -3.9651 & 32.0654 & -0.3951 & -1.7126 \\ 0.0636 & -0.3951 & 0.0056 & 0.0233 \\ 0.2575 & -1.7126 & 0.0233 & 0.0985 \end{bmatrix}$$

and the 95% confidence intervals for the model parameters are given by $b \in [2.6957 \pm 1.7381]$, $\theta \in [9.9133 \pm 11.0988]$, $\lambda \in [0.3852 \pm 0.1475]$ and $\beta \in [19.9365 \pm 0.6153]$.

Based on the values of the goodness-of-fit statistics A^*, W^*, K-S and the p-value of the K-S statistic as shown in Table 3, we conclude that the EHL-TL-LLoGP model performs better than the non-nested models considered in this paper. Figure 5 shows the fitted densities and probability plots for the EHL-TL-LLoGP model. We observe

Table 3 Parameter estimates and goodness-of-fit statistics for various models fitted for guinea pigs survival times data set

Model	Estimates				Statistics							
	b	θ	λ	β	$-2\log L$	AIC	$AICC$	BIC	W^*	A^*	K-S	p-value
EHL-TL-LLoGP	2.6957	9.9133	0.3352	19.9365	780.5	788.5	789.1	797.6	0.1253	0.6963	0.0954	0.5282
	(0.8868)	(5.6626)	(0.0753)	(0.3139)								
	a	b	λ	c								
KOL-LLoG	4.2771	0.3733	0.1683	0.7733	780.9	788.9	789.5	798.0	0.1762	0.9478	0.0982	0.4906
	(3.1783)	(0.3838)	(0.1683)	(0.2867)								
	a	b	λ	α								
BGL	16.0087	0.2035	0.0787	0.1270	782.7	790.7	791.3	799.8	0.2170	1.1680	0.1251	0.2096
	(13.1666)	(0.0883)	(0.0271)	(0.0998)								
	a	b	α	β								
KW	16.0952	0.7353	0.1244	0.5576	780.2	78802	788.8	797.3	0.1427	0.7779	0.0988	0.4831
	(30.5631)	(1.2473)	(0.2849)	(0.5634)								
BW	12.5596	0.5270	0.1108	0.6387	780.1	788.1	788.7	797.2	0.1436	0.7814	0.0982	0.4907
	(16.2845)	(0.6245)	(0.1288)	(0.3799)								
	a	b	α	θ								
TL-WLx	0.5756	0.0480	0.7799	9.2769	780.2	788.2	788.8	797.3	0.1390	0.7596	0.0980	0.4936
	(1.2223)	(0.0975)	(1.4054)	(23.9435)								

the flexibility enjoyed in data fitting from the EHL-TL-LLoGP model compared to the non-nested models.

5.2 Skin Folds Data

The data set was reported by Cook and Weisberg [17] and consists of 12 variables from 100 female and 102 male Australian athletes. We apply the EHL-TL-LLoGP distribution to the sum of skin folds for the 100 female athletes. The observations are 33.8, 36.8, 38.2, 41.1, 41.6, 42.3, 43.5, 43.5, 46.1, 46.2, 46.3, 47.5, 47.6, 48.4, 49.0, 49.9, 50.0, 52.5, 52.6, 54.6, 54.6, 55.6, 56.8, 57.9, 58.9, 59.4, 61.9, 62.6, 62.9, 65.1, 67.0, 68.3, 68.9, 69.9, 70.0, 71.3, 71.6, 73.9, 74.7, 74.9, 75.1, 75.2, 76.2, 76.8, 77.0, 80.1, 80.3, 80.3, 80.3, 80.6, 83.0, 87.2, 88.2, 89.0, 90.2, 90.4, 91.0, 91.2, 95.4, 96.8, 97.2, 97.9, 98.0, 98.1, 98.3, 98.5, 99.8, 99.9, 101.1, 102.8, 102.8, 103.6, 103.6, 104.6, 106.9, 109.0, 109.1, 109.5, 109.6, 110.2, 110.7, 111.1, 113.5, 114.0, 115.9, 117.8, 122.1, 123.6, 125.9, 126.4, 126.4, 131.9, 136.3, 143.5, 148.9, 156.6, 156.6, 171.1, 181.7, 200.8.

The estimated variance-covariance matrix is given by

$$\begin{bmatrix} 0.2746 & 0.0065 & 0.0129 & 0.0183 \\ 0.0065 & 0.0002 & 0.0003 & 0.0004 \\ 0.0129 & 0.0003 & 0.0006 & 0.0008 \\ 0.0183 & 0.0004 & 0.0008 & 0.0012 \end{bmatrix}$$

and the 95% confidence intervals for the model parameters are given by $b \in [2.6665 \pm 1.0272]$, $\theta \in [43.6656 \pm 0.0247]$, $\lambda \in [0.4212 \pm 0.0488]$ and $\beta \in [43.2601 \pm 0.0686]$.

Based on the values of the goodness-of-fit statistics A^*, W^*, K-S and the p-value of the K-S statistic as shown in Table 4, we conclude that the EHL-TL-LLoGP model performs better than the non-nested models considered in this paper. Figures 6 shows the fitted densities and probability plots for the EHL-TL-LLoGP model. We observe the flexibility enjoyed in data fitting from the EHL-TL-LLoGP model compared to the non-nested models.

6 Concluding Remarks

We develop a new and large class of distributions, namely, the exponentiated half logistic-Topp-Leone-G power series (EHL-TL-GPS) class of distributions. Expansion of the EHL-TL-GPS pdf shows that the distribution can be expressed as an infinite linear combination of the Exp-G densities. This property is handy in the derivation of other statistical properties of the EHL-TL-GPS class of distributions. We present some special classes in the new proposed distribution. From the special

Table 4 Parameter estimates and goodness-of-fit statistics for various models fitted for Skin folds data set

	Estimates				Statistics							
Model	b	θ	λ	β	$-2\log L$	AIC	$AICC$	BIC	W^*	A^*	K-S	p-value
EHL-TL-LLoGP	2.6665	43.6656	0.4212	43.2601	974.2	982.2	982.7	992.7	0.0775	0.4857	0.0757	0.6157
	(0.5241)	(0.0126)	(0.0249)	(0.0350)								
	a	b	λ	c								
KOL-LLoG	33.2280	6.5647	0.7529	0.3714	974.4	982.4	982.8	992.8	0.0804	0.4996	0.0772	0.5899
	(1.0842)	(9.5878)	(0.2216)	(0.1164)								
	a	b	λ	α								
BGL	17.9784	1.4713	0.0338	0.2425	974.3	982.3	982.7	992.7	0.0816	0.5040	0.0786	0.5669
	(0.0014)	(2.2374)	(0.0333)	(0.1090)								
	a	b	α	β								
KW	107.8242	9.1641	1.1518	0.2894	974.2	982.2	982.6	992.6	0.0826	0.5068	0.0777	0.5813
	(0.6178)	(9.2570)	(0.8911)	(0.0698)								
BW	25.1070	21.4350	0.0076	0.5268	974.0	982.0	982.5	992.5	0.0859	0.5186	0.0803	0.5393
	(0.0004)	(3.2078×10^{-5})	(0.0004)	(0.0375)								
	a	b	α	θ								
TL-WLx	0.2909	0.1083	2.9210	3.4402	974.3	982.3	982.7	992.7	0.0839	0.5135	0.0792	0.5567
	(0.2387)	(0.2293)	(2.1752)	(2.8310)								

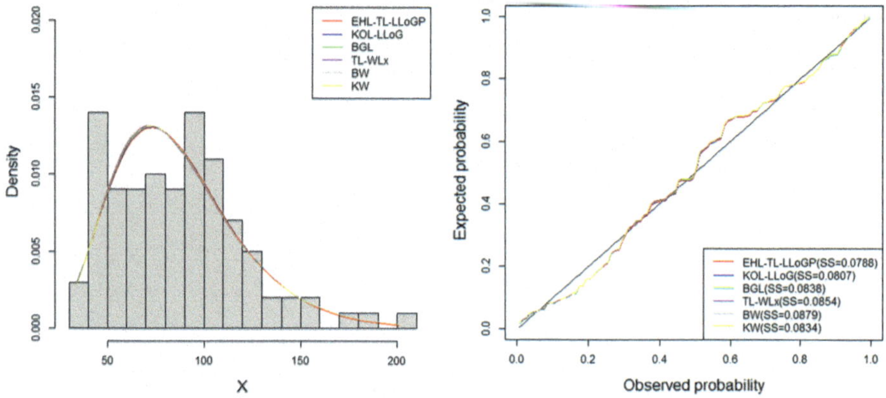

Fig. 6 Fitted pdfs and probability plots for run off data set

cases presented, we observe that the new class of distributions applies to heavy tailed, left or right-skewed data sets and various forms of kurtosis. Structural properties were also derived including moments, distribution of order statistics, Rényi entropy, and maximum likelihood estimates. We conducted a simulation study to evaluate the consistency of the maximum likelihood estimates. Based on the simulation study results, we conclude that the new proposed model produce consistent results. We also applied the EHL-TL-LLoGP distribution to two heavy tailed real data examples to illustrate the usefulness of the new class of distributions. The EHL-TL-LLoGP model performs better than several non-nested models on the selected data sets.

Acknowledgements The authors are grateful to the referees and editor for some very valuable comments which led to this improved version of this manuscript.

A: R-Codes

https://drive.google.com/file/d/1P6pMxPJyKD1kwcxBcs6Xxb1965A8wd7I/view?usp=sharing

References

1. Afify, A.Z., Altun, E., Alizadeh, M., Ozel, G., Hamedani, G.G.: The odd exponentiated Half-Logistic-G family: properties, characterizations and applications. Chilean J. Stat. **8**(2), 65–91 (2017)
2. Ahmad, Z., Elgarhy, M., Hamedani, G.G.: A new Weibull-X family of distributions: properties, characterizations and applications. J. Stat. Distrib. Appl. **5**(1), 5 (2018). https://doi.org/10.1186/s40488-018-0087-6

3. Aldahlan, M.A., Jamal, F., Chesneau, C., Elbatal, I., Elgarhy, M.: Exponentiated power generalized weibull power series family of distributions: properties, estimation and applications. PLoS ONE **15**(3), e0230004 (2019). https://doi.org/10.1371/journal.pone.0230004
4. Alizadeh, M., Tahir, M.H., Cordeiro, G.M., Mansoor, M., Zubair, M., Hamedani, G.G.: The Kumaraswamy Marshall-Olkin family of distributions. J. Egypt Math. Soc. **23**, 546–557 (2015)
5. Anwar, A., Bibi, A.: The Half-Logistic Generalized Weibull Distribution. J. Probab. Stat. **2018**, Article ID 8767826, 12 (2018). https://doi.org/10.1155/2018/8767826
6. Anwar, M., Zahoor, J.: The half-logistic lomax distribution for lifetime modeling. J. Probab. Stat **2018**, Article ID 3152807, 12 (2018). https://doi.org/10.1155/2018/3152807
7. Al-Shomrani, A., Arif, O., Shawky, A., Hanif, S., Shahbaz, M.Q.: Topp-Leone family of distributions: some properties and application. Pakistan J. Stat. Oper. Res. **12**(3), 443–451 (2016)
8. Bantan, R.A.R., Jamal, F., Chesneau, C., Elgarhy, M.: Type II power Topp-Leone generated family of distributions with statistical inference and applications. Symmetry **12**(1), 75 (2020). https://doi.org/10.3390/sym12010075
9. Bjerkedal, T.: Acquisition of resistance in guinea pigs infected with different doses of virulent tubercle bacilli. Am. J. Hygiene **72**, 130–148 (1960)
10. Bourguignon, M., Silva, R.B., Cordeiro, G.M.: The Weibull-G family of probability distributions. J. Data Sci. **12**, 53–68 (2014)
11. Chambers, J., Cleveland, W., Kleiner, B., Tukey, P.: Graphical Methods of Data Analysis. Chapman and Hall (1983)
12. Chen, G., Balakrishnan, N.: A general purpose approximate goodness-of-fit test. J. Q. Technol. **27**(2), 154–161 (1995)
13. Chesneau, C., El Achi, T.: Modified Odd Weibull Family of Distributions: Properties and Applications (2019). ArXivhal-02317235
14. Chipepa, F., Oluyede, B., Makubate, B.: The odd generalized half-logistic Weibull-g family of distributions: properties and applications. J. Stat. Model. Theory Appl. (2020) (in Press)
15. Chipepa, F., Oluyede, B., Makubate, B.: The Topp-Leone-Marshall-Olkin-G family of distributions with applications. Int. J. Stat. Probab. **9**(4) (2020). https://doi.org/10.5539/ijsp.v9n4p15
16. Chipepa, F., Oluyede, B., Makubate, B.: A new generalized family of odd Lindley-G distributions with application. Int. J. Stat. Probab. **8**(6) (2019). https://doi.org/10.5539/ijsp.v8n6p1
17. Cook, R.D., Weisberg, S.: An Introduction to Regression Graphics. Wiley, New York (1994)
18. Cordeiro, G.M., Alizadeh, M., Marinho, P.R.D.: The type I half-logistic family of distributions. J. Stat. Comput. Simul. **86**(4), 707–728 (2016)
19. Cordeiro, G.M., Silva, R.B.: The complementary extended weibull power series class of distributions. Ciéncia e Nat. **36**(3) (2014)
20. Cordeiro, G.M., Ortega, E.M.M., da Cunha, D.C.C.: The exponentiated generalized class of distributions. J. Data Sci. **11**, 1–27 (2013)
21. Cordeiro, G.M., Gomes, A., da Silva, C., Ortega, E.M.M.: The beta exponentiated weibull distribution. J. Stat. Comput. Simul. **38**(1), 114–138 (2013)
22. Cordeiro, G.M., Ortega, E.M.M., Nadarajaah, S.: The Kumaraswamy Weibull distribution with application to failure data. J. Franklin Inst. **347**, 1399–1429 (2010)
23. Flores, J., Borges, P., Cancho, V.G., Louzada, F.: The complementary exponential power series distribution. Brazilian J. Probab. Stat. **27**(4), 565–584 (2013)
24. Gradshetyn, I.S., Ryzhik, I.M.: Tables of Integrals, Series and Products, 6th edn. Academic, San Diego (2000)
25. Jamal, F., Reyad, H.M., Nasir, M.A., Chesneau, C., Shah, M.A.A., Ahmed, S.O.: Topp-Leone Weibull-Lomax distribution: properties, regression model and applications (2019). ArXiv:hal-02270561
26. Johnson, N.L., Kotz, S., Balakrishnan, N.: Continuous Distributions, vol. 1. Wiley, New York, NY (1994)
27. Makubate, B., Moakofi, T., Oluyede, B.: A new generalized Lindley-Weibull class of distributions: theory. properties and applications. Mathematica Slovaka (2020). (In Press)
28. Morais, A.L., Barreto-Souza, W.: A compound class of Weibull and power series distributions. Comput. Stat. Data Anal. **55**(3), 1410–1425 (2011)

29. Oluyede, B.O., Mdlongwa, P., Makubate, B., Huang, S.: The Burr-Weibull power series class of distributions. Austrian J. Stat. **48**, 1–13 (2019)
30. Muhammad, M., Yahaya, M.A.: The half logistic-poisson distribution. Asian J. Math. Appl. **1–15** (2017)
31. Oluyede, B., Chipepa, F., Wanduku, D., Peter, O.P., Makubate, B.: Exponentiated Half Logistic-Topp-Leone-G Family of Distributions (2020). Submitted
32. Oluyede, B.O., Yang, T.: A new class of generalized lindley distributions with applications. J. Stat. Comput. Simul. **10**, 2072–2100 (2015). https://doi.org/10.1080/00949655.2014.917308
33. Rezaei, S., Sadr, B.B., Alizadeh, M., Nadarajah, S.: Topp-Leone generated family of distributions: properties and applications. Commun. Stat. Theory Methods **46**(6) (2016)
34. Reyad, H., Selim, M.A., Othman, S.: The Nadarajah-Haghighi Topp-Leone-G family of distributions with mathematical properties and application. Pakistan J. Stat. Oper. Res. XV, IV, 849–866 (2019)
35. Rényi, A.: On measures of entropy and information. In: Proceedings of the Fourth Berkeley Symposium on Mathematical Statistics and Probability, Vol. 1. The Regents of the University of California (1961)
36. Sangsanit, Y., Bodhisuwan, W.: The Topp-Leone generator of distributions: properties and inferences. Songklanakarin J. Sci. Technol. **38**(5), 537–548 (2016)
37. Shannon, C.E.: Prediction and entropy of printed english. Bell Syst. Tech. J. **30**(1), 50–64 (1951)
38. Silva, R.B., Bourguignon, M., Dias, C.R.B., Cordeiro, G.M.: The compound class of extended weibull power series distributions. Comput. Stat. Data Anal. **58**, 352–367 (2013)
39. Silva, R.B., Cordeiro, G.M.: The Burr XII power series distributions: a new compounding family. Brazilian J. Probab. Stat. **29**(3), 565–589 (2015)
40. Silva, R., Frank Gomes-Silva, F.G., Ramos, M., Cordeiro, G., Marinho, P., De. Andrade, T.A.N.: The exponentiated Kumaraswamy-G class: general properties and application. Revista Colombiana de Estadística **42**(1), 1–33 (2019)

Fixed Points of Multivalued $(\alpha_* - \phi)$-Contractions and Metric Transforms

Basit Ali, Talat Nazir, and Nozara Sundus

Abstract The functions when composed with metrics yield another metric, are known as metric preserving functions in the literature. A particular case of metric preserving functions are metric transforms. Metric transform is a function $I : [0, \infty) \to \mathbb{R}$ which is strictly increasing and concave with $I(0) = 0$. There are metric preserving functions which are not metric transforms. In this chapter, we consider the concept of existence of fixed point sets of multivalued mappings of metric spaces in connection with metric transforms. In this context, we consider $(\alpha - \phi)$ contractions, multivalued $(\alpha_* - \phi)$ contractions, $(\varepsilon - \phi)$ uniform local multivalued contraction and generalized multivalued $(\alpha_* - \phi)$ contractions. Our purpose is to extend some fixed point results for multivalued contractions to the case multivalued $(\alpha_* - \phi)$ contractions. Further, we used the metrics which are sequentially, strong semi sequentially and semi sequentially equivalent to the Hausdorff metric on collection of non-empty, closed and bounded subsets to obtain more general fixed point results. We obtain important fixed point results in the literature as the corollaries of the main theorem in this paper. We present some examples to manifest the applicability, usefulness and generality of our conclusions.

Keywords Metric space · Metric transforms · Sequentially equivalent distances · Strong semi sequentially equivalent distances · Semi sequentially equivalent distances · Multivalued $(\alpha_* - \phi)$ contractions · $(\varepsilon - \phi)$ uniform local multivalued contraction · Multivalued mapping · Fixed point

---

We dedicate this chapter to Ljubomir B. Ćirić (late) for his contributions to the metric fixed point theory.

B. Ali (✉) · N. Sundus
Department of Mathematics, School of Science, University of Management and Technology, C-II Johar Town, Lahore, Pakistan

T. Nazir
Department of Mathematical Science, University of South Africa, Johannesburg, South Africa

J. Singh et al. (eds.), *Methods of Mathematical Modelling and Computation for Complex Systems*, Studies in Systems, Decision and Control 373,
https://doi.org/10.1007/978-3-030-77169-0_15

1 Introduction and Preliminaries

Metric transforms and metric preserving functions are the generalized functions in metric spaces. Blumenthal [5] introduced the concept of metric transform, the notion of metric preserving functions was addressed by Wilson [34] in 1935. These concepts were investigated in detail by many authors (compare: [9–16, 25–27, 32, 33]).

Fixed point theory has always been a very important area of exploration because of its wide span of applications in mathematics and also in different branches of science. The first astonishing theorem in fixed point theory is Banach contraction principle by Banach [6], that relay on iterated function sequences that converges to unique fixed point. In 1969, Banach contraction principle was extended for multivalued mappings in metric spaces by Nadler [31]. He proved the Banach's [6] and Edelstein's [17] results for set-valued mappings. Many authors [1–4, 7, 8, 17–24, 29] generalized the Banach contraction principle in one to many directions.

Kirk and Shahzad [21] gave the proofs of fixed point theorems for metric transform by assuming local radial contractions. They proved some fixed point results for set valued mappings too. They defined a new extended metric than Hausdorff metric H that is sequentially equivalent metric D_1 which was sequentially equivalent to Hausdorff metric, and presented some fixed point results for sequentially equivalent metric.

Many authors extended the Banach's contraction condition for single valued as well as for multivalued mappings. As Samet et al. [30] studied $(\alpha - \psi)$ contraction for single valued mappings and Asl et al. [3] gave more generalized contraction which was $(\alpha_* - \psi)$ contraction for multivauled mappings.

For a metric space (Y, d), we have following notions:

- $CB(Y) = \{C : C \neq \emptyset$ closed and bounded subset of $Y\}$.
- $2^Y = \{C : C \neq \emptyset$ compact subset of $Y\}$.
- $Cl(Y) = \{C : C \neq \emptyset$ is the closed subset of $Y\}$.
- The functions Ψ represent the class of metric transforms.

Let for metric space (Y, d), for $U, V \in CB(Y)$, the Hausdorff metric between U and V is defined as

$$H(U, V) = \max\{\rho(U, V), \rho(V, U)\},$$

where

$$\rho(U, V) = \sup_{y \in U} d(y, V), \quad \rho(V, U) = \sup_{y \in V} d(y, U).$$

Definition 1 ([31]) A mapping $I : Y \to CB(Y)$ is a multivauled contraction mapping if there is a constant $0 < r < 1$ such that

$$H(Iy, Iz) \leq rd(y, z)$$

where $y, z \in Y$.

Nadler [31] proved the following fixed point result for multivauled mappings in complete metric space.

Theorem 2 ([31] *Let for a complete metric space* (Y, d)*, if the mapping* $I : Y \to CB(Y)$ *satisfies for some* $r \in (0, 1)$ *such as*

$$H(Iy, Iz) \leq rd(y, z)$$

where $y, z \in Y$*, then there is a point* $y \in Y$ *such that* $y \in Iy$*.*

Kirk and Shahzad replaced Hausdroff metric H with sequentially equivalent metric D_1 on $CB(Y)$ in [21] and generalized Nadler's Theorem 2.

Definition 3 ([21]) The metric D_1 is said to be sequentially equivalent to Hausdorff metric H if for $U \in CB(Y)$ and $\{U_p\} \subset CB(Y)$, we have

$$\lim_{p\to\infty} D_1(U_p, U) = 0 \text{ if and only if } \lim_{p\to\infty} H(U_p, U) = 0.$$

Theorem 4 ([21]) *Let for a complete metric space* (Y, d)*. Suppose* $I : Y \to CB(Y)$ *satisfies,*

(1) there is a constant $r \in (0, 1)$ such as

$$D_1(Iy, Iz) \leq rd(y, z)$$

for every $y, z \in Y$,

(2) if $y \in Y$ and $z \in Iy$

$$d(z, Iz) \leq D_1(Iz, Iy).$$

Then I has a fixed point.

We extend sequentially equivalent metric to semi sequentially equivalent metric, it is defined as below.

Definition 5 The metric D is semi sequentially equivalent to Hausdorff metric H if for $U \in CB(Y)$ and $\{U_p\} \subset CB(Y)$, we have

$$\lim_{p\to\infty} D(U_p, U) = 0 \text{ implies } \lim_{p\to\infty} H(U_p, U) = 0.$$

Following is the definition of strong semi sequentially equivalent metric, which is stronger than semi sequentially equivalent metric.

Definition 6 The metric D_2 is strong semi sequentially equivalent metric to Hausdorff metric H if for $U \in CB(Y)$ and $\{U_p\} \subset CB(Y)$

$$H(U_p, U) \leq D_2(U_p, U).$$

Samet et al. [30] studied new class of functions Φ, these are the class of nondecreasing functions which are defined as following,

$$\Phi = \left\{ \phi : [0, \infty) \to [0, \infty) : \sum_{p=1}^{\infty} \phi^p (I) < \infty, \text{ for each } I > 0 \right\}$$

where ϕ^p is p-th iterate of ϕ.

Assume for a metric space (Y, d), for mappings $I : Y \to Y$ and $\alpha : Y \times Y \to [0, +\infty)$, the mapping I is said to be α-admissible if for all $y, z \in Y$,

$$\alpha (y, z) \geq 1 \text{ implies } \alpha (Iy, Iz) \geq 1.$$

Samet et al. [30] define $(\alpha - \phi)$ contraction condition for single valued mappings.

Definition 7 ([30]) Let for a metric space (Y, d), the mapping $I : Y \to Y$ be an $(\alpha - \phi)$ contractive mapping if there are two functions $\alpha : Y \times Y \to [0, \infty)$ and $\phi \in \Phi$ is satisfying

$$\alpha (y, z) d (Iy, Iz) \leq \phi (d (y, z))$$

for all $y, z \in Y$.

Samet et al. [30] proved following result by using $(\alpha - \phi)$ contraction for point valued mappings for getting fixed point.

Theorem 8 ([30]) *Let for a complete metric space (Y, d), if the mapping $I : Y \to Y$ satisfies,*

(i) $\alpha (y, z) d (Iy, Iz) \leq \phi (d (y, z))$, *for every* $y, z \in Y$,
(ii) I *is* α*-admissible,*
(iii) there is $y_0 \in Y$ *such as* $\alpha (y_0, Iy_0) \geq 1$,
(iv) either (a) I is continuous or (b) if $\{y_p\}$ *is a sequence in Y such as* $\alpha (y_p, y_{p+1}) \geq 1$ *for all p further* $y_p \to y \in Y$ *as* $p \to +\infty$, *then* $\alpha (y_p, y) \geq 1$ *for all p.*

Then I has a fixed point.

Following notions were given by Hasanzade et al. [3].

- Assume for a metric space (Y, d), the mapping $I : Y \to 2^Y$. Define

$$\alpha_* (Iy, Iz) = \inf \{\alpha (m, n) : m \in Iy, \ n \in Iz\}.$$

- For functions $I : Y \to CB(Y)$ and $\alpha : Y \times Y \to [0, \infty)$ we say that I is α_*-admissible, if $\alpha (y, z) \geq 1$ implies $\alpha_* (Iy, Iz) \geq 1$.

Hasanzade et al. [3] generalized $(\alpha - \phi)$ contraction condition for multivauled mappings and define $(\alpha_* - \phi)$ contraction condition.

Definition 9 ([3]) Let for a metric space (Y, d), a mapping $I : Y \to CB(Y)$ is called an $(\alpha_* - \phi)$ contractive mapping if there are two functions $\alpha : Y \times Y \to [0, \infty)$ and $\phi \in \Phi$ such that

$$\alpha_* (Iy, Iz) H (Iy, Iz) \leq \phi (d (y, z))$$

for all $y, z \in Y$.

Hasanzade et al. [3] used $(\alpha_* - \phi)$ contraction for multivauled mappings for getting fixed point as given below.

Theorem 10 ([3]) *Let for a complete metric space* (Y, d), $\alpha : Y \times Y \to [0, \infty)$ *be a function,* $\phi \in \Phi$, *if mapping* $I : Y \to Cl(Y)$ *satisfying,*

(i) $\alpha_* (Iy, Iz) H (Iy, Iz) \leq \phi (d (y, z))$, *for every* $y, z \in Y$,
(ii) I *is* α_**-admissible,*
(iii) *there is* $y_0 \in Y$ *and* $y_1 \in Iy_0$ *such as* $\alpha (y_0, y_1) \geq 1$,
(iv) *if* $\{y_p\}$ *is a sequence in* Y *such as* $\alpha (y_p, y_{p+1}) \geq 1$ *for all* p *and* $y_p \to y$, *then* $\alpha (y_p, y) \geq 1$ *for all* p,

Then I *has a fixed point.*

Definition 11 ([31]) A metric space (Y, d) is said to be ε-chainable (where $\varepsilon > 0$ is fixed) if and only if $l, m \in Y$ there is an ε-chain from l to m, that is, a finite set of points $y_0, y_1, \ldots y_p \in Y$ such that $y_0 = l$, $y_p = m$, and

$$d (y_{i-1}, y_i) < \varepsilon$$

for all $i = 1, 2, \ldots n$.

Definition 12 ([31]) A mapping $I : Y \to CB(Y)$ is said to be an (ε, r)-uniform local multivauled contraction (where $\varepsilon > 0$ and $r \in (0, 1)$ if

$$d (y, z) < \varepsilon \Rightarrow H (Iy, Iz) \leq rd (y, z).$$

The concept of (ε, c) uniform local multivauled contraction was given by Nadler [31]. Kirk and Shahzad generalized this concept in terms of metric transforms as follows.

Theorem 13 ([21]) *Let for a metric space* (Y, d), *there is metric transform* $\psi \in \Psi$ *and constant* $r \in (0, 1)$, *if the mapping* $I : Y \to CB(Y)$ *satisfies,*

(a) *for every* $y, z \in Y$

$$\psi (H (Iy, Iz)) \leq rd (y, z),$$

(b) *for* $t > 0$ *sufficiently small,there exists* $c \in (0, 1)$ *such that*

$$rt \leq \psi (ct).$$

Then for sufficiently small $\varepsilon > 0$, I *is an* (ε, c) *uniform local multivauled contraction on* (Y, d).

Following is Nadler's result [31] for (ε, r) uniform local contractive multivauled mapping for finding the fixed point.

Theorem 14 ([31]) *Let for a* ε*-chainable complete metric space* (Y, d). *If the mapping* $I : Y \to 2^Y$ *is an* (ε, r) *uniform local contractive multivauled, then* I *has a fixed point.*

In [21] Theorems 13 and 14 are gathered to get the following result.

Theorem 15 ([21]) *Assume for* ε*-chainable, connected and complete metric space* (Y, d). *If mapping* $I : Y \to 2^Y$ *is an* (ε, r) *uniform local contractive multivauled, then* I *has a fixed point.*

Proof For any $\varepsilon > 0$ a connected complete metric space is ε-chainable. ■

Pathak and Shahzad [28] defined a new metric H^+ which is an example of a metric on $CB(Y)$, moreover this metric is metrically equivalent to Hausdorff metric.

Metric H^+ is defined as, for $U, V \in CB(Y)$, we have

$$H^+(U, V) = \frac{1}{2}(\rho(U, V) + \rho(V, U)).$$

It can be easily seen that H^+ is metrically equivalent to the Hausdorff metric:

$$\frac{1}{2}H(U, V) \le H^+(U, V) \le H(U, V).$$

Following is the definition of H^+ contraction given in [28].

Definition 16 ([28]) A multivauled mapping $I : Y \to CB(Y)$ is an H^+ contraction if

(1) there is $r \in (0, 1)$ in such a way

$$H^+(Iy, Iz) \le rd(y, z)$$

for every $y, z \in Y$

(2) for every $y \in Y$ and $z \in Iy$,

$$d(z, Iz) \le H^+(Iy, Iz).$$

2 Extensions by Using Multivalued ($\alpha_* - \phi$) Contractions

In this section, we are going to extend the results related to multivauled mappings, which are discussed in [21], by using ($\alpha_* - \phi$) contractions. Examples are used to verify our extension. Some of the result related corollaries are also discussed here. Moreover, we consider sequentially equivalent metric, strong semi sequentially equivalent, and semi sequentially equivalent metrics, which are more general metrics than Hausdorff metric. Further, we will define ($\varepsilon - \phi$) uniform local multivauled contraction on metric transform, by using this we get some fixed point results.

Let we move on our main result:

Kirk and Shahzad used sequentially equivalent metric in [21] for finding fixed point, moreover they used simple contraction condition to verify their result. We generalize this result (Theorem 4) by using ($\alpha_* - \phi$) contraction, for this purpose we consider metrics D (semi sequentially equivalent), D_2 (strong semi sequentially equivalent) and D_1 (sequentially equivalent).

Theorem 17 *Let (Y, d) be a complete metric space and D (semi sequentially equivalent to Hausdorff metric) be any metric on $CB(Y)$. Suppose the mapping $I : Y \to CB(Y)$ satisfies,*

(i) $\alpha_* (Iy, Iz) D (Iy, Iz) \leq \phi (d (y, z))$, for every $y, z \in Y$,
(ii) If $y \in Y$ and $z \in Iy$
$$d (z, Iz) \leq D (Iz, Iy),$$
(iii) I is α_*-admissible,
(iv) there is $y_0 \in Y$ and $y_1 \in Iy_0$ such as $\alpha (y_0, y_1) \geq 1$.
(v) Either (a) $\{y_p\}$ is a sequence in Y such that $\alpha (y_p, y_{p+1}) \geq 1$ for all p, further $y_p \to y \in Y$ as $p \to \infty$, then $\alpha (y_p, y) \geq 1$ for all $p \in \mathbb{N}$ or (b) I is upper hemi continuous.

Then I has a fixed point y.

Proof By condition (iii) and (iv) there is $y_0 \in Y$ and $y_1 \in Iy_0$ such as

$$\alpha (y_0, y_1) \geq 1 \Rightarrow \alpha_* (Iy_0, Iy_1) \geq 1.$$

If $y_1 = y_0$ then $y_0 = y_1 \in Iy_0$ that is $y_0 \in Iy_0$, which gives y_0 is fixed point of I. Proof is completed. So assume $y_0 \neq y_1$, By condition (i) and (ii)

$$d (y_1, Iy_1) \leq D (Iy_1, Iy_0) = D (Iy_0, Iy_1).$$

If $d (y_1, Iy_1) = 0$ then $y_1 \in Iy_1$ implies that y_1 is a fixed point of I so the proof is completed. Now consider that $d (y_1, Iy_1) > 0$. Given $q_1 > 1$, so there is $y_2 \in Iy_1$ such as

$$0 < d (y_1, Iy_1) \leq d (y_1, y_2) < q_1 d (y_1, Iy_1).$$

That is

$$\begin{aligned}0 &< d(y_1, y_2) < q_1 d(y_1, Iy_1) \leq q_1 D(Iy_0, Iy_1)\\&\leq q_1 \alpha_* (Iy_0, Iy_1) D(Iy_0, Iy_1)\\&\leq q_1 \phi(d(y_0, y_1)).\end{aligned}$$

Hence

$$d(y_1, y_2) < q_1 \phi(d(y_0, y_1)). \tag{1}$$

If $y_2 = y_1$ then we have done, so consider that $y_2 \neq y_1$. As I is α_*-admissible so

$$\alpha(y_1, y_2) \geq \alpha_* (Iy_0, Iy_1) \geq 1.$$

Now put $b_\circ = d(y_0, y_1)$, then from (1)

$$d(y_1, y_2) < q_1 \phi(b_\circ)$$

since ϕ is strictly increasing,

$$\phi(d(y_1, y_2)) < \phi(q_1 \phi(b_\circ)).$$

Put

$$q_2 = \frac{\phi(q_1 \phi(b_\circ))}{\phi(d(y_1, y_2))} > 1.$$

Given $q_2 > 1$ and

$$0 < d(y_2, Iy_2) \leq D(Iy_1, Iy_2).$$

If $y_2 = y_3$ then nothing to prove. Let $y_2 \neq y_3$. Then for $q_2 > 1$ there is $y_3 \in Iy_2$ such as $0 < d(y_2, y_3)$ and

$$\begin{aligned}d(y_2, y_3) &< q_2 d(y_2, Iy_2) \leq q_2 D(Iy_1, Iy_2)\\&\leq q_2 \alpha_* (Iy_1, Iy_2) D(Iy_1, Iy_2)\\&\leq q_2 \phi(d(y_1, y_2)) \leq \frac{\phi(q_1 \phi(b_\circ))}{\phi(d(y_1, y_2))} . \phi(d(y_1, y_2))\\&\leq \phi(q_1 \phi(b_\circ)).\end{aligned}$$

It is clear that $y_3 \neq y_2$,

$$\alpha(y_2, y_3) \geq \alpha_* (Iy_2, Iy_3) \geq 1,$$

and since ϕ is strictly increasing so

$$\phi d(y_2, y_3) < \phi^2 (q \phi(b_\circ)).$$

Now put

$$q_3 = \frac{\phi^2 (q\phi (b_\circ))}{\phi (d (y_2, y_3))}.$$

For given $q_3 > 1$, we obtain

$$0 < d (y_3, Iy_3) \leq D (Iy_3, Iy_4)$$

for $q_3 > 1$ there exists $y_4 \in Iy_3$ such that

$$\begin{aligned} 0 < d (y_3, y_4) &< q_3 d (y_3, Iy_3) \\ &\leq q_3 \alpha (Iy_2, Iy_3) D (Iy_2, Iy_3) \\ &\leq q_3 \phi (d (y_2, y_3)) \leq \frac{\phi^2 (q_1 \phi (b_\circ))}{\phi (d (y_2, y_3))} \phi (d (y_2, y_3)) \\ &\leq \phi^2 (q_1 \phi (b_\circ)). \end{aligned}$$

Resuming in this way, we obtain a sequence $\{y_p\}$ in Y such that $y_p \in Iy_{p-1}$, and $y_p \neq y_{p-1}$, $\alpha (y_p, y_{p+1}) \geq 1$ and

$$d (y_p, y_{p+1}) \leq \phi^{p-1} (q_1 \phi (b_\circ))$$

for all p. As $\phi \in \Phi$, so $\sum_{i=1}^{\infty} \phi^i (l) < \infty$ for all $l > 0$, hence for any $\varepsilon > 0$ there exists $n_1 \in \mathbb{N}$ such as

$$\sum_{i=p+1}^{\infty} \phi^{i-1} (q_1 \phi (b_\circ)) < \varepsilon$$

for all $p \geq n_1$. Now for each $q > p$ using triangular inequality

$$\begin{aligned} d (y_p, y_q) &\leq d (y_p, y_{p+1}) + d (y_{p+1}, y_{p+2}) + \ldots + d (y_{q-1}, y_q) \\ &\leq \phi^{p-1} (q_1 \phi (b_\circ)) + \phi^p (q_1 \phi (b_\circ)) + \cdots + \phi^{q-2} (q_1 \phi (b_\circ)) \\ &= \sum_{i=p}^{q-1} \phi^{i-1} (q_1 \phi (b_\circ)) \leq \sum_{i=p}^{\infty} \phi^{i-1} (q_1 \phi (b_\circ)) < \varepsilon \end{aligned}$$

for all $p \geq n_1$. Hence for any given $\varepsilon > 0$ there is $n_1 \in \mathbb{N}$ such as

$$d (y_p, y_q) = \sum_{i=p+1}^{\infty} \phi^{i-1} (q \phi (b_\circ)) < \varepsilon$$

for all $p, q \geq n_1$. This implies $\{y_p\}$ is a Cauchy sequence., consequently y_p converges to some y in Y as Y is complete. Now assume (a) holds in (V). As $\alpha (y_p, y_{p+1}) \geq 1$ and $y_p \to y$ so $p \to \infty$, so $\alpha (y_p, y) \geq 1$ for all $p \in \mathbb{N} \cup \{0\}$.

As I is α_*-admissible so $\alpha_*\left(Iy_p, Iy\right) \geq 1$ so by (i) we have

$$\begin{aligned} D\left(Iy_p, Iy\right) &\leq \alpha_*\left(Iy_p, Iy\right) D\left(Iy_p, Iy\right) \\ &\leq \phi\left(d\left(y_p, y\right)\right) \end{aligned}$$

this implies

$$\lim_{p\to\infty} D\left(Iy_p, Iy\right) \leq \lim_{p\to\infty} \phi\left(d\left(y_p, y\right)\right).$$

That is $\lim_{p\to\infty} D\left(Iy_p, Iy\right) = 0$ because ϕ is continuous at 0. As D is semi sequentially equivalent to H so

$$\lim_{p\to\infty} H\left(Iy_p, Iy\right) = 0.$$

Consequently, we obtain

$$\lim_{p\to\infty} d\left(y_p, Iy\right) \leq \lim_{p\to\infty} H\left(Iy_p, Iy\right) = 0$$

implies

$$d\left(y, Iy\right) = \lim_{p\to\infty} d\left(y_p, Iy\right) = 0.$$

Hence $y \in Iy$. As $\lim_{p\to\infty} y_p = y$, and $y_{p+1} \in I\left(y_p\right)$, so by upper hemi continuity, $y \in Iy$. ■

It is observed that semi sequentially equivalent metric D is weaker condition than D_1 sequentially equivalent metric and strong sequentially equivalent metric D_2.

Following are the corollaries which are obtained from Theorem 17.

Corollary 18 *Let for a complete metric space (Y, d), sequentially equivalent metric D_1 on $CB(Y)$. Suppose the mapping $I : Y \to CB(Y)$ satisfies,*

(i) $\alpha_*(Iy, Iz) D_1(Iy, Iz) \leq \phi(d(y, z))$, for every $y, z \in Y$,
(ii) If $y \in Y$ and $z \in Iy$

$$d(z, Iz) \leq D_1(Iz, Iy),$$

(iii) I is α_*-admissible,
(iv) there is $y_0 \in Y$ and $y_1 \in Iy_0$ such as $\alpha(y_0, y_1) \geq 1$,
(v) either (a) $\{y_p\}$ is a sequence in Y such that $\alpha\left(y_p, y_{p+1}\right) \geq 1$ for all p, further $y_p \to y \in Y$ as $p \to \infty$, then $\alpha\left(y_p, y\right) \geq 1$ for all $p \in \mathbb{N}$ or (b) I is upper hemi continuous.
Then I has a fixed point.

Corollary 19 *Let for a complete metric space (Y, d), and D_2 (strong semi sequentially equivalent) be any metric on $CB(Y)$. Suppose the mapping $I : Y \to CB(Y)$ satisfies,*

(i) $\alpha_*(Iy, Iz)\, D_2(Iy, Iz) \le \phi(d(y, z))$, for every $y, z \in Y$,
(ii) If $y \in Y$ and $z \in Iy$

$$d(z, Iz) \le D_2(Iz, Iy),$$

(iii) I is α_*-admissible,
(iv) there is $y_0 \in Y$ and $y_1 \in Iy_0$ such as $\alpha(y_0, y_1) \ge 1$,
(v) either (a) $\{y_p\}$ is a sequence in Y such that $\alpha(y_p, y_{p+1}) \ge 1$ for all p, further $y_p \to y \in Y$ as $p \to \infty$, then $\alpha(y_p, y) \ge 1$ for all $p \in \mathbb{N}$ or (b) I is upper hemi continuous.
Then I has a fixed point.

Corollary 20 *Let for a complete metric space (Y, d), if mapping $I : Y \to Y$ satisfies,*

(i) $\alpha(y, z)\, d(Iy, Iz) \le \phi(d(y, z))$, *for every* $y, z \in Y$,
(ii) I is α-admissible,
(iii) there is $y_0 \in Y$ such as $\alpha(y_0, Iy_0) \ge 1$,
(iv) either (a) $\{y_p\}$ is a sequence in Y such that $\alpha(y_p, y_{p+1}) \ge 1$ for all p, further $y_p \to y \in Y$ as $p \to \infty$, then $\alpha(y_p, y) \ge 1$ for all $p \in \mathbb{N}$ or (b) I is continuous.

Then I has a fixed point.

Proof If $I : Y \to CB(Y)$ in Theorem 17 is replaced by $I : Y \to Y$, we get the desired result. ■

Corollary 21 *Let for a complete metric space (Y, d), $\alpha : Y \times Y \to [0, \infty)$ be a function, $\phi \in \Phi$, if the mapping $I : Y \to Cl(Y)$ satisfies,*

(i) $\alpha_*(Iy, Iz)\, H(Iy, Iz) \le \phi(d(y, z))$, *for every* $y, z \in Y$,
(ii) I is α_*-admissible,
(iii) there is $y_0 \in Y$ and $y_1 \in Iy_0$ such as $\alpha(y_0, y_1) \ge 1$,
(iv) assume that if $\{y_p\}$ is a sequence in Y such that $\alpha(y_p, y_{p+1}) \ge 1$ for all p and $y_p \to y$, then $\alpha(y_p, y) \ge 1$ for all p.
Then I has a fixed point.

Proof If $D = H$ in Theorem 17, then result follows. ■

Remark 22 For Corollary 21 we can consider $Cl(Y)$ instead of $CB(Y)$.

Corollary 23 *Let for a complete metric space (Y, d), and let D_1(sequentially equivalent metric) be any metric on $CB(Y)$. If mapping $I : Y \to CB(Y)$ satisfies,*

1. *there is constant $r \in (0, 1)$ such that*

$$D_1(Iy, Iz) \le r d(y, z)$$

for every $y, z \in Y$,

2. *for* $y \in Y$ *and* $z \in Iy$

$$d(z, Iz) \leq D_1(Iz, Iy).$$

Then I has a fixed point.

Proof For $\alpha_*(Iy, Iz) = 1$ and $\phi(b_\circ) = rb_\circ$ in Theorem 17, we obtain the required result. ■

Following example is used to verify Theorem 17.

Example 24 Let $Y = \{y_1, y_2, y_3, y_4\}$. Define the mapping $d : Y \times Y \to \mathbb{R}^+$ as

$$\begin{aligned} d(y_3, y_4) &= d(y_2, y_4) = 6, \\ d(y_2, y_3) &= d(y_1, y_4) = 9, \\ d(y_1, y_2) &= d(y_1, y_3) = 5, \\ d(y, y) &= 0 \text{ and } d(y, y_\circ) = d(y_\circ, y) \end{aligned}$$

for all $y, y_\circ \in Y$. Clearly (Y, d) is a complete metric space Define a mapping $I : Y \to CB(Y)$ by

$$Iy = \begin{cases} \{y_1\} \text{ if } y = y_1, y_2, y_3, \\ \{y_2\} \text{ if } y = y_4, \end{cases}$$

Define $\alpha : Y \times Y \to \mathbb{R}^+$ by $\alpha(y_i, y_j) = 1$ for all $i, j \in \{1, 2, 3, 4\}$. If we set

$$\phi(b_\circ) = \frac{8}{9} b_\circ$$

for $I \in \mathbb{R}^+$, then $\phi \in \Phi$. Now we check the contraction condition of Theorem 17: This condition holds for any $y, z \in \{y_1, y_2, y_3\}$ as

$$H(Iy, Iz) = 0 \leq \phi(d(y, z)) = \frac{8}{9} d(y, z).$$

If $y = y_1$ and $z = y_4$ then

$$H(Iy_1, Iy_4) = 5 \leq \phi(d(y_1, y_4)) < \phi(9) = 8.$$

When $y \in \{y_2, y_3\}$, and $z = y_4$.

$$H(Iy, Iy_4) = d(y_1, y_2) = 5 \leq \phi(d(y, z)) \leq \phi(6) = \frac{48}{9}.$$

As $\alpha(y, z) \geq 1$ implies $\alpha_*(Iy, Iz) \geq 1$ so I is α_*-admissible. Note that eventually constant sequences are the only convergent sequences in (Y, d), hence the condition in Theorems 17 is satisfied, so I has a fixed point.

Kirk and Shahzad [21] provided the single valued version of Theorem 4, after generalizing it we get the following result.

Theorem 25 *Let for a complete metric space (Y, d), ρ (sequentially equivalent to the d) be any metric on Y. Suppose $I : Y \to Y$ satisfies,*

(i) $\alpha(y, z)\,\rho(Iy, Iz) \leq \phi(d(y, z))$ *for every* $y, z \in Y$,

(ii) if $y \in Y$

$$d\left(Iy, I^2y\right) \leq \rho\left(Iy, I^2y\right), \tag{2}$$

(iii) either (a) $\{y_p\}$ *is a sequence in* Y *such that* $\alpha\left(y_p, y_{p+1}\right) \geq 1$ *for all* p, *further* $y_p \to y \in Y$ *as* $p \to \infty$, *then* $\alpha\left(y_p, y\right) \geq 1$ *for all* $p \in \mathbb{N}$ *or (b)* I *is continuous.*

Then I has a fixed point.

Proof Condition (i) and (ii) implies

$$d\left(Iy, I^2y\right) \leq \rho\left(Iy, I^2y\right) \leq \alpha(y, z)\,\rho\left(Iy, I^2y\right) \leq \phi(d(y, Iy))$$

this implies

$$d\left(Iy, I^2y\right) \leq \phi(d(y, Iy))$$

Let $y_0 \in Y$ be any arbitrary point of Y, define

$$y_1 = Iy_0,\ y_2 = Iy_1, \ldots y_p = Iy_{p-1}.$$

That is

$$y_p = Iy_{p-1} = I\left(I\left(y_{p-2}\right)\right) = I^2\left(y_{p-2}\right) = \cdots I^p(y_0).$$

If $y_p = y_{p+1}$ then y_p is a fixed point. Let $y_p \neq y_{p+1}$ for any $p \in N$.

$$\begin{aligned} d\left(y_p, y_{p+1}\right) &= d\left(Iy_{p-1}, I^2y_{p-1}\right) \leq \phi\left(d\left(y_{p-1}, Iy_{p-1}\right)\right) \\ &\leq \phi\left(d\left(y_{p-1}, y_p\right)\right) \leq \phi^2\left(d\left(y_{p-2}, y_{p-1}\right)\right) \\ &\cdots \\ &\leq \phi^p(d(y_0, y_1)). \end{aligned}$$

Now for each $q > p$, we have

$$d\left(y_p, y_q\right) \leq \sum_{i=p}^{q-1} d(y_i, y_{i+1}) \leq \sum_{i=p}^{q-1} \phi^{i-1} d(y_0, y_1).$$

For any $\varepsilon > 0$ there is n_1 such that

$$\sum_{i=p+1}^{\infty} d(y_i, y_{i+1}) < \varepsilon$$

for all $p \geq n_1$. Now by using triangular inequality for all $p, q \geq n_1$

$$\begin{aligned} d(y_p, y_q) &\leq d(y_p, y_{p+1}) + d(y_{p+1}, y_{p+2}) + \cdots d(y_{q-1}, y_q) \\ &= \sum_{i=p}^{q-1} d(y_i, y_{i+1}) \leq \sum_{i=p}^{q-1} \phi^{i-1}(d(y_0, y_1)) \\ &\leq \sum_{i=p}^{\infty} \phi^{i-1}(d(y_0, y_1)). \end{aligned}$$

For all $\varepsilon > 0$ there exists n_1 such that

$$d(y_p, y_q) = \sum_{i=p+1}^{\infty} \phi^{i-1}(d(y_0, y_1)) < \varepsilon$$

for all $p, q \geq n_1$. As $\sum_{i=1}^{\infty} \phi^i < \infty$, so this implies$\{y_p\}$ is a Cauchy sequence. so $y_p \to y$ as $p \to \infty$. If either
(a) As $\alpha(y_p, y) \geq 1$ for all p, now using the triangular inequality

$$\begin{aligned} d(Iy, y) &\leq d(Iy, Iy_p) + d(y_{p+1}, y) \\ &\leq \alpha(y_p, y) d(Iy_p, Iy) + d(y_{p+1}, y) \\ &\leq \phi(d(y_p, y)) + d(y_{p+1}, y). \end{aligned}$$

Letting $p \to \infty$, as ϕ is continuous at $I = 0$, we get $d(Iy, y) = 0$, which implies $y = Iy$. So I has a fixed point. Or
(b) As $\{y_p\}$ is Cauchy sequence. in complete metric space (Y, d), so there is $y \in Y$ such as $y_p \to y$ as $p \to \infty$. As I is continuous, it follows that $y_{p+1} = Iy_p$ converges to Iy as $p \to \infty$. By uniqueness of limit, we get $y = Iy$. That is y is a fixed point of I. ∎

Kirk and Shahzad [21] define (ε, c) uniform local multivauled contraction. We generalize this definition as follows.

Definition 26 A mapping $I : Y \to CB(Y)$ is an $(\varepsilon - \phi)$ uniform local multivauled contraction, where $\varepsilon > 0$ and $\alpha : Y \times Y \to [0, \infty)$ and $\phi \in \Phi$,

$$\text{if } d(y, z) < \varepsilon, \text{ then } \alpha_*(Iy, Iz) H(Iy, Iz) \leq \phi(d(y, z)).$$

Following result is generalized form of [21, Theorem 3.4].

Theorem 27 *Let for a metric space (Y, d), suppose there is a metric transform $\psi \in \Psi$, if mapping $I : Y \to CB(Y)$ satisfies,*

1. *for each* $y, z \in Y$

$$\psi(\alpha_*(Iy, Iz) H(Iy, Iz)) \leq \phi(d(y, z)),$$

2. *there exists* $\upsilon \in \Phi$, *such as for* $l > 0$ *sufficiently small*

$$\phi(l) \leq \psi(\upsilon(l)).$$

Then for $\varepsilon > 0$ *sufficiently small,* I *is an* $(\varepsilon - \upsilon)$ *uniform local multivauled contraction on* (Y, d).

Proof Let $y, z \in Y$ by condition (1) and (2)

$$\psi(\alpha_*(Iy, Iz) H(Iy, Iz)) \leq \phi(d(y, z))$$

and

$$\phi(d(y, z)) \leq \psi(\upsilon(d(y, z))).$$

Then

$$\psi(\alpha_*(Iy, Iz) H(Iy, Iz)) \leq \psi(\upsilon(d(y, z)))$$

as ϕ is strictly increasing so we get

$$\alpha_*(Iy, Iz) H(Iy, Iz) \leq \upsilon(d(y, z)).$$

So I is an $(\varepsilon - \upsilon)$ uniform local multivauled contraction on (Y, d). ■

Corollary 28 *Let for a metric space* (Y, d), *there is a metric transform* $\psi \in \Psi$, *and* $r \in (0, 1)$ *if the mapping* $I : Y \to CB(Y)$ *satisfies,*

1. *for each* $y, z \in Y$

$$\psi(H(Iy, Iz)) \leq rd(y, z),$$

2. *for* $l > 0$ *sufficiently small, there exists* $c \in (0, 1)$ *such that*

$$rl \leq \psi(cl).$$

Then for sufficiently small $l > 0$, I *is an* (ε, c) *uniform local multivauled contraction on* (Y, d).

Proof In Theorem 13 for $\upsilon(t) = rt$ and $\alpha(y, z) = 1$ for some $r \in (0, 1)$, we get the required result. ■

By using the above Corollary 28 and Theorem 14 we get the result as follows.

Corollary 29 *Let* (Y, d) *be a* ε*-chainable, connected and complete metric space Then* I *has a fixed point.*

Proof For any $\varepsilon > 0$ a connected metric space is ε-chainable. ■

3 Generalized ($\alpha_* - \phi$) Multivalued Contractions

In this section, we generalized the fixed point results concerning multivauled mappings by using generalized ($\alpha_* - \phi$) contraction. We extend some theorems which are proved in previous section. The extended results are verified through examples. We prove some other fixed point results, by using them some of the corollaries will be obtained which are exist in literature.

Following Theorem 30 is extended form of Theorem 17 for metric D (semi sequentially equivalent Hausdorff metric). In this result we will use ($\alpha_* - \phi$) contraction condition for generalized metric $\acute{\mathcal{M}}(y, z)$ instead of $d(y, z)$. Where,

$$\acute{\mathcal{M}}(y, z) = \max\left\{d(y, z), d(y, Iy), \frac{d(y, z)\, d(z, Iz)}{1 + d(y, z)}\right\}.$$

Theorem 30 *Let for a complete metric space* (Y, d), *D be semi sequentially equivalent metric on* $CB(Y)$. *If the mapping* $I : Y \to CB(Y)$ *satisfies,*

(i) $\alpha_*(Iy, Iz)\, D(Iy, Iz) \le \phi\left(\acute{\mathcal{M}}(y, z)\right)$, for every $y, z \in Y$,
(ii) If $y \in Y$ and $z \in Iy$

$$d(z, Iz) \le D(Iz, Iy),$$

(iii) I is α_*-admissible,
(iv) there is $y_0 \in Y$ and $y_1 \in Iy_0$ such as $\alpha(y_0, y_1) \ge 1$,
(v) either

- $\{y_p\}$ is a sequence in Y such that $\alpha\left(y_p, y_{p+1}\right) \ge 1$ for all p, further $y_p \to y \in Y$ as $p \to \infty$, then $\alpha\left(y_p, y\right) \ge 1$ for all p,

or

- I is upper hemi continuous.

Then I has a fixed point y.

Proof By condition (iii) and (iv) there is $y_0 \in Y$ and $y_1 \in Iy_0$ such as $\alpha(y_0, y_1) \ge 1$ which implies $\alpha_*(Iy_0, Iy_1) \ge 1$. If $y_1 = y_0$ then $y_0 = y_1 \in Iy_0$ that is $y_0 \in Iy_0$, which implies y_0 is fixed point of I. Proof is completed. So assume $y_0 \neq y_1$, then by using condition (i) and (ii)

$$d(y_1, Iy_1) \le D(Iy_1, Iy_0) \le \alpha_*(Iy_0, Iy_1)\, D(Iy_0, Iy_1) \le \phi\left(\acute{\mathcal{M}}(y_0, y_1)\right)$$

If $d(y_1, Iy_1) = 0$ then $y_1 \in Iy_1$ implies that y_1 is a fixed point of I and the proof is finished. So assume that $d(y_1, Iy_1) > 0$ and

$$\begin{aligned}
0 < d(y_1, Iy_1) &\leq \phi\left(\acute{\mathcal{M}}(y_0, y_1)\right) \\
&= \phi\left(\max\left\{d(y_0, y_1), d(y_0, Iy_0), \frac{d(y_0, y_1)\, d(y_1, Iy_1)}{1 + d(y_0, y_1)}\right\}\right) \\
&\leq \phi(\max\{d(y_0, y_1), d(y_1, Iy_1)\}).
\end{aligned}$$

That is

$$0 < d(y_1, Iy_1) \leq \phi(\max\{d(y_0, y_1), d(y_1, Iy_1)\}).$$

If $\max\{d(y_0, y_1), d(y_1, Iy_1)\} = d(y_1, Iy_1)$ then we have that

$$0 < d(y_1, Iy_1) \leq \phi(d(y_1, Iy_1)).$$

As $d(y_1, Iy_1) > 0$ and $\phi \in \Phi$ and $\phi(l) < l$,

$$0 < d(y_1, Iy_1) < d(y_1, Iy_1)$$

gives a contradiction. Hence

$$0 < d(y_1, Iy_1) \leq \phi(d(y_0, y_1)).$$

We may choose $y_2 \in Iy_1$ and $q_1 > 1$ such that

$$0 < d(y_1, Iy_1) \leq d(y_1, y_2) < q_1 d(y_1, Iy_1).$$

Thus

$$0 < d(y_1, y_2) < q_1 d(y_1, Iy_1) \leq q_1 \phi(d(y_0, y_1)) = q_1 \phi(b_\circ)$$

where $b_\circ = d(y_0, y_1)$. Note that $y_2 \neq y_1$ and $\boldsymbol{\alpha}(y_1, y_2) \geq \alpha_*(Iy_0, Iy_1) \geq 1$. Thus $\alpha(y_1, y_2) \geq 1$ and hence $\alpha_*(Iy_1, Iy_2) \geq 1$. As $\phi \in \Phi$,

$$\phi(d(y_1, y_2)) < \phi(q_1 \phi(b_\circ)).$$

If we set

$$q_2 = \frac{\phi(q_1 \phi(b_\circ))}{\phi(d(y_1, y_2))},$$

then $q_2 > 1$. Now if $y_2 \in Iy_2$ then proof is finished. Let $y_2 \notin Iy_2$ then by similar process we obtain

$$0 < d(y_2, Iy_2) \leq \phi(d(y_1, y_2))$$

and $y_3 \in Iy_2$ such that

$$0 < d(y_2, y_3) < q_2 d(y_2, Iy_2) \leq q_2 \phi(d(y_1, y_2)) = \phi(q_1 \phi(b_\circ)). \tag{3}$$

Note that $y_3 \neq y_2$,

$$\alpha(y_2, y_3) \geq \alpha_*(Iy_1, Iy_2) \geq 1.$$

Thus $\alpha(y_2, y_3) \geq 1$ implies that $\alpha_*(Iy_2, Iy_3) \geq 1$. By 3 we have

$$\phi(d(y_2, y_3)) < \phi^2(q_1(\phi(b_\circ))).$$

If

$$q_3 = \frac{\phi^2(q_1\phi(b_\circ))}{\phi(d(y_2, y_3))},$$

then $q_3 > 1$. Now if $y_3 \in Iy_3$, then the proof is finished. Let $y_3 \notin Iy_3$. Resuming in this way, we get a sequence $\{y_p\}$ in Y and it satisfies $y_{p+1} \in Iy_p$, $y_{p+1} \neq y_p$ and $\alpha(y_{p+1}, y_{p+2}) \geq 1$, such that

$$0 < d(y_{p+1}, Iy_{p+1}) \leq \phi(d(y_p, y_{p+1}))$$

which gives

$$0 < d(y_{p+1}, y_{p+2}) \leq \phi^p(q_1\phi(b_\circ)).$$

Now we prove that $\{y_p\}$ is Cauchy sequence. in Y. Now for each $q > p$, we have

$$d(y_p, y_q) \leq \sum_{i=p}^{q-1} d(y_i, y_{i+1}) \leq \sum_{i=p}^{q-1} \phi^{i-1}(q_1\phi(b_\circ)) < \infty.$$

As $\phi \in \Phi$ for any $\varepsilon > 0$ there exists n_1 such that $\sum_{i=p+1}^{\infty} d(y_i, y_{i+1}) < \varepsilon$ for all $p \geq n_1$. Now by using triangular inequality

$$\begin{aligned} d(y_p, y_q) &\leq d(y_p, y_{p+1}) + d(y_{p+1}, y_{p+2}) + \ldots d(y_{q-1}, y_q) \\ &= \sum_{i=p}^{q-1} d(y_i, y_{i+1}) \leq \sum_{i=p}^{q-1} \phi^{i-1}(q_1\phi(b_\circ)) \\ &\leq \sum_{i=p}^{\infty} \phi^{i-1}(q_1\phi(b_\circ)) \end{aligned}$$

for all $p, q \geq n_1$. Now for all $\varepsilon > 0$ there is n_1 such that

$$d(y_p, y_q) = \sum_{i=p+1}^{\infty} \phi^{i-1}(q_1\phi(b_\circ)) < \varepsilon \text{ for all } p, q \geq n_1.$$

As $\sum_{i=1}^{\infty} \phi^i < \infty$, so this implies $\{y_p\}$ is a Cauchy sequence. so y_p converges to y. As $\alpha(y_p, y_{p+1}) \geq 1$ and $y_p \to y$ so $\alpha(y_p, y) \geq 1$ for all $p \in N$. As I is α_*-

admissible so $\alpha_* \left(Iy_p, Iy\right) \geq 1$ so by (i)

$$\begin{aligned}
D\left(Iy_p, Iy\right) &\leq \alpha_*\left(Iy_p, Iy\right) D\left(Iy_p, Iy\right) \\
&\leq \phi\left(\acute{\mathcal{M}}\left(y_p, y\right)\right) \\
&= \phi\left(\max\left\{d\left(y_p, y\right), d\left(y_p, Iy_p\right), \frac{d\left(y_p, Iy_p\right) d\left(y, Iy\right)}{1 + d\left(y_p, y\right)}\right\}\right) \\
&\leq \phi\left(\max\left\{d\left(y_p, y\right), d\left(y_p, y_{p+1}\right), \frac{d\left(y_p, y_{p+1}\right) d\left(y, Iy\right)}{1 + d\left(y_p, y\right)}\right\}\right).
\end{aligned}$$

On taking limit as p tends to ∞, we have

$$\lim_{p\to\infty} D\left(Iy_p, Iy\right) \leq \lim_{p\to\infty} \phi\left(\max\left\{d\left(y_p, y\right), d\left(y_p, y_{p+1}\right), \frac{d\left(y_p, y_{p+1}\right) d\left(y, Iy\right)}{1 + d\left(y_p, y\right)}\right\}\right).$$

That is $\lim_{p\to\infty} D\left(Iy_p, Iy\right) = 0$ as ϕ is continuous at 0. As D is semi sequentially equivalent to H so

$$\lim_{p\to\infty} H\left(Iy_p, Iy\right) = 0.$$

Consequently, we obtain

$$\lim_{p\to\infty} d\left(y_p, Iy\right) \leq \lim_{p\to\infty} H\left(Iy_p, Iy\right) = 0$$

implies

$$d\left(y, Iy\right) = \lim_{p\to\infty} d\left(y_p, Iy\right) = 0.$$

Hence $y \in Iy$. As $\lim_{p\to\infty} y_p = y$, and $y_{p+1} \in I\left(y_p\right)$, so by upper hemi continuity, $y \in Iy$. ■

By using Theorem 30 the following Corollary can be obtained.

Corollary 31 *Let for a complete metric space (Y, d) and D_1 be sequentially equivalent metric on $CB(Y)$. If the mapping $I : Y \to CB(Y)$ satisfies,*

(i) $\alpha_*(Iy, Iz) D_1(Iy, Iz) \leq \phi\left(\acute{\mathcal{M}}(y, z)\right)$, for every $y, z \in Y$,
(ii) if $y \in Y$ and $z \in Iy$

$$d(z, Iz) \leq D_1(Iz, Iy),$$

(iii) I is α_*-admissible,
(iv) there is $y_0 \in Y$ and $y_1 \in Iy_0$ such as $\alpha(y_0, y_1) \geq 1$,
(v) either (a) $\left\{y_p\right\}$ is a sequence in Y such that $\alpha\left(y_p, y_{p+1}\right) \geq 1$ for all p, further $y_p \to y \in Y$ as $p \to \infty$, then $\alpha\left(y_p, y\right) \geq 1$ for all p, or (b) I is upper hemi continuous.

Then I has a fixed point y.

Corollary 32 *Let for a complete metric space (Y, d), D_2 be strong semi sequentially equivalent metric on $CB(Y)$. If the mapping $I : Y \to CB(Y)$ satisfies,*

(i) $\alpha_*(Iy, Iz) D_2(Iy, Iz) \leq \phi\left(\acute{\mathcal{M}}(y, z)\right)$, for every $y, z \in Y$,
(ii) If $y \in Y$ and $z \in Iy$

$$d(z, Iz) \leq D_2(Iz, Iy),$$

(iii) I is α_*-admissible,
(iv) there is $y_0 \in Y$ and $y_1 \in Iy_0$ such as $\alpha(y_0, y_1) \geq 1$,
(v) either (a) $\{y_p\}$ is a sequence in Y such that $\alpha\left(y_p, y_{p+1}\right) \geq 1$ for all p, further $y_p \to y \in Y$ as $p \to \infty$, then $\alpha\left(y_p, y\right) \geq 1$ for all p, or (b) I is upper hemi continuous.

Then I has a fixed point y.

Corollary 33 *Assume for a complete metric space (Y, d), and D be semi sequentially equivalent metric on $CB(Y)$. If the mapping $I : Y \to CB(Y)$ satisfies,*

(i) $\alpha_*(Iy, Iz) D(Iy, Iz) \leq \phi(d(y, z))$, *for every* $y, z \in Y$,
(ii) If $y \in Y$ *and* $z \in Iy$

$$d(z, Iz) \leq D(Iz, Iy),$$

(iii) I *is* α_**-admissible,*
(iv) there is $y_0 \in Y$ *and* $y_1 \in Iy_0$ *such as* $\alpha(y_0, y_1) \geq 1$,
(v) either (a) $\{y_p\}$ *is a sequence in* Y *such that* $\alpha\left(y_p, y_{p+1}\right) \geq 1$ *for all* p*, further* $y_p \to y \in Y$ *as* $p \to \infty$*, then* $\alpha\left(y_p, y\right) \geq 1$ *for all* $p \in \mathbb{N}$ *or (b)* I *is upper hemi continuous.*

Then I *has a fixed point* y.

Proof If we put $\mathcal{M}(y, z) = d(y, z)$ in Theorem 30 we obtain the required result. ■

Definition 34 Let for a metric space (Y, d), the mapping $I : Y \to CB(Y)$ is a multivauled $(\alpha_* - \phi)$ generalized contractive mapping with two functions $\alpha : Y \times Y \to [0, \infty)$ and $\phi \in \Phi$ such as

$$\alpha_*(Iy, Iz) H(Iy, Iz) \leq \phi(\mathcal{M}(y, z))$$

for all $y, z \in Y$. Where H is the Hausdorff metric and

$$\mathcal{M}(y, z) = \max\left\{\begin{array}{l} d(y, z), d(y, Iy), d(z, Iz), \\ \dfrac{d(y, Iz) + d(z, Iy)}{2}, \dfrac{d(y, Iy) d(z, Iz)}{1 + d(y, z)} \end{array}\right\}.$$

Hassanzade et al. [3] proved fixed point result for Hausdorff metric by using $(\alpha_* - \phi)$ contraction. Following Theorem is extended form of that result.

Theorem 35 *Let for a complete metric space (Y, d), and H be any Hausdorff metric on $CB(Y)$. If the mapping $I : Y \to CB(Y)$ satisfies,*

(i) $\alpha_*(Iy, Iz) H(Iy, Iz) \leq \phi(\mathcal{M}(y, z))$, for every $y, z \in Y$,
(ii) I is α_*-admissible,
(iii) there is $y_0 \in Y$ and $y_1 \in Iy_0$ such as $\alpha(y_0, y_1) \geq 1$,
(iv) either (a) $\{y_p\}$ is a sequence in Y such that $\alpha(y_p, y_{p+1}) \geq 1$ for all p, further $y_p \to y \in Y$ as $p \to \infty$, then $\alpha(y_p, y) \geq 1$ for all p,
or (b) I is upper hemi continuous.

Then I has a fixed point y.

Proof By condition (iii) and (iv) there is $y_0 \in Y$ and $y_1 \in Iy_0$ such as $\alpha(y_0, y_1) \geq 1$ which implies $\alpha_*(Iy_0, Iy_1) \geq 1$. If $y_1 = y_0$ then $y_0 = y_1 \in Iy_0$ that is $y_0 \in Iy_0$. Which implies y_0 is fixed point of I hence proof is completed. So assume $y_0 \neq y_1$, then by using condition (i) we get

$$\begin{aligned} d(y_1, Iy_1) &\leq H(Iy_1, Iy_0) \\ &\leq \alpha_*(Iy_0, Iy_1) H(Iy_0, Iy_1) \\ &\leq \phi(\mathcal{M}(y_0, y_1)). \end{aligned}$$

If $d(y_1, Iy_1) = 0$ then $y_1 \in Iy_1$ implies that y_1 is a fixed point of I and the proof is done. So consider that $d(y_1, Iy_1) > 0$ and

$$\begin{aligned} d(y_1, Iy_1) &\leq \phi(\mathcal{M}(y_0, y_1)) \\ &= \phi\left(\max\left\{\begin{array}{l} d(y_0, y_1), d(y_0, Iy_0), \\ d(y_1, Iy_1), \dfrac{d(y_0, Iy_1) + d(y_1, y_1)}{2}, \\ \dfrac{d(y_0, y_1) d(y_1, Iy_1)}{1 + d(y_0, y_1)} \end{array}\right\}\right) \\ &\leq \phi\left(\max\left\{\begin{array}{l} d(y_0, y_1), d(y_1, Iy_1), \\ \dfrac{d(y_0, Iy_1)}{2}, d(y_1, Iy_1) \end{array}\right\}\right) \\ &\leq \phi\left(\max\left\{\begin{array}{l} d(y_0, y_1), d(y_1, Iy_1), \\ \dfrac{d(y_0, y_1) + d(y_1, Iy_1)}{2} \end{array}\right\}\right) \\ &\leq \phi(\max\{d(y_0, y_1), d(y_1, Iy_1)\}). \end{aligned}$$

That is

$$0 < d(y_1, Iy_1) \leq \phi(\max\{d(y_0, y_1), d(y_1, Iy_1)\}).$$

If

$$\max\{d(y_0, y_1), d(y_1, Iy_1)\} = d(y_1, Iy_1)$$

then we have that

$$0 < d(y_1, Iy_1) \leq \phi(d(y_1, Iy_1)).$$

As $d(y_1, Iy_1) > 0$ and $\phi \in \Phi$ and $\phi(t) < t$,

$$0 < d(y_1, Iy_1) < d(y_1, Iy_1)$$

gives a contradiction. Hence

$$0 < d(y_1, Iy_1) \leq \phi(d(y_0, y_1)).$$

We may choose $y_2 \in Iy_1$ and $q_1 > 1$ such that

$$0 < d(y_1, Iy_1) \leq d(y_1, y_2) < q_1 d(y_1, Iy_1).$$

Thus

$$\begin{aligned} 0 < d(y_1, y_2) &< q_1 d(y_1, Iy_1) \\ &\leq q_1 \phi(d(y_0, y_1)) = q_1 \phi(b_\circ) \end{aligned}$$

where $b_\circ = d(y_0, y_1)$. Note that $y_2 \neq y_1$ and $\alpha(y_1, y_2) \geq \alpha_*(Iy_0, Iy_1) \geq 1$. Thus $\alpha(y_1, y_2) \geq 1$ and hence $\alpha_*(Iy_1, Iy_2) \geq 1$. As $\phi \in \Phi$,

$$\phi(d(y_1, y_2)) < \phi(q_1 \phi(b_\circ)).$$

If we set

$$q_2 = \frac{\phi(q_1 \phi(b_\circ))}{\phi(d(y_1, y_2))},$$

then $q_2 > 1$. Now if $y_2 \in Iy_2$ then proof is finished. Let $y_2 \notin Iy_2$ then by similar process we obtain

$$0 < d(y_2, Iy_2) \leq \phi(d(y_1, y_2))$$

and $y_3 \in Iy_2$ such that

$$\begin{aligned} 0 < d(y_2, y_3) &< q_2 d(y_2, Iy_2) \\ &\leq q_2 \phi(d(y_1, y_2)) = \phi(q_1 \phi(b_\circ)). \end{aligned} \tag{4}$$

Note that $y_3 \neq y_2$,

$$\alpha(y_2, y_3) \geq \alpha_*(Iy_1, Iy_2) \geq 1.$$

Thus $\alpha(y_2, y_3) \geq 1$ implies that $\alpha_*(Iy_2, Iy_3) \geq 1$. By (4) we have

$$\phi(d(y_2, y_3)) < \phi^2(q_1(\phi(b_\circ))).$$

If

$$q_3 = \frac{\phi^2 (q_1 \phi (b_\circ))}{\phi (d (y_2, y_3))},$$

then $q_3 > 1$. Now if $y_3 \in I y_3$, then the proof is finished. Let $y_3 \notin I y_3$. Resuming in this way, we get a sequence $\{y_p\}$ in Y and satisfies

$$y_{p+1} \in I y_p,\ y_{p+1} \neq y_p,\ \alpha (y_{p+1}, y_{p+2}) \geq 1,$$

$$0 < d (y_{p+1}, I y_{p+1}) \leq \phi (d (y_p, y_{p+1}))$$

which gives

$$0 < d (y_{p+1}, y_{p+2}) \leq \phi^p (q_1 \phi (b_\circ)).$$

Now we prove that $\{y_p\}$ is Cauchy sequence. in Y. Now for each $q > p$, we have

$$\begin{aligned} d (y_p, y_q) &\leq \sum_{i=p}^{q-1} d (y_i, y_{i+1}) \\ &\leq \sum_{i=p}^{q-1} \phi^{i-1} (q_1 \phi (b_\circ)) < \infty. \end{aligned}$$

As $\phi \in \Phi$ for any $\varepsilon > 0$ there exists N_1 such that

$$\sum_{i=p+1}^{\infty} d (y_i, y_{i+1}) < \varepsilon$$

for all $p \geq N_1$. Now by using triangular inequality

$$\begin{aligned} d (y_p, y_q) &\leq d (y_p, y_{p+1}) + d (y_{p+1}, y_{p+2}) + \ldots d (y_{q-1}, y_q) \\ &= \sum_{i=p}^{q-1} d (y_i, y_{i+1}) \leq \sum_{i=p}^{q-1} \phi^{i-1} (q_1 \phi (b_\circ)) \\ &\leq \sum_{i=p}^{\infty} \phi^{i-1} (q_1 \phi (b_\circ)) \end{aligned}$$

for all $p, q \geq N_1$. Now for all $\varepsilon > 0$ there exists N_1 such that

$$d (y_p, y_q) = \sum_{i=p+1}^{\infty} \phi^{i-1} (q_1 \phi (b_\circ)) < \varepsilon$$

for all $p, q \geq N_1$. As $\sum_{i=1}^{\infty} \phi^i < \infty$, so this implies $\{y_p\}$ is a Cauchy sequence. so y_p converges to y. Now we want to show that y is a fixed point of I. As $\alpha(y_p, y_{p+1}) \geq 1$ and $y_p \to y$ as $n \to \infty$, so by given condition, $\alpha(y_p, y) \geq 1$ for all $p \in N$. As I is α_*-admissible so $\alpha_*(Iy_p, Iy) \geq 1$. We have

$$
\begin{aligned}
d(y_{p+1}, Iy) &\leq H(Iy_p, Iy) \leq \alpha_*(Iy_p, Iy) H(Iy_p, Iy) \\
&\leq \phi(\mathcal{M}(y_p, y)) \\
&= \phi\left(\max\left\{\begin{array}{l} d(y_p, y), d(y_p, Iy_p), \\ d(y, Iy), \dfrac{d(y_p, Iy) + d(y, Iy_p)}{2}, \\ \dfrac{d(y_p, Iy_p) d(y, Iy)}{1 + d(y_p, y)} \end{array}\right\}\right) \\
&\leq \phi\left(\max\left\{\begin{array}{c} d(y_p, y), d(y_p, y_{p+1}), \\ d(y, Iy), \dfrac{d(y_p, Iy) + d(y, y_{p+1})}{2}, \\ \dfrac{d(y_p, y_{p+1}) d(y, Iy)}{1 + d(y_p, y)} \end{array}\right\}\right).
\end{aligned}
$$

On taking limit we get

$$
d(y, Iy) \leq \phi(d(y, Iy))
$$

implies that $y \in Iy$. As $\lim_{p\to\infty} y_p = y$, and $y_{p+1} \in I(y_p)$, so by upper hemi continuity, $y \in Iy$. ■

Following is the example which is satisfying all the conditions of Theorem 35.

Example 36 Let $Y = \{y_1, y_2, y_3, y_4, y_5\}$. Define the mapping $d : Y \times Y \to \mathbb{R}^+$ by

$$
\begin{aligned}
d(y_2, y_5) &= d(y_3, y_4) = d(y_3, y_5) = d(y_2, y_4) = 6, \\
d(y_2, y_3) &= d(y_1, y_4) = d(y_1, y_5) = 9, \\
d(y_1, y_2) &= d(y_1, y_3) = 5, \\
d(y_4, y_5) &= 1, \\
d(y, y) &= 0 \text{ and } d(y, y_\circ) = d(y_\circ, y)
\end{aligned}
$$

for all $y, z \in Y$. Clearly (Y, d) is a complete metric space. Define a mapping $I : Y \to CB(Y)$ by

$$
Iy = \begin{cases} \{y_1\} \text{ if } y = y_1, y_2, y_3, \\ \{y_2\} \text{ if } y = y_4, \\ \{y_3\} \text{ if } y = y_5. \end{cases}
$$

Define $\alpha : Y \times Y \to \mathbb{R}^+$ by $\alpha(y_i, y_j) = 1$ for all $i, j \in \{1, 2, 3, 4, 5\}$. If we set

$$\phi(t) = \frac{8}{9}t$$

for $t \in \mathbb{R}^+$, then $\phi \in \Phi$. I is α_*-admissible. Note that

$$\alpha_*(Iy, Iz)\, H(Iy, Iz) \le \phi(\mathcal{M}(y, z))$$

where

$$\mathcal{M}(y, z) = \max\left\{\begin{array}{l} d(y, z), d(y, Iy), d(z, Iz), \\ \frac{d(y, Iz) + d(z, Iy)}{2}, \frac{d(y, Iy)\, d(z, Iz)}{1 + d(y, z)} \end{array}\right\}.$$

This condition holds for any $y, z \in \{y_1, y_2, y_3\}$ as

$$H(Iy, Iz) = 0 \le \phi(\mathcal{M}(y, z)) = \frac{8}{9}d(y, z).$$

Now consider the remaining cases:
If $y = y_1$ and $z \in \{y_4, y_5\}$ then

$$\begin{aligned} H(Iy_1, Iz) &= 5 \le \phi(\mathcal{M}(y_1, z)) \\ &= \phi\left(\max\left\{\begin{array}{l} d(y_1, z), d(y_1, Iy_1), d(y_4, Iz), \\ \frac{d(y_1, Iz) + d(z, Iy_1)}{2}, \frac{d(y_1, Iy_1)\, d(z, Iz)}{1 + d(y_1, z)} \end{array}\right\}\right) \\ &= \phi(9) = \frac{8}{9} \times 9 = 8. \end{aligned}$$

When $x = y_2$ and $y = y_4$ then

$$\begin{aligned} H(Iy, Iz) &= d(y_1, y_2) = 5 \le \phi(\mathcal{M}(y_2, y_4)) \\ &= \phi\left(\max\left\{\begin{array}{l} d(y, z), d(x, Ix), d(y_4, Iy_4), \\ \frac{d(x, Iy_4) + d(y_4, Ix)}{2}, \frac{d(x, Iy_2)\, d(x, Ix)}{1 + d(y_2, x)} \end{array}\right\}\right) \\ &= \phi(6) = \frac{48}{9}. \end{aligned}$$

When $y = y_2$ and $z = y_5$ then

$$\begin{aligned} H\left(Iy_2, Iy_5\right) &= d\left(y_1, y_3\right) = 5 \leq \phi\left(\mathcal{M}\left(y_2, y_5\right)\right) \\ &= \phi\left(\max\left\{\begin{array}{l} d\left(y_2, y_5\right), d\left(y_2, Iy_2\right), d\left(y_5, y_3\right), \\ \frac{d\left(y_2, Iy_5\right) + d\left(y_5, Iy_2\right)}{2}, \frac{d\left(y_2, Iy_2\right) d\left(y_5, Iy_5\right)}{1 + d\left(y_2, y_5\right)} \end{array}\right\}\right) \\ &= \phi\left(\max\left\{\begin{array}{l} d\left(y_2, y_5\right), d\left(y_2, y_1\right), d\left(y_5, y_3\right), \\ \frac{d\left(y_2, y_3\right) + d\left(y_5, y_1\right)}{2}, \frac{d\left(y_2, y_1\right) d\left(y_5, y_3\right)}{1 + d\left(y_2, y_5\right)} \end{array}\right\}\right) \\ &= \phi\left(\max\left\{6, 5, 6, \frac{9+9}{2}, \frac{5 \times 6}{1+6}\right\}\right) = \phi\left(9\right) = 8. \end{aligned}$$

Now for $y = y_3$ and $z = y_4$ then

$$\begin{aligned} H\left(Iy_3, Iy_4\right) &= d\left(y_1, y_2\right) = 5 \leq \phi\left(\mathcal{M}\left(y_3, y_4\right)\right) \\ &= \phi\left(\max\left\{\begin{array}{l} d\left(y_3, y_4\right), d\left(y_3, Iy_3\right), d\left(y_4, Iy_4\right), \\ \frac{d\left(y_3, Iy_4\right) + d\left(y_4, Iy_3\right)}{2}, \frac{d\left(y_3, Iy_3\right) d\left(y_4, Iy_4\right)}{1 + d\left(y_3, y_4\right)} \end{array}\right\}\right) \\ &= \phi\left(\max\left\{\begin{array}{l} d\left(y_3, y_4\right), d\left(y_3, y_1\right), d\left(y_4, y_2\right), \\ \frac{d\left(y_3, y_2\right) + d\left(y_4, y_1\right)}{2}, \frac{d\left(y_3, y_1\right) d\left(y_4, y_2\right)}{1 + d\left(y_3, y_4\right)} \end{array}\right\}\right) \\ &= \phi\left(\max\left\{6, 5, 6, \frac{9+9}{2}, \frac{5 \times 6}{1+6}\right\}\right) = \phi\left(9\right) = 8. \end{aligned}$$

When $x = y_3$ and $y = y_5$.

$$\begin{aligned} H\left(Iy_3, Iy_5\right) &= d\left(y_1, y_3\right) = 5 \leq \phi\left(\mathcal{M}\left(y_3, y_5\right)\right) \\ &= \phi\left(\max\left\{\begin{array}{l} d\left(y_3, y_5\right), d\left(y_3, Iy_3\right), d\left(y_5, Iy_5\right), \\ \frac{d\left(y_3, Iy_5\right) + d\left(y, Iy_5\right)}{2}, \frac{d\left(y_3, Iy_3\right) d\left(y_5, Iy_5\right)}{1 + d\left(y_3, y_5\right)} \end{array}\right\}\right) \\ &= \phi\left(\max\left\{\begin{array}{l} d\left(y_3, y_5\right), d\left(y_3, y_1\right), d\left(y_5, y_3\right), \\ \frac{d\left(y_3, y_3\right) + d\left(y_5, y_3\right)}{2}, \frac{d\left(y_3, y_1\right) d\left(y_5, y_3\right)}{1 + d\left(y_3, y_5\right)} \end{array}\right\}\right) \\ &= \phi\left(\max\left\{6, 5, 6, \frac{0+6}{2}, \frac{5 \times 6}{1+6}\right\}\right) = \phi\left(6\right) = \frac{8}{9} \times 6 = \frac{48}{9}. \end{aligned}$$

When $y = y_4$ and $z = y_5$.

$$\begin{aligned} H(Iy_4, Iy_5) &= d(y_2, y_3) = 9 \\ &= \phi\left(\max\left\{\begin{array}{l} d(y_4, y_5), d(y_4, Iy_4), d(y_5, Iy_5), \\ \frac{d(y_4, Iy_5) + d(y_5, Iy_4)}{2}, \frac{d(y_4, Iy_4)\, d(y_5, Iy_5)}{1 + d(y_4, y_5)} \end{array}\right\}\right) \\ &= \phi\left(\max\left\{\begin{array}{l} d(y_4, y_5), d(y_4, y_2), d(y_5, y_3), \\ \frac{d(y_4, y_3) + d(y_5, y_2)}{2}, \frac{d(y_4, y_2)\, d(y_5, y_3)}{1 + d(y_4, y_5)} \end{array}\right\}\right) \\ &= \phi\left(\max\left\{1, 6, 6, \frac{6+6}{2}, \frac{6 \times 6}{1+1}\right\}\right) = \phi(18) = 16. \end{aligned}$$

The sequences which are constant or eventually constant will be happen in discrete metric spaces. So we deduce that all the conditions of Theorem 17 is satisfied so I has a fixed.

Remark 37 By using Example 36 it can be easily seen that Theorem 35 is the generalization of Theorem 17 because if we use this example for Theorem 17 then contraction condition for $y = y_4$ and $z = y_5$ will not be satisfied.

Corollary 38 *Let for a complete metric space (Y, d), $\alpha : Y \times Y \to [0, \infty)$ be a function, $\phi \in \Phi$. If the mapping $I : Y \to Cl(Y)$ satisfies,*

(i) $\alpha_*(Iy, Iz)\, H(Iy, Iz) \leq \phi(d(y, z))$, *for every* $y, z \in Y$,
(ii) *I is α_*-admissible,*
(iii) *there is $y_0 \in Y$ and $y_1 \in Iy_0$ such as $\alpha(y_0, y_1) \geq 1$.*
(iv) *Assume that if $\{y_p\}$ is a sequence in Y such that $\alpha(y_p, y_{p+1}) \geq 1$ for all p and $y_p \to y$, then $\alpha(y_p, y) \geq 1$ for all p.*
Then I has a fixed point.

Proof If $\mathcal{M}(y, z) = d(y, z)$ and in Theorem 35 we get the required result. ∎

Corollary 39 *Let for a complete metric space (Y, d),if the mapping $I : Y \to Y$ satisfies,*

(i) $\alpha(y, z)\, d(Iy, Iz) \leq \phi(d(y, z))$, *for every* $y, z \in Y$,
(ii) *I is α-admissible,*
(iii) *there is $y_0 \in Y$ such as $\alpha(y_0, Iy_0) \geq 1$,*
(iv) *I is continuous.*

Then I has a fixed point, that is, there exists $y \in Y$ such that $Iy = y$.

Proof By using point valued analogous of Theorem 35 and using $\mathcal{M}(y, z) = d(y, z)$ we find the required result. ∎

Corollary 40 *Let for a complete metric space (Y, d), if the mapping $I : Y \to Y$ satisfies,*

(i) $\alpha(y, z)\, d(Iy, Iz) \leq \phi(d(y, z))$, *for every* $y, z \in Y$,

(ii) *I is α-admissible,*
(iii) *there is $y_0 \in Y$ such as $\alpha(y_0, Iy_0) \geq 1$,*
(iv) *if $\{y_p\}$ is a sequence in Y such that $\alpha(y_p, y_{p+1}) \geq 1$ for all p, further $y_p \to y \in Y$ as $p \to +\infty$, then $\alpha(y_p, y) \geq 1$ for all p.*

Then I has a fixed point, that is, there exists $y \in Y$ such that $Iy = y$.

Proof By using point valued analogous in Theorem 35 and using $\mathcal{M}(y, z) = d(y, z)$ we find the required result. ■

Now we extend Theorem 30 by using more general metric $\mathcal{M}(y, z)$ instead of $\acute{\mathcal{M}}(y, z)$ with a new metric D_2 which is strong semi sequentially equivalent to Hausdorff.

Theorem 41 *Let (Y, d) be a complete metric space and D_2 be strong sequentially equivalent to Hausdorff metric on $CB(Y)$. If the mapping $I : Y \to CB(Y)$ satisfies,*

(i) $\alpha_*(Iy, Iz) D_2(Iy, Iz) \leq \phi(\mathcal{M}(y, z))$, for every $y, z \in Y$,
(ii) for $y \in Y$ and $z \in Iy$

$$d(z, Iz) \leq D_2(Iz, Iy),$$

(iii) I is α_*-admissible,
(iv) there is $y_0 \in Y$ and $y_1 \in Iy_0$ such as $\alpha(y_0, y_1) \geq 1$,
(v) if $H(M_p, M) \leq D_2(M_p, M)$,
(vi) either (a) $\{y_p\}$ is a sequence in Y such that $\alpha(y_p, y_{p+1}) \geq 1$ for all p, further $y_p \to y \in Y$ as $p \to \infty$, then $\alpha(y_p, y) \geq 1$ for all p, or (b) I is upper hemi continuous.

Then I has a fixed point y.

Proof By condition (iii) and (iv) there is $y_0 \in Y$ and $y_1 \in Iy_0$ such as

$$\alpha(y_0, y_1) \geq 1 \Rightarrow \alpha_*(Iy_0, Iy_1) \geq 1.$$

If $y_1 = y_0$ then $y_0 = y_1 \in Iy_0$ that is $y_0 \in Iy_0$. Which implies y_0 is fixed point of I. Proof is completed. So assume $y_0 \neq y_1$, then by using condition (i) and (ii)

$$d(y_1, Iy_1) \leq D(Iy_1, Iy_0) \leq \alpha_*(Iy_0, Iy_1) D(Iy_0, Iy_1) \leq \phi(\mathcal{M}(y_0, y_1))$$

If $d(y_1, Iy_1) = 0$ then $y_1 \in Iy_1$ implies that y_1 is a fixed point of I and the proof is done. So assume that $d(y_1, Iy_1) > 0$

$$
\begin{aligned}
0 < d(y_1, Iy_1) &\leq \phi(\mathcal{M}(y_0, y_1)) \\
&= \phi\left(\max\left\{\begin{array}{c} d(y_0, y_1), d(y_0, Iy_0), d(y_1, Iy_1), \\ \frac{d(y_0, Iy_1) + d(y_1, y_1)}{2}, \frac{d(y_0, y_1) d(y_1, Iy_1)}{1 + d(y_0, y_1)} \end{array}\right\}\right) \\
&\leq \phi\left(\max\left\{d(y_0, y_1), d(y_1, Iy_1), \frac{d(y_0, Iy_1)}{2}, d(y_1, Iy_1)\right\}\right) \\
&\leq \phi\left(\max\left\{d(y_0, y_1), d(y_1, Iy_1), \frac{d(y_0, y_1) + d(y_1, Iy_1)}{2}\right\}\right) \\
&\leq \phi(\max\{d(y_0, y_1), d(y_1, Iy_1)\}).
\end{aligned}
$$

That is

$$0 < d(y_1, Iy_1) \leq \phi(\max\{d(y_0, y_1), d(y_1, Iy_1)\}).$$

If $\max\{d(y_0, y_1), d(y_1, Iy_1)\} = d(y_1, Iy_1)$ then we have that

$$0 < d(y_1, Iy_1) \leq \phi(d(y_1, Iy_1)).$$

As $d(y_1, Iy_1) > 0$ and $\phi \in \Phi$ and $\phi(t) < t$,

$$0 < d(y_1, Iy_1) < d(y_1, Iy_1)$$

gives a contradiction. Hence

$$0 < d(y_1, Iy_1) \leq \phi(d(y_0, y_1)).$$

We may take $y_2 \in Iy_1$ and $q_1 > 1$ such that

$$0 < d(y_1, Iy_1) \leq d(y_1, y_2) < q_1 d(y_1, Iy_1).$$

Thus

$$0 < d(y_1, y_2) < q_1 d(y_1, Iy_1) \leq q_1 \phi(d(y_0, y_1)) = q_1 \phi(b_\circ)$$

where $d(y_0, y_1) = b_\circ$. Note that $y_2 \neq y_1$ and $\alpha(y_1, y_2) \geq \alpha_*(Iy_0, Iy_1) \geq 1$. Thus $\alpha(y_1, y_2) \geq 1$ and hence $\alpha_*(Iy_1, Iy_2) \geq 1$. As $\phi \in \Phi$,

$$\phi(d(y_1, y_2)) < \phi(q_1 \phi(b_\circ)).$$

If we set

$$q_2 = \frac{\phi(q_1 \phi(b_\circ))}{\phi(d(y_1, y_2))},$$

then $q_2 > 1$. Now if $y_2 \in Iy_2$ then proof is finished. Let $y_2 \notin Iy_2$ then by similar process we obtain

$$0 < d(y_2, Iy_2) \leq \phi(d(y_1, y_2))$$

and $y_3 \in Iy_2$ such that

$$0 < d(y_2, y_3) < q_2 d(y_2, Iy_2) \leq q_2 \phi(d(y_1, y_2)) = \phi(q_1 \phi(b_\circ)). \quad (5)$$

Note that $y_3 \neq y_2$,

$$\alpha(y_2, y_3) \geq \alpha_*(Iy_1, Iy_2) \geq 1.$$

Thus $\alpha(y_2, y_3) \geq 1$ implies that $\alpha_*(Iy_2, Iy_3) \geq 1$. By 5 we have

$$\phi(d(y_2, y_3)) < \phi^2(q_1(\phi(b_\circ))). \quad (6)$$

If

$$q_3 = \frac{\phi^2(q_1\phi(b_\circ))}{\phi(d(y_2, y_3))},$$

then $q_3 > 1$. Now if $y_3 \in Iy_3$, then the proof is finished. Let $y_3 \notin Iy_3$. Resuming in this way, we get a sequence $\{y_p\}$ in Y and it satisfies $y_{p+1} \in Iy_p$, $y_{p+1} \neq y_p$,

$$\alpha(y_{p+1}, y_{p+2}) \geq 1,$$

$$0 < d(y_{p+1}, Iy_{p+1}) \leq \phi(d(y_p, y_{p+1}))$$

which gives

$$0 < d(y_{p+1}, y_{p+2}) \leq \phi^p(q_1\phi(b_\circ))$$

Now we prove that $\{y_p\}$ is Cauchy sequence. in Y. Now for each $q > p$, we have

$$d(y_p, y_q) \leq \sum_{i=p}^{q-1} d(y_i, y_{i+1}) \leq \sum_{i=p}^{q-1} \phi^{i-1}(q_1\phi(b_\circ)) < \infty.$$

As $\phi \in \Phi$ for any $\varepsilon > 0$ there exists n_1 such that $\sum_{i=p+1}^{\infty} d(y_i, y_{i+1}) < \varepsilon$ for all $p, q \geq n_1$. Now by using triangular inequality

$$
\begin{aligned}
d\left(y_p, y_q\right) &\leq d\left(y_p, y_{p+1}\right) + d\left(y_{p+1}, y_{p+2}\right) + \ldots d\left(y_{q-1}, y_q\right) \\
&= \sum_{i=p}^{q-1} d\left(y_i, y_{i+1}\right) \leq \sum_{i=p}^{q-1} \phi^{i-1}\left(q_1 \phi\left(b_\circ\right)\right) \\
&\leq \sum_{i=p}^{\infty} \phi^{i-1}\left(q_1 \phi\left(b_\circ\right)\right)
\end{aligned}
$$

for all $p, q \geq n_1$. Now for all $\varepsilon > 0$ there exists n_1 such that

$$
d\left(y_p, y_q\right) = \sum_{i=p+1}^{\infty} \phi^{i-1}\left(q\phi\left(b_\circ\right)\right) < \varepsilon \text{ for all } p, q \geq n_1.
$$

As $\sum_{i=1}^{\infty} \phi^i < \infty$, so this implies $\{y_p\}$ is a Cauchy sequence. so y_p converges to y. As $\boldsymbol{\alpha}\left(y_p, y_{p+1}\right) \geq 1$ and $y_p \to y$ so $\alpha\left(y_p, y\right) \geq 1$ for all $p \in N$. As I is α_* admissible so $\alpha_*\left(Iy_p, Iy\right) \geq 1$ so by using condition (i) and (v) we have

$$
\begin{aligned}
d\left(y_{p+1}, Iy\right) &\leq H\left(Iy_p, Iy\right) \leq D\left(Iy_p, Iy\right) \leq \alpha_*\left(Iy_p, Iy\right) D\left(Iy_p, Iy\right) \\
&\leq \phi\left(M\left(y_p, y\right)\right) \\
&= \phi\left(\max\left\{\begin{array}{c} d\left(y_p, y\right), d\left(y_p, Iy_p\right), d\left(y, Iy\right), \\ \dfrac{d\left(y_p, Iy\right) + d\left(y, Iy_p\right)}{2}, \dfrac{d\left(y_p, Iy_p\right) d\left(y, Iy\right)}{1 + d\left(y_p, y\right)} \end{array}\right\}\right) \\
&\leq \phi\left(\max\left\{\begin{array}{c} d\left(y_p, y\right), d\left(y_p, y_{p+1}\right), d\left(y, Iy\right), \\ \dfrac{d\left(y_p, Iy\right) + d\left(y, y_{p+1}\right)}{2}, \dfrac{d\left(y_p, y_{p+1}\right) d\left(y, Iy\right)}{1 + d\left(y_p, y\right)} \end{array}\right\}\right).
\end{aligned}
$$

On taking limit as n tends to ∞ on both sides of above inequality, we have

$$
d\left(y, Iy\right) \leq \phi\left(d\left(y, Iy\right)\right) < d\left(y, Iy\right)
$$

a contradiction. Consequently $y \in Iy$. ■

Definition 42 Let for a metric space (Y, d), the mapping $I : Y \to CB(Y)$ is an $(\alpha_* - \phi)$ generalized contractive mapping on metrically equivalent Hausdorff space with two functions as $\alpha : Y \times Y \to [0, \infty)$ and $\phi \in \Phi$ such that

$$
\alpha_*\left(Iy, Iz\right) H^+\left(Iy, Iz\right) \leq \phi\left(\mathcal{N}\left(y, z\right)\right)
$$

for all $y, z \in Y$. Where H^+ is metrically equivalent to Hausdorff metric and

$$\mathcal{N}(y, y_\circ) = \max\left\{\begin{array}{c} d(y, y_\circ), d(y, Iy), \dfrac{d(y_\circ, Iy_\circ)}{2}, \\ \dfrac{d(y, Iy_\circ) + d(y_\circ, Iy)}{2}, \dfrac{d(y, Iy)\, d(y_\circ, Iy_\circ)}{1 + d(y, y_\circ)} \end{array}\right\}$$

Theorem 43 *Let for a complete metric space (Y, d), H^+ be metrically equivalent to Hausdorff metric on $CB(Y)$. If the mapping $I : Y \to CB(Y)$ satisfies,*

(i) $\alpha_*(Iy, Iz)\, H^+(Iy, Iz) \le \phi(\mathcal{N}(y, z))$, for every $y, z \in Y$,
(iii) I is α_*-admissible,
(iv) there is $y_0 \in Y$ and $y_1 \in Iy_0$ such as $\alpha(y_0, y_1) \ge 1$,
(v) either (a) $\{y_p\}$ is a sequence in Y such that $\alpha(y_p, y_{p+1}) \ge 1$ for all p, further $y_p \to z \in Y$ as $p \to \infty$, then $\alpha(y_p, z) \ge 1$ for all p,

or (b) I is upper hemi continuous.

Then I has a fixed point.

Proof By condition (iii) and (iv) there is $y_0 \in Y$ and $y_1 \in Iy_0$ such as $\boldsymbol{\alpha}(y_0, y_1) \ge 1$ which implies $\alpha_*(Iy_0, Iy_1) \ge 1$. If $y_1 = y_0$ then $y_0 = y_1 \in Iy_0$ that is $y_0 \in Iy_0$. Which implies y_0 is fixed point of I. Proof is completed. So assume $y_0 \ne y_1$, then by using condition (i) we get

$$\begin{aligned} d(y_1, Iy_1) &\le H^+(Iy_1, Iy_0) \\ &\le \alpha_*(Iy_0, Iy_1)\, H^+(Iy_0, Iy_1) \\ &\le \phi(\mathcal{N}(y_0, y_1)). \end{aligned}$$

If $d(y_1, Iy_1) = 0$ then $y_1 \in Iy_1$ implies that y_1 is a fixed point of I and the proof is done. Assume that $d(y_1, Iy_1) > 0$

$$\begin{aligned} 0 < d(y_1, Iy_1) &\le \phi(\mathcal{N}(y_0, y_1)) \\ &= \phi\left(\max\left\{\begin{array}{c} d(y_0, y_1), d(y_0, Iy_0), \dfrac{d(y_1, Iy_1)}{2}, \\ \dfrac{d(y_0, Iy_1) + d(y_1, y_1)}{2}, \dfrac{d(y_0, y_1)\, d(y_1, Iy_1)}{1 + d(y_0, y_1)} \end{array}\right\}\right) \\ &\le \phi\left(\max\left\{d(y_0, y_1), \frac{d(y_1, Iy_1)}{2}, \frac{d(y_0, Iy_1)}{2}, d(y_1, Iy_1)\right\}\right) \\ &\le \phi\left(\max\left\{d(y_0, y_1), \frac{d(y_1, Iy_1)}{2}, \frac{d(y_0, y_1) + d(y_1, Iy_1)}{2}\right\}\right) \\ &\le \phi\left(\max\left\{d(y_0, y_1), \frac{d(y_1, Iy_1)}{2}\right\}\right). \end{aligned}$$

That is

$$0 < d(y_1, Iy_1) \le \phi\left(\max\left\{d(y_0, y_1), \frac{d(y_1, Iy_1)}{2}\right\}\right).$$

If

$$\max\left\{d(y_0, y_1), \frac{d(y_1, Iy_1)}{2}\right\} = \frac{d(y_1, Iy_1)}{2}$$

then we have that

$$0 < d(y_1, Iy_1) \leq \phi\left(\frac{d(y_1, Iy_1)}{2}\right).$$

As $d(y_1, Iy_1) > 0$ and $\phi \in \Phi$ and $\phi(t) < t$,

$$0 < d(y_1, Iy_1) < \frac{d(y_1, Iy_1)}{2}$$

gives a contradiction. Hence

$$0 < d(y_1, Iy_1) \leq \phi(d(y_0, y_1)).$$

We may choose $y_2 \in Iy_1$ and $q_1 > 1$ such that

$$0 < d(y_1, Iy_1) \leq d(y_1, y_2) < q_1 d(y_1, Iy_1).$$

Thus

$$\begin{aligned} 0 < d(y_1, y_2) &< q_1 d(y_1, Iy_1) \\ &\leq q_1 \phi(d(y_0, y_1)) = q_1 \phi(b_\circ) \end{aligned}$$

where $b_\circ = d(y_0, y_1)$. Note that $y_2 \neq y_1$ and

$$\alpha(y_1, y_2) \geq \alpha_*(Iy_0, Iy_1) \geq 1.$$

Thus $\alpha(y_1, y_2) \geq 1$ and hence $\alpha_*(Iy_1, Iy_2) \geq 1$. As $\phi \in \Phi$,

$$\phi(d(y_1, y_2)) < \phi(q_1 \phi(b_\circ)).$$

If we set

$$q_2 = \frac{\phi(q_1 \phi(b_\circ))}{\phi(d(y_1, y_2))}$$

then $q_2 > 1$. Now if $y_2 \in Iy_2$ then proof is finished. Let $y_2 \notin Iy_2$ then by similar process we obtain

$$0 < d(y_2, Iy_2) \leq \phi(d(y_1, y_2))$$

and $y_3 \in Iy_2$ such that

$$0 < d(y_2, y_3) < q_2 d(y_2, Iy_2) \leq q_2\phi(d(y_1, y_2)) = \phi(q_1\phi(b_\circ)). \tag{7}$$

Note that $y_3 \neq y_2$,

$$\alpha(y_2, y_3) \geq \alpha_*(Iy_1, Iy_2) \geq 1.$$

Thus $\alpha(y_2, y_3) \geq 1$ implies that $\alpha_*(Iy_2, Iy_3) \geq 1$. By 7 we have

$$\phi(d(y_2, y_3)) < \phi^2(q_1(\phi(b_\circ))). \tag{8}$$

If

$$q_3 = \frac{\phi^2(q_1\phi(b_\circ))}{\phi(d(y_2, y_3))},$$

then $q_3 > 1$. Now if $y_3 \in Iy_3$, then the proof is finished. Let $y_3 \notin Iy_3$. Resuming in this way, we get a sequence $\{y_p\}$ in Y and it satisfies $y_{p+1} \in Iy_p$ as $y_{p+1} \neq y_p$, and $\alpha(y_{p+1}, y_{p+2}) \geq 1$,

$$0 < d(y_{p+1}, Iy_{p+1}) \leq \phi(d(y_p, y_{p+1}))$$

which gives

$$0 < d(y_{p+1}, y_{p+2}) \leq \phi^n(q_1\phi(b_\circ)).$$

Now we prove that $\{y_p\}$ is Cauchy sequence. in Y. Now for each $q > p$, we have

$$d(y_p, y_q) \leq \sum_{j=p}^{q-1} d(y_j, y_{j+1}) \leq \sum_{j=p}^{q-1} \phi^{j-1}(q_1\phi(b_\circ)) < \infty.$$

As $\phi \in \Phi$ for any $\varepsilon > 0$ there exists N_1 such as $\sum_{j=p+1}^{\infty} d(y_j, y_{j+1}) < \varepsilon$ for all $p \geq N_1$. Now by using triangular inequality

$$\begin{aligned} d(y_p, y_q) &\leq d(y_p, y_{p+1}) + d(y_{p+1}, y_{p+2}) + \ldots d(y_{q-1}, y_q) \\ &= \sum_{j=p}^{q-1} d(y_j, y_{j+1}) \leq \sum_{j=p}^{q-1} \phi^{j-1}(q_1\phi(b_\circ)) \\ &\leq \sum_{j=p}^{\infty} \phi^{j-1}(q_1\phi(b_\circ)) \end{aligned}$$

for all $p, q \geq N_1$, now for all $\varepsilon > 0$ there exists N_1 such as,

$$d(y_p, y_q) = \sum_{j=p+1}^{\infty} \phi^{j-1}(q_1\phi(b_\circ)) < \varepsilon$$

for all $p, q \geq N_1$. As $\sum_{j=1}^{\infty} \phi^j < \infty$, so this implies $\{y_p\}$ is a Cauchy sequence. so y_p converges to z. Now we want to show that z is a fixed point of I. As $\alpha(y_p, y_{p+1}) \geq 1$ and $y_p \to z$ as $p \to \infty$, so by given condition, $\alpha(y_p, z) \geq 1$ for all $p \in N$. As I is α_*-admissible so $\alpha_*(Iy_p, Iz) \geq 1$. Now we have

$$\begin{aligned} d(y_{p+1}, Iz) &\leq H(Iy_p, Iz) \leq 2H^+(Iy_p, Iz) \\ &\leq 2\phi(\mathcal{N}(y_p, z)) \\ &= 2\phi\left(\max\left\{\begin{array}{c} d(y_p, z), d(y_p, Iy_p), \frac{d(z, Iz)}{2}, \\ \frac{d(y_p, Iz) + d(z, Iy_p)}{2}, \frac{d(y_p, Iy_p)\, d(z, Iz)}{1 + d(y_p, z)} \end{array}\right\}\right) \\ &\leq 2\phi\left(\max\left\{\begin{array}{c} d(y_p, z), d(y_p, y_{p+1}), \frac{d(z, Iz)}{2}, \\ \frac{d(y_p, Iz) + d(z, y_{p+1})}{2}, \frac{d(y_p, y_{p+1})\, d(z, Iz)}{1 + d(y_p, z)} \end{array}\right\}\right) \end{aligned}$$

On taking limit as $p \to \infty$ on both sides of above inequality, we have

$$\begin{aligned} d(z, Iz) &\leq 2\phi\left(\frac{d(z, Iz)}{2}\right) \\ &< 2\frac{d(z, Iz)}{2} = d(z, Iz) \end{aligned}$$

a contradiction so implies that $z \in Iz$. As $\lim_{p\to\infty} y_p = z$, and $y_{p+1} \in I(y_p)$, so by upper hemi continuity, $z \in Iz$. ■

Corollary 44 *Let for a complete metric space (Y, d), H be any Hausdorff metric on $CB(Y)$. If mapping $I : Y \to CB(Y)$ satisfies,*

(i) $\alpha_*(Iy, Iz)\, H(Iy, Iz) \leq \phi(\mathcal{N}(y, z))$,

(ii) I is α_*-admissible,
(iii) there exists $y_0 \in Y$ and $y_1 \in Iy_0$ s.t $\alpha(y_0, y_1) \geq 1$,
(iv) either (a) $\{y_p\}$ is a sequence in Y such that $\alpha(y_p, y_{p+1}) \geq 1$ for all p, further $y_p \to y \in Y$ as $p \to \infty$, then $\alpha(y_p, y) \geq 1$ for all p, or (b) I is upper hemi continuous.

Then I has a fixed point.

Proof Using condition

$$H^+(M, N) \leq H(M, N)$$

in Theorem 43 we get the required result. ■

References

1. Agarwal, R.P., El-Gebeily, M.A., O'Regan, D.: Generalized contractions in partially ordered metric spaces. Applicable Analysis **87**(1), 109–116 (2008)
2. Altun, I., Simsek, H.: Some fixed point theorems on ordered metric spaces and application. Fixed Point Theory and Applications **2010**(1), 1–17 (2010)
3. Asl, J.H., Rezapour, S., Shahzad, N.: On fixed points of α-ψ-contractive multifunctions. Fixed Point Theory Appl. **2012**(1), 212 (2012)
4. Beg, I., Butt, A.R.: Fixed point for set-valued mappings satisfying an implicit relation in partially ordered metric spaces. Nonlinear Analysis: Theory, Methods & Applications **71**(9), 3699–3704 (2009)
5. Blumenthal, L.M.: Theory and Applications of Distance Geometry (1970)
6. Banach, S.: Sur les opérations dans les ensembles abstraits et leur application aux équations intégrales. Fund. Math. **3**(1), 133–181 (1922)
7. Butt, A.R., Beg, I.: Fixed point in partially ordered metric spaces with a metric transform. Journal of Nonlinear Analysis and Application **2017**(1), 66–74 (2017)
8. Chifu, C., Petruşel, G.: Fixed-point results for generalized contractions on ordered gauge spaces with applications. Fixed Point Theory Appl. **2011**(1), 979586 (2011)
9. Corazza, P.: Introduction to metric-preserving functions. The American Mathematical Monthly **106**(4), 309–323 (1999)
10. L. Coroian, On some generalizations of Nadler's contraction principle, Stud. Univ. Babes-Bolyai Math., 60 (2015) 123–133
11. Dobos, J.: Metric Preserving Functions. Stroffek, Kosice (1998)
12. J. Dobos, On modification of the Euclidean metric on reals, Tatra Mt. Math. Publ., 8 (1996) 51–54
13. Dobos, J.: A survey of metric preserving functions, Questions and Answers in General. Topology **13**(2), 129–134 (1995)
14. Dobos, J., Piotrowski, Z.: A note on metric preserving functions. International Journal of Mathematics and Mathematical Sciences **19**(1), 199–200 (1996)
15. Dobos, J., Piotrowski, Z.: Some remarks on metric preserving functions. Real Anal. Exchange 317–320 (1993)
16. Das, P.P.: Metricity preserving transforms. Pattern Recognition Letters **10**(2), 73–76 (1989)
17. Edelstein, M.: An extension of Banach's contraction principle. Proceedings of the American Mathematical Society **12**(1), 07–10 (1961)
18. Harjani, J., Sadarangani, K.: Generalized contractions in partially ordered metric spaces and applications to ordinary differential equations. Nonlinear Analysis: Theory, Methods & Applications **72**(3–4), 1188–1197 (2010)
19. Hu, T., Kirk, W.A.: Local contractions in metric spaces. Proceedings of the American Mathematical Society **68**(1), 121–124 (1978)
20. Jachymski, J.: Equivalent conditions for generalized contractions on (ordered) metric spaces. Nonlinear Analysis: Theory, Methods & Applications **74**(3), 768–774 (2011)
21. Kirk, W.A., Shahzad, N.: Remarks on metric transforms and fixed-point theorems. Fixed Point theory and Applications **2013**(1), 106 (2013)
22. Karapinar, E., Samet, B.: Generalized α-ψ contractive type mappings and related fixed point theorems with applications. Abstract Appl. Anal. Hindawi Limited (2012)
23. P. Pongsriiam, I. Termwuttinpong, On metric-preserving functions and fixed point theorems. Fixed Point Theory Appl. **2014**(1), 1-14 (2014)
24. O. Popescu, A new type of multivauled contractive operators, Bull. Sci. Math., 137 (2013) 30–44
25. Petrusel, A., Rus, I.A., Serban, M.A.: The role of equivalent metrics in fixed point theory. Topological Methods in Nonlinear Analysis **41**(1), 85–112 (2013)
26. Z. Piotrowski, R.W. Vallin, Functions which preserve Lebesgue spaces, Commentationes Mathematicae-Prace Matematyczne-Seria 1-Rocz Polsk Towarz Matematycz **43**(2), 249–256 (2003)

27. I. Pokorny, Some remarks on metric-preserving functions of several variables, Tatra Mt. Math. Publ., 8 (1996) 89–92
28. Pathak, H.K., Shahzad, N.: A generalization of Nadler's fixed point theorem and its application to nonconvex integral inclusions. Topological Methods in Nonlinear Analysis **41**(1), 207–227 (2013)
29. Suzuki, T.: A generalized Banach contraction principle that characterizes metric completeness. Proceedings of the American mathematical Society **136**(5), 1861–1869 (2008)
30. Samet, B., Vetro, C., Vetro, P.: Fixed point theorems for $\alpha - \psi$-contractive type mappings. Nonlinear Anal. Theory, Methods Appl. **75**(4), 2154–2165 (2012)
31. Nadler, S.B.: Multi-valued contraction mappings. Pacific Journal of Mathematics **30**(2), 475–488 (1969)
32. Termwuttipong, I., Oudkam, P.: Total boundedness, completeness and uniform limits of metric-preserving functions. Italian Journal of Pure and Applied Mathematics **18**, 187 (2005)
33. R.W. Vallin, Continuity and differentiability aspects of metric preserving functions, Real Anal. Exchange, 25(2) (1999) 849–868
34. Wilson, W.A.: On certain types of continuous transformations of metric spaces. American Journal of Mathematics **57**(1), 62–68 (1935)

Approximation of Signals Via Different Summability Means with Effects of Gibbs Phenomenon

Bidu Bhusan Jena, Susanta Kumar Paikray, and Hemen Dutta

Abstract The paper aims to investigate the notions of the deferred Cesàro, deferred Nörlund and their product summability means of the Fourier series. We estimated the degree of approximation of signal functions belonging to a generalized Lipschitz class by using these notions, and also established some new fundamental approximation theorems in classical sense. Moreover, we introduced the statistical versions of these notions, and demonstrated some Korovkin-type approximation results for trigonometric test functions over a Banach space. Furthermore, in view of our proposed means, we presented some examples demonstrating that the statistical versions of approximation results are stronger than the classical versions. Finally, as regards to the convergence of the Fourier series, the effect of the Gibbs phenomenon has been presented via our proposed means.

Keywords Statistical convergence · Statistical deferred Nörlund summability · Positive linear operators; Sequences of real variables; Banach space · Korovkin-type approximation theorems

2010 Mathematics Subject Classification Primary 40A05 · 40G15 · Secondary 41A36

1 Introduction and Preliminaries

The theory of classical approximation is an extraordinarily wide field covering different areas of applied mathematics. The estimation of trigonometric functions has incredible scientific interests as well as incredible functional significance. As already

B. B. Jena · S. K. Paikray (✉)
Department of Mathematics, Veer Surendra Sai University of Technology, 768018 Burla, Odisha, India
e-mail: skpaikray_math@vssut.ac.in

H. Dutta
Department of Mathematics, Gauhati University, 781014 Guwahati, Assam, India

J. Singh et al. (eds.), *Methods of Mathematical Modelling and Computation for Complex Systems*, Studies in Systems, Decision and Control 373,
https://doi.org/10.1007/978-3-030-77169-0_16

mentioned in [16], the L_p space in general; however, the L_2 and L_1 spaces are specifically assumed as inherent part of the theory of signal functions. It is well-known that the classical approximation theory has started from the well established Weierstrass approximation theory. These approximations have found their wide applications in signal analysis (see [15]), in general and specifically in digital signal processing system. Investigation on signals (or time) has many significant aspects as they pass on data (or characteristics) of some phenomenon. The engineers and scientists use the properties of Fourier approximation for outlining digital filters and signals. Particularly, Psarakis and Moustakides [16] exhibited another L_2 based technique for outlining the finite impulse response digital filters and got comparing optimum approximations having enhanced execution. Recently, Diger *et al.* [5], and Mittal and Singh [10] have obtained numerous nice results on the theory of approximation utilizing Nörlund and Riesz means of summability techniques with monotonicity on the rows of the corresponding matrix T (a digital filter) by using Cesàro mean of summability method presented earlier by Armitage and Maddox (see [2]).

Moreover, the notion of statistical convergence was initially studied by two eminent mathematicians Fast [6] and Steinhaus [20]. Recently, statistical approximation (or statistical convergence) has been an active research area because it is stronger than the classical approximation (or classical convergence) and such a result with its several applications are discussed in the various fields of pure and applied mathematics such as Fourier Analysis, Number Theory and Approximation Theory etc. For more details, see [3, 8, 18, 19].

Let $f(x) \in L_p[0, 2\pi]$ $(p \geqq 1)$ be a 2π periodic signal function and its Fourier series is given by

$$f(x) = \frac{a_0}{2} + \sum_{\mu=1}^{\infty}(a_\mu \cos \mu x + b_\mu \sin \mu x) = \sum \psi_\mu(x). \tag{1}$$

Let $s_n(f)$ be the nth partial sum of the series (1), given by

$$s_n(f) = \frac{a_0}{2} + \sum_{\mu=1}^{n}(a_\mu \cos \mu x + b_\mu \sin \mu x). \tag{2}$$

Now, the integral modulus of continuity of f is defined by

$$\omega_p(f; \delta) = \sup_{0<|h|\leqq\delta} \left\{ \frac{1}{2\pi} \int_0^{2\pi} |f(x+h) - f(x)|^p dx \right\}^{\frac{1}{p}}. \tag{3}$$

If

$$\omega_p(f; \delta) = O(\delta^\alpha) \quad (0 < \alpha \leqq 1)$$

then

$$f \in Lip(\alpha, p) \quad (p \geqq 1).$$

Furthermore, if $p \to \infty$, $Lip(\alpha, p)$ class reduces to the $Lip(\alpha)$ class.

Recalling here the L_p-norm $\|.\|_{L_p}$, we have

$$\|f\|_{L_p} = \left\{ \frac{1}{2\pi} \int_0^{2\pi} |f(x)|^p dx \right\}^{\frac{1}{p}} \quad \left(\because f \in L_p; \quad p \geqq 1\right)$$

and the L_∞- norm of a function f over $\mathbb{R}$ is defined by,

$$\|f\|_{L_\infty} = \sup\{|f(x)| : x \in \mathbb{R}\}.$$

The degree of approximation of a function f over $\mathbb{R}$ by a trigonometric polynomial (t_n) of degree n under the supremum norm $\|.\|_{L_\infty}$ is defined by Zygmund (see [22]) and is given by,

$$\|t_n - f\|_{L_\infty} = \sup\{|t_n - f(x)| : x \in \mathbb{R}\}$$

and the error (E_n) of a function $f \in L_p$ out of the approximation is defined by

$$E_n = \min_n \|t_n - f\|_{L_p}.$$

Now we recall, the formula of Abel's transformation of the following form

$$\sum_{k=m}^{n} u_k v_k = \sum_{k=m}^{n-1} U_k(v_k - v_{k+1}) - U_{m-1} v_m + U_n v_n \quad (0 \leqq m < n), \tag{4}$$

where

$$U_k = u_0 + u_1 + \cdots + u_k \quad (k \geqq 0; \quad U_{-1} = 0).$$

If $v_m, v_{m+1}, \ldots, v_n$ are non-negative and non-increasing, the left hand side of (4) does not exceed

$$\left| 2v_m \max_{m-1 \leqq k \leqq n} U_k \right|.$$

Moreover,

$$\left| \sum_{k=m}^{n} u_k v_k \right| = \max |U_k| \left\{ \sum_{k=m}^{n-1} (v_k - v_{k+1}) - v_m + v_n \right\}$$
$$= 2v_m \max |U_k|.$$

Next recalling the monotonicity, a non-negative sequence (c_n) is almost monotonically decreasing (*resp.* increasing) if there exists a constant $\mathfrak{K} = \mathfrak{K}(c_m)$, depending on the sequence (c_m) only, such that,

$$c_n \geqq \mathfrak{K} c_m \quad (resp.\ c_n \leqq \mathfrak{K} c_m).$$

A non-negative sequence (c_n) which is either almost increasing sequence (AIS) or almost decreasing sequence (ADS) is called an almost monotone sequence (AMS).

2 Definitions and Motivation

Let $(\mu_n)_{n\in\mathbb{N}}$ and $(\xi_n)_{n\in\mathbb{N}}$ be strictly increasing sequences of positive integers such that $(\mu_n) < (\xi_n)$ and $(\xi_n) = \infty$ as $n \to \infty$.

Now we define the deferred Cesàro (DC) summability mean for the sequence of partial sum of the Fourier series as

$$\mathfrak{L}_n(f) = \frac{1}{(\xi_n) - (\mu_n)} \sum_{\upsilon=\mu_n+1}^{\xi_n} s_\upsilon(f), \tag{5}$$

where (s_υ) is the nth partial sum of the Fourier series. Also, it is regular under the usual conditions (see [1]). Moreover, for $(\mu_n) = 0$ and $(\xi_n) = n$, $\mathfrak{L}_n(f)$ mean reduces to the Cesàro $(C, 1)$ summability mean of order 1.

Definition 1 Let (μ_n) and (ξ_n) be strictly increasing sequences of positive integers. A Fourier series $\sum \psi_\mu(x)$ is statistically deferred Cesàro (DC) summable to a function $s(x)$ if, for every $\epsilon > 0$,

$$\{m : m \in \mathbb{N} \quad \text{and} \quad |\mathfrak{L}_m(f) - s(x)| \geqq \epsilon\}$$

has zero natural density, that is,

$$\lim_{n\to\infty} \frac{1}{n} |\{m : m \leqq n \quad \text{and} \quad |\mathfrak{L}_m(f) - s(x)| \geqq \epsilon\}| = 0.$$

In this case, we write

$$\text{stat}_{\text{DC}} \mathfrak{L}_n(f) = s(x).$$

Next, we define the deferred Nörlund (DN) summability mean for the sequence of partial sum of the Fourier series as

$$\mathfrak{L}_n^{**}(f) = \frac{1}{P_n} \sum_{\upsilon=\mu_n+1}^{\xi_n} p_{\xi_n-\upsilon} s_\upsilon(x) \tag{6}$$

where

$$P_n = \sum_{v=\mu_n+1}^{\xi_n} p_v.$$

Definition 2 Let (μ_n) and (ξ_n) be strictly increasing sequences of positive integers and let (p_n) be a positive real sequence. A Fourier series $\sum \psi_\mu(x)$ is statistically deferred Nörlund (DN) summable to a function $s(x)$ if, for every $\epsilon > 0$,

$$\{m : m \leqq n \quad \text{and} \quad |\mathfrak{L}_m^{**}(f) - s(x)| \geqq \epsilon\}$$

has zero natural density, that is,

$$\lim_{n\to\infty} \frac{1}{n} |\{m : m \leqq n \quad \text{and} \quad |\mathfrak{L}_m^{**}(f) - s(x)| \geqq \epsilon\}| = 0.$$

In this case, we write

$$\text{stat}_{\text{DN}} \mathfrak{L}_n^{**}(f) = s(x).$$

Subsequently, we define the deferred Cesàro (DC) and deferred Nörlund (DN) product summability (DCN) mean for the sequence of partial sum of the Fourier series of the form

$$\begin{aligned}
\mathfrak{T}_n^*(f) &= \frac{1}{(\xi_n - \mu_n)} \sum_{v=\mu_n+1}^{\xi_n} \mathfrak{L}_{kn}^{**}(f) \\
&= \frac{1}{(\xi_n - \mu_n)} \sum_{v=\mu_n+1}^{\xi_n} \left[\frac{1}{P_n} \sum_{v=\mu_n+1}^{\xi_n} p_{\xi_n - v} s_v(x) \right] \\
&= \frac{1}{(\xi_n - \mu_n)} \sum_{v=\mu_n+1}^{\xi_n} a_{\xi_n, v} \quad \left(\because a_{\xi_n, v} = 0 \; \forall \; \xi_n > v\right).
\end{aligned} \tag{7}$$

Definition 3 Let (μ_n) and (ξ_n) be strictly increasing sequences of positive integers and let (p_n) be a positive real sequence. A Fourier series $\sum \psi_\mu(x)$ is statistically deferred Cesàro and deferred Nörlund (DCN) product summable to a function $s(x)$ if, for every $\epsilon > 0$,

$$\{m : m \leqq n \quad \text{and} \quad |\mathfrak{T}_m^*(f) - s(x)| \geqq \epsilon\}$$

has zero natural density, that is,

$$\lim_{n\to\infty} \frac{1}{n} |\{m : m \leqq n \quad \text{and} \quad |\mathfrak{T}_m^*(f) - s(x)| \geqq \epsilon\}| = 0.$$

In this case, we write

$$\text{stat}_{\text{DCN}}\mathfrak{T}_n^*(f) = s(x).$$

Many researchers like, Parida et al. [11, 12], Pradhan et al. [13, 14] and a few others used different summability means to determine the degree of approximation of trigonometric polynomials. Recently, Deger et al. [5] and Mittal and Singh [10] used the more general Cesàro summability mean (C_λ) (see, Armitage and Maddox [2]) and established a result on the approximation of signals by trigonometric polynomials in L_p- norm. Very recently, Jena et al. [7] proved an approximation result via general matrix summmability mean under the effect of the Gibbs Phenomenon.

Motivated essentially by the above-mentioned investigations and results, we estimate the degree of approximation of signal functions belonging to the generalized Lipschitz class based on deferred Cesàro, deferred Nörlund and their product summability means of the Fourier series, and accordingly we prove some new fundamental approximation theorems. Moreover, based on our proposed means, we also present the statistical versions of Korovkin-type approximation results for trigonometric test functions of the sequences of positive linear operators over a Banach space, and demonstrate that our statistical versions of approximation results are stronger than the classical versions of approximation theorems. Finally, as regards to the convergence of the Fourier series, the effect of the Gibbs Phenomenon has been discussed here via our proposed means. In particular, the graphs of nth partial sum and $\mathfrak{L}_n(f)$, $\mathfrak{L}_n^{**}(f)$ and $\mathfrak{T}_n^*(f)$ means of the Fourier series are plotted by using the Matlab and are compared to support the proposed investigation.

3 Classical Approximation Theorems

Now we prove the following classical approximation theorems based upon our proposed means.

Theorem 1 *If $f \in Lip(\alpha, p)$ and (p_n) is positive, and also let (μ_n) and (ξ_n) be strictly increasing sequences of positive integers and suppose if one the following conditions hold:*

(i) $p > 1$, $\alpha \in (0, 1)$ *and* (p_n) *is ADS;*
(ii) $p > 1$, $\alpha \in (0, 1)$, (p_n) *is ADS and* $(\xi_n + 1)p_{\xi_n} = O(P_{\xi_n})$,

then

$$\|\mathfrak{L}_n^{**}(f) - f\|_{L_p} = O\left(\frac{1}{(\xi_n)^\alpha}\right),$$

where $\mathfrak{L}_n^{**}(f)$ is the DN mean.

Theorem 2 *If $f \in Lip(\alpha, p)$ and let $(\mu_n)_{n\in\mathbb{N}}$ and $(\xi_n)_{n\in\mathbb{N}}$ be strictly increasing sequences of positive integers and the following condition holds:*

$$p > 1,\ \alpha \in (0,1)\ \ and\ \ (\xi)_n + 1) = O(1),$$

then

$$\|\mathfrak{L}_n(f) - f\|_{L_p} = O\left(\frac{1}{(\xi_n)^\alpha}\right),$$

where $\mathfrak{L}_n(f)$ *is the DC mean.*

Theorem 3 *Let* $f \in Lip(\alpha, p)$, $(0 < \alpha < 1)$. *If the conditions,*

$$\sum_{v=\mu_n+1}^{\xi_n-1} |\Delta_v a_{\xi_n,v}| = O\left(\frac{1}{\xi_n}\right)$$

and

$$(\xi_n + 1)a_{\xi_n,\xi_n} = O(1)$$

hold true, then

$$\|\mathfrak{T}_n^*(f) - f\|_{L_p} = O\left(\frac{1}{(\xi_n)^{1+\alpha}}\right),$$

where $\mathfrak{T}_n^*(f)$ *is the DCN product mean.*

Each of the following Lemmas will be needed in our present work.

Lemma 1 (see [17]) *If* $f \in Lip(\alpha, p)$, *for* $\alpha \in (0,1]$ *and* $p > 1$, *then*

$$\|s_n(f) - f\|_{L_p} = O\left(\frac{1}{n^\alpha}\right). \tag{8}$$

Lemma 2 (see [17]) *If* $f \in Lip(\alpha, p)$, $\alpha \in (0,1)$, *then*

$$\|\mathfrak{L}_n(f) - f\|_{L_p} = O\left(\frac{1}{n^\alpha}\right). \tag{9}$$

Lemma 3 *If* $f \in Lip(\alpha, p)$, $\alpha \in (0,1)$, *and* (p_n) *is AIS (resp. ADS), and suppose*

$$(\xi_n + 1)p_{\xi_n} = O(P_{\xi_n}),$$

then

$$\sum_{v=\mu_n+1}^{\xi_n} v^{-\alpha} p_{\xi_n - k} = O(\xi_n)^{-\alpha} P_{\xi_n}$$

holds.

Proof Let (p_n) be AIS and let $(\xi_n + 1)p_{\xi_n} = O(P_{\xi_n})$ holds, then

$$\begin{aligned}
\sum_{v=\mu_n+1}^{\xi_n} v^{-\alpha} p_{\xi_n-k} &= \sum_{v=\mu_n+1}^{r} v^{-\alpha} p_{\xi_n-k} + \sum_{v=r+1}^{\xi_n} v^{-\alpha} p_{\xi_n-k} \\
&\leqq \mathcal{K} p_{\xi_n-r} \sum_{v=\mu_n+1}^{r} v^{-\alpha} + (r+1)^{-\alpha} \sum_{v=r+1}^{\xi_n} p_{\xi_n-k} \\
&\leqq \mathcal{K} p_{\xi_n-r} \sum_{v=\mu_n+1}^{\xi_n} v^{-\alpha} + (r+1)^{-\alpha} \sum_{v=\mu_n}^{\xi_n} p_{\xi_n-k} \\
&\leqq \mathcal{K} p_{\xi_n-r} (\xi_n)^{1-\alpha} + O(\xi_n)^{-\alpha} P_{\xi_n} \\
&= O(\xi_n)^{-\alpha} P_{\xi_n}.
\end{aligned}$$

Similarly, in other hand we immediately get, if $f \in Lip(\alpha, p)$, $\alpha \in (0, 1)$, and (p_n) is ADS, and suppose that

$$(\xi_n + 1)p_{\xi_n} = O(P_{\xi_n}),$$

then

$$\sum_{v=\mu_n+1}^{\xi_n} v^{-\alpha} p_{\xi_n-k} = O(\xi_n)^{-\alpha} P_{\xi_n}$$

holds. □

Proof **of Theorem** 1 Initially, we wish to prove under conditions (i) and (ii) together, we have

$$\mathfrak{L}_n^{**}(f) - f = \frac{1}{P_{\xi_n}} \sum_{v=\mu_n+1}^{\xi_n} p_{\xi_n-v}\{s_v(x) - f(x)\}. \tag{10}$$

Next, in view of Lemmas 1 and 3, we have

$$\begin{aligned}
\|\mathfrak{L}_n^{**}(f) - f\|_{L_p} &\leqq \frac{1}{P_{\xi_n}} \sum_{v=\mu_n+1}^{\xi_n} p_{\xi_n-v} \|s_v(x) - f(x)\|_{L_p} \\
&= \frac{1}{P_{\xi_n}} \sum_{v=\mu_n+1}^{\xi_n} p_{\xi_n-v} \|s_v(x) - f(x)\|_{L_p} \\
&= \frac{1}{P_{\xi_n}} \sum_{v=\mu_n+1}^{\xi_n} p_{\xi_n-v} (v^{-\alpha}) \\
&= O(\xi_n)^{-\alpha}.
\end{aligned}$$

This completes the proof of Theorem. □

Proof of Theorem 2 The proof of the Theorem 2 is similar to the proof of Theorem 1. We, therefore, choose to skip the details involved. □

Proof of Theorem 3 Using the given conditions we have,

$$
\begin{aligned}
\mathfrak{T}_n^*(f) - f &= \frac{1}{\xi_n - \mu_n} \sum_{v=\mu_n+1}^{\xi_n} a_{\xi_n,v}(s_v(f) - f) \\
&= \frac{1}{\xi_n - \mu_n} \sum_{v=\mu_n+1}^{\xi_n-1} (a_{\xi_n,v} - a_{\xi_n,v+1}) \\
&\quad \cdot \sum_{j=\mu_n+1}^{\xi_n} (s_j(f) - f) + a_{\xi_n,\xi_n} \sum_{v=\mu_n+1}^{\xi_n} (s_v(f) - f) \\
&= \frac{1}{\xi_n - \mu_n} \sum_{v=\mu_n+1}^{\xi_n-1} a_{\xi_n,v} - a_{\xi_n,v+1})(v+1) \\
&\quad \cdot (\mathfrak{L}_n(f) - f) + a_{\xi_n,\xi_n}(1+\xi_n)(\mathfrak{L}_n(f) - f).
\end{aligned}
$$

$$
\begin{aligned}
\|\mathfrak{T}_n^*(f) - f\|_{L_p} \leqq \frac{1}{\xi_n - \mu_n} &\sum_{v=\mu_n+1}^{\xi_n-1} (a_{\xi_n,v} - a_{\xi_n,v+1})(v+1) \\
&\cdot \|\mathfrak{L}_n(f) - f\|_{L_p} + a_{\xi_n,\xi_n}(1+\xi_n)\|\mathfrak{L}_n(f) - f\|_{L_p}
\end{aligned}
$$

$$
\begin{aligned}
&\leqq \frac{1}{\xi_n - \mu_n} \sum_{v=\mu_n+1}^{\xi_n-1} |a_{\xi_n,v} - a_{\xi_n,v+1}|(1+v)^{1-\alpha} \\
&\quad + a_{\xi_n,\xi_n}(1+\xi_n)^{1-\alpha} \ (\because \text{ Lemma 2}) \\
&\leqq \frac{(1+\xi_n)}{(1+\xi_n)^{1+\alpha}} \left(\sum_{v=\mu_n+1}^{\xi_n-1} |a_{\xi_n,v} - a_{\xi_n,v+1}| + a_{\xi_n,\xi_n} \right) \\
&= O\left(\frac{1}{(\xi_n)^{1+\alpha}} \right).
\end{aligned}
$$

This establishes the theorem. □

Next, we establish the following corollary in view of the consequence of the Theorem 3.

Corollary 1 *If $p \to \infty$ and $0 < \alpha < 1$, then the generalized $Lip(\alpha, p)$ class reduces to the class $Lip(\alpha)$, and the degree of approximation of the function f belongs to the $Lip(\alpha)$-class is given by*

$$\|\mathfrak{T}_n^*(f) - f\|_{L_\infty} = O\left(\frac{1}{\xi_n^\alpha}\right).$$

Proof For $p \to \infty$ and $0 < \alpha < 1$, we have

$$\begin{aligned}\|\mathfrak{T}_n^*(f) - f\|_{L_\infty} &= \sup\left\{\left|\mathfrak{T}_n^*(f) - f\right| : 0 \leqq x \leqq 2\pi\right\} \\ &= O\left(\frac{1}{(\xi_n)^\alpha}\right).\end{aligned}$$

This establishes the corollary. □

4 Statistical Korovkin-Type Approximation Results

The Korovkin-type theorems are very useful in the convergence analysis in which the approximation of functions is considered by certain sequences of functions, that is, the continuous functions are approximated by polynomials. Now various researchers developed these concepts by different settings such as measurable convergence, statistical convergence, probability convergence, lacunary convergence, ideal convergence and so on. For current research works in this direction, see [4, 18, 19, 21].

In fact, here we demonstrate some new Korovkin-type approximation results via our proposed statistical summability means.

Let $C_{2\pi}(\mathbb{R})$ be the space of all 2π-periodic real-valued continuous functions defined on $\mathbb{R}$ and suppose that $\mathcal{L} : C_{2\pi}(\mathbb{R}) \to C_{2\pi}(\mathbb{R})$ be a linear operator.

We say that the operator $\mathcal{L}$ is a sequence of positive linear operator provided that

$$\mathcal{L}(f; x) \geqq 0 \quad \text{whenever} \quad f \geqq 0, f \in C_{2\pi}(\mathbb{R}).$$

It is also known that $C_{2\pi}(\mathbb{R})$ is a Banach space. For $f \in C_{2\pi}(\mathbb{R})$, the norm of the function f, denoted by $\|f\|$, is given by

$$\|f\|_{2\pi} = \sup_{x \in [0,2\pi]} |f(x)|.$$

Now we first propose the following result by using the notion of statistical deferred Cesàro summability mean.

Result 1 Let

$$\mathcal{L}_m : C_{2\pi}(\mathbb{R}) \to C_{2\pi}(\mathbb{R})$$

be a sequence of positive linear operators. Then, for all $f \in C_{2\pi}(\mathbb{R})$,

$$\text{stat}_{\text{DC}} \lim_{m \to \infty} \|\mathcal{L}_m(f; x) - f(x)\|_{2\pi} = 0 \tag{11}$$

if and only if

$$\text{stat}_{\text{DC}} \lim_{m\to\infty} \|\mathcal{L}_m(1; x) - 1\|_{2\pi} = 0, \tag{12}$$

$$\text{stat}_{\text{DC}} \lim_{m\to\infty} \|\mathcal{L}_m(\cos x; x) - \cos x\|_{2\pi} = 0 \tag{13}$$

and

$$\text{stat}_{\text{DC}} \lim_{m\to\infty} \|\mathcal{L}_m(\sin x; x) - \sin x\|_{2\pi} = 0. \tag{14}$$

We present below an illustrative example (see Example 1 below) for the sequences of positive linear operators that does not satisfy our classical version of the result, that is, approximation Theorem 2; but it satisfies our statistical version of the result (approximation Result 1). Consequently, our Result 1 is a stronger approach than our usual approximation theorem.

We recall here the *Fejér convolution operators*, let $f \in C_{2\pi}(\mathbb{R})$ and also let the Fourier series of f at $t = x$ be given by the Eq. (1) with its nth partial sum is given by (2), and the deferred Cesàro mean of $(f_n(x))$ is already mentioned in (5), that is,

$$\mathfrak{L}_n(f; x) = \frac{1}{\xi_n - \mu_n} \sum_{v=\mu_n+1}^{\xi_n} s_v(f; x).$$

Furthermore, if $\mu_n = 0$ and $\xi_n = n$, by simple calculation, we obtain

$$\begin{aligned}
\mathfrak{L}_n^*(f; x) &= \frac{1}{2\pi} \int_{-\pi}^{\pi} f(t) \frac{1}{n+1} \sum_{v=0}^{n} \frac{\sin\left[\frac{(2v+1)(x-t)}{2}\right]}{\sin\left[\frac{(x-t)}{2}\right]} dt \\
&= \frac{1}{2\pi} \int_{-\pi}^{\pi} f(t) \frac{1}{n+1} \sum_{v=0}^{n} \frac{\sin^2\left[\frac{(n+1)(x-t)}{2}\right]}{\sin^2\left[\frac{(x-t)}{2}\right]} dt \\
&= \frac{1}{2\pi} \int_{-\pi}^{\pi} f(t)\phi_n(x-t)dt,
\end{aligned}$$

where

$$\phi_n(x) = \begin{cases} \dfrac{\sin^2\left[\frac{(n+1)(x-t)}{2}\right]}{(n+1)\sin^2\left[\frac{(x-t)}{2}\right]} & (\text{x is a multiple of } 2\pi) \\[2ex] n+1 & (\text{x is not a multiple of } 2\pi). \end{cases}$$

The sequence $\{\phi_n(x) : n \in \mathbb{N}\}$ is a positive kernel which is known as the *Fejér kernel* and the corresponding operators $\mathfrak{L}_n^*(f; x)$ are called the *Fejér convolution operators*.

Next, we use the Fejér convolution operators in connection with Example 1 (given below).

Example 1 Let $\mathcal{L}_m : C_{2\pi}(\mathbb{R}) \to C_{2\pi}(\mathbb{R})$ be defined by

$$\mathcal{L}_m(f; x) = [1 + x_m]\mathfrak{L}_n^*(f; x) \quad (f \in C_{2\pi}(\mathbb{R}), \tag{15}$$

where (x_m) is a sequence of the form; for m is odd

$$x_m = \begin{cases} 1 & (m = j^2;\ j \in \mathbb{N}) \\ 0 & (\text{otherwise}). \end{cases} \tag{16}$$

and also, for m is even

$$x_m = \begin{cases} -1 & (m = j^2;\ j \in \mathbb{N}) \\ 0 & (\text{otherwise}). \end{cases} \tag{17}$$

Here, for $a_m = 2m$ and $b_m = 4m$, the sequence (x_m) is not summable in classical sense. However, it is statistically deferred Cesàro summable to 0. It is easily observed that

$$\mathcal{L}_m(1; x) = [1 + x_m]1 = 1 + x_m,$$

$$\mathcal{L}_m(\cos s; x) = [1 + x_m]\frac{m-1}{m}\cos x$$

and

$$\mathcal{L}_m(\sin s; x) = [1 + x_m]\frac{m-1}{m}\sin x,$$

so that we have

$$\text{stat}_{\text{DC}} \lim_{m\to\infty} \|\mathcal{L}_m(1; x) - 1\|_{2\pi} = 0,$$

$$\text{stat}_{\text{DC}} \lim_{m\to\infty} \|\mathcal{L}_m(\cos x; x) - \cos x\|_{2\pi} = 0$$

and

$$\text{stat}_{\text{DC}} \lim_{m\to\infty} \|\mathcal{L}_m(\sin x; x) - \sin x\|_{2\pi} = 0,$$

that is, the sequence $\mathcal{L}_m(f;x)$ satisfies the conditions (12) to (14). Therefore, by Result 1, we have

$$\text{stat}_{\text{DC}} \lim_{m\to\infty} \|\mathcal{L}_m(f;x) - f\|_{2\pi} = 0.$$

Hence, (x_m) is statistically deferred Cesàro summable, but it is not classically summable, so we conclude that our classical version of the result (approximation Theorem 2) is not valid for the operators defined by (15); whereas our statistical version, that is, approximation Result 1 still works for the operators defined by (15).

Next, we propose the following result by using the notion of the statistical deferred Nörlund summability mean with associated example.

Result 2 Let

$$\mathcal{L}'_m : C_{2\pi}(\mathbb{R}) \to C_{2\pi}(\mathbb{R})$$

be a sequence of positive linear operators. Then, for all $f \in C_{2\pi}(\mathbb{R})$,

$$\text{stat}_{\text{DN}} \lim_{m\to\infty} \|\mathcal{L}'_m(f;x) - f(x)\|_{2\pi} = 0 \tag{18}$$

if and only if

$$\text{stat}_{\text{DN}} \lim_{m\to\infty} \|\mathcal{L}'_m(1;x) - 1\|_{2\pi} = 0, \tag{19}$$

$$\text{stat}_{\text{DN}} \lim_{m\to\infty} \|\mathcal{L}'_m(\cos x;x) - \cos x\|_{2\pi} = 0 \tag{20}$$

and

$$\text{stat}_{\text{DN}} \lim_{m\to\infty} \|\mathcal{L}'_m(\sin x;x) - \sin x\|_{2\pi} = 0. \tag{21}$$

Example 2 Let $\mathcal{L}'_m : C_{2\pi}(\mathbb{R}) \to C_{2\pi}(\mathbb{R})$ be defined by

$$\mathcal{L}'_m(f;x) = [1 + x'_m]\mathfrak{L}^*_n(f;x) \quad (f \in C_{2\pi}(\mathbb{R}), \tag{22}$$

where (x_m) is a sequence defined by; for m is odd

$$x'_m = \begin{cases} m+1 & (m = j^2;\ j \in \mathbb{N}) \\ 0 & (\text{otherwise}). \end{cases} \tag{23}$$

and also, for m is even

$$x'_m = \begin{cases} -m & (m = j^2;\ j \in \mathbb{N}) \\ 0 & (\text{otherwise}). \end{cases} \tag{24}$$

Here, $a_m = 2m$ and $b_m = 4m$, the sequence (x'_m) is not classically summable in deferred Nörlund sense; however, it is statistically deferred Nörlund summable to 0. It is easily observed that

$$\mathcal{L}'_m(1; x) = [1 + x'_m]1 = 1 + x'_m,$$

$$\mathcal{L}'_m(\cos s; x) = [1 + x'_m]\frac{m-1}{m}\cos x$$

and

$$\mathcal{L}'_m(\sin s; x) = [1 + x'_m]\frac{m-1}{m}\sin x,$$

so that we have

$$\text{stat}_{\text{DN}} \lim_{m\to\infty} \|\mathcal{L}'_m(1; x) - 1\|_{2\pi} = 0,$$
$$\text{stat}_{\text{DN}} \lim_{m\to\infty} \|\mathcal{L}'_m(\cos x; x) - \cos x\|_{2\pi} = 0$$

and

$$\text{stat}_{\text{DN}} \lim_{m\to\infty} \|\mathcal{L}'_m(\sin x; x) - \sin x\|_{2\pi} = 0,$$

that is, the sequence $\mathcal{L}'_m(f; x)$ satisfies the conditions (19) to (21). Therefore, by Result 2, we have

$$\text{stat}_{\text{DN}} \lim_{m\to\infty} \|\mathcal{L}'_m(f; x) - f\|_{2\pi} = 0.$$

Hence, it is statistically deferred Nörlund summable; but it is not classically summable, so we conclude that our classical part of the approximation result, that is, Theorem 1 is not valid for the operators defined by (22); whereas our statistical part of approximation Result, that is, 2 still works for the operators defined by (22).

Finally, we now propose the following result by using the notion of the statistically product DCN (that is, deferred Cesàro and deferred Nörlund product) summability mean with an illustrated example.

Result 3 Let

$$\mathcal{L}''_m : C_{2\pi}(\mathbb{R}) \to C_{2\pi}(\mathbb{R})$$

be a sequence of positive linear operators. Then, for all $f \in C_{2\pi}(\mathbb{R})$,

$$\text{stat}_{\text{DCN}} \lim_{m\to\infty} \|\mathcal{L}''_m(f; x) - f(x)\|_{2\pi} = 0 \tag{25}$$

if and only if

$$\text{stat}_{\text{DCN}} \lim_{m\to\infty} \|\mathcal{L}''_m(1;x) - 1\|_{2\pi} = 0, \tag{26}$$

$$\text{stat}_{\text{DCN}} \lim_{m\to\infty} \|\mathcal{L}''_m(\cos x;x) - \cos x\|_{2\pi} = 0 \tag{27}$$

and

$$\text{stat}_{\text{DCN}} \lim_{m\to\infty} \|\mathcal{L}''_m(\sin x;x) - \sin x\|_{2\pi} = 0. \tag{28}$$

Example 3 Let $\mathcal{L}''_m : C_{2\pi}(\mathbb{R}) \to C_{2\pi}(\mathbb{R})$ be defined by

$$\mathcal{L}''_m(f;x) = [1 + x''_m]\mathfrak{L}^*_n(f;x) \quad (f \in C_{2\pi}(\mathbb{R}), \tag{29}$$

where (x''_m) is a sequence defined as; for m is odd

$$x''_m = \begin{cases} \frac{m+1}{2} & (m = j^2;\ j \in \mathbb{N}) \\ 0 & (\text{otherwise}). \end{cases} \tag{30}$$

and also, for m is even

$$x''_m = \begin{cases} \frac{-m}{2} & (m = j^2;\ j \in \mathbb{N}) \\ 0 & (\text{otherwise}). \end{cases} \tag{31}$$

Here, $a_m = 2m$ and $b_m = 4m$, the sequence (x''_m) is not classically summable in DCN product sense. However, it is statistically DCN product summable to 0. It is easily observed that

$$\mathcal{L}''_m(1;x) = [1 + x''_m]1 = 1 + x''_m,$$

$$\mathcal{L}''_m(\cos s;x) = [1 + x''_m]\frac{m-1}{m}\cos x$$

and

$$\mathcal{L}''_m(\sin s;x) = [1 + x''_m]\frac{m-1}{m}\sin x,$$

so that we have

$$\text{stat}_{\text{DCN}} \lim_{m\to\infty} \|\mathcal{L}''_m(1;x) - 1\|_{2\pi} = 0,$$

$$\text{stat}_{\text{DCN}} \lim_{m\to\infty} \|\mathcal{L}''_m(\cos x;x) - \cos x\|_{2\pi} = 0$$

and

$$\text{stat}_{\text{DCN}} \lim_{m\to\infty} \|\mathcal{L}''_m(\sin x; x) - \sin x\|_{2\pi} = 0,$$

that is, the sequence $\mathcal{L}''_m(f; x)$ satisfies the conditions (26) to (28). Therefore, by Result 3, we have

$$\text{stat}_{\text{DCN}} \lim_{m\to\infty} \|\mathcal{L}''_m(f; x) - f\|_{2\pi} = 0.$$

Hence, it is statistically DCN product summable; but it is not classically DCN product summable, so we conclude that our classical part of approximation Theorem 3 is not valid for the operators defined by (29); whereas our statistical part of DCN product approximation Result 3 still works for the operators defined by (29).

5 Effects of Gibbs Phenomenon

As regards to the effects of the Gibbs Phenomenon in the following example, we will see how the Cesàro mean $\mathfrak{L}_n(f)$, Nörlund mean $\mathfrak{L}^{**}_n(f)$ and the DCN product $\mathfrak{T}^*_n(f)$ means of the partial sums of Fourier series of a 2π periodic signal is better behaved than the sequence of partial sums $s_n(x)$ itself.
Consider

$$f(x) = \begin{cases} -1 & (-\pi \leqq x < 0) \\ 1 & (0 \leqq x < \pi), \end{cases}$$

be periodic with period 2π. Clearly, it is an odd function. So its Fourier series is given by

$$f(x) = \sum_{n=1}^{\infty} b_n \sin nx,$$

$$b_n = \frac{2}{\pi} \int_0^{\pi} f(x) \sin nx = \frac{2}{\pi} \left(\frac{1-(-1)^n}{n} \right).$$

Thus, the Fourier series of f is,

$$f(x) = \frac{2}{\pi} \sum_{n=1}^{\infty} \frac{1-(-1)^n}{n} \sin nx, \quad x \in [-\pi, \pi]. \tag{32}$$

The nth partial sum $s_n(x)$ of the Fourier series (32) is given by

$$s_n(x) = \frac{4}{\pi}\left(\sin x + \frac{1}{3}\sin 3x + \cdots + \frac{1}{n}\sin nx\right) \tag{33}$$

and the deferred Cesàro mean is

$$\mathfrak{L}_n(f) = \frac{1}{(n+1)}\sum_{k=0}^{n} s_k(f). \tag{34}$$

Next, the deferred Nörlund mean is

$$\mathfrak{L}_n^{**}(f) = \frac{2}{(n+1)(n+2)}\sum_{k=0}^{n}(n-k+1)s_k(f). \tag{35}$$

Finally, the DCN product mean $\mathfrak{T}_n^*(f)$ is given by

$$\mathfrak{T}_n^*(f) = \frac{2}{(n+1)^2(n+2)}\sum_{k=0}^{n}(n-k+1)s_k(f). \tag{36}$$

Now the graphs for the signals, namely graph for nth partial sum $s_n(x)$, deferred Cesàro $\mathfrak{L}_n(f)$, deferred Nörlund $\mathfrak{L}_n^{**}(f)$ and finally for the DCN product sum $\mathfrak{T}_n^*(f)$ are plotted in the following figures.

According to Gibbs Phenomenon, in the neighborhood of discontinuity, the convergence of Fourier series is not uniform and the sequence of partial sum is over estimated the signal by 18%, that is, in the neighborhood of discontinuity overshoots in the peaks of partial sum $s_n(x)$ are noticed closure of the line passing through the point of discontinuity as n- increases.

6 Concluding Remarks and Observations

In the last section of our investigation, we demonstrate different remarks and observations relating the several outcomes which we have presented here.

Remark 1 The product transforms $\mathfrak{T}_n^*(f)$ of the present form plays an important role as a double digital filter in signal theory as well as the theory of Machines in Mechanical Engineering (see [9]).

Remark 2 From Theorem 3, as $1 + \alpha \geqq \alpha$, $\alpha \in (0, 1)$, so it gives still sharper estimates. Thus, as regards to convergence of $f(x)$, the product summability $\mathfrak{T}_n^*(f)$ gives better estimate than the individuals.

Remark 3 Let $(x_m)_{m\in\mathbb{N}}$ be a real sequence given in Example 1. Then, since

$$\text{stat}_{\text{DC}} \lim_{m \to \infty} x_m = 0$$

we have

$$\text{stat}_{\text{DC}} \lim_{m \to \infty} \|\mathfrak{L}_m(f_i; x) - f_i(x)\|_{2\pi} = 0 \quad (i = 0, 1, 2). \tag{37}$$

Thus, by Result 1, we can write

$$\text{stat}_{\text{DC}} \lim_{m \to \infty} \|\mathcal{L}_m(f; x) - f(x)\|_{2\pi} = 0, \tag{38}$$

where

$$f_0(x) = 1, \quad f_1(x) = \cos x \quad \text{and} \quad f_2(x) = \sin x.$$

As we know, (x_m) is not DC summable in the ordinary sense, thus, the associated approximation theorem does not work here for the operators defined by (15). Hence, this application clearly indicates that our Result 1 is a non-trivial generalization of the classical Theorem 2.

Remark 4 Let $(x'_m)_{m \in \mathbb{N}}$ be a real sequence given in Example 2. Then, since

$$\text{stat}_{\text{DN}} \lim_{m \to \infty} x'_m = 0$$

we have

$$\text{stat}_{\text{DN}} \lim_{m \to \infty} \|\mathcal{L}'_m(f_i; x) - f_i(x)\|_{2\pi} = 0. \tag{39}$$

Thus, by Result 2, we can write

$$\text{stat}_{\text{DN}} \lim_{m \to \infty} \|\mathcal{L}'_m(f; x) - f(x)\|_{2\pi} = 0, \quad (i = 0, 1, 2), \tag{40}$$

where

$$f_0(x) = 1, \quad f_1(x) = \cos x \quad \text{and} \quad f_2(x) = \sin x.$$

As we know, (x'_m) is not classically summable in deferred Nörlund the sense, thus, the associated deferred Nörlund classical approximation theorem does not work here for the operators defined by (22). Hence, this application clearly indicates that our Result 2 is a non-trivial generalization of the classical Theorem 1.

Remark 5 Let $(x''_m)_{m \in \mathbb{N}}$ be a real sequence given in Example 3. Then, since

$$\text{stat}_{\text{DCN}} \lim_{m \to \infty} x''_m = 0$$

we have

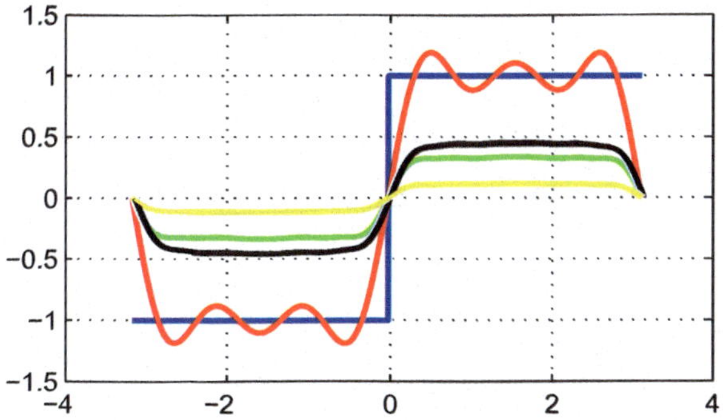

Fig. 1 The signals $f(x) =$ blue, $s_n(x) =$ red, $\mathfrak{L}_n(f) =$ black, $\mathfrak{L}_n^{**}(f) =$ green, $\mathfrak{T}_n^*(f) =$ yellow; for $n = 5$

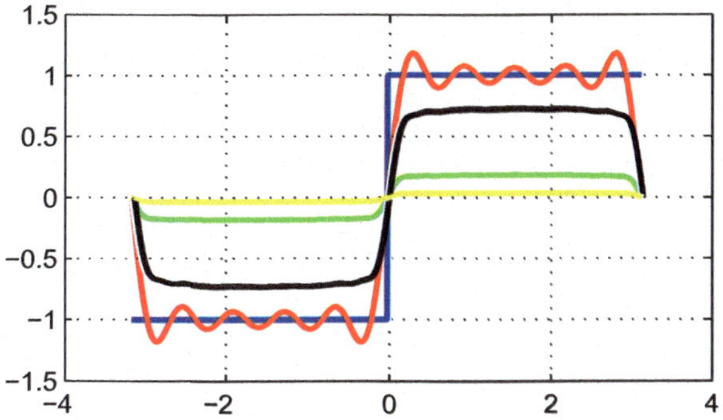

Fig. 2 The signals $f(x) =$ blue, $s_n(x) =$ red, $\mathfrak{L}_n(f) =$ black, $\mathfrak{L}_n^{**}(f) =$ green, $\mathfrak{T}_n^*(f) =$ yellow; for $n = 10$

$$\text{stat}_{\text{DCN}} \lim_{m\to\infty} \|\mathcal{L}''_m(f_i; x) - f_i(x)\|_{2\pi} = 0. \tag{41}$$

Thus, by Result 3, we can write

$$\text{stat}_{\text{DCN}} \lim_{m\to\infty} \|\mathcal{L}''_m(f; x) - f(x)\|_{2\pi} = 0, \quad (i = 0, 1, 2), \tag{42}$$

where

$$f_0(x) = 1, \quad f_1(x) = \cos x \quad \text{and} \quad f_2(x) = \sin x.$$

As we know, (x''_m) is not DCN product summable in the ordinary sense, thus, the associated classical approximation theorem does not work here for the operators defined by (29). Hence, this application clearly indicates that our Result 3 is a non-trivial generalization of the classical Theorem 3.

Remark 6 From the above Figs. 1 and 2, we observe that $\mathfrak{L}_n(f)$, $\mathfrak{L}_n^{**}(f)$ and $\mathfrak{T}_n^*(f)$ converges quickly to $f(x)$ than the sequence of partial sum s_n in the interval $[-\pi, \pi]$. We further notice that in the neighborhood of discontinuity, that is, in the neighborhood of $-\pi$, 0 and π, the graphs of s_5 and s_{10} show overshoots in peaks and move closer to the line passing through the points of discontinuity as n increases, but in the graphs of $\mathfrak{L}_n(f)$, $\mathfrak{L}_n^{**}(f)$ and $\mathfrak{T}_n^*(f)$ at $n = 5$ and 10, the peaks become flatter.

Clearly, the DCN product summability mean of the Fourier series of f overshoots the Gibbs Phenomenon and shows the smoothing effect of the method. Thus, $\mathfrak{L}_n(f)$, $\mathfrak{L}_n^{**}(f)$ and $\mathfrak{T}_n^*(f)$ are better approximates than $s_n(x)$, and the DCN product $\mathfrak{T}_n^*(f)$ summability is better behaved than the individuals $s_n(x)$, $\mathfrak{L}_n(f)$ and $\mathfrak{L}_n^{**}(f)$ summability methods.

References

1. Agnew, R.P.: On deferred Cesàro means. Ann. Math. **33**, 413–421 (1932)
2. Armitage, D.H., Maddox, I.J.: A new type of Cesàro mean. Anal. **9**, 195–204 (1989)
3. Braha, N.L., Srivastava, H.M., Mohiuddine, S.A.: A Korovkin-type approximation theorem for periodic functions via the statistical summability of the generalized de la Vallée Poussin mean. Appl. Math. Comput. **228**, 162–169 (2014)
4. Das, A.A., Paikray, S.K., Pradhan, T., Dutta, H.: Statistical $(C, 1)(E, \mu)$-summablity and associated fuzzy approximation theorems with statistical fuzzy rates. Soft Comput. **24**, 10883–10892 (2020)
5. Deger, U., Dagadur, İ, Küçükaslan, M.: Approximation by trigonometric polynomials to functions in L_p -norm. Proc. Jangjeon Math. Soc. **15**, 203–213 (2012)
6. Fast, H.: Sur la convergence statistique. Colloq. Math. **2**, 241–244 (1951)
7. Jena, B.B., Mishra, L.N., Paikray, S.K., Misra, U.K.: Approximation of signals by general matrix summability with effects of Gibbs phenomenon. Bol. Soc. Paran. Mat. **38**, 141–158 (2020)
8. Jena, B.B., Paikray, S.K., Dutta, H.: On various new concepts of statistical convergence for sequences of random variables via deferred Cesàro mean. J. Math. Anal. Appl. **487**, 123950 (2020)
9. Mishra, V.N., Khatri, K., Mishra, L.N.: Product $(N, p_n)(C, 1)$ summability of a sequence of Fourier coefficients. Math. Sci. **6**, Article ID 38, 1–5 (2012)
10. M. L. Mittal and M. V. Singh, Approximation of signals functions by trigonometric polynomials in L_p -norm. Int. J. Math. Math. Sci. **2014**, Article ID 267383, 1–6 (2014)
11. Parida, P., Paikray, S.K., Dutta, H.: On approximation of signals in $Lip(\alpha, r)$ -class using the product $(\overline{N}, p_n, q_n)(E, s)$ - summability means of conjugate Fourier series. Nonlinear Stud. **27**, 1–9 (2020)
12. Parida, P., Paikray, S.K., Dash, M., Misra, U.K.: Degree of approximation by product $(\overline{N}, p_n, q_n)(E, q)$ summability of Fourier series of a signal belonging to $Lip(\alpha, r)$ -class. TWMS J. App. Eng. Math. **9**, 901–908 (2019)
13. Pradhan, T., Jena, B.B., Paikray, S.K., Dutta, H., Misra, U.K.: On approximation of the rate of convergence of Fourier series in the generalized Hölder metric by Deferred Nörlund mean. Afrika Matematika (14), 1119–1131 (2019). https://doi.org/10.1007/s13370-019-00706-y
14. Pradhan, T., Paikray, S.K., Das, A.A., Dutta, H.: On approximation of signals in the generalized Zygmund class via $(E, 1)(N, p_n)$ ssssummability means of conjugate Fourier series. Proyecciones J. Math. **38**, 1015–1033 (2019)
15. Proakis, J.G.: Digital Communications. McGraw-Hill, New York (1985)
16. Psarakis, E.Z., Moustakides, G.V.: An L_2 -based method for the design of 1-D zero phase FIR digital filters, IEEE Trans. Circuits Syst. I. Fundam. Theor. Appl. **44**, 591–601 (1997)
17. Quade, E.S.: Trigonometric approximation in the mean. Duke Math. J. **3**, 529–542 (1937)
18. Srivastava, H.M., Jena, B.B., Paikray, S.K., Misra, U.K.: A certain class of weighted statistical convergence and associated Korovkin type approximation theorems for trigonometric functions. Math. Methods Appl. Sci. **41**, 671–683 (2018)

19. Srivastava, H.M., Jena, B.B., Paikray, S.K., Misra, U.K.: Generalized equi-statistical convergence of the deferred Nörlund summability and its applications to associated approximation theorems. Rev. Real Acad. Cienc. Exactas Fís. Nat. Ser. A Mat. (RACSAM) **112**, 1487–1501 (2018)
20. Steinhaus, H.: Sur la convergence ordinaire et la convergence asymptotique. Colloq. Math. **2**, 73–74 (1951)
21. Zraiqat, A., Paikray, S.K., Dutta, H.: A certain class of deferred weighted statistical B-summability involving (p, q)-integers and analogous approximation. Filomat **33**, 1425–1444 (2019)
22. Zygmund, A.: Trigonometric Series, 3rd edn. Cambridge University Press, Cambridge (2002)

Printed by Printforce, the Netherlands